INSTRUMENTATION FUNDAMENTALS FOR PROCESS CONTROL

INSTRUMENTATION FUNDAMENTALS FOR PROCESS CONTROL

Douglas O. J. deSá

Senior Applications Engineering Consultant
Retired
Foxboro G. B. Limited

New York   London

Denise T. Schanck, Vice President
Robert H. Bedford, Editor
Catherine M. Caputo, Assistant Editor
James A. Wright, Marketing Director
Albert Ezratty, Marketing Associate

Published in 2001 by
Taylor & Francis
29 West 35th Street
New York, NY 10001

Published in Great Britain by
Taylor & Francis
11 New Fetter Lane
London EC4P 4EE

10 9 8 7 6 5 4 3 2 1

Library of Congress Cataloging-in-Publication Data
deSá, Douglas O. J.
Instrumentation fundamentals for process control / by Douglas O. J. deSá.
p. cm.
Includes index.
ISBN 1-56032-901-7 (alk. paper)
1. Chemical process control—Instruments.   I. Title.
TP155.75 .D4 2000
670.47'7—dc21   00-041189

Contents

PART V
INSTALLATION OF INSTRUMENTATION 335

CHAPTER 18
CONSIDERATIONS IN PROPER INSTALLATION 337

Preface

This book is conceived as a practical guide and introduction to the principles of process measurement and control to those who are entering, or contemplating a career in the instrumentation and control industry. The work is not intended to supersede the formal mathematical treatment of control theory that is, and will continue to be, an essential requirement for a rounded understanding of the subject. It may also serve as a refresher for those practicing the "art" and who feel the need to look again at the basics.

I have tried in general to avoid the essential "heavy" mathematics normally associated with control theory, because a full treatment of this topic can be obtained from the many excellent publications available. The reason for presenting this less mathematical treatment of the subject is that, in the world of applied instrumentation and control, one often does not need to draw on the formal mathematical approach learned in one's university or college course. The practicing control engineer will, of course, occasionally be called upon to carry out a formal theoretical analysis of an application to determine a solution and on these occasions the academic approach will be found to be valuable, if not a necessity.

For the most part, the topics covered in this book were the subjects of a series of lectures and the chapter contents formed the notes handed out to the attendees. It is for this reason that the language is deliberately intimate and the derivation of most of the various formulas given has been covered in detail. Regarding the level of detail in the mathematics used in the text, I would like to state that it is intentional and has been in response to several requests by those attending the talks for a step-by-step explanation. In this regard I ask the readers' forbearance, as I am sure that some may find the solutions somewhat tedious in their execution.

For newcomers, especially those readers who are currently pursuing a course of study in another discipline, many of the explanations and supporting formulas have intentionally reverted to the subject of physics, where they have their foundation. It is hoped that this method of treating the subject will go some way toward answering the age old question, put by students to many teachers of science subjects in schools, "Where will I ever use this?"

Some of the instruments described in the text may be considered old-fashioned, but this has also been intentional, for these designs show very clearly their dependence on the subject of physics. In defense of the stance taken, I have included modern instruments as well, to try and maintain the balance, for even these newer items of equipment depend on principles established a while ago. It has not been my intention to provide an exhaustive treatment on the topics covered; rather, I have sought to present to the reader the breadth of this most fascinating subject that has given me so much pleasure, and a little pain, in acquiring a level of understanding gained over years in the industry.

It is with much admiration and respect that I mention the part played by so many of my colleagues in the Foxboro Company, now a division of the Invensys Group of companies, and in particular Messrs. F. G. Shinskey, P. Badavas, and M. J. Cooper for their inspiration. I further take this opportunity to acknowledge the stoicism of the several people who attended the series of talks and the many kind words offered, and to thank Mr. G. W. Skates of Honeywell Control Systems and Messrs M. J. Cooper and J. Whiting of Foxboro for their great interest and encouragement throughout the entire series of lectures. I also wish to acknowledge the many helpful suggestions made by Messrs. P. Males of Honeywell and P. Robinson of the University of Plymouth and to thank them for their time in reviewing the initial work. My grateful thanks are also due to Messrs. J. Gough of Foxboro, M. Doyle of CISE (North) Ltd., M. Machacek of Apax Computers, and Mr. D. R. Beeton, a good friend and former colleague at Foxboro, for reviewing the present work; and, last, but not least, my wife Halina for her patience and understanding during the prolonged preparation of this book.

Doug deSá

April 2000

Introduction: Concepts in Instrumentation for Process Control

The power of logical reasoning distinguishes humans from all other species, and it was as a consequence of this capability that they made very significant progress. Their development of tools and techniques for exploiting the materials around them stemmed from their ability to analyze a situation and draw conclusions. Discovery of the means of generating fire at will, and the methods to regulate its intensity on demand, caused radical changes in humans' way of life and thinking. The reaction of materials to fire spurred the need to understand their properties and the rules governing their behavior, in order to derive the most benefit and use from them. The ancient alchemist with his potions, chants, and bubbling cauldrons brought advancement of a kind, but people such as this have long since given way to technical innovators like Galileo, Newton, and many others.

With every scientific advance, the material expectations and demand for newer and better products increased. However, when higher prices were demanded for these enhanced products, the public support did not materialize, even for popular items. It fell to scientists and engineers to develop the technology to meet these demands and to provide an expanded and improved range of products, and at affordable prices.

These goals of product improvement and diversification within financial constraints are just as valid today, as they continue bringing the need to increase scientific research, and to develop new or enhance existing technology. Above all, they demand an increased sophistication in measurement and control technology. With an increasing number of new measurements to be made in combination with a narrowing of product tolerances, those scientists, engineers, and technicians involved in the development of measuring techniques and equipment find their role continually expanding. The measuring instrument, control device, and control systems industries are growing and will continue to do so in line with the other branches of science.

FUNDAMENTALS OF PROCESS CONTROL

This book is devoted to providing a working knowledge of process control and its technology, which many will find intellectually stimulating. One of the first concepts we must get to grips with is, what constitutes a process? In order to provide a reasonable answer, we will have to realize that a process can either be a whole manufacturing sequence consisting of a collection of individual stages or comprise just one stage in that sequence. The simplest definition of a process would be a series of manufacturing stages, which could be either mechanical, electrical, physical, chemical, or a combination of all these, that the feed material or materials would have to undergo to be transformed into the desired end products.

To illustrate what has just been said, let us consider just one stage, and for simplicity we shall not consider it to be a materials processing one. Let the "stage" be the living room of a house in which there is a heater of adequate size, fitted with a switch enabling it to be turned on or off. The room has a door and a window, and the insulation is fairly good; as an added feature, let there be a wall-mounted thermometer fitted. Now let us assume that it is winter and a person enters the room, ensures the door and window are closed, turns the heater on, and settles down in a chair to read. Under the action of the heater, the air surrounding the occupant will start to rise in temperature. The actual temperature attained can be read at any time from the wall mounted thermometer. The air temperature will continue to rise until it reaches a value such that the occupant feels compelled to either turn the heater off or open the window or door to limit the increase. Depending on the size of the heater and whether the window and door remain shut, the temperature in the living room could become intolerable if the occupant took no action. In this scenario, we have all the requisites of a *temperature control loop*:

- A *measuring device*: the occupant and validation using the thermometer

- A *controller*: the occupant

- A *controlled device*: the heater with its switch

If we analyze this example further, we see that:

1. In this particular example, *the measuring device is an intrinsic part of the controller*. This is not the normal case; it is more usual to have the measuring and control functions as separate entities even though they can be physically contained within the same housing.

2. The *heat generated* is transmitted to the air in the room by convection currents, and thus there are bound to be areas that will take a while to warm up, resulting in temperature variations in the room. This will certainly affect the reaction of the occupant.

3. The location of the *temperature indicator* is important in order for the reading to give a meaningful value.

4. The *controller* (the occupant), we shall see, has the following "attributes":

   - A *sensor signal proportional to the measurement*: the occupant's ability to detect a change in temperature and validate it via the thermometer.

   - A *desired value*: the occupant's built-in "comfort" temperature value, called a *setpoint*.

   - A *comparator*: the occupant's ability to compare, i.e., to determine the difference between a desired value and the detected temperature, the difference being called the *error*. If the detected temperature is above the desired temperature, the error is considered positive; and if it is below the desired temperature, the error is considered negative.

   - An *output*: The occupant's action of switching the heater will depend on the following conditions:

     a. If the error is negative, then the occupant will allow the heater switch to stay on.

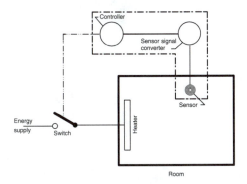

Figure 1: Expanded schematic of control loop.

> b. If the error is zero, then there is a "don't care" situation. However, leaving the heater switch on will raise the temperature, or turning the switch off will allow the temperature to fall.
>
> c. If the error is positive, then the occupant will turn the heater switch off.
>
> Note that *the controller always responds with an action that will tend to bring the measurement and desired value to coincidence.*

5. The *control loop* is entirely self-contained. That is, the controller—the occupant— acts on the measured variable, the air temperature within the room, via the heater to provide comfortable conditions. Such a self-contained loop is called a *closed loop*.

Having given a very simple outline of a control loop, let us now depict it in typical graphical symbols, as in Figure 1, but in this instance let us keep the measuring and control devices separate. For the present, we shall maintain the on/off control strategy; we shall discuss alternatives a little further on. Note that in this example the occupant represented in Figure 1 by the box containing the sensor, the signal converter (a means of changing the sensor signal to one that can be recognized by the controller), and the controller are all actually located within the confines of the room itself. The heater switch should also be within the room. This depiction is for clarity only.

As we have seen from our example, we will never obtain a room temperature that coincides exactly with the desired value. On the contrary, we would obtain a room temperature that was on one side or the other of the desired value. This scattering of temperature values will form a band encompassing the desired value over a period of time. The width of the band will, in very simplified terms, depend on the amount of heat, the rate of heat loss, the time it takes the heater to provide the make up amount of heat to restore the lost heat, and the deadband of the sensor and controller. The *deadband* is defined in Figure 2 as the amount of change, in the direction of the measured variable producing it, that occurs during the interval required to produce the appropriate signal. In this respect the deadband is very akin to the inertia of a body or system.

Since the losses are more or less continuous and are being made up only during those periods when the heater is switched on, it will be appreciated that if the actual room temperature was plotted graphically against time, the trace would fluctuate

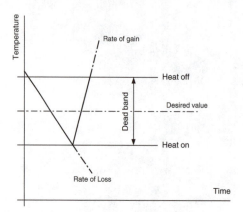

Figure 2: Measurement deadband.

around the desired value. The slope of the part of the curve representing the heat loss might not be identical to the slope of the heat gained part (see Figure 2). The slopes are dependent on the differing rates of heat loss and heat gain; the physical location of the sensor also has a marked effect, because it would not be measuring the temperature of the room, but that in its own vicinity only. In this part of the discussion, remember that the room temperature is a measure of the heat content.

In any control system, process or otherwise, the objective is to achieve coincidence between the measurement and the desired value; anything other than this indicates poor or no control, depending on the amount of separation between the two. Let us now carry the illustration a little further, this time adding instruments to perform the various tasks that humans could do—but not as efficiently—for tiredness and boredom would be the major drawbacks. Before proceeding too far, we should state what we mean by the words *instrument* and *control*, which will be used very often in this text.

For the word *instrument*, we shall generally mean a measuring device. The science behind making the measurement is something we shall be exploring in the first few chapters. These will provide readers with an opportunity to see how the time they devoted to physics, chemistry, and mathematics can be put to use and was not one terrible waste.

A measuring instrument is frequently fitted with a means of directly displaying the measurement made, usually by a pointer moving over a graduated scale; and/or converting the measurement to a corresponding signal, which could be a pneumatic or hydraulic signal or an electric voltage or current that can vary only over a fixed range of values. The latter type of instrument is called a *transmitter*, for this type of instrument is usually employed to transmit the measurement so that it can be read and/or used some distance from the actual point of measurement. In today's world the distance could extend over thousands of kilometers; signals might be beamed to communications satellites, or, conversely, measurements taken by instruments located in space vehicles, manned or unmanned, might need to be read out on earth. In satellite communications, digital signals are used exclusively and are converted to analog variations if and when necessary.

In process control, the main transmitter signals used are pneumatic and/or electric. Pneumatic signals in measurement and control instruments are not as popular now as they once were, although they are still the virtual de facto standard for driving final control devices such as control valves and dampers—if only by virtue of the purchase and installation price of equivalent electronic/electrical equipment alternatives.

The action of a control valve can best be described by an everyday example. The thermostatic valve on a central heating radiator is in reality a self-actuated control valve fitted with a bimetallic temperature sensor that changes shape when exposed to heat—all metals expand when heated, the amount of expansion depending on the coefficient of linear expansion; two different metals are used for stability—and acts on a plunger in the valve body to regulate the flow of the hot water through the radiator. The rotatable dial on these devices allows the user to set the temperature at which the plunger will shut off the hot water flow. In an industrial control valve the plunger is spring-loaded and moved, say, by a piston in a cylinder—the reverse of a bicycle pump—that is driven by a pneumatic signal from a controller to a position dependent on the magnitude of the signal. The position of the plunger determines the amount of fluid that can pass, and thus regulates the flow of hot water through the pipeline and into the radiator. There are methods other than a piston for driving the plunger, and these we shall also discuss in Chapter 16.

A control damper can best be visualized from the following description: a shaped flap that is constrained to move within a large duct so that its movement permits the effective bore of the duct to be varied between fully open and fully closed to the passage of the contained fluid, usually air or gases.

Pneumatic signals are almost universally in the range 3 to 15 lb/in^2, or 0.2 to 1.0 kg/cm^2 or 0.2 to 1.0 bar, the imperial units being more common in the United States and metric in Europe and elsewhere. Electric signals can be either in mA of current or mV of voltage; the ranges associated with electric current signals are almost standardized at 4 to 20 mA, although 0 to 20 mA can sometimes be encountered; when mV voltage signals are used, the most common is 0 to 10 mV. Sometimes microvolt signals are used but these are most often associated with pH, electrical conductivity, or other analytical instruments; microvolt signals are not used for transmitting over a distance. Should this be required then conversion to a milliampere current is needed.

The usefulness of transmitted signals in a process plant is in overcoming losses when the separation distances are large, and they also simplify standardization of the equipment, the control systems, and the communications within and between them. In this respect, milliampere current transmission has the greater advantage over millivolt voltage, as the latter is more susceptible to electrical noise and transmission line losses owing to the resistance of the interconnecting wire. One could argue that the resistance of the transmission line can be reduced, although this can only be achieved by increasing the cross-sectional area of the wire. Increased wire size adds to the costs and to the difficulty in installation due to the greater physical size and weight of the cable.

Every measurement we make must have a minimum and a maximum value; these values could be negative, zero, or positive and are dependent on what is to be measured, or, more correctly, the *parameter* we are measuring. For example, the parameters temperature and pressure could have minimum or maximum values that are either negative, positive, or zero. To be meaningful, it is important that we state the two values (minimum and maximum) in which we are interested—e.g., −30°C minimum and −5°C maximum, or 0°C minimum and 100°C maximum, or −0.1 kg/cm^2 minimum and 2.0 kg/cm^2 maximum. There is a shorthand way of writing this, which is −30 to −5°C, 0 to 100°C, and −0.1 to 2.0 kg/cm^2, respectively, and it is always called the *measurement range*. The minimum value is called the *lower range value (lrv)* and the maximum value the *upper range value (urv)*. Be aware that there are other measurables that cannot have negative values, for example, the rotational speed of a motor or the speed of a conveyor belt; in this example during normal operation negative values usually have little real meaning; however, there may be occasions where the conveyor drive needs to be reversed slowly under operator control to eliminate a problem. Subtracting the lower range value from the upper range value (urv − lrv) will give the *span* of the measurement. For the examples given, the measurement spans are therefore 25°C, 100°C, and 2.1 kg/cm^2, respectively.

ILLUSTRATION: A SIMPLE CONTROL SYSTEM

We shall try to make our previous example a little more industrially realistic and at the same time retain its simplicity. To do this we shall replace the person in that example with transmitters and controllers having a continuously variable output, which is proportional to the error (i.e., the output will be proportional to the difference between what we measure and what we want the value to be). Every automatic (nonhuman) controller is designed to operate with:

- A desired value that can be set by the process operator or by some other device, which could be another controller, or can be computed by a software routine in a computer. When set by the operator, the desired value—the *setpoint*—is termed the *local* setpoint. When set by another controller or a computer, it is termed the *remote* setpoint, sometimes also called the *cascade* setpoint.

- An output that results from computations (analog or digital) within the confines of the controller itself is deemed to be *automatic*; alternatively, the process operator can drive the output *manually* to a value judged to be necessary. These ways of producing the output are called the operating *modes* of the controller. The modes are selectable by the operator and can only be invoked one at a time. Changing the controller mode from Auto to Manual has the effect of "freezing" the last value of the output, or allowing it to be adjusted by the operator; while reverting to Auto unlocks the output, which will change to a computed value based on the error.

COMPONENTS OF THE SYSTEM

In the example that follows, some liberties have been taken in the interest of simplification to show how the process works. It is possible in the food industry for juices, syrups, and milk to be thickened in specialized vessels called evaporators, of which there are different types, such as the *falling film* used in the concentration of milk or glucose. In practice there is usually more than one vessel involved, each vessel being called an *effect*, the process of concentrating the raw product being spread over several effects. This collection of effects is called a *multiple-effect evaporator*, the actual number of vessels used depending on the material being processed.

Let us assume that the instrumentation is analog and a cylindrical vessel, as shown in Figure 3 that has a heating jacket partially covering its lower portion, replaces the room in the previous example. The jacket forms what can best be described as a second skin. The vessel forms one part of a system used for reducing—by evaporation—water-rich, fresh, edible juice to a thicker fluid called *concentrate*. To retain the taste, the evaporation of the excess water is carried out at a pressure that is lower than the ambient so that the temperatures involved are not high enough to cook the fruit. Heat is imparted to the process fluid (juice) by steam, in a way that allows the juice maximum exposure to the heat source. Hence, steam enters through a connection on the top part of the outer jacket, and exits as hot water—when steam

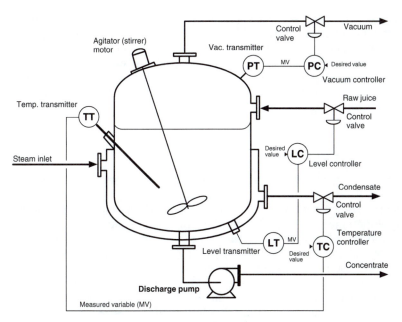

Figure 3: Simplified edible juice concentrating control scheme.

gives up its heat it will condense back to water, in which state it is called *condensate*—from a connection at the lower part of the jacket cavity. Such an arrangement in a real process vessel is called a *steam heating jacket*, but there are other methods of steam heating that involve introducing the steam either through a *steam heating coil* fitted within the vessel, or directly into the liquid, when the technique is called *live steam injection*. Both of these latter methods are unsuitable for this particular application. If used here, steam heating coils can become clogged with the concentrate, permitting bacterial growth, which is a health hazard; or live steam injection will dilute the juice and therefore be counterproductive.

The vessel is fitted with the following:

- A level transmitter and its associated control device, which acts through a control valve in the supply line delivering the raw juice.

- A temperature transmitter and its control device, which acts through a control valve in the condensate line and not in the steam line, as some might have expected, for economic and heat conservation reasons.

- A vacuum transmitter and its control device, which acts through a control valve in the vacuum line.

The three control valves are used to regulate material flow through the pipelines. The whole arrangement is shown in Figure 3.

In Figure 3, each function has been shown separately, as is normal for this type of diagram. For convenience, the measurement (i.e., MV, or measured variable) and the desired value for each controller have been labeled. As we stated earlier, the measuring and control devices are analog instruments; hence, the monitoring and the control functions will be continuous—i.e., their signals will trace smooth curves with respect to time. Now if the control valves are also continuously variable, then they will respond to the variations of the controller output and as a result will, in time, change the measurement. A suitable analogy of what is happening here would be the accelerator pedal and speedometer on an automobile. The driver varies the travel of the pedal continuously to suit the road conditions and also to stay within the speed limit. However, from experience we know that there is always a delay between a change in pedal position and the change in speed of the vehicle. In control parlance, this is an *output lag*.

OPERATION AND REGULATION

The sequence of actions we shall describe next typifies the actions of a process operative in control of a single evaporator in the middle of a multiple-effect evaporator processing system. The process operator vents the vessel, sets the required vessel level as the setpoint on the level controller, and sets the mode to Auto. This allows the control valve in the raw juice line to open and start to fill the still vented vessel. The controller monitors the level, regulates the juice flow, and, when the measurement coincides with the setpoint, closes the vent and supply control valves. With this system, the operator will have to change the mode of the juice controller to Manual to keep the control valve shut to allow the contained juice to be concentrated.

The stirrer motor is switched on at a predetermined juice level to avoid stratification; because juice is a solution of salts, solids, and water, separation will occur naturally in time. The other important and very beneficial effect of stirring (agitating) the juice is to ensure a uniform temperature through its whole mass.

The process operator then sets the required juice temperature on the temperature controller—as mentioned earlier this has to be much lower than the boiling point of the juice at atmospheric (ambient) pressure to preserve the taste—and sets the controller mode to Auto to allow it to admit and regulate the heating steam to the jacket by allowing some condensate to leave the jacket. When the juice temperature reaches the setpoint, the controller closes the condensate valve fully. The process operator leaves the temperature controller mode in Auto to maintain the temperature in the vessel. Note that what has just been described is deliberately very simplified, since what is required is the admission of some heat (steam) to replace that which is lost due to evaporation and the surroundings.

While the juice is attaining the required temperature, the process operator sets the vacuum required by gently raising the value on the vacuum controller (for reasons given later in this section). Be aware that a lower pressure will increase the observed liquid level, and if the level controller had been left on Auto, this would drive the valve open and upset the required thickening process. At this point the vacuum controller is still in the Manual mode from the previous operation, which holds the vacuum valve shut. When the desired temperature throughout the mass has been reached, the process operator changes the vacuum controller mode to Auto, which will drive the vacuum control valve open thereby reducing the pressure in the vessel. The juice starts to boil, but at a much lower temperature than would be the case at ambient pressure, and forcing some of the water in the juice to change to steam which is either vented or removed up the vacuum line for further use.

Since the juice is giving up part of its water content, it will become more dense (undergo a density change) and thicker (more viscous), and the level in the vessel will fall commensurably. The object in most liquid food concentrating processes is to reduce the inlet volume to about one-third the initial value. The process operator monitors the level in the vessel and changes the vacuum controller mode to Manual when either the desired level or concentration is attained. There is normally a vent line on the vessel (for simplicity, this has been omitted on the drawing), which is slowly opened by the process operator to allow the vessel to attain ambient pressure. When ambient pressure is reached, the discharge pump is started to empty the vessel to make it ready for the next "charge" of juic; this is typically a *batch process*.

Two important points have been smoothed over in the preceding description, and these should be explained before we progress too far. First, it is very important that a controller mode be changed from Manual back to Auto only when setpoint and measurement are aligned, for this will avoid a *bump*—a rapid change in the measured condition—in the process when the control valve has to react suddenly to a change from its initial position to that demanded by the controller. The severity of the bump will depend on how far apart the setpoint and measurement are at the time of transfer. Modern controllers are designed to balance the measurement and the setpoint automatically by what is called *bumpless Auto/Manual transfer*. The second point is that the opening of the vacuum control valve should be gentle. It should not be too difficult to imagine the drastic effect that a sudden exposure of a pressurized vessel to a vacuum could have on the process fluid; it will tend to all froth up and rush out immediately. To avoid this, the setpoint on the vacuum controller should be slowly increased to the desired value, or, in other words, there should be a *ramped setpoint*, the idea being to have the measurement and setpoint change in unison.

This extremely simplified semiautomatic control scheme on a single evaporator vessel will show that instrumentation and control allow an ever vigilant check on the more mundane parts of the process with minimal operator intervention. Consistently repeatable product quality will inevitably follow, provided that the operator makes the changes of controller mode at the appropriate times.

MODERN INSTRUMENTATION AND ITS EVOLUTION

Even from this brief illustration it should not be too difficult to visualize the possibility of a more fully automated process, still manipulated by means of small individual control loops, but using the refinements of modern control equipment. However, this is not the end of the control story, it is only the beginning. To obtain a much more appealing product, the temperature rise initially imparted to the raw juice should not be detrimental to those enzymes and volatile components that stimulate the taste buds, and this will involve processing the juice in more than one vessel, for we can then preserve the quality needed by not adding heat, but reducing the boiling point temperature in small stages as a result of progressively increasing the vacuum in each succeeding process vessel. In terms of energy saving, there is every reason to utilize the steam extracted in the first evaporator vessel as the heat source for succeeding evaporator vessel(s), or even to warm up the incoming raw juice. There is a limit, which lies in the efficiency and running cost of the vacuum system, that dictates how far this can be carried through. The hot condensate can also be used to, say, wash the raw produce from which the juice was obtained.

Our experience of the real world tells us it is highly unlikely that there is such a thing as a single independent process parameter; most are to some extent interactive. In some cases, because of the complexities of taking *all* the interactions into account for very minimal actual results, we employ simpler but effective control schemes that have a consistent and desired effect. Herein lies the fascination and the challenges of the science and technology of control and instrumentation.

Manufacturing processes are becoming ever more complex. For instance, we no longer distill crude petroleum for just providing automobile gasoline, a few aromatics, kerosene, some lubricating oils, wax, and tar. Now the products derived from the same raw material, crude petroleum, have been diversified to such an extent that there are, to mention a few only, several grades of gasoline to cover aviation and automobile requirements, jet fuel, heating oils, a whole range of products that form the basis of the petrochemical industry, dyestuffs, a lubricating and cutting oil industry, a range of waxes, bitumen, and some pharmaceuticals. The same story is repeated in the pulp and paper, food and drugs, water and sewage treatment, and gas and power generation industries. To achieve this kind of development, chemists, physicists, and chemical, mechanical, and electrical engineers have been fully and ably supported by instrument and control engineers, for it is due to these latter technologists that we are able to obtain in quantity and quality the whole range of products that we take for granted.

To focus the discussion, it is difficult to envisage a business environment of the future, whether it be short or long term, that will be successful when using the methods and technology of the past. Concern, in any manufactured product industry at the present and for the foreseeable future, is with the multiple requirements of producing the goods for the:

• Minimum manufacturing cost

• Maximum return on financial investment

• Minimum consumption of nonrenewable raw material resources

• Minimum demand on energy and maximum use of that consumed

• Greatest possibility for recycling at the end of product life

• Minimum waste

To achieve all of these is a task that will demand much ingenuity and skill of all the personnel involved. In any case, how well the objectives are met will involve good, careful design and close monitoring at every stage during the manufacture of the product. Out of overall process plant capital costs, instrumentation and control systems account for approximately 3 to 9% for *continuous processes* and up to 15% for *batch processes*. A *continuous* process is one that takes a constant stream of raw material, performs some work on it, and obtains the final product(s) at the output of the process. A *batch* process is one that is very complex and takes in specific quantities of several raw or processed materials or combinations of both at very specific points in the manufacturing cycle to obtain a (comparatively small) quantity of product that is usually technically sophisticated and expensive; pharmaceutical products are an example. It is not a shortened version of a continuous process.

These financial investments are a necessary and vital part of the strategy to attain the aforementioned economic goals. Instruments provide a "window" through which the manufacturing process can be observed, both by those who operate it and by the control systems that provide the means of manipulating it. The early individual, large, totally analog indicating and control instruments were at one time the only such equipment available. These instruments had to be accessible to those operating the process and therefore had to be mounted on a plate, which became what was called an "easel-type" panel. As process complexity grew, there were a greater number of measurements to be made, and therefore many more instruments that had to be mounted; necessary steps were taken to reduce instrument size and format but without sacrificing readability. The easel panel gave way to the "cubicle" and "control console" types of panels, for reasons of stability and compactness, scheme complexity, and numbers of indicators, controllers, lamps, and switches involved. In spite of the reduced physical size of the instruments, control panels could in some instances have a length of tens of meters. To assist the operator and reduce the mental burden, fixed pictorial representations of the process, or *mimic diagrams* were added to the panel. These pictures would be, say, of a process vessel with the main material feed and process pipelines, each painted in a different color for easy recognition of their function and the material flowing in them. In some instances the actual associated instruments were mounted in the correct relative position in the lines to provide a vivid visual display. Once again, process complexity and sheer number of measurements forced large-scale pictures of the vessels to be abandoned and replaced with symbolized versions containing lamps of different colors that illuminated to depict the "health" or status of each pipeline or vessel. The associated instruments were located on the panel, but away from the picture. From this developed the highly symbolized and stylized pictures of the process and what are today called the *process graphics*.

The physical size of the control panels meant that the control rooms within which they were located also had to be large. Coupled to this was the fact that each item in a control loop had to be individually piped or wired. If computations were involved, then the number of components in the loop increased, and correct scaling of the signals was vitally important for correct results from the signal manipulation. Corrections or modifications to the control room and the control loop were an expensive affair in terms of both money and time. These difficulties were the incentive to search for a solution, which eventually resulted in large pneumatic-driven systems becoming less prevalent, and the application of the semiconductor and the emerging integrated circuit to increasingly provide the answers. The transistor and its developments brought very rapid transition in their wake from analog-to-digital techniques, which in turn gave the world *microprocessor-based individual controllers* and the *distributed control system* (*DCS*). New microprocessor-based individual controllers have very largely superseded the old style "moving pointer" display instruments. These new devices are hybrid, in that they still use analog input and output signals that are usually in the milliampere current range (4 to 20 mA) but internally carry out the functions digitally much like the pocket digital calculator. There are no moving parts, and the instruments are fully programmable and provided with facilities enabling them to be interrogated and commanded by a digital computer if required.

DCS systems are discussed further on in this book. Briefly, in these control systems the process measurements and the controlled outputs are physically separated from the control and computation functions, but all are interconnected as required by cabling and software links. The control, computation, and logic capabilities are held as *algorithms*—predefined sequences of software routines for calculation and/or logic—that can be invoked by the control system designer on demand to perform specific control, computation, or logical functions. The system includes a *keyboard* (a standard computer keyboard) for giving instructions to the system; a *printer* for outputs and messages from the system; and a *visual display unit* (*VDU*) or monitor, through which the process operator is able to view and manipulate the whole process, usually only one section at a time, via the keyboard and/or a mouse or trackball. The process engineers can additionally change parameters, measurement ranges, and *alarm points*, i.e., points in the measurement range where the process parameter is considered to be approaching a hazard. The foregoing process data accessible to the process engineer is *protected* in that only authorized personnel are allowed to make changes to it to provide security of the formulation and ensure consistent product quality. The DCS arrangement allows very large control panels to be dispensed with; and since all internal interconnections are made by software and not by physical wiring, it enables modifications and rearrangement to be implemented relatively quickly and without undue difficulty.

It is a fact that the lifeblood of all modern control systems is data and information manipulation; this aspect has been given a great deal of prominence in recent times, but why should it be? Data and information are nothing new; it has always been there and available. The truth is that only now is the enabling technology and expertise at our disposal to begin to exploit the vast amounts of information being continually generated. It is possible to propose a scenario where allocation of the size (quantity) of production runs, which is normally based on process management personnel's best guess of sales and marketing demand, can be dynamically modified by the management supervisory computer system to reflect values based on the actual volume of product sales. More *intelligent measuring and transmitting instruments*, some of these new instruments already with us and in use, provide measurements to limits hitherto impossible, are capable of being reranged remotely through a keyboard or a handheld configurator, and can include measurement data validation and many more features.

As a direct consequence, the whole business operation—administration, procurement, manufacturing, sales, and product development—will be made to run at a level most advantageous to the improvement of cash flow and profitability of the corporation, and will become more adept at conserving our nonrenewable natural resources. The bedrock of such an enterprise is professionally qualified personnel with a high level of business acumen and technical ability in management, manufacturing, and product development. Assisting these persons in their decision making will be very powerful, easily configurable computer control systems providing up-to-the-minute reporting on throughput, quantity, and quality of both raw and finished materials. The combination of vision and fact set out in the scenario is gradually taking shape; it will be the way forward. Without capable and innovative instrument and control engineers, however, the chances of achieving this scenario are slim. It is hoped that this book will be a start for the reader to a fulfilling and exciting future.

SUMMARY

1. A process can be defined as a series of manufacturing stages, which could be either mechanical, electrical, physical, or chemical, or a combination of all these, that the feed materials would have to undergo to be transformed into desired products.

2. Flow, pressure, level, temperature, and similar quantities are called *parameters* of the process variable.

3. A control loop in general comprises a measuring device, a controller having the desired value (setpoint) that can be set by the process operator, and a controlled device. However, when the setpoint of one controller is set by another controller or a computer it is termed the *remote* setpoint, sometimes also called the *cascade* setpoint; such a combination is called a *cascade loop*.

4. The separation between the minimum and maximum values of a measurement is called the range, and the difference between lower and upper range values is the measurement span.

5. A controller must include all the following components: a measuring unit, a setpoint, and a comparator to determine the difference between setpoint and measurement to generate the error; a control unit that operates on the error and

produces an output; and an output unit to drive the correcting device. A correctly operating controller always responds with an action that will tend to bring the measurement and desired value to coincidence.

6. The controller setpoint can be local (generated within the controller) or remote (generated outside the controller). The setpoint can be ramped (gradually changed upward or downward) or can follow a predefined profile with different rates of change.

7. The controller has two operating modes: Auto and Manual. In Auto the controlled output is directed to the Control Device without operator intervention. In Manual the controlled output is bypassed, and the Control Device is driven directly by the operator.

8. A mimic or process graphic is a pictorial representation of the process. It can be static, showing the picture only, or dynamic when it includes live process parameter data.

9. A distributed control system (DCS) is a microprocessor-based control system with centralized hard wired inputs and outputs that are software-connected and configured to provide control and computation. To communicate with the system and the process, a DCS is always provided at the minimum with a visual display unit (VDU), plus a computer keyboard and printer.

ROAD MAP THROUGH THIS BOOK

This book has been arranged to provide a structure that in general replicates a typical control loop, in that it starts with the methods of obtaining a *measurement*, progresses to *control*, and then moves on to *applications* and *final control devices*. The last several chapters discuss the installation of measuring instruments, modern control systems, system communications, and safety.

Two tables of notation used in this book precede Chapter 1. The first lists symbols used in Chapters 1 through 5 on measurement. The second shows symbols used in Chapters 6 through 15 on control.

PART I: PRIMARY DEVICES AND MEASUREMENTS

Without a knowledge of what the process is doing at any instant, any desire to control it may as well be abandoned. Therefore, any endeavors to obtain the best possible results in measurement will make the quality of control that much better. Chapters 1 through 5 discuss the reasons behind the various methods of making a process measurement, each chapter being devoted to a specific topic so that the following can be found:

Chapter 1: allocated to flow measurement

Chapter 2: pressure measurement

Chapter 3: temperature measurement

Chapter 4: level measurement

Chapter 5: analytical measurements

PART II: PROCESS CONTROL

Chapters 6 through 10 explain the theory behind the control of the process variable once the measurement is obtained. The topics are as follows:

Chapter 6: gain

Chapter 7: lags (delays)

Chapter 8: control actions

Chapter 9: control equations—the mathematics of control (algorithms)

Chapter 10: tuning—methods by which controllers are adjusted for fast and accurate response to any error or change of measurement or requirements

PART III: APPLICATIONS

Chapters 11 through 15 discuss a few typical industrial applications of instrumentation and control.

PART IV: FINAL CONTROL DEVICES

Chapters 16 and 17 give an insight into how controlled devices operate. They discuss the final control devices—i.e., control valves, dampers, power cylinders—and give the mathematical formulas used by valve makers for sizing control valves. Actuators that produce large forces to move heavy final controlled devices are also discussed.

PART V: INSTALLATION OF INSTRUMENTATION

Chapter 18 outlines some of the problems that can be encountered in real process plants, and it also gives some of the practices to be followed together with the dangers that should be avoided when making an installation.

PART VI: MODERN CONTROL SYSTEMS

Chapters 19 and 20 discuss distributed control systems (DCS) and their implementation with examples of some real systems currently being used.

Chapters 21 through 23 discuss systems communication, safety, and process alarms.

Chapter 24 discusses Boolean or combinational logic, which is used frequently when dealing with circuit switching.

Chapter 25 discusses sequential logic, the bedrock of batch processing operations—a branch of control engineering that is fast gaining in status in pursuit of competitiveness and growth.

Notations Used in This Book

NOTATIONS USED IN CHAPTERS ON MEASUREMENTS

The symbolic notation shown here is used in Part I (chapters 1 through 5). Because the notation changes in succeeding chapters viz. Control, a second list follows this one.

A	Area	R	Resistance
a	Acceleration	$\Re$	Universal gas constant
B	Magnetic flux density	r	Radius (assigned by subscripts)
C	Specific heat	T	Torque, temperature (as applicable)
c	Capacitance	t	Time
D	Major diameter	u	Initial velocity
d	Minor diameter	V	Volume
E	Electropotential	v	Final velocity
E_λ	Energy radiated	W	Weight
e	Dielectric constant (used with subscripts where necessary)	z	Compressibility factor (for a gas)
		α	Mean temperature coefficient
F	Force	β	Coefficient of cubic expansion
f	Frequency	γ	Specific heat ratio
g	Gravitational acceleration	Δ	Change
h	Height	δ	Rate of change
K	Constant (assigned by application)	θ	Angular displacement
k	Constant (assigned by application)	κ	Specific conductance
l	Length	λ	Wavelength of light
M	Molecular weight	ρ	Density
m	Mass	σ	Boltzmann constant
p	Pressure	Ω	Moment of oscillation
Q	Quantity (assigned by subscripts per application)	ω	Angular velocity

NOTATIONS USED IN CHAPTERS ON CONTROL

The following symbolic notation is used in Part II (chapters 6 through 10) and in Part III (chapters 11 through 15). Some symbols that are similar to those used in the chapters on measurements are omitted to minimize the length of the notation listing. Not all the subscripts used in the text are listed here, but the meaning of any unlisted symbols should be apparent from the example set here.

A	Amplitude	E	Total emissivity (radiation)
b	Bias	Er	Error
C	Absolute velocity	e	Exponential constant (2.718)
c	Capacity	G	Gain

H	Head		t	Time
I	Current		u	Initial velocity
K	Constant (assigned by application)		u_r	Relative velocity
k	Constant (assigned by application)		V	Output volume voltage (as applicable)
L	Inductance			
l	Length (distance)		v	Final velocity
M	Instantaneous corrected value		W	Heat, weight, work watts [Power] (as applicable)
m	Mass (flow rate)			
N	Speed (rotational)		Wq	Heat transfer
n	Polytropic index		z	Compressibility factor
P	Proportional band		α	Cross-sectional area
p	Pressure		γ	Ratio of specific heats
Q	Flow rate (quantity)		θ	Measured variable
q	Fractional flow rate		τ_d	Dead time
R	Derivative action time		τ_0	Periodic time
R_v	Compression ratio		ϕ	Natural period of the process
$\Re$	Universal gas constant			
r	Resistance		**Subscripts**	
S	Integral action time		ad	Adiabatic
s	Specific heat		avg	Average
Sp	Setpoint		d	Differential
T	Torque, temperature (as applicable)		m	Molecular weight
			Th	Thermal
			vap	Vapor

PRIMARY DEVICES AND MEASUREMENTS

CHAPTER 1

Flow Measurement

In almost all process plants, *flow* is a parameter of prime importance, for the simple reason that most processes involve moving material from one part of the plant to another or from one piece of processing equipment to another. Each movement involves doing work on the material so as to reduce or increase its mass, alter one or several of its properties, add heat, or otherwise change its condition. In this chapter we shall learn the relationship between the volume flow and the differential head created when a restriction such as an orifice plate or a venturi section is placed in the flow stream. Although the use of meters based on this principle is diminishing, there are still a large number in operation today.

The magnetic flowmeter uses the principle of voltage induction, and the vortex meter uses vortex shedding as a means of measuring volumetric flow, and these instruments are gradually replacing the orifice plate and venturi meters. Measuring the flow of a viscous fluid is not possible with the meters just described; hence, we shall also look at methods using a target in the flow stream and a meter having gears forced to rotate by the fluid itself.

Mass flow has always been an essential process measurement but it was only very recently that the means to measure it directly via the Coriolis meter have been available. We shall investigate how this has been achieved. The Doppler effect and the time-of-flight methods are measurements used to determine flow rate, and both of these are discussed. High-accuracy measurements of low-viscosity fluids can be made with turbine meters, and we shall discuss these instruments later in this chapter.

The topics presented in this chapter are not exhaustive; they are only a collection of methods that cover some of the older and newer techniques and give a flavor of the subject.

BASIC THEORY OF FLOW

Any material in motion possesses energy, and this energy is called *kinetic energy*, from the Greek word for movement. Let us now consider the method of determining the kinetic energy possessed by a body in motion. **Note:** In this discussion, the mass of the body m is given as W/g, because direct measurement of mass is not generally easily made, whereas the weight of a body can be determined very simply. This statement has to be modified slightly in the light of modern technology, where the mass of liquid fluids can be measured directly using the Coriolis meter, which will be discussed later.

Suppose a body has a weight W and this body attains a velocity v from rest in a time t under the influence of a force F, which causes it to move a distance l. The product of the applied force and the distance over which it moves gives the work done:

Work done $= Fl$

We know that force has been defined as mass times acceleration:

$$F = ma$$

$$= \frac{W}{g}a$$

where g = acceleration due to gravity and a = acceleration of the body. We also know that the final velocity v of a body equals its initial velocity plus the product of its acceleration and the duration through which the acceleration is applied:

$$v = u + at$$

from which we can say:

$$t = \frac{v - u}{a}$$

where u is the initial velocity.

We know that the distance traversed under constant acceleration is the product of the average velocity and the time it took to cover the distance; mathematically:

$$l = \left(\frac{u + v}{2}\right)t$$

Substituting the expression for v we obtained above into this equation, we have:

$$l = \left(\frac{2u + at}{2}\right)t$$

$$= ut + \frac{1}{2}at^2$$

Since

$$l = \frac{u + v}{2}t$$

$$= \left(\frac{u + v}{2}\right)\frac{v - u}{a}$$

$$= \frac{v^2 - u^2}{2a}$$

we can also say

$$v^2 = u^2 + 2al$$

If we consider the initial velocity to be zero for the body starting from rest, then

$$l = \frac{v^2}{2a}$$

and substituting this into the equation for work done, we have

$$\text{Work done} = \frac{W}{g}a\frac{v^2}{2a}$$

$$= \frac{Wv^2}{2g}$$

which is the *kinetic energy* of the body (conforming to the familiar general relationship that energy is proportional to the square of the velocity). We can apply this equation to determine the kinetic energy of a fluid. For the purposes of this discussion, we shall consider a liquid, and let this liquid be a common one: water. If

W = weight of water flowing

v = velocity of flow

then as given previously the kinetic energy possessed by this flowing water is given by

$$\frac{Wv^2}{2g}$$

which is another way of expressing the general form of kinetic energy of any body:

$$\text{Kinetic energy} = \frac{1}{2}mv^2$$

where m is the mass and v the velocity.

The mass of the body in the flow form of the equation is shown by W/g; so, to make the flow form similar in appearance to the general form, we can write it as:

$$\frac{1}{2}\left(\frac{W}{g}\right)v^2$$

Hence, in general terms we can say the velocity head, or kinetic energy per unit weight, of the water in physical dimensions is given by

$$\frac{v^2}{2g} = \frac{\text{length}^2/\text{time}^2}{\text{length}/\text{time}^2} = \text{length}$$

which is the pressure head due to the fluid velocity or the *velocity head*.

Since the mass of fluid flowing is constant, any change in flow rate is dependent on the fluid velocity alone. This can be observed when one uses a hosepipe with a jet nozzle fitted to it; the more restricted the nozzle, the faster and further the jet will go.

In order for the water to flow, there must be a force applied to it, and the force necessary is derived from the applied pressure to give the water the acceleration needed to make it flow. We all know that *pressure* is defined as *force per unit area*. The required force will vary depending on the type of fluid involved, and more specifically upon the *fluid density*. Hence, we can redefine pressure in terms that include the density of the fluid. To do this, we shall continue to use water as the fluid and consider a column of water. If we let the column have a cross-sectional area of one square unit, then the volume of the water will be directly proportional to the height of the column, the *constant of proportionality* being the area of the cross section, in this case, unity, or 1.0. Since water has a density (mass per unit volume), it will exert a force on the base of the column, but we have already said that pressure is force per unit area, and therefore the pressure exerted by the column of water on the base is given by the height of the column multiplied by the density of the water. Thus, if p is the pressure, h is the height of the column, and ρ is the density, then

$$p = \rho h$$

and this, then, is the pressure energy of the water, or the pressure head.

We all know from experience that if we elevate a tank of water, then the water acquires energy due to its position, and this energy is called the *potential energy* of the water. Then, the *total energy* of the water in the general case will be the summation of all the energies involved:

Total energy = kinetic energy + pressure energy + potential energy

or, using symbols:

$$\text{Total energy} = \frac{Wv^2}{2g} + \frac{pW}{g\rho} + \frac{Wh}{g}$$

Since we transmit water from one place to another by pipe, it is easy to calculate the energy available at any point. Remember, pipes are not always laid on the same level, but the same pipeline could be run at several different levels. Steps must therefore be taken in the computations to account for all the changes in energy that are involved. Hence, in the general ideal case (no losses), we can say:

$$\frac{Wv_1^2}{2g} + \frac{p_1W}{g\rho} + \frac{Wh_1}{g} = \frac{Wv_2^2}{2g} + \frac{p_2W}{g\rho} + \frac{Wh_2}{g}$$

where subscripts 1 and 2 are any two locations.

If we consider real-world cases, then there are going to be some losses, and this relation above must be modified to take account of them. If the flow direction is from location 1 to location 2, the equation will then give

$$\frac{Wv_1^2}{2g} + \frac{p_1W}{g\rho} + \frac{Wh_1}{g} = \frac{Wv_2^2}{2g} + \frac{p_2W}{g\rho} + \frac{Wh_2}{g} + \text{losses}$$

This is Bernoulli's law, and it is no more than a restatement of the law of conservation of energy and mass.

PRIMARY DIFFERENTIAL-PRESSURE DEVICES

THE VENTURI METER

Having now arrived at an understanding of the theoretical aspects of flow, how do we make use of the information? To explain this further, making use of all that we have seen so far, we shall consider a *primary element*, another way of saying a device that is inserted into the line to create a pressure difference between specific points at that particular location in the flow stream. The device we shall consider for illustration is called a *venturi meter* or *tube*, after G. B. Venturi and shown diagrammatically in Figure 1.1.

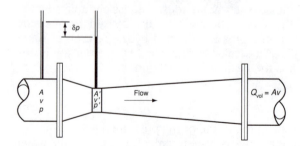

Figure 1.1: Schematic of a venturi meter.

In Figure 1.1 the convergent and divergent parts are suitably tapered to minimize the turbulence, with the internal surfaces made as smooth as possible to reduce the losses to a minimum. Let the meter be horizontal to eliminate the potential energy effect; then, from the sketch we can say that, if Q_{vol} is the volumetric flow rate of the water, A is the area of the entrance, A' is the area of the throat, v is the velocity at the entrance, and v' is the velocity at the throat, then:

$$Q_{vol} = \text{area} \times \text{velocity}$$

$$= Av \qquad \text{for the entrance}$$

$$= A'v' \qquad \text{for the throat}$$

$$Av = A'v'$$

that is since the quantity entering must be the same as that leaving the meter. Therefore

$$v = A'v'/A \tag{1.1}$$

Applying Bernoulli's law, we have:

$$\frac{p}{\rho g} + \frac{v^2}{2g} = \frac{p'}{\rho g} + \frac{v'^2}{2g}$$

$$\frac{v'^2 - v^2}{2g} = \frac{p - p'}{\rho g}$$

where p is the pressure of the liquid in the inlet pipe and p' is the pressure in the throat. Substituting Equation (1.1) into this, we have:

$$\frac{v'^2}{2g} - \left(\frac{A'}{A}\right)^2 v'^2 = \frac{p - p'}{\rho g}$$

$$\frac{v'^2}{2g}\left(1 - \frac{A'^2}{A^2}\right) = \frac{p - p'}{\rho g}$$

But

$$\frac{p - p'}{\rho g} = \delta p$$

which is the differential pressure head, and

$$1 - \frac{A'^2}{A^2} = \frac{A^2 - A'^2}{A^2}$$

Therefore

$$\frac{v'^2}{2g}\left(\frac{A^2 - A'^2}{A^2}\right) = \delta p$$

$$v'^2 = 2g\,\delta p\left(\frac{A^2}{A^2 - A'^2}\right)$$

Taking the square root of both sides, we have:

$$v' = \frac{A\sqrt{2g\,\delta p}}{\sqrt{A^2 - A'^2}}$$

The quantity of liquid flowing is

$$Q_{vol} = Av$$

Therefore

$$Q_{vol} = A \left(\frac{A' \sqrt{2g \, \delta p}}{\sqrt{A^2 - A'^2}} \right)$$

In a real meter, A and A' are constants which, when combined, give another constant, k; hence, we can say:

$$Q_{vol} = k \sqrt{2g \, \delta p}$$
$$= K \sqrt{\delta p}$$

if we regard $\sqrt{2g}$ as a constant as well, to give an overall constant K.

   This last equation is valid for all primary devices using a differential head as a means of flow measurement.

OTHER PRIMARY DIFFERENTIAL-PRESSURE DEVICES

The best known differential-pressure creating devices are:

Orifice plates

Venturi meters/tubes

Dall tubes (see Figure 1.2)

Venturi nozzles (see Figure 1.2)

Venturi flumes, of which there are two types: standard and Parshall; see Figure 1.3

A *pipe bend* has also been used as a differential-pressure generator, although this method has a major problem in that there is always considerable turbulence associated with the device. The reason being a deliberate and rapid change in flow direction, and that always causes a great amount of noise to be impressed on the measurement signal.

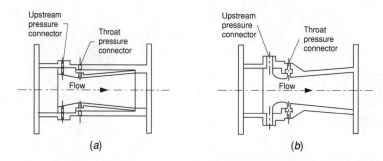

Figure 1.2: (*a*) General arrangement of a Dall tube. (Original design by Kent Instruments Ltd.) (*b*) General arrangement of a venturi nozzle. (Design principle following BS 1042.)

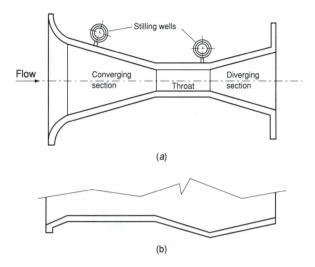

Figure 1.3: (*a*) Plan view of Parshall flume. (*b*) Contour of the base of the Parshall flume.

Venturi Flumes

Flow measurements of very large quantities of water like rivers and streams and also very common in the water and sewage treatment industries cannot be handled by the venturi meter primary device we discussed earlier; other methods have to be used. For these measurements, the standard venturi flume is employed instead. This device is capable of handling very large flow rates, usually given in millions of cubic units. The relative ease of maintenance of the measuring system is another reason for its popularity. What does one of these devices look like? In plan view the device looks, much like the Parshall flume without the contoured base, but the cross section is usually rectangular, the inlet shorter, and there is a gentle taper on the outlet similar to the venturi meter of Figure 1.1. To get an idea of the size difference, the pipe Venturi meter would usually have relatively small dimensions, frequently in millimeters/inches, whereas in the Venturi flume these would be measured in meters/feet. In these devices it is always the practice to perform the measurement of the differential pressure as the difference in levels in *stilling wells*, so called because of the effect they have in reducing the turbulence effects on the measurement. Air reaction or float and cable systems, which we shall discuss in Chapters 2 and 4, make the measurement itself. Because of their size, it is usual for these meters to be open to the surroundings just like a river or stream and therefore this type of measurement is referred to as *open-channel flow measurement*. Parshall flumes are very much more efficient than the standard flumes and generate a standing wave to effect a measurement.

Orifice Plates, Dall Tubes, and Flow Nozzles

At one time it was almost universal to find a flow orifice plate on a process line in which a flow measurement was required. The reason for this was that the device was relatively inexpensive to purchase, was easy to install, required virtually no maintenance, and offered an acceptable level of accuracy and repeatability. The situation is not the same today, for we have outgrown the capabilities offered by the

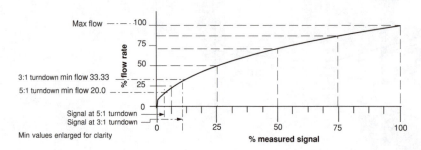

Figure 1.4: Turndown ratios for differential-pressure instruments.

orifice plate. Process engineers are now demanding much greater *turndown ratios* on the primary device, where turndown ratio is defined as the ratio of the maximum measured flow to the minimum measured flow at a given accuracy:

$$\text{Turndown} = \frac{\text{maximum flow}}{\text{minimum flow}}$$

Suppose the maximum flow was 9000 units and the minimum flow was 3000 units; then from the foregoing we can say:

$$\text{Turndown} = \frac{9000}{3000}$$

That is, the turndown ratio is 3:1, in which a specific case can be identified in Figure 1.4.

This low turndown ratio is a property of almost all differential pressure creating devices and is clearly shown on the graph of Figure 1.4. From this diagram it can be seen that the larger the turndown, the more cramped the measurement signal will be at low flow rates. At best the turndown ratio of an orifice plate is about 5:1 but more typically it is of the order of 3:1. Compare this to a magnetic flowmeter with a turndown of 40:1, or a Coriolis meter, where the ratio is quite often 100:1; these are straight through (obstructionless) flow devices, which in general have low loss characteristics and we shall discuss both these instruments later in this chapter. It is important to realize that, apart from the difficulty of reading the scale, for the lower flow rates the accuracy of the measurement is also much reduced (remember, we are considering only the measured variable itself as related to the linear signal produced by the transmitter). It is quite usual for some instrument manufacturers to make accuracy statements for linear scales in terms of full-scale values, i.e., statements like ±0.5% FSD (full-scale deflection). What does this mean? An explanation is as follows. If the measurement signal corresponds, say, to a maximum flow with a span of 100 units, then the transmitted or indicated value at maximum span will be accurate to within ±0.5%. If now the measurements were to be reduced to 50%, then the instrument would be only ±1.0% accurate and at 10% of measurement the instrument would be only ±5% accurate—the computed accuracy being derived from

$$\frac{\text{Full-span accuracy}}{\% \text{Measured span}}$$

Therefore, at, say, 40% span, it would be $0.5/40 = \pm1.25\%$ accurate, which shows that, with accuracy statements like that given, the lower the measured value, the less accurately the measurement is made.

To address the low turndown ratio limitations imposed when using differential-pressure cells for flow measurement, the *smart differential pressure cell* was devised. This new instrument was possible only with the advent of the microprocessor chip.

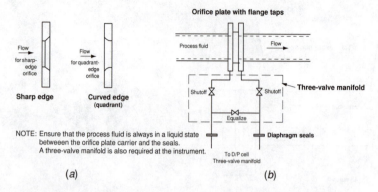

Figure 1.5: (*a*) Sharp-edge and quadrant-edge orifice plates. (*b*) Installation of a differential-pressure cell with chemical seals.

The "smart" instrument allowed recalibration of the device without removing it from the process line. The effect of this was an immediate increase in the turndown ratio capability of the device, but only with calibrations within the range of the sensor (D/P capsule, Bourdon, etc.; these are pressure elements that will be discussed in Chapter 2). In all of this do not lose sight of the fact that the orifice bore still imposes a limitation on the measurement, for its dimension is unchangeable. The range limitation would now be that of the orifice plate itself. Furthermore, the pressure loss across an orifice plate is very high and almost totally unrecoverable, increasing the pumping "overhead" both pressurewise and financially.

To handle flow rates that are larger than the capabilities of an orifice plate with an acceptable differential pressure loss across it, Dall tubes and flow nozzles were invented. These primary devices have specially contoured inlets, as shown in Figure 1.2, that guide the flow stream smoothly towards the calculated bore that creates the required differential pressure to effect the flow measurement. Unlike the venturi meters there is no contoured exit, the element terminates at the calculated bore; this arrangement results in a "chamber" in the vicinity of the bore where a low pressure is generated. The difference between the pressure of the approaching fluid and that generated in the chamber is the differential pressure of the measurement. Size for size these elements have the capability of handling much larger flows with a lower pressure head loss than a comparable orifice plate.

There are two types of orifice plates: (1) the sharp edge and (2) the curved or quadrant edge, the sharp-edge device being used for fluids of low viscosity and therefore more common, with the curved-edge device specified for viscous fluids. Particularly in the latter case, the process fluid must be prevented from entering the measuring device, which is usually a differential-pressure meter/transmitter. To achieve this, *chemical seals* are employed to isolate the process fluid from the measuring instrumentation; the pipe work or, more correctly, the capillary tubes between the seals and the measuring instruments, are filled with an appropriate fluid, usually silicone oil. See Chapter 2 for a discussion of chemical seals. Figure 1.5 illustrates some of the things we have been discussing.

MEASURING DEVICES FOR VISCOUS LIQUIDS

If the process fluid is very viscous, as are heavy fuel oils, then even curved-edge orifice plates are not suitable, and one has to resort to other means of measurement such as the target meter or oval-gear meters. A point worth remembering is that

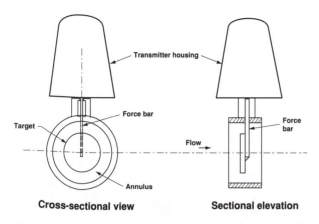

Figure 1.6: Schematic diagram of a target meter.

heavy fuel oils cannot be combusted in their normal state, since the oil would not flow through the burner nozzles. The oil has to be heated up to reduce the viscosity; steam-heated coils usually carry out this heating and when suitable the oil is *atomized* (forced into a fine vaporous form), also using steam for the purpose. Quite often it will be found that in fuel oil systems the burners are of the spill type. This means that not all the oil supplied for combustion is actually burned. Therefore, in these cases, it is important to measure the flow of both the supply and the return lines (recirculation system) to determine the quantity burnt and to use this value to manipulate the fuel/air ratio.

The Target Meter

The *target meter*, as its name implies, presents a physical target to the fluid flow. The actual instrument is shown diagrammatically in Figure 1.6. The instrument works as follows. The fluid being measured impinges on the circular target and is forced to give up some of its momentum (resulting in a pressure difference across the target); this loss of momentum is reflcted in minimal deflection of the force bar for instruments that use it as a feedback source and at the same time forces the fluid to flow through the annular space between it and the enclosing pipe body. The size of the target is calculated so that the size of the annulus created is such that the quantity of fluid displaced is the maximum obtainable for the given pressure (driving force). The force bar constrains the target to a specific position with respect to the enclosing pipe body; the transmitting mechanism is zeroed for no force applied to the bar, i.e., no fluid flow. The deflection of the force bar under the action of the fluid flow is proportional to the applied force and therefore proportional to the square of the volumetric flow rate. Some targets are sharp edged.

The mathematics of the fluid involved are as follows. The liquid is brought to rest, and its pressure is thereby increased. The force F on the target is given by:

$$F = \frac{k \rho v^2 A_t}{2} \text{(Newton)}$$

where ρ = density (mass/unit volume), kg/m^3
A_t = target area, m^2
v = fluid velocity, m/s
k = constant

From this we can obtain the velocity in terms of flow:

$$2F = k\rho v^2 A_t$$
$$= k\rho v^2 \frac{\pi d^2}{4}$$
$$8F = k\rho v^2 \pi d^2$$

and therefore

$$v = \sqrt{\frac{8F}{k\rho\pi d^2}}$$

where d is the diameter of the target.

If D is the internal diameter of the pipe, then the area of the annulus is given by

$$A_{annulus} = \pi\left(\frac{D^2 - d^2}{4}\right) \qquad m^2 \ (ft^2)$$

The volumetric flow rate in m^3/s (ft^3/s) is given by

$$Q = Av$$
$$= \frac{\pi(D^2 - d^2)}{4}\sqrt{\frac{8F}{k\rho\pi d^2}}$$
$$= \frac{K(D^2 - d^2)}{d}\sqrt{\frac{F}{\rho}}$$

where K is the collective constant.

Positive-Displacement Meter (Oval-Gear Type)

Oval-gear meters as the name implies, uses oval-shaped gears as the measuring system; see Figure 1.7. These meters are positive-displacement devices, i.e., they will deliver a specific quantity of fluid per rotation and are used to measure the flow of viscous fluids, often replacing the target meters.

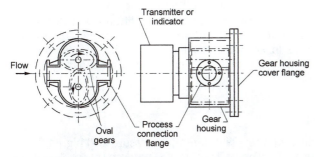

NOTE: The process fluid provides the motive power to drive the gears. The gears are magnetically coupled to the instrument mechanism.

Figure 1.7: Schematic arrangement of an oval-gear meter.

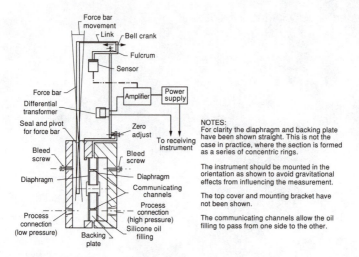

Figure 1.8: Schematic arrangement of a differential-pressure cell.

DIFFERENTIAL-PRESSURE-MEASURING INSTRUMENTS

This section describes several instruments that measure differential head.

DIFFERENTIAL-PRESSURE CELLS

Differential-pressure cells are the "workhorse" of modern differential-pressure measurements and are manufactured in a large variety of forms; see Figure 1.8. Each variant is designed to obtain the required measurement at the optimum accuracy. The design is very modular and basically comprises a differential-pressure capsule (see Figure 2.3 in Chapter 2) clamped between two end pieces that provide the means of connecting the two pressure lines from the primary element (orifice plate, venturi meter, etc.) to the measuring capsule. The capsule itself consists of two diaphragms welded one to each side of a backing plate, used for overpressure limiting, with the intervening spaces between the diaphragm and backing plate in communication with each other and filled with a fluid, normally silicone oil. Alternative fluids are used for special services. When this coupled constant-volume diaphragm pair is subjected to a different pressure across each diaphragm, the capsule assembly is subjected to a net force dependent on the pressure difference. A force bar system converts this feedback-balanced force into a movement and applies it via a mechanism to a transducer designed to produce a signal representing the difference between the two pressures. The actual movement is of the order of a few micrometers or a few thousandths of an inch. The transduced/transmitted signal is either an electronic or a pneumatic one of the appropriate range that is directly proportional to the pressure differences between the two process connections.

U-TUBE MANOMETERS

U-tube manometers are divided into low differential-pressure and high differential-pressure devices. The liquid fill accounts for the distinction between low-range instruments, which use water or a light oil, and high-range instruments, which use mercury. As U-tube manometers per se, they serve only as indicators.

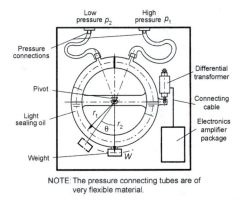

NOTE: The pressure connecting tubes are of very flexible material.

Figure 1.9: Schematic arrangement of a ring balance transmitter.

RING BALANCES

Ring balances are another version of the U-tube manometer, but in this form can be made to work as ring tube transmitters; see Figure 1.9.

The rotating moment is equal to $(p_1 - p_2)Ar_1$, where A is the cross-sectional area of the bore, r_1 the ring mean radius, and $p_1 - p_2$ the difference in pressure. The restoring moment is $r_2 W \sin \theta$, where W is the counterbalance weight and r_2 the radius away from the center of rotation. Thus,

$$(p_1 - p_2)Ar_1 = r_2 W \sin \theta$$

when in equilibrium, or

$$p_1 - p_2 = \left(\frac{r_2}{r_1}\right)\left(\frac{W}{A}\right) \sin \theta$$
$$= k \sin \theta$$

For small angular movements, the angle of rotation is proportional to the differential pressure (and thus to the square of the flow rate through the associated primary element). These instruments are always low-range devices using a very light oil as the fill fluid.

BELL-TYPE METERS

Bell-type meters again use the manometer principle and a light oil as the fill liquid. As its name implies, each measuring unit is a calibrated one-end-open cylinder, the open end being immersed in a trough of oil that is designed not to completely fill the cylinder. There are two such cylinders balanced on either end of a pivoted beam; see Figure 1.10. Pressure from the point of measurement is directed into the space above the oil via a tube that projects above the surface of the oil in the trough. With the same pressure in each cylinder, the beam is balanced. Any change in pressure will cause the beam to pivot, the amount it pivots being proportional to the differential pressure. These instruments can be used as transmitters, recorders, and indicators.

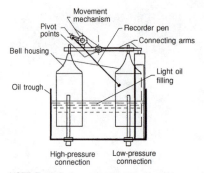

NOTE: For clarity the case and chart backing plate are not shown.

Figure 1.10: Schematic arrangement of a bell-mechanism recorder.

SLACK DIAPHRAGM

In slack diaphragm instruments, a slack neoprene diaphragm partitions the measuring chamber and provides a good pressure seal; see Figure 1.11. Inlet ports are provided to each side of the diaphragm unit, and these ports are in communication with the pressures being measured. Through one side of the chamber slides a slender shaft that works through another pressure seal to the pointer/output mechanism. With equal pressures to each port, the diaphragm is in equilibrium; a change in one side or the other will cause the diaphragm to move in the direction of the applied pressure against the spring shown. Therefore, with unequal pressures applied to the two ports, the diaphragm will take up a position of equilibrium that is proportional to the difference of the two pressures. These instruments are used as indicators, recorders, and transmitters.

VIRTUALLY OBSTRUCTIONLESS FLOW-MEASURING INSTRUMENTS

There are further methods of measuring flow using principles other than differential pressure, and we shall look at some of the more recent and important ones and discuss the operating principles and details in each case.

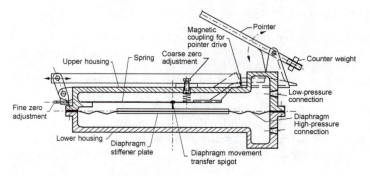

Figure 1.11: Schematic arrangement of a slack diaphragm indicator.

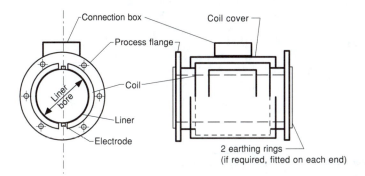

Figure 1.12: Magnetic flowmeter.

THE ELECTROMAGNETIC FLOWMETER

Let us first examine the *magnetic flowmeter*. This instrument operates on the law proposed by Michael Faraday, which can be paraphrased as, "When an electrical conductor moves in a direction that is perpendicular to a magnetic field, an emf is set up on the conductor; the magnitude of the induced emf is proportional to the strength of the magnetic field, the length of the conductor, and the speed at which the conductor moves through it." Or, in symbols,

$$\text{emf} \propto Blv$$

where B is field strength, l is conductor length, and v is velocity of the traverse (perpendicular) component. If we make B and l constant and call this product k, then:

$$\text{emf} = kv$$

In a real-life magnetic flowmeter, the magnetic field is developed by an electric coil in two halves fixed on the outside of a pipe diametrically opposite each other, the size of the pipe being selected for the required flow; see Figure 1.12. The coils are formed to follow the contour of the outside of the pipe and arranged to leave equal gaps on either side. Electrodes are passed through the gaps so that they are in intimate contact with the process fluid, which itself *must not* be electrically nonconducting.

The instrument works as follows. The coils produce an intense uniform magnetic field. The distance between the electrodes represents the length of the conductor, the fluid itself being the conductor. Since the magnetic field and the distance between the electrodes can be held constant, they can be seen as being equivalent to k in the foregoing equation. When fluid flows, the electrodes will pick up the induced voltage, which is proportional to the velocity of the fluid. Because the bore of the pipe to which the coils and electrodes are fitted is known, the volume of fluid flowing can be measured, or symbolies:

$$\frac{\pi d^2}{4} v = \text{area} \times \text{velocity} = \text{volume per unit time}$$

Most of these meters are powered from the ac supply, but the waveforms are nonsinusoidial to avoid transformer effects (quadrature/phase) that produce interfering voltages and zero drift. Pure dc excitation is avoided because it polarizes the electrodes, and because of the electrochemical effect that can sometimes occur.

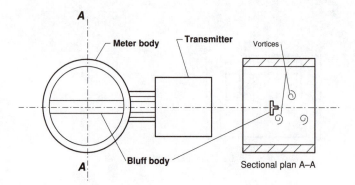

Figure 1.13: Vortex flowmeter—wafer type.

THE VORTEX FLOWMETER

The vortex flowmeter, shown in Figure 1.13, is one of the newer flow-measuring devices that have been gaining much ground in recent years and is fast replacing the orifice plate/differential-pressure transmitter arrangement in many process plants. The reason for the popularity is that a single in-line device is all that is necessary to produce a measurement, coupled with the advantage that the turndown obtained is very much higher and of the order of 40:1, as the instrument is fundamentally a linear-response device.

The instrument works as follows. The *bluff body* presents an obstruction to the fluid flow, causing the flow stream to divide, and in so doing causes a change in the velocity profile of the stream, the lowest velocities being near the bluff body. A simple, but a little crude, analogy is the situation that occurs when a stream of people moving along a street is suddenly forced to divide by an inserted barrier and the people go through two gates; those nearest the gateposts could find themselves not moving at all for a little while. From the edges of the bluff body, vortices are generated due to the separation of the flow stream, and reverse flow occurs below the line of zero velocity with a pressure difference as the result. The number of vortices shed by the body is proportional to the fluid flow rate. It is reasonably easy to visualize the eddies that are formed in the wake traveling in the direction of flow and therefore away from the bluff body or vortex shedder. There is a pressure difference between the main flow and the eddy, and it is this pressure difference that is sensed as a means of detecting the presence of an eddy; if we count the number of eddies in a unit time period, then we can evaluate the flow rate. In some designs, the sensors are mounted on the side of the bluff body behind the surface that causes the stream to divide, while other designs have the sensors mounted away from the bluff body, but in a position where the eddies can be detected.

The instrument shown in the diagram is usually called a wafer-type instrument, which means that it has to be fitted between two Process pipeline flanges, with through bolts between the two flanges retaining the device in its position. This is not the only type of instrument available, for there are instruments with flanged ends or "sanitary" end fittings. The sanitary end connections are highly specialized fittings that inhibit the retention of any process fluid at the joints; they are mandatory in process lines that move food, pharmaceuticals, or products in which bacterial growth is not permitted to occur. While on the subject of fitting primary devices into pipelines, perhaps it may be well to state that in general the following installation connection methods are available for most instruments:

Flanged ends

Screwed (threaded) ends

Welded ends

Sanitary ends

Selecting the correct connection will depend on the process condition and is largely influenced by the pressure and temperature of the fluid involved. The material of construction of the meter body, sensor, and process connections will also be determined by the compatibility of the material with the process fluid. There are tables of materials and their reaction to chemicals that give an idea of the suitability, but it is recommended that expert advice be sought in difficult circumstances. As a general rule, the pipeline specification will be a good guide to the type of process connections and material of construction. However, be aware that instrument makers do not, or, more correctly, cannot make equipment in every different material, mainly for economic reasons. Hence, they (instrument makers) always choose materials and process connections that have the broadest applicability to the process fluids usually encountered.

A word of caution on blindly using the pipe specification as a guide: In some instances the material—perhaps say, 316L stainless steel—for the pipeline is chosen even though it may not be the most compatible (i.e., the effect of the process fluid that does erode it over time is considered to be minimal), but because it is the most economic. It is accepted that under the circumstances the amount of material taken off the pipe wall is such that the reduction in wall thickness does not seriously affect the integrity and safety of the piping system over the design lifespan of the line. In this situation, because of the aggressiveness of the process fluid if a differential-pressure transmitter with a 316L stainless steel capsule assembly were to be used, the limited thickness of the diaphragm itself will make the diaphragm be liable to failure under chemical attack, and an alternative diaphragm material will have to be used to provide the measurement. This assumes that the alternative material is available as an option; otherwise chemical seals of the appropriate material will need to be provided on the instrument to measure the flow. For details of chemical seals, see Chapter 2.

THE TURBINE FLOWMETER

Prior to the vortex meter, the need to obtain a stable and accurate measure of flow, especially in applications involving fluids subject to excise duty, or aircraft fuel systems, was satisfied by the turbine flowmeter. Turbine flowmeters are constructed around a set of unpowered and freely rotatable hydro- or aerofoils held like an aircraft or marine propeller on the central longitudinal axis of the pipeline. The type of foil used is dependent on the fluid being measured. Since the foil is made to rotate solely by the action of the flowing fluid its rotation is used to determine the flow rate. Magnets are embedded in the rotor housing, and a suitable pickup coil, isolated from the fluid, is placed outside the periphery of the rotor blades. The rotating magnets induce a voltage pulse in the coil each time they pass by the coil. The pulse frequency is proportional to the velocity of the fluid, and since the diameter of the pipe containing the rotor is uniform, the generated pulse frequency is proportional to the volume flowing. The measurement signal is inherently digital and thus capable of high accuracy using digital counting circuits. The meter design is both mechanically and electronically sophisticated, the support bearing friction, the lift and drag forces on the foil, and the axial thrust that the rotating arrangement generates being allowed for and the counting circuits designed to be capable of high resolution, frequency, and responsiveness (sensitivity). Analog output signals of flow are possible using elegant digital-to-analog conversion to minimize the inevitable errors involved.

It is important that on the upstream side, any pipeline device such as valves placed nearby, does not cause the fluid to accelerate beyond the normal (use turbulence and flow straighteners to minimize these effects), as this will cause excessive rotational speed and have serious consequences on the life of the meter.

THE CORIOLIS MASS FLOWMETER

Before the advent of the Coriolis mass flowmeter, the only way to determine true mass flow was to obtain the volumetric flow and multiply this by the fluid density. On-line densitometers are not the easiest of instruments to maintain, for the simple reason that the methods used require frequent calibration to keep them performing correctly. The versatile differential-pressure (D/P) transmitter can be used as a means of measuring density. When the D/P transmitter is used in this application, the vertical distance between the two measuring points depends on the measurement span of the density, and the connections must therefore be accurately positioned. The scheme will work only if the fluid velocity is very low; because of this it is not suitable for most process lines, but it is acceptable for process tanks where the level does not change rapidly.

The French physicist G. de Coriolis was the first to observe the effect of an apparent force that caused bodies (when viewed from an imaginary point above the North Pole) to deflect to the right in the Northern Hemisphere and to the left in the Southern Hemisphere, due solely to the rotation of the earth—the washwater in the basin and plughole phenomena. The magnitude of the force is related to the mass of the body and the rotational speed of the earth. In recognition, the force was named the Coriolis force.

The Coriolis mass flowmeter works on the same principle by causing the fluid to "rotate," although the rotation is not literal; but the pipe containing the fluid is vibrated at a predetermined frequency—and we all know that circular rotation can be converted mathematically into an oscillation (or vice versa) resulting in a sinusoidal waveform. This has the effect of generating the Coriolis force, which is measured by sensors mounted near the pipe.

There are a few specific points worth considering when applying this instrument. The first is personnel and plant protection in the event of measuring tube rupture due to stress cracking of the welds or the tube wall itself if the wall thickness is insufficient; one or both can be brought about by the forced vibration of the measurement technique, but the provision of a secondary containment enclosure can avoid spillage until a replacement is effected. Others are ensuring an equal split of the flow stream in those meter designs having two parallel tubes, and the exclusion of air from the process fluid, since it seriously affects the measuring accuracy. Figure 1.14 shows the geometry of some of the meter pipe work used.

For those who wish to understand the principle of operation of the Coriolis meter, we shall now consider the device in more detail. For this purpose we shall use a tube of a very simple geometry, as this will reduce the mathematics involved but nonetheless give a very good idea of how the meter works. Figure 1.15 shows the meter tube we shall work with for the discussion.

Let the fluid have a mass m and a velocity v. Let the fluid enter the tube that is rotating with an angular velocity ω about the axis O-O. As mentioned earlier, a

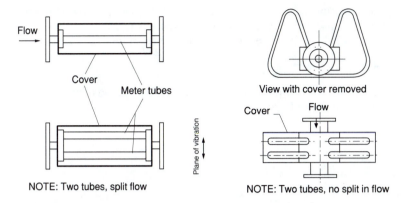

Figure 1.14: Coriolis mass flowmeter. The plane of vibration is common to both instruments.

sinusoidal vibration can simulate this rotation. The Coriolis force induced by the flow is given by:

$$F_c = 2m\omega x v \qquad (1.2)$$

where F_c is the Coriolis force, m is the mass of the fluid in the pipe, v is the velocity of the fluid, ω is the angular velocity of the tube, and x is the vector cross product operator (conversion factor) that aligns the angular with the linear velocity. Note that the equation just given is equivalent to the well-known Newtonian equation for force:

$$F = ma$$

the difference being that it has a rotational frame of reference. The inlet and outlet velocities are equal, but opposite in direction.

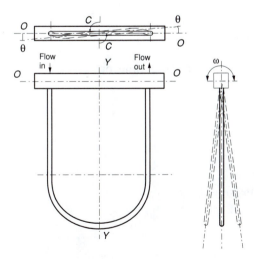

Figure 1.15: Geometry of a Coriolis meter tube. The tube shown is only one of the many tubes available.

Considering the elevation view at the top of Figure 1.15, we find that since the tube vibrates about the axis O-O, the Coriolis force creates an oscillatory moment Ω about the central axis Y-Y of the tube, and this moment can be expressed by:

$$\Omega = F'r' + F''r'' \tag{1.3}$$

where r is the radius. But $F' = F''$ and $r' = r''$, so if we write F as the common force and r as the common radius, we have:

$$\Omega = 2Fr$$

from which

$$F = \frac{\Omega}{2r}$$

We can now rewrite Equation (1.2) in these terms to give:

$$\frac{\Omega}{2r} = 2m\omega xv$$

Therefore

$$\Omega = 4m\omega xvr \tag{1.4}$$

Now mass equals volume times density and if we let ρ be the density, A be the cross-sectional area, and l be the length of the meter tube, then:

$$\text{Mass} = Al\rho$$

But the fluid is moving at a velocity v which can be written as:

$$v = \frac{l}{t}$$

Therefore

$$t = \frac{l}{v}$$

If Q_{mass} is the mass flow rate, which is really mass per unit time, or

$$Q_{mass} = \frac{m}{t}$$

then, combining, we have:

$$Q_{mass} = \frac{mv}{l} \tag{1.5}$$

Using this last relationship, we can rewrite Equation (1.4) as:

$$\Omega = 4\frac{Q_{mass}l}{v}\omega xvr$$

$$= 4Q_{mass}l\omega xr$$

This equation defines the force that acts against the stiffness of the meter tube material that produces the deflection θ in the tube. In this respect the tube behaves exactly as a helical spring, and for any helical spring the torque is given by:

$$T = k\theta$$

where T is the torque, k is the spring stiffness, and θ is the angle through which the force is applied. Thus, we can say $T = \Omega$ and rewrite the last equation as:

$$k\theta = 4Q_{mass}l\omega xr$$

or we can rearrange the terms to give:

$$Q_{mass} = \frac{k\theta}{4\omega xrl}$$

In a real meter design all the terms on the right-hand side can be obtained, so that measuring mass flow by the Coriolis forces involved is a physical reality.

ULTRASONIC FLOW MEASUREMENT

We conclude by stating that ultrasonics are also used in flow measurement, and for this a source of the ultrasound and a receiver is needed. An important aspect to be considered is the effect on sound transmission when entrained particles in the fluid are involved. In one method of measurement there is active dependence on entrained particles, and in another the design is such that the dependence on discontinuities is avoided altogether. Both these methods of flow measurements will be discussed using the following instruments.

The Doppler-Effect Ultrasonic Flowmeter

The first type of instrument we consider is one that makes use of the frequency shift whenever there is relative motion between the frequency source, the receiver, and/or the carrier medium. The resulting change in frequency is termed the *Doppler effect*. Figure 1.16 shows the arrangement; with this instrument the entrained particles in the fluid are used to provide the measurement.

If f_1 is the transmitted frequency, f_2 is the reflected frequency, v is the fluid velocity, and θ is the angle of the transmitted ultrasonic beam relative to the flow stream, then:

$$f_2 = f_1 \pm 2v(\cos\theta)\frac{f_1}{C}$$

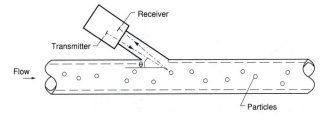

Figure 1.16: Schematic arrangement of a Doppler meter.

where C is the velocity of sound in the fluid. Rearranging, we have:

$$v = \frac{C(f_2 - f_1)}{2 f_1 \cos\theta}$$

which shows that the velocity is proportional to the frequency change and depends on the discontinuities of the particles in the fluid. This means that there must be entrained particles of some kind contained in the fluid, for it is not possible to use this technique to measure the flow of clear fluids. The pipe itself must also be capable of transmitting the ultrasonic signals. Since the dimensions of the pipe are constant, the cross-sectional area will be constant and therefore the fluid flow rate can be determined. This technique is also very dependent on the presence of entrained solids in the fluid and is not suitable for, say, clear liquids.

The Time-of-Flight Ultrasonic Flowmeter

The next method of determining flow rate also uses ultrasonics but does not depend on the existence of particles or discontinuities in the fluid as required by the Doppler method; rather, it works on a "time-of-flight" principle. In this arrangement there are two transmitters/receivers placed on opposite sides of the pipeline, as shown in Figure 1.17. The transmitters put out ultrasonic waves of a predetermined frequency, once in the direction of flow and then in the opposite direction, the difference in frequencies being proportional to the average fluid velocity and hence the flow rate.

Let

t_1 = pulse transit time downstream

t_2 = pulse transit time upstream

l = length of path between the two transmitters/receivers within the pipe

C = velocity of sound in the fluid

v = fluid velocity

θ = angle of pulse stream to the fluid

Then, since elapsed time is given by distance traversed divided by velocity, we can

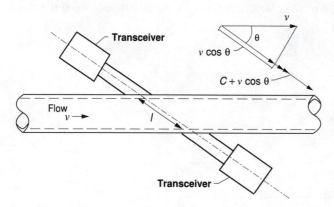

Figure 1.17: Schematic arrangement of a time-of-flight meter.

say (where $t_1 > t_2$)

$$t_1 = \frac{l}{C + v \cos \theta}$$

$$t_2 = \frac{l}{C - v \cos \theta}$$

$$t = t_1 - t_2 = \frac{2lv \cos \theta}{C^2 - v^2 \cos^2 \theta} = \frac{2lv \cos \theta}{C^2}$$

Since v^2 is very small in comparison to C^2, we can ignore it. For convenience, we shall develop a relationship that removes the dependency on the speed of sound and use the signal frequency instead. If f_1 is the effective signal frequency of the downstream pulse as received by the appropriate element and f_2 is the signal frequency of the initiating upstream pulse, and if $f_1 = 1/t_1$ and $f_2 = 1/t_2$, then the average velocity is given by:

$$\bar{v} = \frac{v}{\cos \theta}$$

or we can say:

$$f_1 - f_2 = \frac{(2\bar{v} \cos \theta)}{l}$$

which removes the dependence on the velocity of sound C, while path length l and θ are parts of the calibration constant.

HOT-WIRE ANEMOMETERS

Hot-wire anemometers are frequently used to measure the flow of gases. This device is a resistance wire carrying a current causing it to heat up. Under no-flow conditions the wire will attain a temperature dependent on the resistance and the value of the current, i.e., the $I^2 R$ effect. As flow starts, then the wire temperature will drop depending on the flow velocity and the thermal conductivity of the gas itself, and this will be manifest in the change in the value of the resistance, which will in turn represent the flow rate via the appropriate instrument calibration.

DATA REQUIRED TO PROVIDE A MEASUREMENT OF PROCESS FLOW

This section identifies the general required data relating to the containment of process fluids relevant to flow conditions. The list of data corresponds to what would appear on a checklist for a suitable flow measurement system.

PROCESS FLUID

Name

Characteristic   Gaseous or liquid

Chemical action   Aggressive or nonaggressive (**Note:** Aggressiveness is defined as the ability of the fluid to attack chemically or physically the containment or contact materials)

PROCESS PARAMETERS

For the following, the appropriate engineering units are required.

Flow rate

Maximum operating

Normal operating

Pressure

Maximum operating

Normal operating

Temperature

Maximum operating

Minimum operating

Normal operating

Viscosity

Normal operating

Density

Flowing (kinematic)

PHYSICAL LOCATION

In the following, give the standards with which the material must comply—e.g., BS, ANSI, etc. In the case of process connections, the pressure rating must also be given.

Straight pipe runs are defined as lengths of pipe without any pipe fittings (tees, elbows, reducers, valves, etc.) included.

Process line	Nominal size
	Pipe schedule
	Material of construction
Process connections	Screwed
	Flanged
	Welded
	Hygienic (exclusively for food and drugs)
Straight pipe runs	Example: upstream—min 15 to 20 pipe diameters; downstream—min 7 to 10 pipe diameters

PRIMARY DEVICE

The type of primary device required must be specified, since it is to be mounted in the process line. The maximum pressure drop that can be tolerated must also be specified.

The standards to which the element design is to conform—e.g., BS 1042 (British Standards) or ASME (American Society of Mechanical Engineers)—must be stated, together with the formulas by which the sizing calculation is to be carried out, e.g., BS 1042 ASME, or those given in *Flow Measurement Engineering Handbook*, by R. W. Miller published by McGraw-Hill Book Company. The last publication mentioned provides a single reference source for all closed-conduit flow measurement and has replaced the very well known and widely used work by L. K. Spink published by the Foxboro Company (Foxboro, Ma., U.S.A.).

APPLICATION

Indication	Local or remote
Recording	Local or remote
Integration	Local or remote (flow totalization)
Control—local	On/Off or modulating
Remote	On/Off or continuous (**Note:** Local is defined as having the measurement or control function at or near the point of penetration of the sensor or transducer into the process. Remote is defined as having the measurement or control function at some distance from the point of penetration of the sensor or transducer into the process.)

Do not be too anxious about the control aspects at this stage. They are stated here for completeness only and will be discussed in further detail in later chapters.

CRITERIA FOR SELECTING A SUITABLE FLOW-MEASURING INSTRUMENT

SENSING ELEMENTS

Differential-pressure sensors can generally be divided into two categories:

1. Force-balance differential pressure. All the modern D/P cells fall into this category.

2. Motion-balance differential pressure. The ring balance, bell-type, bellows-type, and slack diaphragm instruments fall into this category.

OPERATING RANGE

The operating range and static pressure rating are very important when selecting a measuring instrument. The static pressure rating is the maximum pressure that can be applied to the instrument. In the case of the differential-pressure cell it is the maximum pressure that the body and backing plate for the cell diaphragms can withstand without rupture.

For differential-pressure sensing instruments, these two criteria can roughly be broken down into the following:

Low range. Used typically on airflow measurements on furnaces, or those applications where the static pressure is relatively low—e.g., slightly above atmospheric pressure—and the differential is approximately within a span of 12.5 mm (0.5 in) H_2O or a span of 25 mm (1.0 in) H_2O (typical spans offered by manufacturers).

When used on these low spans, the static pressure rating is of the order of 5 lb/in^2.

Medium range. Used typically on applications where the differential is approximately within a span of 2540 mm (100 in) H$_2$O or a span of 5080 mm (200 in) H$_2$O. The static pressure in these applications can be relatively high. The instruments can be made with a static pressure rating of 40 bars (600 lb/in^2), or other values up to 400 bars (6000 lb/in^2).

High range. Used typically on applications where the differential is approximately within a span of 3040 mm (120 in) H$_2$O or a span of 19,050 mm (750 in) H$_2$O. The instruments made can have static pressure ratings of 100 bars (1500 lb/in^2), or 200 bars (3000 lb/in^2) or 400 bars (6000 lb/in^2).

AGGRESSIVE FLUIDS

Materials of Construction

Where the compatible material for aggressive fluids is the same as one from which the sensor can be manufactured, then the instrument as made will be suitable. However, it is more usual for the user to specify a suitable material and leave the manufacturer to comply with it. If none of the materials of the instrument are suitable, then provision of chemical seals between the process and the measuring instruments may be the alternative.

Chemical Seals

As mentioned earlier, chemical seals are physical barriers installed to separate the process fluid from the measuring device. There are several manufacturers of such equipment; two of the more well known are Ashcroft and Wika. The former is a US manufacturer, while the latter is European.

For more details on chemical seals, please refer to Chapter 2 on pressure measurement.

SUMMARY

1. Bernoulli's law is no more than a restatement of the law of conservation of energy and is given by:

$$\frac{Wv_1^2}{2g} + \frac{Wp_1}{g\rho} + \frac{Wh_1}{g} = \frac{Wv_2^2}{2g} + \frac{Wp_2}{g\rho} + \frac{Wh_2}{g} + \text{losses}$$

where v is the velocity of flow, p is the pressure, h is the height of the column, ρ is the fluid density, g is the acceleration due to gravity, and subscripts 1 and 2 stand for two locations.

2. For a venturi meter having an entrance area A and a throat area A',

$$Q_{\text{vol}} = A\left(\frac{A'\sqrt{2g\,\delta p}}{\sqrt{A^2 - A'^2}}\right)$$

which, for a real meter, can be reduced to:

$$Q_{\text{vol}} = k\sqrt{2gh}$$
$$= K\sqrt{h}$$

where Q_{vol} is the quantity of liquid flowing and K is a combined constant that includes the entrance and throat dimensions, and the numerical constant 2 times the acceleration due to gravity.

3. All differential-head instruments when used as means of measuring flow use the foregoing equation for Q_{vol}—typically, orifice plates, venturi meters, Dall tubes, venturi nozzles, and venturi flumes.

4. Orifice plates do not have any pressure head recovery capability; almost all the head developed across the device is lost. Venturi meters, on the other hand, do allow most of the differential-pressure heads across the two measurement tappings to be recovered. This is due to the geometrical shape of the meter.

5. Turndown ratio is defined as the ratio of the maximum expected flow to the minimum expected flow for a given accuracy:

$$\text{Turndown} = \frac{\text{maximum flow}}{\text{minimum flow}}$$

For an orifice plate, the best turndown ratio is about 5:1; most often, it is actually 3:1. For magnetic flowmeters, the turndown ratio is usually 40:1, and for Coriolis meters, usually 100:1.

6. A sharp-edge orifice plate should never be used to measure high-viscosity fluids, e.g., heavy fuel oil; a curved-edge orifice plate is much better. Target meters and oval-gear meters are specifically designed to measure the flow of high-viscosity fluids.

7. If a differential-pressure instrument or transmitter is used to measure the flow of viscous or aggressive fluids, use chemical seals to protect the measuring element. Ensure that the process fluid between each seal and the process line is always free to transmit the pressure variations to the seal.

8. Magnetic meters operate on Faraday's law of induced voltage:

$$\text{emf} \propto Blv$$

where B is the magnetic field strength, l is the conductor length, and v is the velocity of traverse. In a real meter, the field strength and conductor length are constant, and hence this gives the volumetric flow rate from:

$$\text{emf} = kv$$
$$= k_1 \frac{\pi d^2}{4} v$$

9. Vortex meters operate on the basis of counting the number of vortices (eddies) generated when a body—called a *bluff* body—is inserted in a stream of flowing fluid. The number of vortices over a fixed period of time—usually one second—is directly proportional to the volumetric flow rate.

10. Turbine flowmeters are constructed around a set of unpowered and freely rotatable hydro- or aerofoils held like an aircraft or marine propeller on the central longitudinal axis of the pipeline. The type of foil used depends on the fluid being measured. Since the foil rotates solely by the action of the flowing fluid, it is used to determine the rate of flow. Magnets embedded in the rotor assembly induce voltage pulses in a suitable pickup coil placed outside the periphery of the rotor blades. The pulse frequency is proportional to the fluid velocity, and also to the volume flowing since the diameter of the pipe containing the rotor is uniform.

11. Coriolis mass flowmeters operate on the basis of the Coriolis force, the force that causes an unrestrained body to deflect to the right in the Northern Hemisphere and to the left in the Southern Hemisphere due solely to the earth's rotation. The magnitude of the force is directly related to the mass of the body and its velocity. The meter is a direct mass-measuring instrument. Several different geometries of the flow tube are available.

12. Instruments that determine flow rate by a change in frequency whenever there is relative motion between the frequency source, the receiver, and/or the carrier medium are called Doppler-effect flowmeters. The fluid velocity is given by:

$$v = \frac{C(f_2 \pm f_1)}{2f_1 \cos\theta}$$

where v is fluid velocity; C is the velocity of sound in the fluid; f is frequency, with subscripts 1 and 2 indicating transmitted and reflected, respectively; and θ is the angle of the pulse stream to the fluid. Flow rate can be determined for the particular pipe size.

13. Time-of-flight flowmeters operate on the principle that, when a sound is transmitted through a moving fluid, the time it takes the sound to travel in the same direction as the fluid flow is less than the time it takes it to travel the same distance in the opposite direction. There is also a difference in frequencies proportional to the average fluid velocity and hence the flow rate. For the *time* difference,

$$t = t_1 - t_2 = \frac{2lv\cos\theta}{C^2 - v^2\cos^2\theta} = \frac{2lv\cos\theta}{C^2}$$

The average velocity is given by:

$$\bar{v} = \frac{v}{\cos\theta}$$

where v is the fluid velocity and θ is the angle of the pulse stream to the fluid. For the frequency difference, if f_1 is the signal frequency of the downstream (received) pulse and f_2 is the signal frequency of the upstream (initiating) pulse, $f_1 = 1/t_1$ and $f_2 = 1/t_2$,

$$f_1 - f_2 = \frac{2\bar{v}\cos\theta}{l}$$

where path length is l and θ the angle of the transmitted ultrasonic beam, both of which are parts of the calibration constant for the instrument.

14. Hot-wire anemometers are frequently used to measure the flow of gases. The device is a resistance wire that carries a current that causes it to heat up. Under no-flow conditions, a temperature dependent on the resistance and the magnitude of the current (I^2R effect) will be attained. As flow starts, the resistor temperature will fall to a value dependent on the flow velocity and thermal conductivity of the gas. The value of the resistance will represent the flow rate.

CHAPTER 2

Pressure Measurement

Atmospheric pressure exerts its effect simultaneously over the entire surface of bodies that are exposed to it. Every single thing on earth is exposed to it continuously. Creatures that inhabit the world would not be able to survive if the ambient pressure surrounding them was not held nearly constant at approximately 1 bar, although we seldom realize it. It is because of this that measurements referred to ambient conditions have a "normal" zero value that is actually higher than the "true" zero.

Pressure can be used inferentially to measure other variables such as flow and level. Pressure is intimately bound up with the parameter flow. Generally speaking, it is not possible to have a flow of fluid or other material without pressure, and in this respect the pressure appears to behave as the "motive power" behind flow. In the case when flow occurs between two points under the influence of gravity, we do have a case of a high- and a low-pressure head due solely to the positional difference of the two points in question, for increasing the positional difference between the two points (i.e., pressure) will cause the material to be moved further and faster than it would under a lower value of this parameter.

As we have just seen, pressure is intimately involved with the parameter flow; it also exerts an influence on other physical parameters. For example, it plays a major part in determining the boiling point of a liquid, and judicious use of this fact helps to preserve the qualities of some raw materials or partly manufactured products to give many of the items that we use everyday. Hence, there is also an intimate link with the parameter temperature, for it was by the combined effects of pressure and temperature on once-living organisms that some of the raw materials we use for further processing were formed—e.g., crude oil and coal, to name but two. In geological terms these two parameters affect many inanimate materials as well. In this chapter we shall be introduced to the concepts of the *Gauge* and *Absolute* measurement scales of pressure, and some of the techniques of measuring these parameters. The Bourdon tube is one of the most commonly found devices for measuring pressure and will be described in detail. Strain gauge and resonant-type sensors are described. The technique for determining level using the back pressure generated when air is forced through a tube immersed in a vessel containing a liquid is described. Chemical seals are discussed, with the reasons for their inclusion in a measuring system. Typical requirements for a pressure measurement are shown, and the criteria applying when selecting an instrument are given. Where manufacturers' names are given in this section, these do not represent a recommendation; the names are few among several that supply actual equipment for inclusion in any scheme. The actual application of any specific device(s) must be discussed with the appropriate manufacturer for suitability.

BASIC THEORY AND DEFINITIONS

All material on the earth is subject to a force exerted by the surrounding atmosphere, the total force depending on the surface area exposed to the atmosphere, the relationship being given by

Pressure = force/area

It is well known that all liquids exhibit three easily demonstrable characteristics:

1. They try to find the lowest level. This means that if a quantity of liquid contained in a vessel is raised above a chosen datum and then allowed to run free, all the freed liquid will flow back toward the datum, provided that the path, which need not necessarily be fixed, is not absorbent and there are no cavities to intercept the released liquid volume located between the datum and the point of release.

2. They take the shape of the containing vessel. Depending on the volumes of the container and the liquid, the liquid may partially or completely fill or even overfill the vessel.

3. From any single point at a chosen level, the liquid exerts pressure on the containing vessel equally and in all directions.

Strictly speaking, weight is a force caused by the effect of gravitational acceleration acting on the mass of a material. The most important idea coming out of this is that in order to define a force we *must* give its magnitude and direction, and thus force is a *vector* quantity.

Gases behave in much the same way as liquids, the main difference being that a volume of air or any gas provides no boundary of its own and will totally occupy the containing vessel irrespective of the container volume and the initial volume of the gas. Heavy gases are defined as those that are heavier than air. Combinations of air and heavier gases separate out, with the heavier gases occupying a portion of the containing vessel. This combination of gases will completely occupy the containing vessel. Like liquids, gases also exert pressure on the walls of the containing vessel equally, and in all directions from a single point.

Before we go any further we must define *density*, which is given as mass per unit volume:

Density = mass/volume

Pressure is commonly quoted as being *absolute* or *gauge*. What is the difference? The easiest way of thinking about it is as follows. If we consider the pressure due only to the fluid itself, then the measuring units will be *absolute* which results in:

Some fluid = some pressure = some *absolute* pressure

No fluid = no pressure = *zero absolute*

If, on the other hand, we also consider the pressure due to the fluid plus that of the surrounding atmosphere, then the measuring units will be gauge, and we will have:

Fluid pressure + atmospheric pressure = some *gauge pressure*

No fluid + atmospheric pressure = *zero gauge pressure*

from which follows:

Gauge pressure − atmospheric pressure = pressure due to the fluid itself

= *absolute fluid pressure*

To indicate whether the units are absolute or gauge, the unit of pressure is appended with the letter A or the abbreviation Abs, or with the letter G or g—e.g., 10 lb/in^2 g or 20 lb/in^2 A.

PRESSURE-MEASURING DEVICES

THE U-TUBE MANOMETER

Let us now consider some of the methods used to measure the parameter of pressure. The simplest is the *U-tube manometer*, a device that is most familiar as a glass tube of uniform bore having both ends open, and formed into a U shape, hence the name. The tube with the arms vertically upward is partially filled with a liquid and clamped to a backing plate; and a scale (ruler) graduated in units of length (in or mm) is fitted to the plate between the two arms of the U. The scale with adjustable vertical positioning to allow zero adjustment for the amount of liquid fill is fitted alongside the end that will not be connected to the pressure source and usually called the reference limb or pressure, the other is the active limb or pressure. The instrument is (always) a differential-pressure measuring device, the limb-limb column height difference giving this value. Zero adjustments are made with both ends of the tube open to the ambient, thus ensuring that the pressures in both limbs of the tube are identical, and indicated by the same liquid level in each limb, the measurement zero is then set. For gauge pressure measurement the limb nearest the scale is left open to the atmosphere, and the opposite limb the U tube is connected to the source whose pressure is to be measured. When the measured pressure, if greater than atmospheric, is applied, the liquid is forced down the connected limb and up the other until a point of stability is reached when no liquid movement is observed. At this point the pressure in each limb of the tube is the same i.e., the sum of the atmospheric pressure and that of the column of liquid exactly balance the source pressure. Readings of the liquid level are made with respect to the scale, which, if the bore is uniform, will show the rise to be the same as the fall of the liquid in the other limb. However, if the bore in one arm of the tube is not the same as that in the other (e.g., the well-type manometer) then the rise and fall of the liquid will not be the same, but the pressure in each limb will be identical, and the scale will have to be calibrated to suit. The chosen liquid determines the range of the instrument, for example low ranges usually use water whereas high ranges use mercury. The physical dimension of the measurement is length, expressed in units of mm (in) H_2O and mm (in) Hg, respectively. There are several variations on this theme. For instance, with very low ranges the U tube is not suitable; instead, an inclined tube is used. By this method, a much greater liquid movement is obtained for a small change in pressure. Remember that the readings in all cases are taken from the center of the liquid meniscus and not from the edges. A very important consideration with regard to using manometers for pressure measurements when subject to temperature variation is that the density of the liquid will change with temperature and these changes must be accounted for. This is accomplished by the following two approximate relationships:

$$\rho = \frac{\rho_0}{1 + \beta(T - T_0)}$$

Alternatively:

$$\rho = \rho_0[1 - \beta(T - T_0)]$$

where ρ = density of the U-tube liquid at ambient temperature
            ρ_0 = density at base condition
            β = coefficient of cubic expansion
            T = ambient temperature
            T_0 = temperature at base condition

THE BOURDON TUBE SENSOR

The most common method of measuring pressure is the *Bourdon tube*. What is a Bourdon tube? As its name suggests, it is a tube, but it is not one with an annulus for a cross section. The starting point in the manufacturing process is a circular tube, which is progressively flattened to present as large a surface area to the pressure source as practicable, as long as the wall thickness remains sufficient to contain the pressure range being measured. The internal surface in relation to the wall thickness and the applied pressure area must also be of such a value to produce the driving force to actuate the instrument's mechanism. Bourdon tubes are made in the following configurations, depending largely on the static pressure and instrument span:

- C type

- Helical type

- Spiral type

For pressure measurement, one end of the tube is sealed tight, while the other end is exposed to the pressure being measured. In all the configurations named, it is easy to visualize the sealed end moving in a manner to straighten out the tube. The amount by which the tube straightens is proportional to the applied pressure, provided that the elastic limit of the tube material is not exceeded.

The movement of the free end of the Bourdon is inversely proportional to the wall thickness and is dependent on the shape of the cross section of the tube. As will be appreciated from Figure 2.1, showing a C-type Bourdon tube, there will be some "backlash" between the drive quadrant and the pinion, and to minimize this, fine

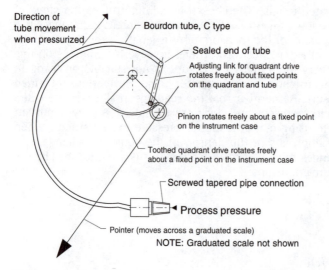

Figure 2.1: Typical Bourdon tube pressure indicator.

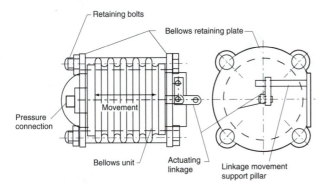

Figure 2.2: Schematic arrangement of a bellows unit.

hair springs are inserted on the quadrant and pointer pivots, the effect being to keep the quadrant and pinion in intimate contact at all times.

Spiral and helical Bourdon tubes provide the means of measuring low and high pressures. Spiral Bourdon tubes are used on relatively low pressures, while helical Bourdon tubes are used on high pressures, the selection being dictated by the opposing requirements of high deflection and thick walls. A linkage mechanism, with one end anchored directly to the free end of the pressure element and the other end fitted to the spindle of the pointer, avoids the backlash that is inevitable when geared drives are involved as they are in Figure 2.1. The material most commonly used for the tube is NiSpan C, since it is compatible with many process fluids and retains its form extremely well. There are instances when 316 stainless steel is used, but the pressure range is fairly high in these cases, as would be expected.

THE BELLOWS-TYPE SENSOR

In those cases where the pressure range is so low that a Bourdon tube is unsuitable, other pressure-sensing elements have to be found. In such instances, use is made of bellows or diaphragms, for the simple reason that they offer potentially a much greater surface area to the measured pressure, and hence produce a much greater driving force and movement for the instrument mechanism. The new device mentioned here is the *bellows*; what is it? As its name implies, the device is constructed from a thin-wall closed-end tube formed into a series of rings that collapse or expand under the influence of the applied pressure; see Figure 2.2. The manufacture of the bellows commences with a deep-sided "cup" of the appropriate material and of the correct uniform thickness. A collapsible mandrel or former of the correct shape and size is inserted into the cup; the assembly is then put into a vacuum chamber and the mandrel interior is subjected to a high vacuum (almost zero absolute). Under this condition the cup collapses around the mandrel, the vacuum is released, and the newly formed bellows and mandrel are withdrawn. Since the mandrel is collapsible, it is then removed to leave just the bellows unit. This technique is called *vacuum forming*.

THE DIAPHRAGM CAPSULE

Figure 2.3 shows the diaphragm capsule that lies at the heart of the differential-pressure cell discussed in Chapter 1, with the force bar positioned by the action of the capsule assembly (see Figure 1.8).

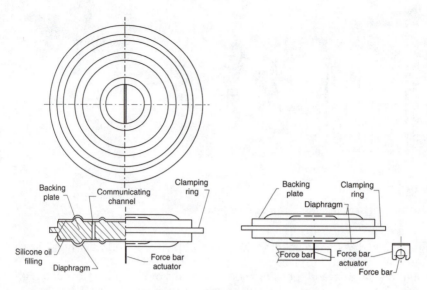

Figure 2.3: Schematic of a diaphragm capsule.

MEASURING ABSOLUTE PRESSURE

So far we have discussed the measurement of gauge pressure only, but how do we measure absolute pressure? The answer should be fairly obvious, as the way would be to measure against a reference vacuum. To do this practically and on an industrial process using Bourdon tubes would require a matched pair of tubes. One of the tubes would be evacuated to as near absolute zero pressure as practicable and then permanently sealed off (this one is now the reference). The other tube would then be linked to the reference in such a way that its movement under the action of the measured pressure opposes the reference. The two forces will equate at the absolute zero pressure of the source being measured. This basic method is also applicable to the bellows and diaphragm units of Figures 2.2 and 2.3, and, in fact, most industrial vacuum measuring instruments work on this principle, although laboratory instrumentation uses other techniques.

THE STRAIN GAUGE SENSOR

We shall now come more up to date in that we shall consider newer types of pressure-sensing devices. The first one we shall look at is the passive *strain gauge* type in which a resistive strain gauge rosette is bonded to a stress-sensing element and is connected in a conventional Wheatstone bridge arrangement (now dated and replaced by more recent detecting systems), as shown in Figure 2.4. With no pressure applied to the strain gauges, the bridge is adjusted so that the output is zero. When a pressure is applied to the unit, it causes the resistance of the strain gauge elements to change, thereby unbalancing the bridge, the amount of imbalance being proportional to the applied pressure. These types of device have many advantages in that they are very robust, have no moving parts, are inherently very stable, and require almost no maintenance. One must be aware that the process fluid is never in contact with the sensor itself, but works via a diaphragm or very small-volume sensor tube. Strain gauge sensors/bridges are provided with facilities for compensating the measurement for temperature variations,

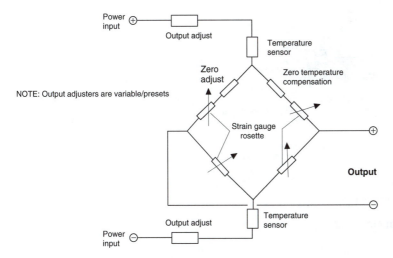

Figure 2.4: Strain gauge pressure sensor.

although they are prone to errors if either the process or ambient temperatures change rapidly. Twice or three times the rated pressure capacity can safely be accommodated without any detrimental effect; but be aware that high-pressure transients can, and do, damage the sensor permanently. Pressure "snubbers"—discussed later in this section and in Chapter 18—will greatly assist in these cases. Most strain gauge pressure sensors are insensitive to mounting orientation in their installation.

THE PIEZORESISTIVE SENSOR

The strain gauge idea has also been updated, for it is now possible to have the Wheatstone bridge as part of the "fabric" of a silicon chip. The process involves diffusing, very precisely, another suitable material into the crystal lattice of the silicon in order to form the bridge elements resistive material. The chip is etched, on the face of the side opposite to that exposed to the pressure source, to leave a very definite thickness of silicon material, against a "masked" pattern that creates all four bridge-connected arms with the correct resistance values of the bridge circuit. Carrying out this delicate manufacturing operation permits the measured pressure to deform the silicon and thereby give rise to a piezoresistive effect. Electrical connections to external circuits from the integrated circuit (IC) device are via "leadouts" of very fine gold wires welded to the appropriate points of the microstructure. As before, the silicon IC is not exposed directly to the measured parameter, as there is the usual diaphragm and oil-filled intermediate arrangement.

THE RESONATING-MICROSTRUCTURE SENSOR

Another recent development is the resonant pressure sensor; in this device a metallic microstructure is caused to resonate. If a force is applied to alter the dimension of the resonating member of the structure, then the frequency will change in proportion to the applied force—the "violin string effect." Figure 2.5 shows diagrammatically the principle of operation for this type of sensor.

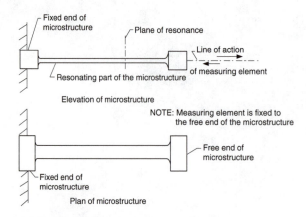

Figure 2.5: Resonating-microstructure pressure sensor.

As suggested before, the sensor shown is only typical; actual units could differ considerably in their design. The great advantage of such designs is that basically the sensor is digital in character and therefore does not require frequency-to-digital converters when used with modern control systems. Additionally, temperature compensation for the sensor, via microminiature resistance temperature detectors, can be very easily accomplished. Instruments of this type are already on the market and are proving robust and extremely accurate.

The relationship between an applied pressure p and the oscillating frequency is given in the following equation

$$p = k \left(1 - \frac{f_0}{f} \right) - k^1 \left(1 - \frac{f_0}{f} \right)^2$$

where f_0 is the resonant frequency for zero pressure, f is the resonant frequency for the applied pressure, and k and k^1 are calibration constants.

THE DEFORMING-MICROSTRUCTURE CAPACITIVE SENSOR

When a quartz crystal is subjected to a compressive force, a "piezoelectric effect" is produced due to the stress in the crystal, the electrical capacitance of the material changes, and the change being used to provide a measure of the force applied. The force is applied along one of the *neutral axes* of the material, and an electrical charge is developed on the surfaces at right angles (along the polar axis) to the line of action of the applied force. A neutral axis is defined as a line parallel to a neutral surface in which the crystals or fibers of the material are unstressed; this means that, because the material has physical dimensions, the stress diminishes in the direction from the surface toward the center and forms a neutral surface at the boundary of the stressed crystals or fibers of the material. The magnitude of the charge developed is related to the dimensions of the crystal, and since these dimensions of the sensing material are known, measuring the magnitude of the charge will determine the pressure involved. This effect is shown in Figure 2.6. The charge is transduced and then applied to a transmitter for use in an instrumentation application.

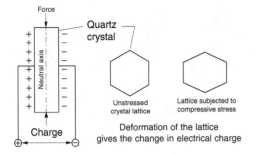

Figure 2.6: Schematic of piezoelectric sensor.

USING PRESSURE MEASUREMENT TO DETERMINE LIQUID LEVEL

Earlier, we defined pressure as force per unit area or: pressure = force/area. If we are dealing with liquids, we can modify this to read:

$$\text{Pressure} = \text{depth} \times \text{density} \times g$$

where density is defined as mass/volume. If we convert these two relationships to dimensional units, we shall find them to be identical, so we can take advantage of this fact and use pressure-measuring devices to ascertain liquid level in tanks. In these applications, the positioning of the pressure sensor is vital, for the measurement of level is made explicitly with respect to the position of the sensor and relies most importantly on the fact that the *liquid density is constant*. Let us now look at the different ways in which we can use pressure to determine the level of liquids in vessels.

See Chapter 4 for a fuller treatment of this parameter.

OPEN-TOP AND CLOSED VESSELS

We shall first consider open-top vessels, i.e., vessels that contain a liquid having the surface exposed to the atmosphere. If indication alone were required, then a straightforward pressure gauge would be most suitable, the range of the instrument being determined from the maximum depth of the liquid and its density. A word of warning here: In all calculations it is vital that the units of measurement be consistent; and it is very easy in hydraulics to make dimensional errors. Do not ever mix imperial and metric units, keep the same units—e.g., length in meters or feet throughout. Also do not mix meters and millimeters, even though they are both metric units of length.

If transmission of the measurement is involved, then a pressure transmitter of the correct range is important. In some cases of very low range, it may be necessary to use a differential-pressure transmitter, in which cases it must be ensured that the low-pressure connection is open to the atmosphere.

In determining the liquid pressure or level in closed vessels with a single instrument, the choice is restricted to differential-pressure systems only. The reason is that the lower connection will have not only the pressure due to the liquid, but also the pressure of the vapor, or other gas, on the liquid surface as well. The liquid pressure in this case will be the difference between the combined total pressure (liquid + vapor or gas) and the pressure due to the vapor or gas alone; hence, a connection from the vapor space will have to be made to the low-pressure connection of the differential-pressure instrument. Another word of warning here: If the vapor is one that condenses to a liquid when cooled, then over a period of time the piping between the vapor connection and the instrument will fill with liquid, and when this is the case the accuracy of the measurement will be in question. To overcome this, the connecting pipe must be deliberately filled with either liquid condensate or another liquid of identical density and compatibility with the process fluid *before* the instrument is put into service. This "standing head" of the liquid in the low-pressure connection must be taken into account when determining the measurement range when buying the instrument and allowed for in any associated indication. Remember to "zero out" this standing head of liquid when commissioning the instrument. The alternative, depending on the vapor pressure, is to fit a pressure repeater, which will eliminate the effect of the standing head, as no liquid is involved in producing the effective vapor pressure for the differential measurement. The repeater is an instrument that produces a pneumatic output of the same magnitude as the measured pressure, which is applied to the low-pressure input of the differential-pressure instrument. The lower vessel connection is the high-pressure connection on the instrument.

In all cases where there is connecting pipe work between the points of measurement and the instrument, it is vital that there be no entrapped gas or air. Every differential-pressure instrument is provided with a bleed screw on the meter body for this purpose. Always install instruments with an isolating valve between the process and the instrument itself, as this makes removal of the instrument possible without shutting the process down. Although this last statement would seem obvious, it is surprising how often it is forgotten.

THE SUBMERSIBLE DIAPHRAGM SENSOR

Another way of using pressure to determine level in vessels is with a submersible diaphragm assembly; this device is shown in Figure 2.7. In the interest of clarity, this figure shows only the equipment used to obtain the measurement; the mechanism of the measuring element (Bourdon tube, bellows assembly, etc.) has been omitted. The system shown is quite suitable for open-top vessels, but proper care must be given in sizing the capillary length and the necessary protection, mechanical and environmental, must be provided.

AIR REACTION METHOD: THE BUBBLE TUBE

While we are discussing the use of pressure for liquid level measurements, let us consider the method of air reaction that was mentioned in Chapter 1 on flow. All

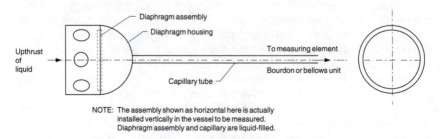

Figure 2.7: Submersible liquid-filled pressure diaphragm assembly.

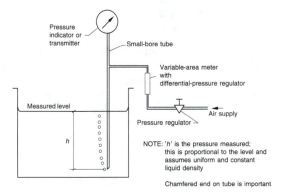

Figure 2.8: Pressure measurement to determine liquid level: bubble-tube method.

the methods that are to be looked at now balance the pressure head of the fluid being measured against the pressure head of the *reaction medium*, which in the common case of water or liquids of similar viscosity (provided the fluids are compatible with air) is air. For this method to work, we shall need the following items: a small-bore tube (say, 12 mm) of material compatible with the process fluid and of a length sufficient to measure the maximum depth, a variable area (V/A) flowmeter with differential-pressure regulator, an air pressure regulator, a pressure gauge or transmitter (as appropriate to the application), and appropriate tube fittings. Figure 2.8 shows the setup.

The system works as follows. As it is normal for instrument air supply pressure in a plant to be approximately 100 lb/in^2 g or the equivalent metric (kg/cm^2) value, this supply pressure must be reduced before it can be used in this application; hence the requirement for the pressure regulator shown in Figure 2.8. The actual pressure required for the measurement is normally only a few lb/in^2 g (kg/cm^2). Clearly, the regulated pressure must be very slightly greater than the maximum liquid head to be measured. With the pipe work connected up as shown, the small-bore tube, also called the *dip tube*, will have to be supported within the vessel; this aspect has not been shown on the diagram, for much will depend on the actual on-site conditions. With no air applied, the system is then vented to allow the entrapped air within the tube to escape, and the liquid to rise up within the dip tube to a height equal to h, the immersion depth (venting, though not essential, is a good starting point for the measurement). The venting arrangement on the tube must then be sealed off, and the V/A meter needle valve opened slowly; the air supplied will then gradually push the liquid out of the dip tube. The amount of air supplied should be such that the escaping air bubbles are single and separated by a small period of time; this is necessary to ensure that the air pressure is only just slightly greater than the liquid head. A typical V/A meter to use would be the Williams & James Pneumastat or the equivalent, which has been specifically designed for this application and includes the differential-pressure regulator, the needle valve, and a viewing tube to enable the air bubbles to be seen. Once set for maximum liquid level, the air pressure within the dip tube will vary for any change in level because more air will escape when the liquid level is reduced; these changes in pressure are proportional to the liquid level. The differential-pressure regulator is very important for ensuring constant airflow at the operating condition, and it does this by maintaining constant differential across the float of the variable-area meter.

USING PRESSURE MEASUREMENT TO DETERMINE SLURRY LEVEL OR SLURRY FLOW: LIQUID REACTION METHOD

The principle just described can also be used for measuring the level and flow of some slurries by substituting a compatible liquid for the air supply, but be aware that the purgemeters (these consist of a V/A meter and a differential-pressure regulator as before, but this time suitable for liquid service, which will involve some changes in the sizing of the instrument) will be different. Primary devices for slurry flows, when measured by this method, will be restricted to large venturi tubes and flumes, with a purgemeter set required for each of the two connections.

DATA REQUIRED TO PROVIDE A MEASUREMENT OF PROCESS PRESSURE

This section identifies the general required data relating to the containment of process fluids relevant to pressure conditions. The list of data corresponds to what would appear on a checklist for a suitable pressure-measuring instrument.

PROCESS FLUID

Name

Characteristic   Gaseous or liquid

Chemical action   Aggressive or nonaggressive (**Note:** Aggressiveness is
                        defined as the ability of the fluid to attack chemically
                        or physically the containment or contact materials.)

PROCESS PARAMETERS

For the following, the appropriate engineering units are required.

Pressure

Maximum operating

Minimum operating

Normal operating

Temperature

Maximum operating

Minimum operating

Normal operating

PHYSICAL LOCATION

In the following, give the standards with which the material must comply—e.g., BS, ANSI, etc. In the case of process connections, the pressure rating must also be given.

Process Line

Nominal size

Pipe schedule

Material of construction

Process Vessel

Process connections	Screwed
	Flanged
	Welded
	Hygienic (exclusively for food and drugs)

APPLICATION

Indication	local or remote
Control—local	On/Off or PID
Remote	On/Off or PID (**Note:** Local is defined as the ability of having the measurement or control function at or near the point of penetration of the sensor or transducer into the process. Remote is defined as the ability of having the measurement or control function at some distance from the point of penetration of the sensor or transducer into the process.)

Do not be too anxious about the control aspects at this stage. They are stated here for completeness only, and will be discussed in further detail in later chapters.

CRITERIA FOR SELECTING A SUITABLE PRESSURE-MEASURING INSTRUMENT

OPERATING RANGE

The operating range is a very important consideration when selecting a measuring instrument. Ranges roughly fall into the following general categories:

1. Extra low range of positive pressures usually below 0.25 in of water column gauge (say, 5 to 6 mm H_2O, i.e., 5 to 6 mm wg, which is the abbreviation for water column gauge).

2. Low range, usually from vacuum to pressures measured in mm or inches water column. Vacuum ranges can be only from atmospheric pressure down to full vacuum (−1 bar gauge). Full vacuum, i.e., absolute zero, is not possible; we can approach very close, but never achieve it.

3. Medium range, usually in pressure ranges of tens to hundreds of bars, kg/cm^2, or lb/in^2.

4. High range, usually in pressure ranges of thousands of bars, kg/cm^2, or lb/in^2.

SENSING ELEMENTS

The sensors can be divided into the following categories:

1. For low-range measurements, slack diaphragms are the usual elements used. This element is also used for small negative pressures (typically furnace stack draft) where the measurement is very slightly below atmospheric pressure. Ranges that

cover both positive and vacuum (negative) values result in what is termed a *compound-range* instrument scale. Compound ranges with slack diaphragms have a minimum range value of 0.25 in H_2O g with a minimum span of 0.5 in H_2O g or the equivalent in metric units.

An alternative to the slack diaphragm, but for positive pressures only, is obtained by using a low-range differential-pressure device. This instrument provides the measurement when the low-pressure connection is left open to the atmosphere. Inverted bell-type D/P meters can be used for local indication and recording, but they are large instruments. On the other hand, if differential-pressure cells are used, then a transmitter signal will be produced, and this means that the measurement can be displayed either locally or at some point away from the point of measurement. The advantage of using a differential-pressure cell to obtain the measurement is the ability of the device to withstand static pressures that are much higher than the maximum being measured without problems, a design feature of the instrument, at the same time as they are able to measure low pressures in the mm H_2O g range.

Extra low ranges can be indicated only using an inclined manometer where the discrimination necessary in the measurement is provided, but transducing this measurement for long-distance transmission is not possible.

2. For medium pressure, the elements available are the C-type Bourdon, spiral Bourdon, and thin-wall helical Bourdon tubes; the C-type elements are the most commonly used. Spiral elements can be used for pressure and differential pressure. When differential pressure is to be measured, a pair of spiral elements will be required. Since the process fluid must enter the sensing element, it is important that the materials used for the element's construction are compatible with the fluid to avoid chemical attack, which could lead to calibration shift or, even, leakage and injury to personnel. Common materials for the elements are NiSpan C, beryllium copper, and 316 stainless steel. Spiral and helical elements provide greater physical movement for a given pressure.

3. High-pressure elements are usually fabricated from thick-wall 316 stainless steel and are in helical form to ensure adequate movement. Especially in applications when the range is in tens of thousands units, the only sensing devices with adequate sensitivity are the strain gauge and vibrating-microstructure types of transducers.

Note: For instruments used in the food- and drug-processing industries, it is a requirement that no mercury be used in any part of the instrument. Furthermore, the installed instrument and/or pipe work must have a physical arrangement that will not permit process fluid to stagnate, to avoid the possibility of bacterial growth, which could contaminate the whole process. All instruments must be provided with patented process connections, the most well known ones being manufactured in Europe by APV Engineering.

PRESSURE PULSATION

Pulsations in the process fluid are usually present when a reciprocating pump is the prime mover of the fluid. The pulsating effect can be minimized by having a receiving vessel in the process line, thus making the arrangement analogous to a capacitance in an electrical circuit, which results in a process damping. However, the biggest advantage is that it smooths the raw fluctuating waveform of the measured variable. It must be appreciated that the inclusion of a receiver is a decision that has to be taken by the owners or users of the process. The alternative to the foregoing is the provision of a "pressure snubber." This in effect introduces a capacity of very minimal size and is included at the process instrument

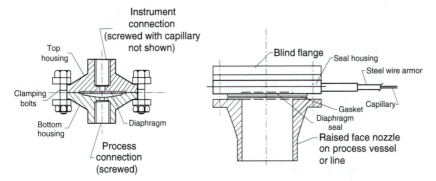

Figure 2.9: Schematic arrangement of typical diaphragm seals. The seals shown are only two of many different types available.

connection (for details and sketch of a pressure snubber, see Chapter 18, on installation); the mass involved is minute in comparison with the mass of the process fluid itself.

AGGRESSIVE FLUIDS

Materials of Construction

Where the sensor is available in a material compatible with the process fluid, then the instrument as made will be suitable. However, it is more usual for the user to specify a suitable material and leave the manufacturer to comply with it. If none of the materials of the instrument are suitable, then provision of a chemical seal between the process and the measuring instrument may be the alternative.

Chemical Seals

As mentioned earlier, chemical seals are physical barriers that are placed to separate the process fluid from the measuring device. There are several manufacturers of such equipment; two of the more well known are Ashcroft and Wika, the former is a US manufacturer and the latter European. Figure 2.9 shows the arrangement of typical seals. It is important that the appropriate unit be selected for the measurement to be made.

Material of the Seal

Wika provides the following materials for construction of the seal:

Stainless steel: 316L (two grades), 321 and 316Ti

Hastelloy: B2, C4, and C276

Incolloy: 825

Inconel: 600

Monel: 400; K500

Nickel: >99.2% purity

Tantalum, titanium

Zirconium

PTFE polytetrafluorethylene

Note: Be aware that the housing and diaphragm materials from other manufacturers will generally differ.

Considerations for Chemical Seals

It is important that the volumes of the instrument and the seal itself be in balance. As a general rule, the volume of the seal must at least be equal to the control volume (to be defined shortly) of the pressure instrument. Furthermore, the variations due to thermal expansion and liquid compressibility must also be considered. This requirement can be summarized as:

Working volume (seal) > control volume of instrument

+ volume due to temperature change + volume due to compression of fluid

The *working volume*, sometimes also referred to as the *displacement volume*, of the seal is defined as the maximum displaced volume of the seal fluid that is equal to the change in volume when the diaphragm fully deflects within the limits of its characteristic. The *control volume* is defined as the amount of fluid displaced when the diaphragm responds to a full deflection of the pressure-measuring element. For example, a Bourdon tube will expand when subjected to a pressure, and this will change the internal volume commensurate with the applied pressure; the control volume for the tube being that volume obtained for 100% change of the sensor calibrated span and represented by an amount of fluid displaced. It is this volume that will have to be taken into consideration when determining the control volume of the seal.

Temperature effects can be allowed for, but there are considerations that must be taken into account when applying this method. The effect of temperature on the measurement is given by:

$$\Delta p = \frac{\Delta V_T}{k}$$

where Δp is the maximum change of pressure, ΔV_T is the volume due to the temperature, and k is the gradient of the diaphragm characteristic (Wika has graphs for the effects of temperature).

Remember that thermal expansion is proportional to the total volume of the filling, and the smaller the diaphragm the less flexible it is and the less it is capable of absorbing the volumetric increase due to thermal fluctuations; larger diaphragms are much better in these instances. This means that systems that need long capillaries, or low-pressure measurements, will be severely affected by thermal variations. This gives rise to an important empirical rule: *the lower the pressure measurement range and the higher the associated process temperature of the fluid being measured, the larger must be the diameter of the diaphragm of the seal.*

Differences in mounting levels will result in the creation of static heads, and these must be zeroed out after the instrument has been installed, i.e., at the commissioning stage.

The response time is dependent on the volume displacement required relative to the total volume of the system. In other words, the smaller the volume, the faster the response.

Fill Fluids for Wika Seals

Fill fluid	Code	Process temperature range		Notes
		$<1\ bar_{abs}$ (°C)	$\geq 1\ bar_{abs}$ (°C)	
Silicone oil	KN2	—	−20 to 200	Standard
Silicone oil	KN17	−40 to 80	−90 to 100	Standard
High-temperature oil	KN3.1	−10 to 100	20 to 300	Standard
High-temperature oil	KN3.2	−10 to 200	−20 to 400	Standard
Halocarbon	KN21	−40 to 80	−40 to 175 (max 160 bars)	Oxygen and chlorine service
Glycerine	KN7	—	+17 to 230	Food-compatible
Glycerine/H_2O	KN12	—	+10 to 120	Food-compatible
Vegetable oil	KN13	−10 to 120	−10 to 250	Food-compatible
Esso Marcol	KN62	−30 to 170	−30 to 250	FDA-approved

Note: For halocarbon the maximum pressure is 160 bars.

SUMMARY

1. Pressure is the force per unit area:

$$\text{Pressure} = \frac{\text{force}}{\text{area}}$$

When using chemical seals, the effect of temperature on the measurement is given by:

$$\Delta p = \frac{\Delta V_T}{k}$$

where Δp is the maximimum change of pressure, ΔV_T is the volume due to the temperature, and k is the gradient of the diaphragm characteristic (most manufacturers have graphs or data for this).

2. Density is mass per unit volume:

$$\text{Density} = \frac{\text{mass}}{\text{volume}}$$

3. Gauge pressure = fluid pressure + atmospheric pressure

   Zero gauge pressure = zero fluid pressure + normal atmospheric pressure

   Absolute pressure = gauge pressure − atmospheric pressure

4. Very low pressures are measured by water-filled inclined-tube manometers, which give large movement of the meniscus for small changes of pressure. Mercury-filled manometers, because of the higher density of the mercury, can measure vacuum and higher pressures. Mercury is rarely used nowadays, as it is considered a health hazard.

5. The density of the liquid in a manometer will change with temperature. The change must be accounted for by the following relationship:

$$\rho = \frac{\rho_0}{1 + \beta(T - T_0)}$$

Alternatively:

$$\rho = \rho_0[1 - \beta(T - T_0)]$$

where ρ = density at ambient temperature
ρ_0 = density at base condition
β = coefficient of cubic expansion
T = ambient temperature
T_0 = Temperature at base condition

6. Bourdon tubes are the most commonly used element for measuring pressure. There are three main configurations of the tube: C type, helical type, and spiral type. All Bourdons tend to straighten when subjected to pressure. The movement of the free end is inversely proportional to the thickness of the tube wall and dependent on the shape of the cross section.

7. In resonant sensors the relationship between an applied pressure p and the oscillating frequency is given by:

$$p = k\left(1 - \frac{f_0}{f}\right) - k^1\left(1 - \frac{f_0}{f}\right)^2$$

where f_0 is the frequency for zero pressure, f is the frequency for the applied pressure, and k and k^1 are calibration constants

8. Pressure can be used as means of determining the level of a liquid in a vessel. For vessels open to the atmosphere with fluid of known constant density, but no air supply available, the instrument to use will be a differential-pressure instrument with the low-pressure connection open to the atmosphere.

9. For vessels open to the atmosphere with fluid of known constant density, air supply available, and indication of level only required, the instrumentation will be a dip tube of the correct length (it must cover the entire range level to be indicated), a variable area meter fitted with a D/P regulator, and a low-range pressure indicator such as a slack diaphragm element. For similar process conditions, but no air supply available, use a submersible liquid-filled diaphragm element coupled to a low range indicator.

10. For closed vessels and single instrument requirement, only differential pressure instruments or transmitters should be used to measure level, with due consideration to the following points:

   - If the vapor is of the condensing type, then in time the pipe work between the vapor connection and the instrument will fill with liquid, and the accuracy of the measurement will be in question.

   - To overcome this problem allow the low-pressure leg to fill naturally before commissioning, or fill the connecting pipe work with liquid condensate or another liquid of identical density and compatibility with the process fluid *before* the instrument is put into service. Instrument engineers refer to this as the "wet leg."

   - Take account of the standing head of the liquid in the low-pressure connection when using the foregoing techniques, and zero this out.

   - Bleed all the air or gas trapped in the interconnecting pipes—sometimes called impulse lines—to ensure correct measurement, as gas compression effects must be avoided.

- An alternative to a wet leg is to install a pneumatic pressure repeater at the vapor phase connection and apply the pneumatic signal to the low-pressure connection on the differential-pressure instrument.

11. Chemical seals are physical barriers that are placed to separate the process fluid from the measuring device. As a general rule, the volume of the seal must at least be equal to the control volume of the pressure instrument. Furthermore, the variations due to thermal expansion and liquid compressibility must also be considered. The requirement can be summarized as:

Working volume (seal) > control volume (instrument)

+ volume (thermal expansion) + volume (liquid compressibility)

CHAPTER 3

Temperature Measurements

This chapter is concerned with the parameter temperature, which, like the parameter pressure, affects all material on the earth. It is a means of determining the heat content of a body for a given condition, and therefore is also indicative of its energy content. We shall initially look at methods using mechanical means of measuring temperature; these rely on the physical changes solids and liquids undergo when they are exposed to a source of heat or cold. In this category fall the bimetallic, liquid, and mercury instruments (the last becoming rarer due to the harmful effect mercury has on humans, animals, and the environment) that are still in use, even today. The mechanical types of instruments can be found mounted within the process plant at locations where the information can be conveniently read by the plant operators as an immediate check on the behavior of parts of the process assigned to their care, and in this respect they are a very useful tool for the continuing and safe operation of the equipment and the process itself.

We shall also review and study the thermoelectric elements such as thermocouples, resistance bulbs, and optical pyrometers that are so very widely used in most process plants, since these provide the means that permit the control system, usually located some distance from the point of measurement, to obtain a continuous view of the process conditions prevailing in the plant and to make corrections to ensure the product is "within specification."

THE TEMPERATURE SCALES

Temperature is a means of determining the amount of heat that a body contains for a given condition; it is *not* a direct measure of the heat of the body. Once again, as for the parameter pressure, there are two scales of measurement: the absolute scale and an "indicated" scale, which is relative to a specific reference. There are also two *systems* of measurement units, Fahrenheit and Celsius (the old name for the latter in the UK and the United States was *centigrade*), usually abbreviated, as F and C. The Fahrenheit system is the imperial system for temperature measurement, while the Celsius system is the metric. The imperial system originated in Britain; it is no longer used there, although it is still in use extensively in the United States, while the metric system is predominant in Europe.

The easiest way to determine which scale is referred to is to look at the definitive values of the freezing point and boiling point of water under normal conditions. These two points are usually referred to as the *fixed points*, and the temperature span between them is called the *fundamental interval*. In the Fahrenheit system the fixed points are 32° and 212°, respectively, whereas on the Celsius system these are 0° and 100°, respectively. It must be remembered that the prevailing *conditions*— e.g., pressure—when determining the fixed points have to be held constant at defined reference or standard values and are to be universally applicable; i.e., results obtained at any geographical location on the earth will be repeatable.

In view of the fact that fixed points are used in the definition of the range, conversion from one to the other is possible in the following way:

To convert °C to °F, multiply °C by 9/5 and add 32:

$$^\circ F = \left(^\circ C \times \frac{180}{100} \right) + 32$$

To convert °F to °C, subtract 32 from °F and multiply remainder by 5/9:

$$^\circ C = (^\circ F - 32) \times \frac{180}{100}$$

The absolute zero on the Fahrenheit scale has a value of 459.57°F below the freezing point. On the corresponding absolute scale the units of measurement are degrees Rankine, abbreviated °R.

The absolute zero Celsius is a value 273.15°C below the freezing point and on this scale the units of measurement are kelvins, abbreviated K. Hence,

To convert °F to °R, add 459.57 to the °F value:

$$^\circ R = {}^\circ F + 459.57$$

To convert °C to K, add 273.15 to the °C value:

$$K = {}^\circ C + 273.15$$

This K scale is also called the absolute thermodynamic scale; it was devised by Lord Kelvin who related the temperature to the amount of mechanical work done by a reversible heat engine working between the two temperatures, thus eliminating the dependence on the properties of the materials involved. In fact, the value of the unit (degree) is the same for both K and °C.

Sometimes the Réaumur scale is used in which the freezing point is 0 and the boiling point is 80; in these instances, conversions are as follows:

To convert °C to R, multiply °C by 4/5

$$R = {}^\circ C \times \frac{4}{5}$$

To convert °F to R, subtract 32 from °F and multiply remainder by 4/9:

$$R = (^\circ F - 32) \times \frac{4}{9}$$

To convert R to °C, multiply R by 5/4:

$$^\circ C = R \times \frac{5}{4}$$

To convert R to °F, multiply R by 9/4 and add 32:

$$^\circ F = \left(R \times \frac{9}{4} \right) + 32$$

The Réaumur scale is quite rarely used, but when it is, great care is required when using the abbreviation R, for it could stand for Rankine or Réaumur. A convention now in use that avoids ambiguity is to represent the Réaumur temperature with a plain R, and this convention has been followed in the foregoing conversions.

TEMPERATURE MEASUREMENT SYSTEMS—MECHANICAL

Temperature can be measured in several ways, but the methods in general fall broadly into two categories, *mechanical* and *electrical*. Under the category of mechanical methods are instruments based on the principles of thermal expansion of solids, liquids, and gases. These will be the subject of this section of the chapter. Methods based on the effects of temperature on the electrical properties of materials will be discussed later in the chapter.

SOLIDS: THE BIMETALLIC THERMOMETER

We consider first the bimetallic thermometer as shown in Figure 3.1. As its name implies, this instrument uses a long strip of two different metals, one of which has a greater expansion coefficient than the other, bonded to each other and wound (unbroken) as two helices that fit coaxially one inside the other, thus presenting a continuous multiturn helix. The reason for this construction of multiple turns is to provide adequate pointer movement and obviate any lateral movement of the pointer spindle in one helix with respect to another. The effect of this is to eliminate the need for jeweled bearings, keep the spindle central, and enable the instrument to absorb mechanical shocks. One end of the helix is permanently fixed to the instrument case and the other end to the stem carrying the pointer when the element is housed within it as shown in the figure. However, when the source of temperature being measured is located away from the instrument dial as in instances where the element has to be inserted to reach the source to be measured; the free end of the element is attached to a shaft (spindle) that is free to rotate. The spindle, supported if necessary at a point along its length, projects through a circular temperature graduated scale plate carries a pointer to provide the indication.

Typical materials used are invar and a nickel-molybdenum alloy. As the temperature increases, the metals begin to expand, but not by the same amount or rate; their differential expansion causes the helices to uncoil. Since the pointer is attached to the spindle, to which the free end is fixed, any movement of the free end results in a pointer rotation. It is important to understand that all changes in length are linear; therefore the pointer movement will also be linear.

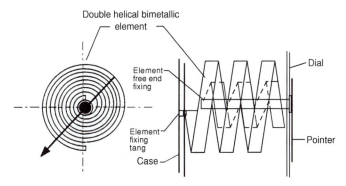

Figure 3.1: Bimetallic thermometer.

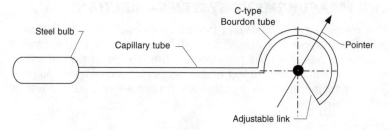

Figure 3.2: Typical liquid-filled thermometer.

LIQUIDS: MERCURY-IN-STEEL AND AROMATIC PETROLEUM LIQUID SYSTEMS

The next device we shall consider is the mercury-in-steel thermometer. Even though, strictly speaking, mercury is a metal in a liquid state and therefore behaves like a metal in some ways, it is considered here. This instrument is the industrial version of the clinical mercury-in-glass instruments usually found in laboratories and hospitals. The main reason for avoiding the glass version in industry is its fragility, and hence its inability to withstand rough handling both in the process and by personnel. In industrial instruments the mercury is contained in a steel bulb joined to a mechanism exactly like the pressure Bourdon tube (spiral or helical), which actuates the pointer movement as described earlier (see Figure 3.2). In this case the mercury must completely fill the bulb, interconnecting capillary tube, and the pointer actuating mechanism with no entrained air in the system. Once again we are dealing with a material that has a linear rate of change, since the coefficient of volumetric expansion in the bulb is uniform. One thing to remember is never to use mercury-in-steel devices in the food industry, for mercury is poisonous and its use is prohibited. These industrial instruments are fitted with the usual types of process connections—i.e., flanged or screwed (threaded).

We shall now consider other liquid-filled thermal systems for which Figure 3.2 is also valid. For these to function correctly the liquid must totally fill the bulb, capillary tubing, and mechanism as before. When this is achieved, the instrument works exactly as the mercury-in-steel thermometer, with the distinction that the operating range will be determined by the type of liquid used as the filling. The following table gives some examples of the liquids used and the operating ranges.

	Operating range	
Liquid	*°F*	*°C*
Alcohol	−110 to +160	−80 to +70
Toluene	−110 to +212	−80 to +100
Pentane	−330 to +80	−200 to +30

The bulbs of other liquid-filled systems can be made of various metals dependent on the application. In this respect they differ from the mercury-filled devices, which are forced to use steel or other mercury-compatible material. The scales of other liquid-filled thermometers are also linear, and these *can* be used in the food industry.

GAS- AND VAPOR-FILLED SYSTEMS

To get down to lower temperatures, vapor-filled systems are employed; the arrangement in Figure 3.2 is also valid in this instance. These systems also use a liquid for the filling, but the liquids used have a low boiling point and do not completely fill the whole system. They work on the principle that, as the liquid absorbs heat, it gives off greater amounts of vapor, thereby exerting more pressure on the indicating mechanism. The pressure exerted depends on whether the temperature of the capillary and Bourdon tube is above or below the temperature of the bulb.

When the temperature(s) of the capillary and Bourdon tube is (are) above the temperature of the bulb, then the capillary and Bourdon tube are full of vapor only, and vapor pressure drives the Bourdon tube. An increase in the temperature surrounding the capillary and Bourdon tube tends to increase the pressure within these, thus causing the vapor in the bulb to condense until the capillary/Bourdon pressure rebalances the saturated vapor pressure of the liquid in the bulb.

When the temperature of the capillary and Bourdon tube is below the temperature of the bulb, the vapor in the capillary and Bourdon tube will condense to a liquid, and this liquid will drive the Bourdon tube under the influence of the pressure in the bulb. If now there is an increase in the temperature surrounding the capillary and Bourdon tube causes the liquid to expand and temporarily increases the pressure within these, thus causing the vapor in the bulb to condense until the pressure balances the saturated vapor pressure of the liquid in the bulb.

Because the vapor is a compressible fluid, the scales of such instruments are temperature-dependent only and nonlinear, the graduations at the lower end of the scale being very close together. From the operation of the system described it will be clear that these instruments must never be used to make measurements of the process that are close to the surrounding ambient temperature. Vapor-pressure thermometers always suffer a time lag every time the temperature of the bulb crosses the ambient temperature of the connecting capillary. By comparison with all other instruments that use a filling, the vapor-pressure thermometer uses the smallest (physical size) bulb to achieve the results.

The last device we consider in this category is the gas-filled system. These instruments operate on the well-known Boyle/Charles principle:

$$k = \frac{pV}{T}$$

from which

$$T = K(pV)$$

where $K = 1/k$. The volume of gas sealed in the measuring system remains constant, and therefore when it is subjected to a change in temperature the pressure will increase or decrease dependent on the direction of the change. The change in pressure per degree of temperature is uniform, and therefore the scale will be linear. In all other respects the gas-filled thermometer is constructed in the same way as the other devices we have been considering so far.

Correcting Filled Systems

As will be appreciated, subjecting a bulb to a heat source while the interconnecting capillary is influenced by its surroundings at a different temperature will give rise to inaccuracies in the indicated values, and there are two classes of correction that can be applied:

Class A. This is a fully compensated system for liquid- and vapor-filled systems and involves the use of a second filled capillary running the full length of the actual system capillary but not connected to the bulb, and right into the instrument case, where it terminates in a second actuating mechanism that carries no pointer. This second "pointer drive" mechanism is arranged to adjust, via an appropriate mechanical movement of gears and levers, the indicating pointer of the instrument for ambient temperature variations of the capillary, i.e., to "back off" the effective capillary temperature variation.

Class B. Under this class, only the instrument case temperature is used to correct the pointer movement, and a bimetallic strip is the usual method employed here.

TEMPERATURE MEASUREMENT SYSTEMS—ELECTRICAL

We shall now consider thermoelectric temperature-measuring devices, which are being used more frequently for process control. The reason for their popularity is that they are more accurate, can be used to make measurements some considerable distance from where they are actually viewed (recorded, indicated, etc.), and are eminently suitable for use with modern control systems, being directly compatible with electronic circuits.

THERMOCOUPLES

The first of the thermoelectric devices we consider are thermocouples, which depend for their operation on the electromotive force (emf) generated when two dissimilar metals are joined together as illustrated in Figure 3.3*a* and subjected to a source of heat. This phenomenon was observed by Seebeck and named as the *Seebeck effect* in his memory. For this to be generated, there must be a temperature difference between the heated end and the end at which the observation is made. Also, Peltier showed that when an electric current flowed across the junction of two dissimilar metals, heat was absorbed when the current flow was in one direction and heat was released when the flow was reversed, the amount of heat being proportional to the quantity of electricity crossing the junction. The amount of heat released or absorbed when a unit of current passes in a unit time is called the *Peltier coefficient*. Reasoning on the basis of the existence of the Peltier effect, the junction of the two dissimilar metals must be the point at which the Seebeck effect originates. The magnitude of the emf depends on the junction temperature. Now, in real, practical terms (once the thermocouple is connected in a circuit), a single junction cannot exist in isolation, and there will always be at least two such junctions in a circuit. If two junctions are connected (as in Figure 3.3*b*) and both are at the same temperature, there will be no net emf in the loop; if we can assume there is an "ideal" measuring device included in the loop, it will now show no reading. Now, starting from the point where both junctions are at the same temperature, if one of the junctions is then heated, the emf across the heated junction will be greater than that across the unheated one; the resultant net emf will cause an instrument indication dependent on the temperature difference between the two junctions, now termed "hot" and "cold" junctions.

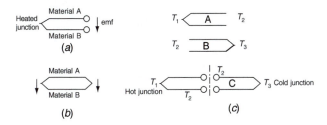

Figure 3.3: Thermocouple fundamentals. (a) Hypothetical single junction thermocouple. (b) The Seebeck effect. Arrows indicate the direction of emf. (c) Law of intermediate temperatures.

The *law of intermediate temperatures* is very useful in calculating the effect of any change in temperature of the cold junction, and makes it possible to correct the indicated value of emf or temperature for these changes. If, as in Figure 3.3c, a thermocouple has one junction at temperature T_1 and the other junction at T_3, then this would be equivalent to having two thermocouples of the same materials with junctions at T_1 and T_2 for one thermocouple and junctions at T_2 and T_3 for the other. There are tables that detail the corrections that must be applied to cold junctions at various temperatures for different thermocouple materials.

The *law of intermediate metals* has an important bearing on the application of thermocouples. Provided that all the apparatus for measuring the thermoelectric emf in the circuit at the cold junction is kept at the same temperature, the presence of any number of different metals (in the series loop or circuit) will not affect the overall emf in the circuit. While this is absolutely true theoretically, it is important in practice that the number of breaks in the circuit be minimal to avoid the problems of the nonuniformity of the cold junction temperature. Furthermore, since the coefficient of resistivity of different metals is also different, the combined effect of temperature and resistance variations would be severe if the measuring or sensing loop were not a true emf sensor. It would be enormously expensive in terms of installation, running, and maintenance costs to provide thermostatically controlled cold junction compensating boxes at each break in the circuit and this arrangement also increases the uncertainty of the measurement by adding more possibilities for failure and inaccurate measurement. It is therefore advisable, in the interests of obtaining the best and most consistent results in practice, that either the same materials as the thermocouple or the appropriate compensating cable pair be used as the extension (connecting) cable back to the indicating or control instrument, ensuring that each member of the cable pair is connected to the correct polarity of the thermocouple terminals. Compensating cables are made from materials compatible with the measuring thermocouple in that no overall thermoelectric effect is introduced at the interface; they are certainly recommended for extension cables on rare metal couples to minimize the cost involved. There are tables that state the type of compensating cable to use in any particular application.

Figure 3.4 shows a typical industrial thermocouple assembly with an arrangement that has a loose process flange to permit adjustment of the depth of insertion into the thermowell or duct. This arrangement is applicable only to installations where thermowells are inserted in the line to effectively seal off the process liquid or gas, or without thermowells when fast response is necessary on applications involving nonnoxious gases, e.g., air only; in these latter instances precautions for personnel

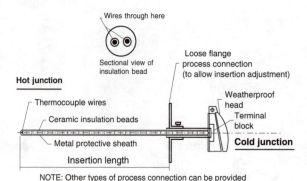

Figure 3.4: Typical industrial thermocouple assembly.

protection must be taken. It is suggested that thermowells be used in instances where it is necessary to contain the process fluid within the pipe or vessel, as this facilitates on-line withdrawal of the unit. The pressure rating of the line or vessel will determine whether the process connections are to be flanged or screwed (threaded). If the gas (air) is at an elevated temperature, then it is advisable to use a thermowell to prevent heat loss and personnel injury. The sheaths, though normally made from steel, can also be made from ceramic, and these latter coverings are used in applications where the process temperatures are in excess of that which can be tolerated by steel. In some other applications, especially in flue gas temperature measurement, it is important to specify sheaths that are coated with a ceramic. This has two benefits: first, it allows high temperatures to be measured; and second, it makes the steel resistant to attack from the sulfur dioxide that is usually present in the exhaust gases, thus enhancing the life expectancy of the sensor.

Tables 3.1 and 3.2 are a tabulation of various thermocouples and typical maximum temperatures for which they are suitable. The choice will also relate to characteristic thermo-emf in each case and the proposed detecting instrumentation (plus

TABLE 3.1
Base metal thermocouples

All temperatures shown are in °C.

Thermocouple	*Temperature*
Copper-constantan	400 continuous, 500 spot
Iron-constantan	850 continuous, 1100 spot
Chromel-constantan	700 continuous, 1000 spot
Chromel-alumel	1100 continuous, 1300 spot
Nicrosil-Nisil	1250 continuous
Tungsten-molybdenum	2600 continuous, 2600 spot—never use below 1250

Notes:
1. The last thermocouple in the tabulation uses pure metal wires only, not alloys.
2. The term *continuous* means that, when necessary, the thermocouple can monitor the stated value without interruption or deterioration.
3. The term *spot* means that the thermocouple can make very short-duration measurements of the stated value value without damage. Continuous exposure to these values will be detrimental to the device.

TABLE 3.2
Rare metal thermocouples

All temperatures shown are in °C.

Thermocouple	Temperature
Platinum–rhodium/platinum 100%; 10%/90%	1400 continuous, 1600 spot
Platinum-rhodium/platinum 100%; 13%/87%	1400 continuous, 1700 spot

costs, distances, and their compensation and chemical resistance in the case of direct insertion)

THERMAL RESISTIVE SENSORS

Another way of measuring temperature is by detecting the change of resistance of a stable resistor when subjected to a change of temperature. Platinum is the most common material for making resistance thermometers and may be used for temperatures up to 600°C. The reason for its popularity is that it is not easily oxidized, although it must be protected from contamination, especially by silica. The usual range of this type of temperature element is from −200 to +600°C. The temperature measured by a platinum bulb is referred to as the *platinum scale*, which approximates, but is not identical with, the *gas scale* over the range of 0° to +100°C.

The Gas Scale

The gas scale was developed from work done by the French physicist Gay Lussac, who gave us the two laws that relate the parameters of pressure and volume of a given mass of gas to its absolute temperature:

1. *The volume of a given mass of gas at constant pressure is directly proportional to its absolute temperature.*

2. *The pressure of a given mass of gas at constant volume is directly proportional to its absolute temperature.*

These laws can be easily visualized and verified from Figure 3.5. Let the volume of a gas at the temperature of melting ice be V_0, and let its volume be V_{100} when its temperature is raised to that of steam produced at standard conditions (the boiling point of water at 1 atm). This data has been plotted in Figure 3.5a and separated by 100 equal units on the x axis. Let the two volumes V_0 and V_{100} be denoted by points A and B, respectively, as shown in the figure. Suppose we join A and B by any curve as shown in the figure. Then, based on this arbitrary curve, for an intermediate volume V_1 there could be three different values, 1, 2, and 3 possible on the x axis. If now the units of the x axis were to be units of temperature in °C instead, then, for a single volume V_1 of the gas, there would simultaneously be three different temperatures, and this is an impossibility. This proves that the curve joining points A and B can only be a straight line, as shown in Figure 3.5b. If now the straight line AB were to be extrapolated back beyond A, then this extended line would intersect the temperature axis at approximately −273°C, as shown in Figure 3.5c. The value −273°C is the absolute zero of the gas (note: for precision work, a value of −273.16 is used) and is given the measurement unit K (for Kelvin), sometimes called °K. In actual practice it

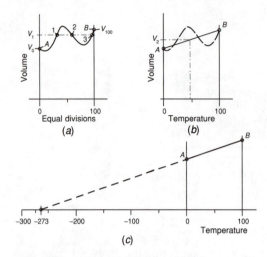

Figure 3.5: Absolute zero of temperature.

is rare to obtain the value for V_{100}, because the value of the atmospheric pressure will not be exactly equal to the barometric column of mercury at standard reference, and hence there is a difficulty in determining the temperature of melting ice. Corrections have to be made for the barometric height, and the steam temperature has to be calculated accordingly.

This gives the *gas* or *thermodynamic temperature scale*, a fundamental scale to which all temperature measurement must be referred.

The Platinum Scale

The platinum scale is obtained by using a platinum resistance element in a thermoresistive effort to replicate the gas scale. There are inevitable variations when this is done and corrections have to be applied.

The resistance at a particular temperature within the range of 0° to +100°C is given by:

$$R_T = R_0(1 + \alpha T)$$

where R_T = resistance in ohms at temperature T (°C)
  R_0 = resistance at 0°C—also called the *ice point*
  α = mean temperature coefficient of resistance of Pt between
    0° and 100°C

The coefficient α is calculated as:

$$\alpha = \frac{R_{100}/R_0 - 1}{100}$$

where R_{100} is the resistance at 100°C—also called the *steam point*. The difference between the resistances at 0° and 100°C is called the *fundamental interval* which, for 100 ohm bulbs, is 39.02 ohms—i.e., $\alpha = 0.003902$ ohms per ohm per degree C ($\Omega / \Omega /°C$).

The temperature measured on the platinum scale is computed from:

$$T_{\text{plat}} = \left(\frac{R_T - R_0}{R_{100} - R_0} \right) 100$$

For example if $R_T = 119.51\,\Omega$, then $T = 50°C$, which conforms to the value of α given previously.

The relationship developed here is a simple one, applicable for temperatures between 0° and 100°C and is known as the platinum temperature. If the formula is used for temperature ranges above 100°C, then the result will be numerically less than that given by the gas scale.

International Practical Scale of Temperature (IPST)

In 1927 the 7th General Conference of Weights and Measures adopted an easily reproducible practical scale to overcome the difficulties in temperature measurement on the thermodynamic scale and called it the International Temperature Scale (ITS). In 1948 the ITS was revised to the knowledge then available; and has been further amended mainly in that the temperatures are expressed in Celcius and now called the IPST.

For temperatures in the range of 0° to 630.5°C (the freezing point of antimony), the resistance of a standard platinum resistance thermometer is usually given by

$$R_t = R_0(1 + \alpha T + K_1 T^2)$$

This equation is usually referred to in the temperature standards (IPST) where it is shown as follows:

$$R_t = R_0(1 + AT + BT^2)$$

The coefficient K_1 is obtained from the measurement of the resistance of the platinum wire at 444.60°C, which is the boiling point of sulfur, also called the *sulfur point*; or at 419.505°C, the freezing point of zinc. The freezing point of zinc is the recommended value to be used.

The preceding equation defines the IPST for values in the range of 0° to 630.5°C. For ranges below 0°C, the following is used:

$$R_T = R_0[1 + \alpha T + K_1 T^2 + K_2(T - 100)T^3]$$

This equation is usually referred to in the temperature standards (IPST) where it is shown as follows

$$R_T = R_0[1 + AT + BT^2 + C(T - 100)T^3]$$

The coefficient K_2 is obtained from the measurement of the resistance of the platinum wire at $-182.97°C$, which is the boiling point of oxygen. All other coefficients are obtained as before. Remember the values of T will now be negative.

Using the platinum scale equation to calculate the temperature above 100°C will result in values that are lower than those derived by the IPST. The difference can be calculated from the following equation:

$$T_I - T_{\text{plat}} = k\left[\left(\frac{T_I}{100}\right)^2 - \frac{T_I}{100}\right]$$

where T_I = temperature determined by the IPST
T_{plat} = temperature determined by platinum scale
k = constant associated with purity of the platinum wire, 1.494 for the purest

Wire-wound resistance thermometers are manufactured in the following way. The resistance is a length of platinum wire that is wound on a ceramic bobbin, the length of wire being such that when the bulb is subjected to a temperature of 0°C its resistance is approximately 100 ohms. Small differences are made up by inserting "ballast resistors" usually in series, to make the value exactly 100 Ω at 0°C. The bobbin is supported on an insulating rod of length to cover the insertion required. The windings are particularly important in that they must be noninductive to avoid pickup, and it is usual to achieve this by having them bent hairpinwise and then wound on a twin helical groove in the bobbin. The whole assembly is covered in a ceramic glaze to inhibit contamination during use, but the rest of the mechanical construction is very similar to that shown for the thermocouple in Figure 3.4. One thing worth pointing out is that the use of platinum for resistance bulbs is fairly standard in Europe, whereas nickel was the more common in the United States, but less so now anywhere. The nickel bulbs have to conform to the SAMA specification; in Europe, to the DIN specification (DIN 43760 and DIN EN 60751); and in the UK usually to BS 1904:1984. Compensating for voltage drops on the current carrying cables is achieved by having 3 or 4 wire connections from the resistance element to the measuring bridge in the receiving instrument.

There are newer resistance elements that are constructed on a relatively thick ceramic substrate; one manufacturer, Heraeus Sensor-Nite, uses an aluminum oxide corundum (Al_2O_3), as the substrate onto which a thin-film platinum element is photolithographically structured. The element is protected with glass layers through which the connection leads to the resistance element are brought out.

Since the resistance thermometer forms part of a Wheatstone bridge circuit, it is subjected to the electric current applied to the bridge. This current will cause the element to heat up and result in an error in the measurement, the magnitude of which is determined by:

$$T_{error} = I^2 R K$$

where R is the resistance of the element, K is the self-heating coefficient provided by the element manufacturers, and $I^2 R$ is the power dissipated.

RADIATION PYROMETRY

We shall now discuss radiation pyrometry. The instruments for this branch of thermometry are called *radiation pyrometers*. The IPST is used as the basis for the measurement, as the ranges involved are usually above the freezing point of gold (1063°C, the *gold point*). This is a very important area for consideration, as direct methods for temperatures above 1400°C are quite useless, because the materials used would melt, or be contaminated beyond use. The main advantage for radiation pyrometers is that they do not need to be in contact with the object whose temperature is sought, since they are based on the measurement of the radiant energy emitted by the body.

We know that there are three methods by which heat transfer is effected—conduction, convection, and radiation, but only the last of these is independent of the presence of matter. Radiation is the method by which solar heat is transmitted to us here on earth. All forms of radiation are associated with electromagnetic waves, which are propagated through space with a common velocity of 299,774 km/s. There are many forms of radiation, but for the purposes of this discussion we shall restrict ourselves to UV (ultraviolet) and IR (infrared) radiations, i.e., to wavelengths from 0.1 μm (micron) in the UV band to approximately. 10 μm the IR band. **Note:** A wavelength of 1 μm is 10^{-6} m (1 micrometer), and 1 Å (Angstrom unit) is 10^{-10} m (or 10^{-8} cm, which is more often used). Within these bands, the radiation behaves in the same way as light in that it travels in straight lines, it can be reflected or refracted, and the amount of radiant energy falling upon a unit area is inversely proportional to the square of the distance between the radiating source and the receiver.

The calibration of radiation pyrometers enables them to measure radiation correctly when sighted on a *blackbody*. What, then, is a blackbody? A blackbody is a body, which, at all temperatures, absorbs all radiation falling on it without transmitting or reflecting any. When cold it would appear totally black, since it would absorb all light falling on it and reflect none. It is also found that a blackbody is a perfect radiator, and radiates more energy than any other body at the same temperature. How does one achieve a blackbody in practice? Having a large enclosure whose inner surfaces are not perfectly reflecting, i.e., only partially, and maintained at a uniform temperature does this. One of the surfaces of the enclosure has a small opening, and, provided that the opening is small enough, the interior of the enclosure appears perfectly black. If we consider a single ray of radiation falling on the opening, it passes through to the inside of the enclosure, where it is reflected by the "not perfectly reflecting" interior surfaces of the enclosure, so that it "bounces" from one side to another (losing energy with every bounce) so many times, that for all practical purposes, it is completely absorbed before it can reach the opening again. The only test of a blackbody is that there should be no temperature change in the body for radiations falling on it, which means that the *emissive power* must equal the *absorptive power*. The enclosure just described is used to calibrate radiation pyrometers by comparing the emissivity at a given temperature to that of the blackbody at the same temperature.

The *Stefan Boltzmann law for blackbody radiation* states that the total energy radiated per second per square centimeter is proportional to the absolute temperature raised to the fourth power, or:

$$E = \sigma T^4$$

where E = total energy (heat flux) radiated/s/cm^2
σ = Stefan-Boltzmann constant 5.670×10^{-12} W/cm^2·K^4 or 5.670×10^{-8} W/m^2·K^{-4}
T = absolute temperature, K

Figure 3.6 shows a diagrammatic arrangement of a typical total radiation pyrometer. The sensor shown as a thermocouple could alternatively be a *thermopile* as the effect will be the same, with the advantage of a higher output from the sensor.

(A thermopile, illustrated in Figure 3.7, is an assembly of miniature thermocouples connected in series, having all the hot junctions falling within a very small target area. The thermocouples are made in extremely narrow strips using very fine wires. One of the easier layouts to obtain the small target is to have the thermocouples arranged radially. The series connection is important, for it enables very small changes to be detected because the emf generated by each thermocouple is additive. In use,

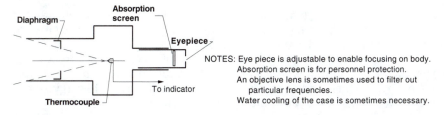

Figure 3.6: Diagrammatic arrangement of total radiation pyrometer.

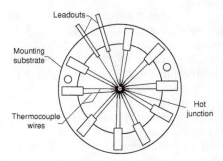

Figure 3.7: Schematic arrangement of a thermopile.

the thermopile is housed in a protective enclosure that is temperature-controlled, which makes the assembly immune to the atmosphere being measured and extremely stable thermodynamically, for the hot junction is rarely above a few hundred °C.)

When used, for instance, in measuring furnace temperature, the sighting hole on the furnace will radiate as a blackbody, if its contents are in equilibrium and provided that there are no flames or gas that absorbs or radiates any wavelength preferentially. This means that the entire contents are at the same temperature and all demarcation lines disappear. If now under these conditions a cooler object is introduced to the furnace, it (the cooler body) will absorb more energy than it radiates, and there will no longer be blackbody radiation; any radiation will depend on the emissivity of the furnace walls.

Optical pyrometers are generally unsuitable for recording or controlling temperature. They are nonetheless most useful in giving an accurate indication of temperature in the range 600° to 3000°C and are therefore also used in checking the calibration of total radiation pyrometers. Optical pyrometers are based on the measurement of radiant energy of one wavelength emitted by a body. Wien's law enables the calculation of the amount of radiant energy at a particular wavelength, emitted from a body at a particular temperature. It has been found that the wavelength of the radiation becomes shorter as the temperature rises; the relation expressed in Wien's displacement law as:

$$\lambda_m T = \text{constant}$$

where λ_m is the wavelength corresponding to the radiation of maximum energy (intensity) and T is the absolute temperature in K.

Wien's second law gives the actual value of the total energy radiated at the wavelength:

$$E_{\lambda_m} = T^5 \cdot \text{constant}$$

It is very important to understand that this equation gives the total energy at one particular wavelength only, whereas the total energy for all wavelengths is given by the Stefan-Boltzmann law and is proportional to T^4.

The Wien formula for the radiation, in the range of wavelengths λ to $(\lambda + \delta\lambda)$, emitted per unit solid angle per unit area of a small aperture in a uniform temperature enclosure in a normal (perpendicular) direction for the distribution of energy among the various wavelengths in the radiation of a hot body is

$$E_\lambda \, \delta\lambda$$

where

$$E_\lambda = \frac{K_3}{\lambda^5 \left(e^{K_4/\lambda T}\right)}$$

where T is the temperature in K, K_3 and K_4 are constants, and e is the base of the natural logarithms.

It was found that Wien's formula gave good results only up to temperatures of 1550°C, so Max Planck later refined it based on his quantum theory. Planck's expression was:

$$E_\lambda = \frac{K_3}{\lambda^5 \left(e^{K_4/\lambda T} - 1\right)}$$

where $K_3 = 3.740 \times 10^{-16}$ (J×m^2)/s, $K_4 = 1.44 \times 10^{-2}$ m·K, and all other symbols are as defined previously.

DATA REQUIRED TO PROVIDE MEASUREMENT OF PROCESS TEMPERATURE

The list of data under this heading corresponds to the information that would appear on a checklist for selecting a suitable temperature-measuring instrument.

PROCESS FLUID

Name

Characteristic	Gaseous or liquid
Chemical action	Aggressive or nonaggressive (**Note:** Aggressiveness is defined as the ability of the fluid to attack chemically or physically the containment or contact materials.)

PROCESS PARAMETERS

For the following, the appropriate engineering units are required.

Temperature

Maximum operating

Minimum operating

Normal operating

Pressure

Maximum operating

Minimum operating

Normal operating

PHYSICAL LOCATION

In the following, give the standards with which the material must comply—e.g., BS, ANSI, etc., pressure rating of process connections must be given.

Process line

Nominal size

Pipe schedule

Material of construction

Line/vessel connections

Screwed

Flanged

Welded

Hygienic (exclusively for food and drugs)

APPLICATION

Indication	Local or remote
Control—local	On/Off or PID
Remote	On/Off or PID (**Note:** Local is defined as having the measurement or control function at or near the point of penetration of the sensor or transducer into the process. Remote is defined as having the measurement or control function at some distance from the point of penetration of the sensor or transducer into the process.)

Do not be too anxious about the control aspects at this stage. They are stated here for completeness only and will be discussed in later chapters.

CRITERIA FOR SELECTING A SUITABLE TEMPERATURE-MEASURING INSTRUMENT

OPERATING RANGE

The operating range and type of function are the important considerations when selecting a measuring instrument. Roughly, these define the following categories:

1. Local reading instruments are fitted with either bimetallic, filled-system temperature-sensing elements, or thermoelectric temperature sensors. Scale graduations are as follows:

   • Bimetallic sensors have uniformly graduated scales, as the relationship between pointer movement and temperature is linear.

   • Vapor-filled systems have a nonlinear relationship between actual temperature and the pointer movement. The scale markings are usually close together at the low values of temperature, opening out for higher values. Vapor-filled systems should never be used for measurements near the ambient temperature.

- Liquid-filled systems have linear graduated scales. These include those filled with mercury that, although it is a metal, is in the (usual) liquid state over the temperature range considered.

- Thermoelectric temperature-measuring devices are defined as thermocouples, resistance thermometers, thermopiles, infrared temperature detectors, and thermistors. Of these, resistance thermometers and infrared detectors have linear scales, while all the others are nonlinear. The derived/ultimate electrical signal, which could be analog or digital, is available for indication and/or transmission as required.

2. Remote reading instruments are fitted with either filled-system sensors or elements that operate on thermoelectric principles:

- Filled systems used for this manner of readout are limited on the separation distance involved. Long distances have a requirement for a large temperature-sensing bulb so that the motive power for the instrument mechanism can be obtained. The matter of compensation for ambient temperature on long capillary runs must also be taken into account. The comments regarding the characteristics of the indicating scales made earlier apply equally here. If required, the measurements can be converted to signals that are either pneumatic or electronic (current) for transmission.

- Thermoelectric temperature-measuring devices have a much better capability for remote readout. Into this category of elements fall thermocouples, resistance thermometers, thermopiles, infrared temperature detectors, and thermistors. Direct connection of the sensing element and the remote display instrument is usually confined to the thermocouple and resistance temperature detectors, all the others have the measurement converted to a standard signal usually in the 4–20 mA range.

SENSING ELEMENTS

For the fastest measurement response, the sensing element itself should ideally be directly exposed to the process fluid, although this is not always possible, for one or more of the following reasons:

- Contamination of the sensing element by the process fluid
- Contamination of the process fluid by the sensing element
- Aggressive attack on the sensing element by the process fluid
- High pressure in the line or vessel containing the process fluid
- A process fluid that is life-threatening or dangerous to handle

The following notes are applicable to the food and drug industries:

1. Do not under any circumstance use a meter containing mercury to make a measurement of a process variable.

2. Process connections used must be of the sanitary type made by specialist manufacturers; APV Engineering and TriClamp® are two well known in Europe.

3. Installation of the instrument must not in any way be such as to permit the process fluid to stagnate; this is to avoid any possibility of bacterial growth, which could lead to contamination of the whole process.

To meet the requirement that the sensing element should avoid contamination of the process, most elements are located within wells or pockets. This serves a twofold purpose in that it separates the sensor and the fluid, as well as providing the means of containing the fluid pressure under all circumstances. It will be appreciated that pressure containment requires a material thickness sufficient to meet the forces involved. This will inevitably mean that the response time for the measurement will be affected, and thermal gradients will develop, which will give rise to steady-state errors. To minimize the effect of this lag, the space between the sensor and the bore of the well is sometimes filled with a heat transfer medium, although great care must be exercised when this is done to avoid process contamination.

NOTE: A thermistor is a semiconductor made of mixed metal oxides. These devices have an electrical resistance that falls with rising temperature; they are thus said to have a negative temperature coefficient. They are not as accurate as resistance thermometers, for it is difficult to control the material composition during manufacture, and they are very nonlinear. Hence, the resolution will vary across the usable span of measurement. A single thermistor will not be able to cover the total measuring range of this type of device, which is −100° to +300°C. Typical resistance values are 10 kΩ at 0°C and 200 Ω at 100°C. Physically, the thermistor is available in a number of different versions, the smallest being encapsulated in rods of 1 to 6 mm diameter, up to 50 mm long; or in 1 to 2.5 mm diameter glass beads. Metal casing similar to that of resistance thermometers is also available. The current-handling capability of the larger devices is excellent, and they may be used directly (without amplifiers) to perform some control functions, provided that due account is taken of the inherent nonlinearity.

LIQUID FILLING

The listing, though not exhaustive, shows the operating temperature ranges of some selected fluids and, with due care taken to avoid process contamination, the fluids can be used in measuring devices to be installed in the food and drug industries.

	Operating range	
Fill fluid	°C	°F
Alcohol	−80 to +70	−110 to +160
Toluene	−80 to +100	−110 to +212
Pentane	−200 to +30	−330 to +80

VAPOR-PRESSURE ELEMENTS

The vapor-pressure sensing elements must never be used for measurements of temperatures around the ambient, as there is insufficient power to drive a mechanism to indicate the measured variable. The scale is very nonlinear in the ambient temperature region, and therefore the resolution and accuracy are questionable.

THERMOCOUPLES

These temperature-sensing elements along with the resistance thermometers are employed today to provide most of the day-to-day temperature measuring requirements in a process plant. They are accurate and repeatable, provided they are not contaminated by exposure to unsuitable process conditions. The listing shows the most often used thermocouples, along with their conductor insulation color code. A few specialized thermocouples have been included to indicate the high-temperature capability of such elements.

Base Metal

Items marked with an asterisk are very specialized elements

Element, +/−	Type	Continuous °C	Spot °C	Conductor/color code +/−		
				UK	US	IEC*
Copper/constantan	T	400	500	wh/bl	bl/r	br/wh
Iron/constantan	J	850	1100	yl/bl	wh/r	blk/wh
Chromel/constantan	E	700	1000	br/bl	mag/r	mag/wh
Chromel/alumel	K	1100	1300	br/bl	yl/r	gr/wh
Nicrosil/Nisil	N	1250		or/bl	or/r	pnk/wh
Tungsten/molybdenum*		2600	2600	(never use below 1250)		
Tungsten/tungsten 26% rhenium*		2300				
Tungsten 5% rhenium/ tungsten 26% rhenium*		2300				

*IEC International Electrotechnical Commission.

Rare Metal

Element, +/−	Type	Continuous °C	Spot °C
Platinum/rhodium platinum 100%; 10%/90%	S	1400	1600
Platinum/rhodium platinum 100%; 13%/87%	R	1400	1700
Platinum rhodium/platinum rhodium 70%/30%; 94%/6%	B	1750	1800

% shown below the thermocouple is the metallic composition of the wire material used.

On base metal thermocouples, the conductors have an overall insulating cover that is also color-coded to the standard specified by the individual authority, but is not shown in this table; it is recommended that this coding be referred to also. The polarity of the emf is indicated in the relevant order for each conductor shown.

RESISTANCE TEMPERATURE DETECTORS

The standard elements use platinum wire for the resistance element, and are designed to have a resistance of 25, 50, 100, 500, and 1000, Ω at 0°C. In Europe the 100 Ω elements must conform to DIN EN 60751, DIN 43760, and BS 1904:1984. The foregoing elements are abbreviated to Pt 25, Pt 50, Pt 100, Pt 500, and Pt 1000, respectively.

A very large European manufacturer, Heraeus, incorporating the large US manufacturer Electro-Nite and now operateing in Europe as Heraeus Sensor-Nite, publishes the following data for its resistance elements:

Type	Construction	Range (°C)	Pt reference	Application
M-FK	Thin film	−70 to +500	100, 500, 1000	Unhoused
M-FK 220	Thin film	−70 to +500	100	Unhoused
M-FC	Thin film	−70 to +200	100, 500, 1000	Unhoused
KN	Ceramic	−200 to +600	100	Industrial
K	Ceramic	−200 to +800	100	Industrial
KE	Ceramic	−200 to +850	100	Industrial
M-FR	Ceramic	−70 to +500	100, 500, 1000	Industrial
M-G	Glass	0 to +600	100	Industrial
M-GX	Glass	−220 to +400	100	Industrial/laboratory

The foregoing is only an abbreviated summary of the devices, and manufacturers' specifications must be consulted for full details.

AGGRESSIVE FLUIDS

Materials of Construction

The user must define suitable material for the thermowell; otherwise, a choice will have to be made and user approval obtained before any supply or installation is undertaken.

SUMMARY

1. There are two engineering unit scales that can be used for temperature measurement: the Fahrenheit scale (now mainly used in the United States); and the Celsius scale (formerly called centigrade in the UK).

2. The temperature ranges between the fixed points of the freezing and boiling points of pure water at a pressure of 762 mm Hg for the respective scales are 32° to 212°F and 0° to 100°C.

3. There are absolute temperature scales also. Zero on the absolute scale when referred to the Fahrenheit and Celsius scales results in negative values. That is, the absolute zeros are below the freezing-point values of the respective scales. Absolute zero for each scale is:

   - Fahrenheit: −459.57°F or 0°R (Rankine)
   - Celsius: −273.15°C or 0 K (Kelvin)

The Kelvin scale is called the absolute thermodynamic scale.

4. Conversions from the various temperature scales are carried out as follows:

To convert °F to °C	$°C = (°F - 32) \times \frac{180}{100}$
To convert °C to °F	$°F = \left(°C \times \frac{180}{100}\right) + 32$
To convert °F to °R (Rankine)	$°R = °F + 459.57$
To convert °C to °K (Kelvin)	$K = °C + 273.15$
To convert °F to R (Réaumur)	$R = (°F - 32) \times \frac{4}{9}$
To convert °C to R (Réaumur)	$R = °C \times \frac{4}{5}$
To convert R (Réaumur) to °F	$°F = \left(R \times \frac{9}{4}\right) + 32$
To convert R (Réaumur) to °C	$°C = R \times \frac{5}{4}$

5. Bimetallic thermometers have two different metals bonded together and wound as continuous helices, one end being fixed and the other fitted with a pointer constrained to move across a graduated sector or circular scale.

6. Mercury-in-steel thermometers are the industrial version of the clinical mercury-in-glass instrument commonly used in hospitals and laboratories. They must never be used in food or drug processes, because of the possibilities of mercury poisoning. Mercury-in-steel thermometers must never be exposed to high temperatures over prolonged periods. The boiling point of mercury at atmospheric pressure is 357°C.

7. Liquid-filled thermometers can be used in those applications where mercury is unsuitable. Typical filling fluids are alcohol, toluene, and pentane.

8. Vapor-pressure thermometers should never be used for ambient temperature measurements. This type of instrument always suffers a time lag every time the bulb temperature crosses the ambient temperature of the capillary. They use the smallest bulb to achieve the measurement range.

9. Gas-filled thermometers are based on the gas laws and have very large bulbs in comparison with the Bourdon tube actuator and the connecting capillary involved. This size difference is to minimize the effects of ambient temperature variations.

10. Thermocouples are formed by joining two wires of different metals together. When the junction is exposed to a heat source, a small emf is generated and this is the Seebeck effect. The junction that is exposed to the heat source is called the hot junction; the other end of the wires is called the cold junction. The cold junction is connected to a device capable of measuring millivolts.

11. When an electric current flows across the junction of two dissimilar metals, heat is absorbed when the current flow is in one direction and heat is released when the current flow is reversed. The amount of heat is proportional to the amount of electricity that crosses the junction. This is the Peltier effect.

12. The amount of heat released or absorbed when a unit of current passes for a unit time is called the Peltier coefficient for the materials of the junction.

13. The law of intermediate temperatures states that if a thermocouple has one junction at temperature T_1 and the other junction at T_3, then this would be equivalent to having two thermocouples of the same materials with junctions at T_1 and T_2 for one thermocouple and junctions at T_2 and T_3 for the other.

14. The law of intermediate metals states that provided all the apparatus for measuring the thermoelectric emf in the circuit at the cold junction is kept at the same temperature the presence of any number of different metals will not affect the generated emf in the circuit.

15. In resistance thermometry, the temperature measured by a platinum resistance bulb is referred to as the platinum scale, which approximates but is not identical with the gas scale over the range of 0° to +100°C. The resistance at a particular temperature within this range is given by:

$$R_T = R_0(1 + \alpha T)$$

where R_T = resistance at temperature T
    R_0 = resistance at 0°C
    α = mean temperature coefficient of Pt (0 to 100)

$$\alpha = \frac{R_{100}/R_0 - 1}{100}$$

$(R_{100} - R_0)$ is the fundamental interval.

16. The Stefan-Boltzman law for blackbody radiation: The total energy radiated per second per square centimeter is proportional to the absolute temperature raised to the fourth power:

$$E = \sigma T^4$$

where E = total energy radiated/s/cm^2
    σ = Stefan Boltzmann constant, 5.670×10^{-12} W/cm^2·K^4 or
        5.670×10^{-8} W/m^2·K^{-4}
    T = absolute temperature, K

17. Wien's law enables the calculation of the amount of radiant energy at a particular wavelength emitted from a body at a particular temperature. The wavelength of the radiation becoming shorter as the temperature rises and expressed in Wien's displacement law as:

$$\lambda_m T = \text{constant}$$

The actual value of the total energy radiated at the wavelength is given by Wien's second law

$$E_{\lambda_m} = T^5 \cdot \text{constant}$$

18. The Wien formula for the radiation, in the range of wavelengths λ to $\lambda + \delta\lambda$, emitted per unit solid angle per unit area of a small aperture in a uniform temperature enclosure in a normal (perpendicular) direction for the distribution of energy among the various wavelengths in the radiation of a hot body is

$$E_\lambda \, \delta\lambda$$

where

$$E_\lambda = \frac{K_3}{\lambda^5 \left(e^{K_4/\lambda T}\right)}$$

where T is the temperature in K, K_3 and K_4 are constants, and e is the base of the natural logarithms. Good results can be obtained from this equation up to temperatures of 1550°C.

19. Planck's refinement of the Wien formula based on his quantum theory is expressed as:

$$E_\lambda = \frac{K_3}{\lambda^5 \left(e^{K_4/\lambda T} - 1 \right)}$$

where $K_3 = 3.740 \times 10^{-16}$ ($J \times m^2$)/s, $K_4 = 1.44 \times 10^{-2}$ m·K, and all other symbols are as defined previously.

CHAPTER 4

Level Measurement

The parameter *level* is an important one in process control, for it is one of the indicators by which the process engineers, process operators, and plant management know how much raw material there is in stock, how much material is in the manufacturing process, and how much finished product is ready for the market. As will doubtless be gathered, it is also an indirect but nevertheless important way of determining the amount of money that has been invested and is tied up in the product from a materials point of view. Economists today are continuously and increasingly referring to *cash flow* as an indicator of a company's financial health. It is therefore not difficult to appreciate that holding large quantities of raw, semimanufactured material and product will seriously affect the financial viability of the organization and therefore it's balance sheet.

Level measurement is important both from a process and fiscal viewpoint. The latter takes on a greater importance when the fluid contained in the tanks is final product and therefore represents inventory (capital investment), or when it is substances that need to be bonded. Bonded products are those that come under the jurisdiction of the customs and taxing authorities—e.g., gasoline, and both industrial alcohols and consumer spirits (whiskey, gin, etc.). Meeting the stringent requirements of this type of application requires specialized and highly accurate instrumentation. Most float-operated servodrive and radar and some hydrostatic tank gauges have been designed for this purpose.

Because many of the materials used in a manufacturing process are liquid, this chapter will discuss ways of determining the level of liquids in vessels and storage tanks using:

1. The "contact" methods of floats, buoyant force, electrical capacitance, and differential pressure.

2. The "noncontact" methods of radar, ultrasonics, and radiometrics.

Some of the methods, although discussed in the context of liquid level, are also suitable for a continuously variable measurement of solid material, but those suitable for solid materials are usually the noncontact methods, especially when discrete electrical switching is involved.

DEFINITION OF LEVEL

The level of a liquid is basically a measurement of *length* in the vertical direction with respect to the surface of the earth. The numerical value is determined from a measurement of the height of the free surface of the liquid from a fixed datum or reference. In many cases use is made of parameters other than the direct measurement of length (height), and therefore the level is *inferred*, which means that it is calculated from the value of other parameters.

The measurement of level was touched upon when we considered the parameter of pressure, but we shall investigate the subject in more detail in this chapter. We can all appreciate that the parameter level is usually associated with solids and liquids; but since air, and gaseous mixtures containing air, always occupy the entire volume of the containment vessel, it will be meaningless to measure the level of gaseous materials.

DEVICES FOR MEASURING LEVEL

THE BUOYANCY SENSOR

The first method of measurement we shall discuss is the *buoyancy method* of determining the level of a liquid in a containing vessel. The method depends on a principle of physics defined by Archimedes many centuries ago and requires a "body" upon which the forces act. In level instruments of this type it is usual for the body to take the form of a closed cylindrical tube that is allowed to be partially immersed in the liquid to be measured. The cylinder is called the *displacer* and is usually weighted with lead shot to a value that allows it to remain partially immersed in the liquid of the highest density that will be encountered in operation. Hence, each float is calibrated for the application, i.e., the liquid, density, and level range over which it will be used. In view of this, an instrument that is calibrated for one set of conditions is therefore not universally applicable for all level measurement, but needs to be recalibrated for the next application. From what has been said, it will be appreciated that, for the possible range of level measurements, the length of the displacer and the density of the liquid to be measured will govern the system *span*. The instrument, when installed on the vessel, is fitted in a position about a normal operating level. Figure 4.1 shows diagrammatically how the instrument operates. For convenience the connections are shown as top and bottom in Figure 4.2; but alternative process connection arrangements are available.

The instrument shown in Figure 4.2 has a *cage*, i.e., a separate chamber within which is located the displacer. The cage assembly is externally fitted to the process vessel at a position that represents the *normal liquid level* (NLL) or working level of the process vessel. It should also be pointed out that the displacer does not have to

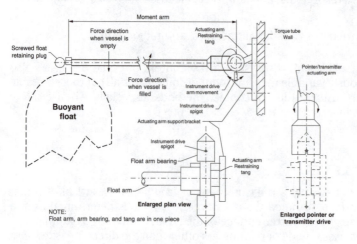

Figure 4.1: Diagram of displacer: instrument arm geometry.

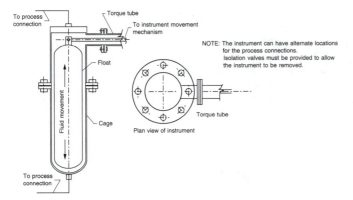

Figure 4.2: Diagrammatic arrangement of bouyancy-level transmitter.

be held within a cage as shown, but could just as well be left submersed uncaged within the vessel itself, either with the torque tube on the side of the vessel, or with the torque tube made to emerge through the roof of the vessel. In the latter case the connecting link will have to be a long rod, and therefore care should be taken to minimize the effect of the inertia and flexing of the rod itself. For internally mounted displacers, precautions must also be taken to minimize the effect of any possible violent fluctuations of the free surface, due to filling, as these will be sensed by the displacer resulting in a (misleading) noisy signal.

The instrument in Figure 4.2 works in the following way. The process liquid enters the cage through the lower process connection for the instrument and fills it to the NLL. In this respect the cage assembly and main vessel are configured and act like a U-tube manometer. The immersed displacer produces an upthrust brought about by the fluid; the downward force of the displacer weight (mass) itself counteracts this upthrust. The magnitude of the upward force is proportional to the height to which the liquid rises on the outer surface of the displacer. In accordance with Archimedes' principle, the displacer will apparently lose weight as the liquid level rises, and gain weight as it falls; this apparent change in displacer weight, in response to each change in fluid level, results in a change in the force applied to the actuating arm, causing a turning moment to be applied; but because the arm movement is constrained, a tortional stress is set up in the tube to which the actuating arm is fitted. The amount of stress in the tube causes the pointer/transmitter actuating arm to rotate a proportional amount and balances the force applied; this balanced position will be the level of the fluid. The rotation is made to drive a pointer, a pen arm, or a transmitting mechanism, and hence produces the level measurement required. When there is no liquid, the whole weight of the displacer is taken by the arm that connects the displacer to the instrument mechanism.

THE FLOAT-ACTUATED SENSOR

The next sensor we consider is a float-operated device, but in this case the float actually rests on or in the liquid surface and is not submerged (not fully immersed). The connection between the float and the instrument drive mechanism is either a counterbalancing weight with just enough bias or a spring-loaded windup wheel with just sufficient tension for the connecting cable or strip never to become slack but to always allow the float to be in contact with the liquid surface. Since the float rests on the liquid surface, the linkage to the instrument drive has to be flexible, but means have to be provided to prevent the float from wandering around the inside of the tank. The two most common methods used are:

1. A *stilling tube* or *well*—a vertical metal tube with holes on the lower portion and of length sufficient to cover the maximum level the tank contents will be expected to take. The internal diameter of the stilling tube, which is fixed permanently to the base of the tank, is just large enough to allow the float to move freely within the tube. The holes in the lower part of the stilling tube allow the process liquid to enter the tube, and this arrangement causes any turbulence on the liquid surface to be damped out and isolated from the measuring instrument while still allowing a steady or mean level to be determined. Manual "dipping" of the tank can also be carried out through the stilling well.

2. A *guide wire* passes through each of the two "eyes" or loops fixed on the float, diametrically opposite to each other. One end of the guide wire pair is fixed permanently to the base of the tank, or to a very heavy block that is lowered into the tank to rest at the bottom, the other end of the pair being permanently fixed to the roof of the tank. Turnbuckles provided on the roof connections allow the guide wires to be stretched taut. The "heavy block" method is usually used when the measuring system is installed after the tank has been in use, as it avoids the necessity of draining the contents. This arrangement constrains the float to move vertically without wandering, but does not eliminate the effects on the measurement due to fluctuations (common to all level measurements) of the liquid surface brought about by filling.

Because of the mechanics of the system, the float-actuated type of instrument is used to determine the level in tall vessels—e.g., oil storage tanks—and where the measurement span may be considerable. The level measurement could be displayed on directly coupled mechanical indicators, or transmitted using transmitting synchros or a slide-wire potentiometer. This technique is widely used by instrument companies such as Whessoe-Varec, which specializes in this method of level measurement.

In all float-and-cable devices, it is important that there be minimal slippage between the float cable and the windup pulley. In many cases the cable is not really a cable at all but a very thin and flexible flat strip of stainless steel with a series of precisely equidistant holes punched out on the centerline of the strip. These holes engage in a series of sprockets or pins on the pulley, very much like the paper feed mechanism on a logging printer. The sprocket pulley is spring-loaded so that the slack on the float cable is always taken up when the float changes position due to a change in level. To minimize the friction losses that are inevitable with such measuring systems, it is best to keep the number of pulleys and the cable length to a minimum. It is therefore usual when measurements have to be transmitted, for the transmitting instrument to be mounted at the top of the tank, as this will require the shortest cable length and only one idling sprocket between the float and the instrument drive mechanism. Allowing the tank level to be read locally and at ground level by a process operator would involve the use of two or more idling sprockets and long cables. This latter arrangement avoids a long climb to the top of the tank by the operator that will be inevitable with the minimum option just described. The accuracy claimed by manufacturers of these devices is of the order of 2 to 3 mm of the measured level. Figure 4.3 shows a typical arrangement of a float-and-cable level installation.

Displacers and floats can be used to locate the interface between two liquids of different densities. In this application the forces acting on a totally immersed

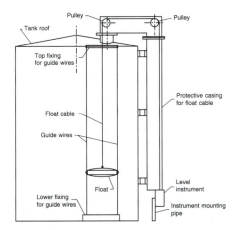

Figure 4.3: Float-and-cable level measurement. This diagram shows the readout instrument at ground level. An alternative position could be on the roof of the tank. This method is used for large storage tanks.

body at the interface will depend on the density of each fluid in the vessel, and it is important to remember, when specifying instruments for this type of application, that the density for each fluid must be known or available for second-order corrections. Displacer-type instruments are made by Fisher Controls and by Masoneilan.

THE CAPACITANCE LEVEL SENSOR

The next type of level measurement we consider is the *capacitance* level measurement method. In a simple system an electrode, which acts as one plate of a capacitor, is introduced into the fluid whose level is sought, the other plate being the vessel itself. An electrical potential is applied to the electrode system, with the fluid acting as the dielectric. The amount of charge developed, and thus the effective capacitance, will be a function of the area of the plate, the dielectric constant, and the distance between the two plates. The capacitance of the system is given by:

$$c = e \left(\frac{A}{d} \right) e_0$$

where A is the plate area covered by the liquid, d is the distance between the plates, e is the relative dielectric constant of the material, and e_0 is the dielectric constant of vacuum ($\cong$ the dielectric constant of air). Note that for air the dielectric constant is 1.0.

The electrodes could be rigid, as shown in Figure 4.4, or could be made of a flexible steel cable attached to a tensioning weight; the type used will depend on the application. Manufacturers of this type of instrumentation are Endress + Hauser, Kent, Eurogauge, Magnetrol, and many others.

The measuring circuit for many capacitance-type level instruments is a capacitance bridge. The Endress + Hauser instrument, however, uses the capacacitance generated as described and applies it to a frequency generator, the calibrated output of which determines the level being measured. This type of measurement is also suitable for solids storage. Care in this type of application must be given to avoid

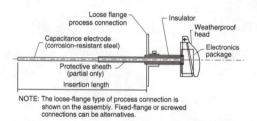

NOTE: The loose-flange type of process connection is
      shown on the assembly. Fixed-flange or screwed
      connections can be alternatives.

Figure 4.4: Typical industrial capacitance level
probe.

"ratholing," or cavity formation around the electrode. The actual precaution to be
taken is to mount the electrode at an angle that is slightly greater than the angle of re-
pose of the material itself, which will have the effect of making the rathole self-filling.

THE DIFFERENTIAL-PRESSURE CELL AS A LEVEL SENSOR

The differential-pressure meter can be, and is frequently, used to measure the
level in tanks and other process vessels. While it is possible to use the standard
screwed process connection instrument when it is reasonably close coupled to the
vessel e.g., Figure 4.6 *Case 1* open vessel level-measuring applications, the flanged
process connection device shown in Figure 4.5a, 4.5b, and 4.5c is more often selected. A screwed pipe connected instrument as shown in Figure 4.6 *Case 1* traps
process fluid (for a fuller treatment of this aspect, see details following on flanged devices) and can produce a static head due to mounting misalignment; supporting the
instrument is also difficult. The flanged instrument location provides a fixed datum

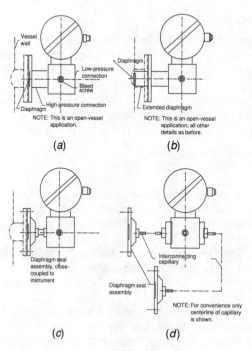

Figure 4.5: Level measurement with D/P cells.

on the vessel that is relevant to the particular measurement. For level applications several manufacturers provide the following options for the instrument as shown in Figure 4.5:

1. Measuring diaphragm flush with the face of the process flange.

2. Measuring diaphragm extended beyond the face of the process flange. The amount of extension is usually 4 in (100 mm) or 6 in (150 mm) to facilitate a diaphragm that is fully flush with the interior wall of the vessel when fitted to a standard 4- or 6-in flanged pipe stub or nozzle

3. A chemical seal can be fitted to a standard screwed process connection on the instrument as shown in Figure 4.5c or via an armored capillary tube as shown in Figure 4.5d. If the vessel is a closed one, i.e., not open to the atmosphere, then chemical seals will have to be fitted to both process connections. (Seals and their application are discussed in more detail in Chapter 2.)

When using the differential-pressure cells to measure level, be aware of the following. In all instances when instruments are fitted with measuring diaphragms as in Figure 4.5b, make sure that the diaphragm is flush with the inside surface of the vessel wall. If a nozzle on the vessel is used to provide a mounting for the instrument, then there is going to be a "dead leg" unless the extended diaphragm equals the overall nozzle length—i.e., the process fluid within the length of the nozzle would otherwise provide a hydraulic coupling or a stagnant volume that would not be continuously replenished or flushed. This can be a problem with crystallizing fluids, since in time the crystal growth will accumulate and inhibit diaphragm movement. Such a method of installation is particularly unacceptable in the food-manufacturing industry, since it encourages bacterial growth.

Instruments with chemical seals provide a solution to the problems of hygiene, crystal formation, and isolation of corrosive fluids. These elements can be provided with special hygienic screwed or quick-release clamp connectors suitable for the food industry, and/or purge connections to permit flushing of the diaphragm face with a suitable solvent. In instances where the process fluid is toxic, consideration should be give to the provision of a "restriction" orifice plate between the vessel and the instrument to limit the amount of toxic release in the event of a leakage.

When using D/P cells fitted with chemical seals, be aware that a variety of different types are available, each for a specific application. Make sure that the one selected is relevant for the measurement in measuring range, pressure rating, and material(s) in contact with the process fluid, which instrument manufacturers usually refer to as the material of the "wetted parts." Furthermore, when seals are installed as a pair, in the event of replacement it will have to be as a matched pair; this is necessary because the volumes of the fill fluid have to be identical, otherwise the determination of pressure differences will not be correct with varying ambient temperature.

When D/P cells are fitted with chemical seals that have long capillaries, be aware that such systems behave like "filled-system" thermometers and as such could give erroneous readings. One solution (especially if the level being measured is in a heated vessel) is to run the capillary very close to the outside wall of the vessel to keep the temperature differences to the absolute minimum.

In some instances where the vessel is closed and capillary-connected diaphragms are envisaged as being suitable, the following problems could be encountered. Be aware that there are some manufacturing limitations imposed on the capillary length, as well as the thermal effect we have discussed. One way out of a situation where capillary lengths would be exceeded due to the height of the vessel, or where the

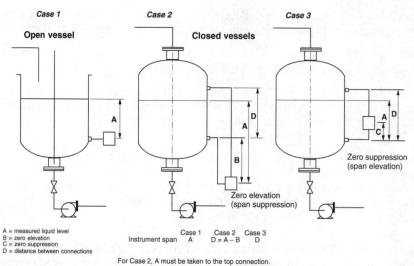

Figure 4.6: Vessel level measurement by differential-pressure cell.

lagging on the vessel prohibits the mounting of capillaries close to the vessel skin, is to have a pressure repeater on the upper (low-pressure) connection. The repeater is, in fact, a 1:1 pneumatic transducer that measures the vessel pressure and produces a pneumatic (air) signal that has direct correspondence to the measurement. This pneumatic signal is then connected to the low-pressure side of the D/P cell. Note that when pressure repeaters are used, there is a maximum limit to the pressure that can be sensed and repeated in the pneumatic signal generated. This maximum pressure rating of the repeater will have to correspond to the maximum internal operating pressure of the vessel. The use of a pressure repeater on the low-pressure connection of a differential-pressure cell can also provide a solution to the problem of measuring the level of volatile liquids like LPG (liquefied petroleum gas), as it eliminates the variations that would occur in the upper pipe connection that is not in contact with the liquid (static leg) brought about by the process vapor liquefying within it. (This is similar to the case, described in Chapter 2, of measuring pressure in a closed vessel.)

To assist in understanding the application of differential-pressure transmitters for determining the level in process vessels, Figure 4.6 shows the three possible methods of mounting the instrument. Always choose the mounting arrangement that will allow the easiest access for installation and maintenance. While the open-vessel arrangement shows the instrument level with the lower connection, this is not an absolute requirement, as either of the other two arrangements can be used. The important points to remember in the case of vessel level measurement are:

• For vented vessels, the low-pressure connection must be open to the atmosphere.

• Dependent on the instrument location, the measurement span and range must be calculated as shown.

• An adjustment kit or (calculation) algorithm has to be provided on the instrument to "elevate" or "suppress" the zero of the measurement, as will be explained shortly.

• It is difficult to obtain an installation where the connections on the vessel and instrument are truly in line as shown in *Case 1* in the figure; the instrument zero will have to be adjusted to suit.

The terms *zero elevation*, *zero suppression*, *span elevation*, and *span suppression* are very often used and a frequent cause of confusion by engineers will now be explained. Let us take *Case 1* to start with. Here we have a situation where the instrument is mounted at the lower process connection point on the vessel; therefore, the measurements will always be made with respect to this point. Liquid level rising above this connection will be measured up to the point that is the upper range value (the maximum height the liquid is allowed to reach) of the calibrated measurement span (span = upper range value − lower range value). The lower range value in this case is zero, since any liquid level below the lower connection will not be measured.

Let us now assume that we still need to measure the liquid level with respect to the lower vessel connection; the vessel is still open, but the instrument is mounted below the lower process connection on the vessel—in some respects, the situation in *Case 2*. Here, even when the level in the vessel goes below the lower process connection, there is a column of liquid always standing at the measurement connection of the instrument. If this situation were not taken care of, the level we would be measuring would be with respect to the location of the instrument connection where the true zero of the measurement is now located rather than the zero of the lower vessel connection, as required. The true measurement span of the arrangement would be the upper range value minus the true zero (lying some distance below the lower vessel connection), which makes the span greater than that required. To overcome this, what we have to do is "artificially" push the instrument connection up to the location of the lower connection on the vessel; in other words, we have to *elevate the zero* of the measurement, and as a direct result we have reduced the true measurement span, or, in instrument parlance, carried out a "span suppression." Elevating the zero always results in a change in the lower range value, and for the case in question, the range will certainly not be zero-based.

Let us now consider the situation where the instrument is located (for practical convenience, say) above the lower process connection. In this case the true zero, as before, will be at the position of the instrument connection, and therefore the true zero will have to be artificially lowered to the position of the process connection. In other words, we have to *suppress the zero* and as a result increase or elevate the span.

A useful memory aid is, zero elevation results in span suppression, and zero suppression results in span elevation. In closed vessels, the pressure above the liquid face must always be accounted for in the calculations for instrument range and span.

THE ULTRASONIC LEVEL SENSOR

So far we have considered devices that are in contact with the process fluid at all times in order to determine its level, but let us now look at some of the noncontact methods of determining liquid level. The first of these will be one of the ultrasonic devices available today. For these instruments to operate, there must be a source of ultrasound and a receiver. In operation (for installations of the transceiver above the liquid) the instrument transmits a burst of sound, typically in the range of 21 kHz to 160 kHz, which travels to the surface of the liquid, from which it is reflected and picked up by the receiver. The time interval between these pulses is a measure of the location of the liquid surface and hence of the level. Since the technique uses sound as the basis for the measurement, it is obvious that there will be some reflection (echoes) from the sides of the containing vessel causing an interference to the signal if precautions are not taken in the design of the installation. The echoes will be most severe in an empty vessel, although most ultrasonic level-measuring instruments are provided with facilities for "ignoring" by electronic time gating most of the remnant/spurious echoes and concentrating only on those that emanate directly from the surface of the fluid. It is by this means of *positive selection* that the unwanted signals from the vessel itself and any other hardware that may be fitted within the vessel can be accounted for and excluded from the measurement indication. Furthermore, since temperature influences the velocity of sound, the ultrasonic level detector is provided with the means of sensing this parameter and accounting for it in the measurement. In this context, recall that the velocity of sound in a gas is given by:

$$v = \sqrt{\frac{\gamma \Re T}{M}}$$

where $\Re$ is the universal gas constant, γ is the ratio of specific heats, T is the absolute temperature, and M is the molecular weight. Therefore, for a given gas:

$$v \propto \sqrt{T}$$

There are application constraints for this type of measurement, the most important being that the liquid should be free from any froth or foam. If froth or foam is present, then some of the transmitted sound will be absorbed and erroneous results will certainly be obtained, e.g., from multiple reflections from the complex surface. Humidity will affect the velocity of sound, since vapors alter the transmission of the ultrasound signal. Some manufacturers, e.g., Endress + Hauser, make their instruments insensitive to condensate on the sound transmitter/receiver. There are steps that can be taken to avoid some of the problems, but these have to be considered individually for each application.

Ultrasonic level sensors are manufactured by Endress + Hauser, as well as several other companies.

THE RADAR LEVEL SENSOR

The next type of device we consider is the radar level detector; this instrument uses a radar signal instead of the ultrasonic one just discussed. The advantage of this device is that it can be focused more sharply than the ultrasonic instrument because of the much higher frequency, and as a result it can be used for large tanks with a very high level of accuracy—error of the order of 1 mm is claimed. Saab (the automobile and aircraft manufacturer) and others have devices on the market using the radar technique to determine the level in large tanks and combine this with density and the tank "strapping table" to produce a signal that represents the mass of the material. A strapping table is a tabulated list of level values obtained in large storage vessels that relate the barreling effect of the containment vessel on the liquid level. The Saab mass meter competes with the hydrostatic tank gauges made by Foxboro, Rosemount, and Honeywell.

LEVEL SWITCHES

When continuous indication of fluid level in a vessel is not required, but a requisite amount is to be available at all times, a level sensor with a two state device or switch is appropriate. These devices have adjustable high and/or low operating limits. Fluid ingress is permitted when the level is below, and inhibited when above the high limit, and/or fluid egress is inhibited when the fluid is at or below the low the limit.

Level sensors that operate on the basis of a switching action can be divided into two main groups: (1) contacting and (2) contactless. Under the first group will be all the float, rotating paddle, vibrating probes, and capacitance-operated devices. This first group is so named because the material being measured must be in 'contact' with the switch-actuating mechanism, and the second group will contain all devices that do not require any part of the switch to be in contact with the material being measured. Into this latter category falls the inductive-type proximity and microwave devices for level measurement. Microwave switches are very useful for detecting the level of electrically nonconductive solid materials. The contactless switches can be used to determine the level in solid storage hoppers and silos, but since these devices are switches they can be located only at predetermined heights on, for instance, a solids hopper; they will not provide any continuous (analog) level measurement or indication. They are virtually indispensable in applications where limiting and safety are a concern, in batch and sequential process operations.

THE RADIOMETRIC (RADIOACTIVE SOURCE) LEVEL SENSOR

We shall now consider the radiometric (also called nucleonic) method of level measurement. For this we need a radioactive source. Cesium137 and cobalt60 are the most commonly used; they are quite safe provided that the required procedures are implicitly followed. In the United Kingdom and most European countries, the site will have to be open for inspection by the relevant safety authority; and appropriate restrictions and regulations also apply elsewhere. The radiation from the source is directed over the full range of level to be measured as indicated in Figure 4.7. The detector senses a reduction in received radiation due to the presence of material in the vessel. A microprocessor handles the tasks of compensation for the loss of natural activity of the gamma source, linearization of the measured value, calibration data, alarms, temperature, etc. The radiation loss to the material of the containment vessel or measured material is claimed to be negligible, making it acceptable for food-manufacturing processes. This instrument can provide an analog (continuously variable) measurement that is proportional to the height of the material in the vessel. There are several manufacturers of this type of equipment, among which is Endress + Hauser.

It must be pointed out that radiometric level measurement is totally non-invasive—i.e., no part of the instrument has to be inserted into the process. Figure 4.7 shows a typical arrangement of a radiometric level installation. The detector

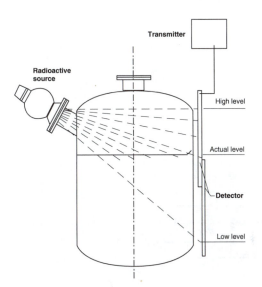

Figure 4.7: Radiometric liquid-level measurement.

comprises a scintillator, a photomultiplier, and a control unit built into a corrosion-resistant steel pipe, held in place by fixing collars.

THE HYDROSTATIC TANK GAUGE (HTG)

Hydrostatic tank gauge (HTG) measuring systems have been designed to meet the needs for very accurately measuring the contents of large storage tanks. The phrase *measuring system* has been used in this instance for the simple reason that the overall output is the combined result computed from more than one measuring instrument covering different parameters or liquid properties. Hydrostatic principles of measurement are not new, but the determination of liquid mass received a great impetus with the more recent technological advancement in pressure sensing. The output from the measuring system is the *total static mass* of the material in the tank, obtained directly from the pressure measurements and the *tank strapping* (capacity) table. Figure 4.8 shows the system arrangement.

In these systems the instrumentation consists of a maximum of three very highly accurate pressure transmitters—on the Foxboro TankExpert® the pressure transmitters have resonant wire sensors with no moving parts, but with temperature compensation and an accuracy of the order of 0.02% of the sensor range—and an RTD (resistance temperature detector described in Chapter 3) temperature transmitter. If we call the three pressure transmitters P1, P2, and P3, we can arrange to mount P1 near the bottom of the tank, P2 at a fixed vertical distance from P1 on the vertical side of the tank shell, and P3 on the roof of the tank. Not all three pressure transmitters will be required for each application; the actual number that will be necessary depends on the type of tank involved. With this arrangement it is possible to calculate the results with a specialized microprocessor in which are stored the constants appropriate to the tank and its strapping tables; each of the following calculations must be combined with the relevant tank constants and tank capacity values to give the information required:

1. Density from the pressure difference between P1 and P2

2. Mass from the pressure difference between P1 and P3, the tank effective area, and stored tank constants

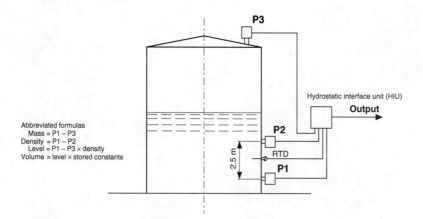

Figure 4.8: Hydrostatic tank gauging system, based on Foxboro TankExpert®.

3. Level from the pressure difference between P1 and P3 and the density

4. Observed volume from level and the tank strapping table

5. Temperature of the fluid from an RTD and the head-mounted temperature transmitter

6. Volume at reference temperature from level, density, temperature, and known liquid parameters

Applications of the Hydrostatic Tank Gauge

The number and kind of instruments involved in using the hydrostatic tank gauge depend on the system arrangements. The following describes the necessary setup in each of four different situations to provide for highly accurate and automatic tank gauging; calculations of mass, level, volume, and density can be carried out and recorded at predefined intervals, or on demand, and an inventory management system giving reliable, accurate data for material balances and product loss control can be supported.

1. In applications having pressurized tanks—i.e., the vessels are totally closed—and requiring density measurement, three pressure transmitters and a temperature transmitter are needed.

2. In applications having tanks that are vented to the atmosphere and requiring density measurement, only two pressure transmitters (P1 and P2) and the temperature transmitter are needed to give the density at the operating temperature.

3. In applications having pressurized tanks and a known density, only two pressure transmitters (P1 and P3) and the temperature transmitter will be required.

4. In applications having tanks that are vented to the atmosphere and a known density, only one pressure transmitter (P1) and the temperature transmitter will be required.

Since the output from the Foxboro TankExpert® HTG system for mass is highly accurate (of the order of better than 0.1% of the urv), it is suitable documentation for the taxing authorities in assessing inventory taxes. The equipment must be inspected and calibration-witnessed, and the tank strapping table must be certified to ensure that the government neither loses revenue nor gains it at the expense of the producer. In addition, the system is arranged to give the fluid level accurate to 2% and the fluid density accurate to better than 0.5%. It should be noted that fiscal measurements by reference volume have been made by servo and radar for a very long time.

In tank gauging, there are always comparisons between float-operated equipment and the more modern approach of hydrostatic measurements, and arguments as to which is the better of the two types of equipment. The hydrostatic measurement system has advantages for the reason that there are no moving parts and there is very little to go wrong, since there are no cables, pulleys, or friction to worry about; and because of this the calibration has a greater propensity for remaining stable for much greater periods. It is also usual for the hydrostatic equipment to use intelligent pressure transmitters that have been especially designed for this type of measurement, thus ensuring very high stability and accuracy with only an adjustment for zero provided for the transmitters. Manufacturers of this type of system are Foxboro, Rosemount, Honeywell, and some others. Some other instrument makers also provide hybrid tank gauging equipment of other types.

DATA REQUIRED TO PROVIDE A MEASUREMENT OF PROCESS LEVEL

The list of data under this heading corresponds to the information that would appear on a checklist for selecting a suitable level-measuring instrument.

PROCESS FLUID

Name

Characteristic   Liquid or solid

Chemical action   Aggressive or nonaggressive (**Note:** Aggressiveness is defined as the ability of the fluid to attack chemically or physically the containment or contact material)

PROCESS PARAMETERS

For the following, appropriate engineering units are required.

Pressure

Maximum operating

Minimum operating

Normal operating

Temperature

Maximum operating

Minimum operating

Normal operating

Density or Specific Gravity (At stated or reference conditions)

PHYSICAL LOCATION

In the following the standards—e.g., BS, ANSI, etc., with which the material of construction must comply, and the pressure rating of process connections must be given.

Process Vessel

Height

Diameter

Other section (state shape required for volume, strapping table data, HTG system, etc.)

Material of Construction

Process Connections

Screwed

Flanged

Welded

Hygienic (exclusively for food and drugs)

APPLICATION

Indication	Local or remote
Control—local	On/Off or PID
Remote	On/Off or PID (**Note:** Local is defined as having the measurement or control function at or near the point of penetration of the sensor or transducer into the process. Remote is defined as having the measurement or control function at some distance from the point of penetration of the sensor or transducer into the process.)

Do not be too anxious about the control aspects at this stage. They are stated here for completeness only and will be discussed in later chapters.

CRITERIA FOR SELECTING A SUITABLE LEVEL-MEASURING INSTRUMENT

OPERATING RANGE

The operating range is the most important criterion when selecting a measuring instrument. This can be specified in:

1. Units of length (metric or imperial)

2. Units of mass (normally associated with HTG)

SENSING ELEMENTS

The sensors used for level-measuring instruments can be divided into the following categories:

1. Differential-pressure cells.

2. Buoyancy type. **Note:** If the sensor is to be located within the vessel, the displacer length must be stated. If the sensor is to be the external cage type, then the distance between the connections on the vessel must be given (together with the orientation—i.e., top/bottom, bottom/side, side/side, top/side).

3. Float and cable type. **Note:** The location of the readout instrument or the transmitter must be stated, as this will permit the best pulley and sprocket arrangement to be provided.

4. Capacitance type. **Note:** The length of the capacitance probe must be stated, together with the location with respect to the vessel wall.

5. Ultrasonic type: details of any internal fittings (e.g., agitators).

6. Radiometric type.

Note: For instruments used in the food-and-drug processing industries, no mercury is allowed to be used in any part of the instrument. Furthermore, installation of the instrument must not be carried out in any way that will permit the process fluid to stagnate; this is to avoid any possibility of bacterial growth, which could lead to contamination of the whole process. All instruments must be provided with the patented process connections, the most well known ones being manufactured by APV Engineering.

In the UK and some countries in Europe, there are some food manufacturers who do not accept radiometric-type sensors; but, if they are used, be aware that any specific factory inspectorate requirements will also have also to be complied with.

AGGRESSIVE FLUIDS

Materials of Construction

Where the sensor or instrument is available in a material compatible with the process fluid, then the instrument as made will be suitable. However, it is more usual for the user to specify a suitable material and leave the manufacturer to comply with it. If none of the materials of the instrument are suitable, then provision of a chemical seal between the process and the measuring instrument may be the alternative.

Chemical Seals

As mentioned earlier, chemical seals are physical barriers that are placed to separate the process fluid from the measuring device. There are several manufacturers of such equipment; two of the more well known are Ashcroft and Wika. The former is a US manufacturer and the latter is European.

For more details on chemical seals, refer to Chapter 2.

SUMMARY

1. Level is basically a measurement of length in the vertical direction with respect to a fixed reference, e.g., the bottom of a vessel. The numerical value is determined from a measurement of the height of the free surface of the material from a fixed datum or reference.

2. The level in a vessel can be measured using a buoyancy level transmitter. This method allows the measurement to be made over a limited range governed by the length of the float, when the density of the fluid is known and fixed. The instrument is most applicable where the vessel will always contain some process fluid; it measures variation in level about the normal.

3. Calibrating the buoyancy-type instrument requires knowing the density of the process fluid and the range over which the instrument is to operate. Since the forces involved are liquid-density–dependent, the instrument is capable of measuring the interface between two different fluids that could be contained in a vessel.

4. The buoyancy type of instrument can be provided with a variety of process connection arrangements to suit the installation and location.

5. The displacer of a buoyancy-type instrument can be cage-enclosed, or it can be located within the vessel itself.

6. Large storage tanks can be measured using a float and cable attached to a synchronous repeater. A change in the position of the float resting on the surface of the liquid in the tank is reflected in a corresponding change of the position of the synchro.

7. Float-and-cable systems are mechanical devices that are capable of measuring the full range of the tank to which they are fitted. Care must be exercised to ensure that the float is not subject to violent fluctuations of the fluid surface (a "stilling tube" will prevent this), and that the float is prohibited from wandering and is constrained by guide wires or a stilling tube to move only in the vertical direction.

8. Since float-and-cable systems are mechanical, they are subject to the effects of friction and inertia, which will be reflected in the accuracy of the measurement.

9. Ultrasonic level measurement uses a transmitter to direct a burst of ultrasound at the fluid surface, from which it is bounced back and picked up by a receiver. The time interval between transmission and receipt of the sound signal pulses is a measure of the position of the fluid surface.

10. Since sound is the medium through which an ultrasonic measurement is made, echoes will be most severe in an empty vessel. Careful design of the installation to avoid interference is vital. Ultrasonic level-measuring instruments are provided with facilities for ignoring by electronic time gating most of the spurious echoes and concentrating only on those that emanate directly from the surface of the fluid.

11. Temperature influences the velocity of sound. The velocity of sound in a gas is given by:

$$v = \sqrt{\frac{\gamma \Re T}{M}}$$

where $\Re$ is the universal gas constant, T is the absolute temperature, M is the molecular weight, and γ is the ratio of specific heats. Therefore, for a given gas:

$$v \propto \sqrt{T}$$

12. Froth or foam will absorb transmitted sound and will produce erroneous results in ultrasonic level measurement. Humidity, too, affects the velocity of sound, since vapors affect the transmission of sound signals.

13. The radar level detector uses a radar signal instead of an ultrasonic signal, the advantage being that it can be focused more sharply than ultrasonic probes or detectors, and thus can be used for large tanks to give a very high level of measurement accuracy.

14. Radiometric (or nucleonic) level measurement is totally noninvasive; i.e., no part of the instrument has to be inserted into the process. The radiation from the source is directed over the full range of level to be measured. A reduction in radiation due to the presence of the vessel contents is sensed by the detector.

15. The radioactive source most usually used is Cesium137 or Cobalt60. It is quite safe, provided that the required procedures are implicitly followed. In the United

Kingdom the site will have to be open to inspection by the relevant safety authority, and appropriate restrictions and regulations apply elsewhere as well.

16. In a simple capacitance level-measurement system, an electrode that acts as one plate of a capacitor is introduced into the fluid whose level is sought, the other plate being the vessel itself. An electrical potential is applied to the electrode system with the fluid acting as the dielectric. The charge developed will be a function of the area of the plate, the dielectric constant, and the distance between the two plates. The capacitance of the system is given by:

$$c = e \left(\frac{A}{d} \right) e_0$$

where A is the plate area covered by the liquid, d is the distance between the plates, e is the dielectric constant, and e_0 is the dielectric constant of vacuum ($\cong$ dielectric constant of air).

17. The differential-pressure transmitter can be used in level measurement in tanks and other process vessels if the fluid density is known and fixed. When used in level-measuring applications, flanged process connections are usually selected for the device.

18. For level applications, several manufacturers provide the following options for the instrument: (a) measuring diaphragm flush with the face of the process flange, (b) measuring diaphragm extended beyond the face of the process flange, and (c) a chemical seal fitted to a standard screwed process connection via an armored capillary tube.

19. In using differential-pressure instruments for level measurement, remember:

- For vented vessels, the low-pressure connection must be open to the atmosphere.

- Dependent on the instrument location, the measurement span/range has to be calculated to allow for the pressure variations that this entails.

- An adjustment kit has to be provided on the instrument to elevate or suppress the zero of the measurement.

- The instrument zero will have to be adjusted to align the differences in level between the process connection and the instrument.

- It is necessary to minimize the error due to the process and ambient temperature effects that will ensue when chemical seals with long capillary tubes are attached to the instrument.

20. A pneumatic pressure repeater can be used to replace the wet leg (upper connection) in a differential pressure method of level measurement of a closed vessel containing condensing vapor. Care must be exercised to ensure that the minimum and maximum internal pressures of the vessel fall within the range capability of the repeater.

21. Hydrostatic tank gauging (HTG) is a measuring system whereby the total static mass of the material in a tank is obtained directly from the combined measurements of up to three pressures, one temperature, and the tank strapping (capacity) table.

22. The direct measurement of the static or inventory as opposed to transfer mass obtained from the HTG system is as accurate a measure—of the order of better than 0.1% of urv—of the tank contents as is possible at the present time. In addition, the system is arranged to give the fluid level accurate to 2% of reading and the fluid density accurate to better than 0.5% of reading.

23. In an HTG system with three pressure transmitters, of which one is mounted on the roof of the tank, two mounted a fixed vertical distance apart near the bottom of the tank, and a temperature transmitter, it is possible to calculate: *density* from a pressure difference of the lower pair, *mass* from the pressure difference and the tank capacity table, *level* from the pressure difference and the density, *volume* from level and the tank strapping table, and *volume at reference temperature* from level, temperature, and known liquid parameters. The fluid temperature is obtained from the temperature transmitter.

CHAPTER 5

Analytical Measurements

CHEMICAL ANALYSIS IN PROCESS CONTROL

The quality of the manufactured product can only be assessed by conducting tests or by measuring one or several constituents of the completed product. In addition, many of the manufactured products demand alterations to the chemical makeup of the basic raw materials as part of the process. In view of this requirement it is important that the changes taking place in the raw material be monitored and controlled if necessary so that the final results are within the limits of acceptability and hence of quality. This idea of working to a limit of acceptance is generally referred to as bringing the product "within specification," or, in the parlance of people in manufacturing, as being "on spec." From this brief statement of the situation, one can understand that there are a number of different parameters that will have to be monitored during the processing of the material in order to ensure that the final manufactured product is entirely suitable for its intended use.

The field of analytical instrumentation is so large and highly specialized that this chapter will not attempt to cover every aspect in detail. In most process industries, it has been the experience that the more esoteric analytical methods available today are rarely used routinely in the plant for maintaining product specification or control of the process. However, these sophisticated measuring techniques are most often to be found in the product control and research laboratories, where the influence of subtle variations in the chemistry may be more evident and significant to the difference between success and failure in the production process.

In line with these comments, we shall limit ourselves to the more "common" analytical measurements made in process plants, and perhaps mention in passing those other exotic methods that are used by the process chemists in their laboratories. It must be stressed that the sophisticated measurements do have a relevance and importance in the real manufacturing world, for without them, it would not be possible to even contemplate a simplification of the process and maybe the instrumentation and control requirements that could bring with them the economies of scale for the benefit of us all.

In this chapter we shall devote our attention to the parameters of pH, conductivity, oxygen content (using the paramagnetic technique) and the methods of chromatography. The measurements covered are the ones that are the most common and widely used in the process industry. Even restricting ourselves to the parameters noted, it must be stated that the actual analytical measurements made on plants are so many and varied that it will not be practicable to cover them all.

HYDROGEN ION CONCENTRATION OR pH

The first measurement we shall consider is pH, which is an abbreviated way of saying the *hydrogen ion concentration*. What does pH mean to those of us who are not chemists or chemical engineers? For the answer to this question we shall have to go back to the fundamentals.

ATOMS AND IONS

When we consider a specific material, we know that at the molecular level it is composed of a combination of atoms, and then looking at the atomic level we find there a collection of particles, some carrying electrical charges. The three important particles from our point of view are the positively charged *protons*, the negatively charged *electrons*, and the chargeless *neutron*. The neutron and protons form the nucleus, which is orbited by the electrons. The number of protons exactly equals the number of electrons, resulting in an electrically neutral atom. The atomic number is the number of protons in the nucleus of the atom of the element, but the number of electrons varies with the ionization.

We have used the word *orbited* to describe the path taken by the electrons around the nucleus. Readership knowledgeable of physics may take exception to it, for it is understood that the electrons do not have a single path, but a multiplicity of paths resulting in what can best be described as a "cloud." Maintaining the simplistic view initially taken, the electron paths can be visualized as a miniature solar planetary system having the nucleus as the sun. There are seven such paths, or electronic "shells," each at a different energy level that are designated consecutively K through Q, with K being nearest the nucleus and Q the furthest. Not all atoms have all these seven shells; they vary from element to element. Tables are available that give an ordered list of the elements and the corresponding number of electrons in each electronic shell. The minimum number of electrons in K is 1, and the maximum 2; the intermediate shells vary between 8 and 32, the most tightly packed shells being M, N, and O; the outermost shell of an atom shows a tendency or preference to form a complete 8-electron combination. This tendency is called *valence*. Hydrogen is unique in its simplicity in that it has only one electron in the K shell, which also explains its unique chemical character that bears no resemblance to any other.

It is possible, within limits, for atoms to gain electrons from, or lose electrons to, neighboring atoms. The electrons involved in the exchanges are called *free* or valence electrons. When an atom *gains* a negatively charged electron, it becomes more *negatively* charged and is then called an *anion*, and the atom that *loses* an electron will become more *positively* charged and is then called a *cation*. The charge carried by an ion is equal to the number of electrons gained or lost by the original atom. This provides us with the fundamental unit of electrical charge, $-e$, equal to about 1.6×10^{-19} coulomb (C), as demonstrated by R. A. Millikan in his "oil drop" experiment, to date the smallest quantity of charge discovered. The process of changing neutral atoms into ions is called *ionization*. Materials that can be split into ions are called *electrolytes*, but energy is always required to do this, and the process of splitting is called *electrolysis*.

THE NORMALITY OF SOLUTIONS

Frequently we are confronted with a material that is a *solution*. There are, broadly speaking, two types of solutions: (1) solid solutions—e.g., all metallic alloys and steels; and (2) liquid solutions—e.g., sugar in water, brine, etc. We shall confine ourselves to a consideration of liquid solutions only, and for the moment a binary (two-component) solution. What is such a liquid solution? This liquid solution is a liquid in which another substance has been dissolved; for this to happen the substance added must be soluble in the liquid. If this were not so, we would have a *mixture* of the two. The liquid is called the *solvent* and the added substance is called the *solute*. Since water—the Latin word being *aqua*—is the most common solvent, *aqueous* solutions are of particular importance.

The concentration of a solution of a substance can be expressed in several ways; the three most common are:

- Percentage by weight

- Number of gram-molecular weights of substance in 1 liter of solution

- In terms of *normal* concentration.

The third of these is by far the most commonly used and bears some resemblance to the second.

A solution has a *normal* concentration when 1 gram-equivalent weight of the substance is dissolved in 1 liter of solution. If 2 gram-equivalent weights of the substance are dissolved in 1 liter of solution, the solution is said to be twice normal, and there is a shorthand way of writing this:

- A normal solution is written as N $(= 1N)$

- A twice normal solution is written as $2N$.

- A one-tenth normal solution (decinormal) is written as $0.1N$.

It is difficult to give a clear definition of equivalent weight (gram equivalent) to cover all reactions. The equivalent weight of an acid is that weight of the acid that contains one replaceable hydrogen, i.e., 1.008 g of hydrogen. The equivalent weight of a base (alkali) is that weight containing one replaceable hydroxyl group, i.e., 17.008 g of ionizable hydroxyl (17.008 g hydroxyl is equivalent to 1.008 g of hydrogen). In general, for a salt, it is defined as its gram molecular weight divided by the total positive or total negative valence and for oxidizing or reducing agents, it is the gram molecular weight divided by the total change in valence of all atoms in the molecule that change valence. Modern chemistry uses the mol equivalent instead of the equivalent weight to simplify the calculations, for under identical conditions equal volumes of gases will contain an equal number of molecules (Avogadro's law).

THEORY OF ELECTROLYTIC DISSOCIATION

Let us look at an example. Common salt has the chemical formula NaCl, which means that a sodium atom (Na) has combined with a chlorine atom (Cl). In this combination the following has occurred: the sodium atom loses an electron to the chlorine atom. As a result of this the sodium atom becomes positively charged and the chlorine atom in gaining the electron becomes negatively charged; they are no longer atoms but are ions, or, symbolically:

$$NaCl \leftrightarrow Na^+ + Cl^-$$

which is represented pictorially as in Figure 5.1. Ions do not behave in the same way as atoms, for a sodium (Na) atom will react with water, whereas a Na^+ ion will not react with water in any way and therefore can exist in water.

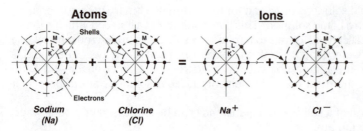

Figure 5.1: Ionic combination of sodium and chlorine. For clarity, only electrons are shown.

Electrolytic substances always dissociate; the extent to which a substance will dissociate depends to an extent on the substance itself, and partly on the dilution of the solution, which determines the ion concentration. The greater the dilution, the larger the percentage of molecules of electrolytes that will dissociate. Hence, with very dilute solutions, almost all the molecules dissociate, but the solution of electrolytes will still remain electrically neutral.

If we consider another example, barium chloride whose formula is $BaCl_2$, dissociates as follows:

$$BaCl_2 \leftrightarrow B^{++} + 2Cl^-$$

From this we can see that when barium chloride is dissolved in water, one barium ion and two chlorine ions are produced from one molecule of the salt; but now the barium ion will carry two positive charges, and this will have the effect of keeping the solution electrically neutral.

ELECTROLYSIS

Once again we shall take an example and see what happens. Let the sample be a container of hydrochloric acid. Symbolically, we have

$$HCl \leftrightarrow H^+ + Cl^-$$

If we now make a solution of hydrochloric acid in water, the result of this is also that the water is ionized to a very slight extent symbolically represented as

$$H_2O \leftrightarrow H^+ + OH^-$$

The OH^- is indicative of the hydroxyl ion, which, along with the hydrogen ions, will be in constant motion in the solution, but in no fixed direction, with the result that there is no overall net flow of electrical current.

If we now immerse two electrodes into the solution and connect them to the terminals of a battery as shown in Figure 5.2, bubbles of gas are seen to be given off from each electrode with rather more from the cathode, an effect of Faraday's *second law of electrolysis* that states, "The masses of various products liberated in electrolysis by the passage of the same quantity of electricity are in the exact ratio of their chemical equivalents." In the example, the gram equivalent of chlorine is 35.5 g and let this contain n atoms. Then 1.008 g of hydrogen must also contain n atoms. Now, since 96,500 coulombs are required to liberate 35.5 g of chlorine, the same amount of electricity will be required to liberate 1.008 g of hydrogen. If we collect

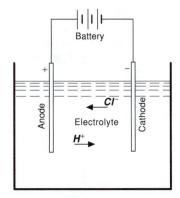

Figure 5.2: Electrolysis of hydrochloric acid. Arrows show direction of ion travel.

and test the gases thus produced, we shall find:

- The gas from the cathode to be hydrogen

- The gas from the anode to be chlorine

What really happens is this: The hydrogen ions from the acid and some from the water each gain an electron from the cathode and become neutral hydrogen atoms, two hydrogen atoms combining to form hydrogen gas (H_2). The chlorine ions and some hydroxyl ions from the water reach the anode, where the chlorine ions part with an electron to the anode more readily than the hydroxyl ions and are discharged, leaving behind chlorine atoms, which combine in pairs to form molecules of chlorine gas (Cl_2). Remember that ions of all metals and hydrogen are always positively charged and will always move to the cathode to be discharged there, whereas the ions of nonmetals are always discharged at the anode.

The process just described is the basis for electroplating, in which the electrolytes are solutions made up of salts of metals such as silver, copper, or nickel.

All acids dissociate when added to water, and in so doing produce hydrogen ions. The extent of this dissociation varies from acid to acid, and increases with the dilution, until almost all the acid is dissociated in very dilute solutions. There are strong and weak acids; the strong ones are those that produce a large number of hydrogen ions when they dissolve, while the weak ones do not produce many hydrogen ions when they dissolve. Typical strong acids are hydrochloric, sulfuric, and nitric acids; and typical weak acids are acetic and carbonic acids.

The dissociation constant k indicates the strength of a weak acid and is defined by

$$k = \frac{[A^-][H^+]}{[HA]}$$

where $[A^-]$ is the molecular concentration of the acidic ions, $[H^+]$ is the hydrogen ion concentration, and $[HA]$ the concentration of undissociated acid. The dissociation constant approximates a unique value that is roughly proportional to the square root of the dilution at a given temperature.

Alkalis, or *bases*, have a similar relationship when in solution, and there are strong and weak alkalis, too. Typical strong alkalis are the hydroxides of potassium and sodium and a typical weak one is ammonium hydroxide. In the case of weak alkalis, the ions are hydroxyl ions and have a dissociation constant also, with a similar relationship:

$$k = \frac{[B^+][OH^-]}{[BOH]}$$

where $[B^+]$ is the concentration of basic ions, $[OH^-]$ is the concentration of hydroxyl ions, and $[BOH]$ is the concentration of undissociated base.

Strong electrolytes do not have a dissociation constant.

THE pH SCALE

We have now seen that the hydrogen ion concentration is indicative of the strength of an acid or a base. This relationship, therefore, is the foundation of pH measurement—the p in the abbreviation standing for *power*, or, in mathematical terminology, the *exponent*.

A $0.1N$ solution of hydrochloric acid is almost completely ionized and contains 0.1 or 10^{-1} g H^+/liter, but a $0.1N$ solution of sodium hydroxide contains 0.000000000000086 g H^+/liter, or 8.6×10^{-14} g H^+/liter. Working with numbers of such a size constitutes an inconvenience and a good opportunity for errors. Sören Peter Sörensen, realizing the trouble dealing with these awkward numbers, devised a simpler workable set of figures that is called the pH scale, based on common logarithms, i.e., logarithms to the base 10. This scale is based on the following relationship:

$$pH = \frac{\log_{10} 1}{H^+ \text{ concentration}}$$

From this we find that for a $0.1N$ HCl solution

$$pH = \frac{\log_{10} 1}{10^{-1}}$$
$$= \log_{10} 10$$
$$= 1.0$$

and for a $0.1N$ NaOH solution

$$pH = \frac{\log_{10} 1}{8.6 \times 10^{-14}}$$
$$= \log_{10} 1.682 \times 10^{13}$$
$$= 13.07$$

This gives the following table (**Note:** Linear increases from 4 to 6, 8 to 10 and 13 pH are omitted):

0	1.0 g H^+/liter	Acidic
1	0.1 g H^+/liter	
2	0.01 g H^+/liter	
3	0.001 g H^+/liter	
7	0.0000001 g H^+/liter	Neutral
11	0.00000000001 g H^+/liter	
12	0.000000000001 g H^+/liter	
14	0.00000000000001 g H^+/liter	Alkaline

Intermediate values are expressed in decimals, and an increase of 0.1 pH represents a 20.57% decrease in H^+ concentration on the pH immediately preceding it.

MEASURING pH

THE STANDARD HYDROGEN REFERENCE ELECTRODE

The basis of measuring pH is the *hydrogen electrode*, which is used as the reference. A practical hydrogen electrode consists of a platinum wire covered with a finely powdered form of platinum, called *platinum black*. When hydrogen gas is bubbled over this electrode, it is absorbed into the surface, and the electrode then behaves as a hydrogen electrode. To see how it works, let us imagine a simple Daniell voltaic cell in which the hydrogen electrode is used as one of the two electrodes, the other being an electrode made from a material to give the appropriate electrical polarity compatible with the electrolyte. If everything around the cell is kept constant, then the potential difference between the electrodes will be a function of the concentration of the hydrogen ions in the solution that is in contact with the hydrogen electrode. The standard *hydrogen potential* is that obtained when an electrode is in contact with a solution of unit hydrogen ion concentration. HCl with an actual concentration of $1.228N$ has an effective activity of hydrogen ions that is unity. The hydrogen electrode is the standard against which all other electrodes are compared.

The electrical potential E in pH volts developed by a hydrogen electrode is related to the pH of the solution being measured, and is given by the following:

$$E = E_0 - 0.0001984T \qquad \text{pH volts}$$

where T is the temperature in K, and E_0 is the potential of the electrode in a solution with a pH value of 0 (zero) and is constant at this value.

THE pH ELECTRODE

There are three types of measuring electrodes used in determining the pH of a fluid: the quinhydrone, the antimony, and the glass electrode. Of the three, the first mentioned (quinhydrone) is not widely used, for it is suitable only for measurements below 8 pH. In the presence of pH values greater than 8, the quinhydrone undergoes partial dissociation and produces errors as large as 2 pH. Furthermore, it is found that quinhydrone contaminates the solution being measured.

The antimony electrode has a useful range of 2 to 12 pH, but also has a major disadvantage, in that the electrode cannot be used in solutions with ions of metals that are electropositive with respect to antimony. Under these circumstances the metal in the solution will be deposited onto the electrode itself and destroy the emf relationship, thus upsetting the measuring system. The main use of this electrode is in viscous fluids and heavy sludges, for antimony is a fairly strong material mechanically and in those applications where the measurement accuracy is not required to be better than ±1 pH. When used for applications where the fluid and measurement criteria conform to those described, the electrode must be mechanically cleaned to permit continuous operation.

The glass electrode is by far the most useful of the three and can be made to cover the whole of the pH range. The other great advantage is that it is unaffected by most chemicals, with the exception of hydrofluoric acid, which is well known to attack glass. The electrode consists of a membrane of a special low-electrical-resistance glass bonded to a tubular stem of relatively nonconducting glass. The tube is filled with a solution of constant pH. If the process liquid has a pH that is different from that in the electrode tube, hydrogen ions will diffuse through the glass membrane from the solution of higher hydrogen ion concentration to that of lower hydrogen ion concentration. This results in a loss of positive charge by the solution of greater concentration, making it more negative. The potential difference will continue to increase until the electrical forces opposing the transfer of ions exactly neutralize the influence of the concentration difference. The quantity of hydrogen ions transferred is very small indeed and can be neglected, leaving the original concentration substantially unaltered. The potential difference developed across the glass membrane is proportional to the difference in the pH values of the two solutions and is measured by putting electrodes into each solution.

Figure 5.3 shows a typical arrangement of a pH-measuring system. The pH of the fluid is determined by measuring the potential difference between the measuring and reference electrodes. To complete this electrical circuit, a second contact with the measured solution is required. This second point of contact is provided by another electrode called the reference electrode, which must fulfill the following requirements:

- The potential difference developed by the electrode and the buffer solution used must be constant.

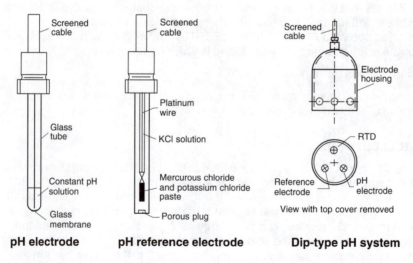

Figure 5.3: A pH-measuring system.

- It should be independent of temperature changes, or vary in a known manner.
- It should be independent of the pH value of the solution.
- It should remain stable over very long periods.

The best reference electrode is the hydrogen electrode, but this is impractical, for three main reasons. (1) It requires a continuous supply of pure hydrogen. (2) It is easily poisoned by substances in the process fluid. (3) Maintenance is a skilled and difficult operation. Hence, a metal/metal salt solution type is used, of which there are two:

1. The calomel (mercury/mercurous chloride) electrode.

2. The silver/silver chloride electrode that is used as the internal reference electrode in glass electrodes.

CONDUCTIVITY OF LIQUIDS

The next parameter we shall consider is *conductivity*, which is defined as the ability of a solution to conduct an electrical current. In terms of measurement units, conductance has a unit that is the reciprocal of resistance, i.e., 1/ohms; because of this the unit was previously called the *mho* (ohm spelled backwards). The modern name for the unit is the *siemens*. The conductivity of a material is defined as the resistance between opposite faces of a cube of the conductor of side 1.0 m length or any object whose length in m is numerically the same as its cross-sectional area in m^2. If κ (kappa) is the specific conductance and R is the specific resistance, then:

$$\kappa = \frac{1}{R}$$

A change in the concentration of a solution affects its specific conductivity in two ways. Dilution, i.e., adding more solvent, (1) will increase the degree of dissociation and, because there is more solvent, (2) will tend toward reducing the number of ions per cm^3. The respective effects are (1) by solvent addition to increase the value of κ; and (2) by increasing the solvent to reduce the value of κ. Because the second effect is stronger, the overall net result is a reduction in the value of κ.

MEASURING CONDUCTIVITY

To measure the conductivity of a solution it is vital that an alternating current supply be used to excite the electrodes in order to eliminate the polarizing effect of dc. To explain this polarization, we shall have to consider what happens when an electric current flows between two electrodes immersed in an electrolyte. The passage of current will cause electrons to be given up by the cathode and acquired by the anode and in the process liberate gas or deposit metal on the electrode surfaces. The effect of this is to produce a surface charge on the electrode, which depends on the metal deposited or the gas liberated. When the electrodes are in this state, they are considered to be polarized, and their surfaces, together with the electrolyte, form a voltaic cell that has an emf called the polarization emf.

The conductance of a liquid between the electrodes is given by:

$$K_{conductance} = \frac{\kappa A}{l} \qquad \text{or} \qquad \kappa = \frac{l}{A} K_{conductance}$$

where κ is the specific conductance, A is the effective cross-sectional area of the path between the electrodes, and l is the distance between the electrodes. The factor l/A is the *cell constant*.

To simplify the measurement of resistance, it is usual to maintain the resistance of conductivity cells between 10.0 and 100,000.0 (10.0×10^4) Ω. This results in a requirement to have cells with a range of cell constants from 0.10 to 100.0. The area of the electrode is limited by the permissible polarization error, which in turn depends on the frequency of the exciting ac current. Conductivity cells are superficially very much like pH electrodes in appearance. One of the important applications of conductivity measurement and control is in boiler feedwater in order to minimize corrosion and coating the tubes with "fur"—actually, lime scale. In practice, the conductivity cell must be fully immersed in the fluid being measured and at least 0.25 in away from any pipe or vessel wall. Furthermore, care must be taken to prevent air from being trapped within the cell. Since conductivity cells involve intimate contact with the process fluid, they will be subjected to the process pressure and temperature. There are limits to the pressure and temperature that the cells can be exposed to, and these are approximately 14 bars (200 lb/in^2) and 95°C. When a cell is fitted to a process line, there is sometimes a modified valve through which the cell is inserted and makes contact with the fluid. This arrangement allows the cell to be withdrawn for maintenance or replacement from the process while the system is under pressure. To remove the cell it is necessary to unscrew it and pull it back just past the shutoff valve arrangement, and then to close the valve to retain the process pressure prior to withdrawing the cell completely. A similar arrangement is also often used for other probes such as thermocouples and resistance bulbs.

CHROMATOGRAPHY

We shall now consider chromatography, but before proceeding further it will be useful to define what the science is all about. Chromatography is a technique used to separate out components in a mixture, using physical or physicochemical means. The separation is on the basis of the molecular distribution between two immiscible phases. One phase is stationary and in a finely divided state, which provides a very large surface area relative to volume. The other phase is mobile and carries the components over the stationary phase.

In gas chromatography, the mobile phase is a gas called the *carrier gas*, and the stationary phase is either a granular solid, or a granular solid coated with a nonvolatile liquid. The first case is referred to as *gas-solid chromatography*, and the second *gas-liquid chromatography*. Since the components of the mixture are carried in the gaseous phase, it is limited to the separation of mixtures of components of significant vapor pressures. The latest versions of these instruments can handle quite high temperatures. In gas-solid chromatography, separation is effected by the different adsorption characteristics of the components onto the solid phase. *Adsorption* is the process of adhesion of molecules of gas, liquid, or dissolved substance to a surface. In gas-liquid chromatography, the separation involves the distribution of the components, in the sample, between the gas and the liquid phases. Figure 5.4 is a block diagram of a complete system.

The instrument used to determine the components in a fluid is called a *chromatograph*; it consists of a tube called a *column* containing the stationary phase that is uniformly packed into it, such tubes being referred to as *packed columns*. The whole assembly is held in an environment capable of being controlled and thus permitting

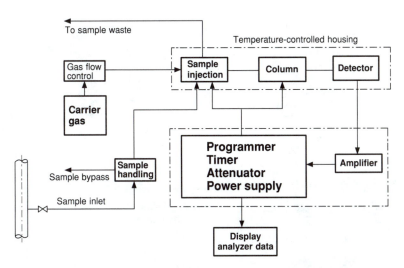

Figure 5.4: Simplified block diagram of chromatograph system.

the environmental conditions to be altered—i.e., heated or cooled. Experience has proved that a capillary tube uniformly coated with the solid or liquid phase gives the best results; hence, such techniques are called *capillary chromatography*. The carrier gas is passed through the system at a controlled rate, and facilities for injecting the gas mixture (at controlled rates) into this carrier gas stream are provided. Suitable sensors that respond to the components sought are provided downstream of the column.

To analyze a sample, an aliquot (an aliquot is defined as a part that is contained an exact number of times in the whole—division without remainder—e.g., 1/10, 1/50) of a known volume is introduced into the carrier stream. The components of the sample interact characteristically with the stationary phase, causing them to pass through the column at different rates. The component that is least adsorbed comes out of the column first; this process is called *elution*. The actual process of component separation is very complex. If a pen recorder is assigned to the sensor output, then the trace obtained will consist of a series of peaks that define the components and their value (% concentration). It is important to understand that a chromatograph gives a series of spot analyses of each of several components in a single aliquot virtually simultaneously, based on regular discrete, rather than continuous, sampling of the process stream. All components of the sample must be eluted or backflushed before a new sample can be introduced for analysis.

For the purpose of this section we shall only name several types of detectors (sensors) available and discuss only some of the simplest ones in detail. The following detectors are used: katharometers, flame ionization, photo ionization, helium ionization, flame photometric, electron capture, ultrasonic, catalytic (pellistor), and semiconductor. In all chromatographic systems the design and construction of the sampling system play a most vital part and is usually a specialist's job.

THE KATHAROMETER

Of the detectors just named, the katharometer is perhaps the most simple to understand. The katharometer is based on Maxwell's kinetic theory of gases and

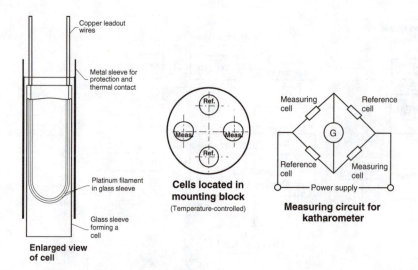

Figure 5.5: The Shakespear katharometer.

works on the basis of the heating effect of an electrical current in the presence of a known gas, the amount of current drawn being dependent on the amount of the known gas present, and hence on the thermal conductivity of the measured gas. The instrument consists of a set of four cells (see Figure 5.5), each containing a platinum wire-measuring element, shaped as a hairpin for strength. In one type the platinum wire is enclosed and sealed within an evacuated glass capillary to ensure that the gas does not contaminate the platinum element. The element is sealed into an open-ended cylindrical glass sheath with an overall metal tube that is made gas-tight with a low-melting-point metal filler. The construction of each element is identical. The element assemblies are fitted into a cell mounting block designed to accept four cells and made of metal, compatible with the sample, to minimize the thermal effects. Two of the cells are called *reference cells* and the other two *measuring cells*. The open end of each measuring cell is exposed to the gas being measured, while the open end of each reference cell is exposed to the purest form of the gas to be detected. Both the measured and reference gases are made to enter the cells only by molecular diffusion, so that the instrument is independent of the rate of flow of the gas. The four cells are arranged to have one measuring and one reference cell in each side of a Wheatstone bridge as shown in Figure 5.5. With the process fluid applied to the cells and the bridge powered up from a source of constant current, the following can occur: if the column effluent does not have the same purity of gas as that in the reference cell, then the resistances in each arm of the bridge will be different and cause the galvanometer to deflect, the deflection being related to the amount of the measured component.

THE ULTRASONIC DETECTOR

The ultrasonic detector works on the principle that the velocity of sound in a gas is inversely proportional to the square root of the molecular weight of the gas. In the case of binary gases, measuring the speed of sound will enable the composition to be deduced. When this type of detector is used in chromatography, it provides good sensitivity and a wide linear dynamic range. However, it is temperature-sensitive, and therefore precise temperature control of the detector is mandatory. In order

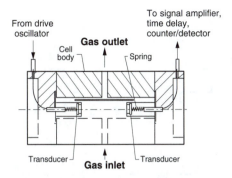

Figure 5.6: Schematic diagram of ultra-sonic detector.

to obtain the efficient transfer of sound between the gas and the transducers, the detector must operate above atmospheric pressure, the sample gas being regulated between 1 and 7 bars g (typically), with the actual value of pressure depending on the gas being measured. Figure 5.6 is a schematic diagram of the detector.

THE PELLISTOR DETECTOR

The pellistor detector, a detector for combustibles, is a catalytic type of device that measures the heat output from the catalytic oxidation of a flammable gas to produce CO_2 and water at a solid surface. The two most suitable materials to promote oxidation of gases containing $C-H$ bonds are platinum and palladium. The sensor, or *pellistor* is a platinum coil, mounted on leadout wire supports. The coil is embedded in porous ceramic, usually alumina, and formed into a 1-mm-long bead. The outer surface of the bead is impregnated with a catalyst of either palladium or palladium oxide. The pellistor is fitted to a carrier base and surrounded with a protective sleeve. A heater is included in the sensor, and when the flammable gas passes over the heater, it is oxidized with a release of heat. The temperature of the sensor is monitored. The detector is arranged to be included in one arm of a Wheatstone bridge, where any temperature variation is indicated, the amount of change being proportional to the amount of flammable gas present. Figure 5.7 is a schematic of the device.

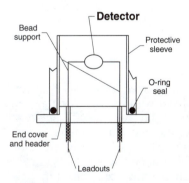

Figure 5.7: Schematic diagram of a pellistor.

NOTE: For clarity the light source has been omitted

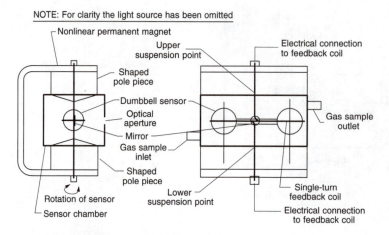

Figure 5.8: Simplified arrangement of a paramagnetic oxygen analyzer.

THE PARAMAGNETIC OXYGEN ANALYZER

One of the more common gases to be measured in the process industries is oxygen, and we shall discuss one of the instruments and techniques used, the paramagnetic oxygen analyzer. Faraday showed that many substances are influenced by a magnetic field. In his experiments he suspended solid specimens by fine wires between the poles of an electromagnet and found that some arranged themselves parallel to the field while others positioned themselves at right angles to the field. Those that set themselves parallel to the magnetic field he called *paramagnetic*, and the ones at right angles *diamagnetic*. Other scientists later found that oxygen behaves just like a piece of steel in a magnetic field, and it is this behavior of oxygen in the presence of an intense magnetic field that is exploited as a means of determining the amount present.

A simplified mechanical arrangement of the analyzer is shown in Figure 5.8.

Oxygen has a very high magnetic susceptibility and exhibits paramagnetic behavior in the presence of a magnetic field. It is this property that is exploited in the paramagnetic oxygen analyzer. The instruments made by Servomex and Maihak are mechanically very alike and operate in exactly the same manner. Instruments of this type can have measuring spans of 1.0 to 100% oxygen, with a resolution of 0.01%. The zero of the instrument can be suppressed in 0.01% steps to a maximum of 99.99%. In measuring range terms, this makes for a very flexible instrument. Figure 5.9 shows an electrical schematic of how the measurement is made.

Instrument Construction

The instrument diagrammed in Figure 5.8 is constructed as follows. The very-small-volume dumbbell-shaped glass container is filled with nitrogen (N_2) gas and permanently sealed. A fine strong platinum or platinum alloy wire of sufficient length is used to provide the sensor and the suspension for the N_2-filled container. Starting with the middle of the length of wire and the center of gravity of the glass container, and moving in opposite directions, the wire is wound around the lengthwise surface of the dumbbell to form the sensor as a single-turn electrical coil, called the *feedback coil*. The excess of wire from both ends is then used to form the suspension

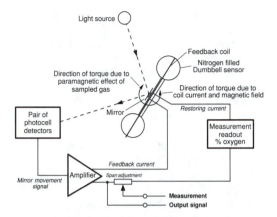

Figure 5.9: Simplified schematic circuit of paramagnetic oxygen analyzer.

for the dumbbell-coil assembly, as well as the electrical connections to the coil. A very light and small flat mirror is permanently attached to the center of the dumbbell coplanar with the suspension. This arrangement allows the dumbbell to freely rotate virtually friction-free about the points of suspension. A gas-tight measuring chamber completely encloses the dumbbell unit. The chamber is designed to permit the suspension/connection wires to pass through without gas leakage. The chamber has inlet and outlet connections to allow the gas to be measured to flow through it. The measuring chamber fits between the pole pieces of the permanent magnet. An optical window is provided such that the mirror attached to the dumbbell is visible. A light source and optics outside the measuring chamber provide a well-defined beam of light to be shone through the window onto the mirror to permit detecting any dumbbell movement. The reflected light is detected by a pair of photocell detectors connected to an autobalancing amplifier, which drives the feedback coil in a direction to counteract the detected movement. The feedback/restoring current provides a measure of the oxygen concentration. A very strong permanent magnet fitted with shaped pole pieces causes a nonuniform magnetic field to lie perpendicular to the sensor. The dumbbell suspension wires pass through and are secured outside the magnet so that the electrical connections can be made to it. The whole sensing and measuring system is temperature-controlled for good accuracy.

Working Principle of the Paramagnetic Oxygen Analyzer

Since nitrogen is dimagnetic, the dumbbell will rotate out of the magnetic field to take up a position that is related to the susceptibility of oxygen in the measuring chamber. As the dumbbell rotates the mirror moves with it, thereby changing the angle of reflection of the beam of light. This change is detected and used to alter the current passing through the coil wound around the dumbbell. However, initially the instrument zero is calibrated with nitrogen in the measuring chamber. This causes the dumbbell to rotate, because of the dimagnetic effect of the nitrogen, and, take up a position due to it; the zero is adjusted until no current flows in the feedback coil, which also establishes the zero for the measurement.

When a calibration gas containing a known amount of oxygen is made to flow through the measuring chamber, the dumbbell will try to rotate by an amount brought about by the paramagnetic constituents moving to congregate at the strongest part of the magnetic field, thus upsetting the balance. As soon as the dumbbell moves, the optical detecting system "sees" the movement and drives the amplifier to change the current flow to the feedback coil to maintain the dumbbell at its zero position. This current, proportional to the volume magnetic susceptibility of the gas mixture in the measuring chamber, is displayed on an indicator having a scale calibrated in terms of percent oxygen. The measured gas causes the same behavior of the sensor. Because the measurement produced is proportional to the partial pressure of the oxygen present in the gas sample, it is directly proportional to the absolute value at atmospheric pressure. Provided that they are not excessive, the instrument is fairly insensitive to other gases in the sample, to sample flow, and to tilt. These analyzers have applications in inerting, fermentation, blast furnaces, chemical production, biotechnology and physiology. (*Inerting* is a procedure used to describe the replacement of air in a process vessel by an inert gas; the replacement gas is usually nitrogen. The "atmosphere" within the vessel is continuously monitored for the presence of oxygen and more nitrogen added if required to eliminate the oxygen.)

THE NONDISPERSIVE INFRARED (NDIR) GAS ANALYZER

Another type of gas analyzer is the nondispersive infrared gas analyzer usually abbreviated NDIR. This instrument is most often used to determine the amount of a single gas component in a process stream. Recent development of this type of analyzer has enabled instruments to be used to measure more than one component of a process gas. Some instruments incorporate an oxygen measurement. When used for multiple components, there are applicable limitations that are associated with the spans of the various components. The Maihak multicomponent NDIR analyzer permits a choice of any three from the following as a standard:

Carbon dioxide	CO_2
Carbon monoxide	CO
Sulfur dioxide	SO_2
Methane	CH_4
Nitric oxide	NO

Other components can also be measured.

NDIR analyzers utilize a Luft-type detector, which is a single-beam infrared instrument. Figure 5.10 is diagrammatic and represents the Maihak single-beam instrument. The description is for a single-component gas and the operation is as follows: The instrument comprises a container section called a *cuvette* that is divided into two cells, one of which, called the *measuring cell*, carries the gas being measured and the other of which, called the *reference cell*, carries the reference gas, which is pure nitrogen. The cuvette has windows of calcium or barium fluoride at each end, and is of a length that is dependent on the measuring range. Immediately following the measuring cuvette is the detector. Infrared radiation from the single source is split by a chopper rotating at a constant speed and interposed between the infrared source and the measuring cuvette. This arrangement causes the beam to be "chopped" into discrete pulses of energy. A splitter causes the pulse of energy to pass, in phase, through the reference and measuring cells of the cuvette and on to the detector.

The Luft detector also consists of two parts separated by windows of calcium or barium fluoride filled with a sample of the gas that is to be analyzed. The front part is called the *front absorption volume* and the other, the *rear absorption volume*. The two parts of the detector are each connected to separate pressure chambers

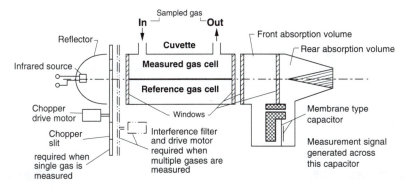

Figure 5.10: Simplified arrangement of an NDIR gas analyzer. (This design is based on the Maihak instrument.)

formed by dividing a single chamber with a flexible membrane separator that is, in fact, a variable capacitor. The front volume of the detector is in communication with one side of the pressure chamber, while the rear volume is in communication with the other side of the pressure chamber. This detector is a very sensitive device, but is susceptible to ambient temperature and external vibration effects. The sensor is temperature-controlled to minimize the effects of temperature variations.

The radiation energy passing through the cuvette to the detector is absorbed by both the front and rear absorbers, giving rise to a heating effect and thereby causing the pressures to increase. If the gas in the measuring side were the same as that in the reference side, then the energy absorption would be the same for both sides and the pressures would be the same. However, if the gas in the measuring side contains some of the components being measured, there will be an imbalance in energy and a corresponding imbalance in pressure. The pressure changes on opposite sides of the membrane capacitor produce a varying capacitance dependent on the magnitude of the pressures involved. The change in capacitance is sensed, amplified, and used as an indicator of the amount of the component being measured.

When multiple components in the gas samples are involved, then a second intermittently rotating disk driven by a stepping motor is included in the design. This disk has interference filters, which permit several components to be measured in turn. This is necessary to enable each of the components to be analyzed. The component is measured during the period a filter is stationary in the optical path to the detector and before the stepping motor switches to the next filter.

The subjects treated in this chapter represent about the most common analytical measurements that one will encounter, but there are several other analytical instruments in use today. Many of the instruments and techniques are highly specialized and these can be studied when the occasion arises.

DATA REQUIRED TO PROVIDE AN ANALYTICAL MEASUREMENT IN A PROCESS

The list of data shown under this heading is typical of the information that would appear on a checklist for selecting a suitable analytical measuring instrument. The requirements of different types of analytical measurement are quite varied, so that the catagories of data considered will be unique to the application.

PROCESS FLUID

Name	
Characteristic	Gaseous or liquid
Chemical action	Aggressive or nonaggressive (**Note:** Aggressiveness is defined as the ability of the fluid to attack the containing vessel chemically)
Number of components to be analyzed	

PROCESS PARAMETERS

For the following, appropriate engineering units are required. Items marked with an asterisk are necessary when chromatographs are required.

Pressure

Maximum operating

Minimum operating

Normal operating

Phase

Liquid*

Vapor*

Bubble point

(At operating pressure)*

Dew Point

(At operating pressure)*

Temperature

Maximum operating

Minimum operating

Normal operating

Water

Minimum*

Normal*

Maximum*

Particulates

Minimum*

Normal*

Maximum*

PHYSICAL LOCATION

In the following, give the standards with which the material must comply—e.g., the BS, ANSI, etc., pressure rating of process connections must be given.

Process line

Nominal size

Pipe schedule

Material of construction

Vessel process connections

Screwed

Flanged

Welded

Hygienic (exclusively for food and drugs)

APPLICATION

Indication	Local or remote
Control-local	On/Off or PID
Remote	On/Off or PID (**Note:** Local is defined as having the measurement or control function at or near the point of penetration of the sensor or transducer into the process. Remote is defined as having the measurement or control function at some distance from the point of penetration of the sensor or transducer into the process.)

CRITERIA FOR SELECTING A SUITABLE ANALYTICAL MEASURING INSTRUMENT

TYPE OF MEASUREMENT REQUIRED

State the measurement parameter required—e.g., pH, conductivity, gas analysis. If the process is to be sampled, then all information regarding the materials to be used in the sampling system should be given, if possible. Any precautions to be taken should also be stated.

It is quite normal for process analyzers to be located in cabins also called analyzer houses, so as to provide the maintenance personnel facilities to perform cleaning and calibration of the equipment. The cabins also allow gas cylinders for the chromatographs to be stored, and details of the location and area classification must be made known. When analyzers are located in hazardous areas of a plant, it could be necessary for the house to be continuously purged with air or nitrogen to minimize any explosion risk. If nitrogen is used for the purge, then every precaution must be taken to ensure that personnel are protected, for nitrogen is a lethal gas. All purged analyzer houses must have interlocked alarms to ensure continuous safety.

OPERATING RANGE

The operating range is the most important criterion when selecting a measuring instrument. The required range must be given together with the high and low limits that may be possible.

AGGRESSIVE FLUIDS

Materials of Construction

Where material has to be compatible with the process fluid, then it is important that this is stated to avoid the possibility of injury to personnel or plant failure.

SUMMARY

1. For specific materials, when we are considering the material at the atomic level, the amount of the positive charge exactly equals the amount of the negative charge. This results in an electrically neutral atom. It is possible for atoms to gain electrons from or lose electrons to neighboring atoms.

2. The electrons that are involved in exchanges between neighboring atoms are called free or valence electrons.

3. In an atom of a specific material, the nucleus is composed of a collection of positively charged protons and chargeless neutrons; the effect is that the nucleus is positively charged. The number of protons in the atom gives the element its atomic number. Electrons orbit the nucleus and always carry a negative charge and, the numbers exactly balances the number of protons making the atom electrically neutral.

4. When an atom gains an electron it becomes more negatively charged and is then called an anion. When the atom loses an electron the atom will become more positively charged and is then called a cation.

5. The charge carried by an ion is equal to the number of electrons gained or lost by the original atom. The process of changing electrically neutral atoms into ions is called ionization.

6. Materials that can be split into ions are called electrolytes. The process of splitting is called electrolysis. Energy is always required to carry out electrolysis.

7. There are, in the main, two types of solutions: (1) solid solutions (e.g., all metallic alloys and steels) and (2) liquid solutions, such as sugar in water or brine. For solutions to occur, we can say that one substance must be soluble in the other.

8. In a liquid solution, the liquid is called the solvent and the added substance is called the solute. Water-based solutions are called aqueous solutions and are of particular importance.

9. A solution's concentration is given in terms of its normality. A normal solution is written as N, twice normal solution is written as $2N$, and so forth.

10. The extent to which a substance will dissociate depends on the substance itself, and the dilution of the solution. The greater the dilution, the greater the dissociation. With very dilute solutions, almost all the molecules dissociate. The solution of electrolytes will remain electrically neutral.

11. All acids dissociate when added to water and in so doing produce hydrogen ions. The extent of dissociation varies from acid to acid and increases with the dilution. In very dilute solutions almost all the acid is dissociated. The dissociation constant k indicates the strength of a weak acid and is defined by

$$k = \frac{[A^-][H^+]}{[HA]}$$

where $[A^-]$ is the molecular concentration of the acidic ions and $[H^+]$ is the hydrogen ion concentration. The dissociation constant approximates a unique value that is roughly proportional to the square root of the dilution at a given temperature. Alkalis, or bases, have a similar relationship when in solution. The ions are hydroxyl ions and have a dissociation constant also, the relationship for a weak alkali (base), being

$$k = \frac{[B^+][OH^-]}{[BOH]}$$

Strong electrolytes do not have a dissociation constant.

12. The hydrogen ion concentration can be indicative of the strength of an acid or a base. This relationship is the foundation of pH measurement, the p standing for power or exponent.

13. The pH scale is based on logarithms to base 10, and on the following relationship:

$$pH = \frac{\log_{10} 1}{H^+ \text{ concentration}}$$

14. The pH scale is given in the following table (**Note:** Linear increases from 4 to 6, 8 to 10 and 13 pH are omitted):

0	1.0 g H^+/liter	Acidic
1	0.1 g H^+/liter	
2	0.01 g H^+/liter	
3	0.001 g H^+/liter	
7	0.0000001 g H^+/liter	Neutral
11	0.00000000001 g H^+/liter	
12	0.000000000001 g H^+/liter	
14	0.00000000000001 g H^+/liter	Alkaline

Intermediate values are expressed in decimals and an increase of 0.1 pH represents a 20.56% decrease in H^+ concentration from the pH immediately preceding it.

15. The hydrogen electrode is the standard against which all other electrodes are compared. The electrical potential in pH volts developed by a hydrogen electrode is related to the pH of the solution being measured, and is given by the following:

$$E = E_0 - 0.0001984T \text{ pH volts}$$

where T is the temperature in K, and E_0 is the potential of the electrode in a solution with a pH value of 0 (zero) and is constant at this value.

16. The three types of electrodes used in pH measurement are; the quinhydrone, the antimony, and the glass electrodes. The glass electrode is by far the most useful of the three and can be made to cover the whole of the pH range.

17. The glass electrode consists of a membrane of a special low-resistance glass bonded to a tubular stem of relatively nonconducting glass. The tube is filled with a solution of constant pH. The constant-pH solution is therefore separated from the process liquid itself. If the process liquid has a pH that is different from that in the electrode tube, hydrogen ions will diffuse through the glass membrane from the solution of higher hydrogen ion concentration to that of lower hydrogen concentration. The quantity of hydrogen ions transferred is very small indeed and can be neglected; this leaves the original concentration unaltered.

18. To complete the electrical circuit for pH measurement, another contact with the measured solution is required; a reference electrode, which must fulfil the following requirements, provides it:

    • The potential difference developed by the electrode and the buffer solution must be constant.

    • It should be independent of temperature changes or vary in a known manner.

    • It should be independent of the pH value of the solution.

    • It should remain stable over very long periods.

    Common reference electrodes are of a metal/metal salt solution type, either the calomel (mercury/mercurous chloride) electrode, or the silver/silver chloride electrode that is used as the internal reference electrode in glass electrodes.

19. Conductivity is defined as the ability of a solution to conduct an electrical current. The measurement unit of conductance is a unit that is the reciprocal of the unit of resistance. The modern name for the unit is the siemens.

20. The siemens is obtained from a measure of the resistance between opposite faces of a cube of the conductor of side 1.0 m length or any object whose length in meters is numerically the same as its cross-sectional area in m^2. If κ (kappa) is the specific conductance, and R is the resistance in ohms, then:

$$\kappa = \frac{1}{R}$$

21. Conductivity cells have to be in intimate contact with the process fluid and will be subjected to the process pressure and temperature. The limits to the pressure and temperature to which they can be exposed are approximately 14 bars (200 lb/in^2) and 95°C.

22. Chromatography is a technique used to separate out components in a mixture using physical or physicochemical means. The separation is on the basis of the molecular distribution between two immiscible phases. One phase is normally stationary and in a finely divided state; the other phase is mobile and carries the components over the stationary phase.

23. In gas chromatography the mobile phase is a gas called the carrier gas, and the stationary phase is either a granular solid or a granular solid coated with a non-volatile liquid. The first case is referred to as gas-solid chromatography and the second gas-liquid chromatography.

24. The instrument used to determine the components in a fluid is called a chromatograph, and consists of a tube containing the uniformly packed stationary phase and called a packed column. This column is held in an environment capable of being controlled (heated or cooled). The carrier gas is passed through the column

at a controlled rate, and facilities for injecting the gas mixture at controlled rates into this carrier gas stream are also provided. Suitable sensors that respond to the components sought are provided downstream of the column.

25. To analyze a sample by chromatography, an aliquot of a known volume is introduced into the carrier stream. The components of the sample interact with the stationary phase, causing them to pass through the column at different rates this process is called elution.

26. The following detectors are used in chromatography: katharometers, flame ionisation, photoionization, helium ionization, flame photometric, electron capture, ultrasonic, catalytic (pellistor), and semiconductor.

27. The paramagnetic oxygen analyzer is an instrument based on the effects observed by Faraday when he found that some solid materials when placed in a strong magnetic field arrange themselves parallel to the field while others position themselves at right angles to it. The ones that set themselves parallel to the magnetic field he called paramagnetic, and the ones at right angles diamagnetic. It was later found that oxygen behaves just like a piece of steel in a magnetic field.

28. Paramagnetic oxygen instruments can have measuring spans of 1.0 to 100% oxygen, with a resolution of 0.01%.

29. The nondispersive infrared (NDIR) analyzer is most often used to determine the amount of a single gas component in a process stream. Recent development of this type of analyzer has enabled instruments to be used to measure more than one component of a process gas.

30. The NDIR instrument comprises a container section, also called a cuvette, divided into two sections: the measuring cell, carrying the measured gas and the reference cell, containing a reference gas of pure nitrogen. The cuvette has a length dependent on the measuring range and has windows of calcium or barium fluoride at each end. The infrared radiation from the single source is split into discrete pulses of energy by a chopper rotating at constant speed, interposed between the infrared source and the measuring cuvette. A splitter causes the pulse of energy to pass, in phase, through the reference and the measuring cells of the cuvette and on to the detector.

31. The Luft detector used in the NDIR analyzer is a single-beam infrared instrument. It consists of two parts separated by windows of calcium or barium fluoride filled with a sample of the gas that is to be analyzed. The front part is called the front absorption volume and the other, the rear absorption volume. The detector is a very sensitive device susceptible to ambient temperature and external vibration effects. The sensor is temperature-controlled to minimize the effects of temperature variations. The two parts of the detector are each connected to two pressure chambers formed by dividing a single chamber with a flexible membrane separator that is in fact a variable capacitor.

32. When multiple components in the gas samples are involved, the NDIR analyzer has a second intermittently rotating disk driven by a stepping motor included in the design. This disk has interference filters, which permit several selected components to be measured and is necessary to enable each of the components to be analyzed. The component is measured during the period a filter is stationary in the optical path to the detector and before the stepping motor switches to the next filter.

PART I APPENDIX

As stated in the preface to this book, practicing instrumentation and control engineers are quite often, in the course of their working day, presented with problems that require solving. However, as stated before, the available data is at most times not clear cut, but contained in a mass of other information. The engineer's first task is to extract the "real" information and then set about providing a workable solution. In order to give the reader a feel for the challenges and hence the excitement involved, the following series of questions have been deliberately constructed in terms that one would not unreasonably expect to encounter in real life. When the engineer is employed by an instrumentation and system vendor, the complexity in determining a solution increases; because of the commercial sensitivity of the manufacturing process, in some instances the process data is made available in as general terms as possible. Under these circumstances, it is up to the engineer to size up the situation and ask suitable questions of the data provider in order to make appropriate recommendations.

Having read in the foregoing chapters details of some of the instrumentation at the engineer's disposal for use in a process plant, it would be advantageous and instructive to determine for one's own benefit the usefulness of the time and effort spent in doing so. Remember, there is no single universal solution; each solution will have its own suitability, and the one that offers the most advantages both technically and financially will be the most successful. Be aware also that "least cost" recommendations are not always the best solution, for advanced technology always comes at a price.

To give an idea of what could be involved in this process, let us consider the following example and go through the decision making step by step.

> **Example:** The level of a corrosive liquid has to be determined. The containing vessel is lined and located at an elevation of approximately 3 m above grade (measured from the underside of the vessel). The overall dimensions of the vessel are 5 m in height and 2 m in diameter. Flanged nozzles are located in the vertical section of the vessel sufficiently removed from the "tan lines" (the joints where the dished ends and the cylinder of vessel meet) to make a good installation. The level of the liquid must be maintained at a height of 0.5 m above the lower nozzle, and the measuring instruments have to be located in the vicinity. Specify the measuring instruments, installation, and any precautions that you would take to ensure a representative measurement and personnel safety.

A good starting point is to produce a sketch of the requirements and systematically proceed as in Figure A.1. We take it for granted that the lining used in the vessel itself is suitable, and the designers have paid due attention to it; we make this assumption clear to the data provider in order to absolve us from any responsibility in this regard. There are however, other points that require our attention. For example, what is the process fluid? Is it an acid or an alkali? Is it otherwise hazardous? Under what conditions is it delivered or to be stored? If, say, it is sulfuric acid of 90% concentration, delivered at a temperature between 15° and 20°C (59° and 68°F) and a pressure of 2.812 kg/cm^2 (40 lb/in^2), we might look in *Perry's Chemical Engineers Handbook* under material compatible with sulfuric acid for concentrations between 40% and 100% and temperatures up to 100°F. We would find that polyethylene exhibits complete resistance to attack. We can now approach instrument and accessory vendors for their views and recommendations.

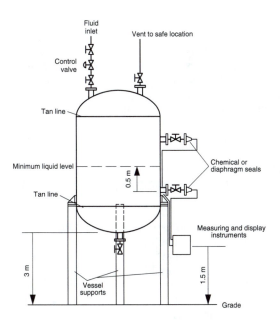

Figure A.1: Sketch of example requirements (not to scale).

In determining the measurement procedure, we first have to acknowledge that the fluid is corrosive and will be fuming because of its concentration and therefore must be kept away from all personnel, with special care for those involved with its maintenance. This means providing a physical separation between the fluid and personnel. The choice is therefore between:

- D/P cells with an extended diaphragm
- D/P cells with diaphragm and capillary assemblies
- D/P cells with chemical seals
- Radiometric level detectors
- Ultrasonic level detectors

From this availability of instruments for the application, serious consideration will have to be given to the financial commitment involved. Of the devices named the radiometric and ultrasonic instruments are the more expensive. Additionally, the radiometric option also involves high installation cost, maintenance, and spares for the radioactive components. The spares will certainly not be commonly used elsewhere in the plant and would entail a traceable, securely protected storage location separated from all other spares holding. The need for health and safety inspections by both onsite and government personnel, along with the accompanying paperwork restricts the radiometric option.

With the D/P cell options, one has to consider the actual measurement range versus capability of the cell diaphragm, especially if the corrosion resistance is obtained with a plastic coating on a metallic base. Discussions with suitable vendors must be entered into to determine the viability of the instrument. The extended-diaphragm

D/P cell will usually be fitted only to the vessel connection that provides the measurement, in this case, the lower nozzle. Since this specific vessel is closed, another instrument with similar corrosion resistance must provide a measurement of the vapor pressure or the system will not give correct results. Protect the capillaries from wide temperature fluctuations, as they will act as filled-system thermometers.

A point worth remembering is that the chemical seal manufacturers usually only provide the seals; the associated instrumentation is manufactured and supplied by others. Quite often an instrument manufacturer has agreements with a number of chemical seal makers to provide the instrument user with suitable equipment for such applications.

If the measurement is a vital one—one on which the manufacturing process is dependent—then serious consideration will have to be given to providing a backup in the event of a failure to avoid process interruption. This will involve finance for both plant and measurement instrumentation.

As far as the installation is concerned, it is important that the instruments be located in a position that is accessible and reasonably clear of obstruction, so that personnel are able to reach the process connections and perform calibration and maintenance checks on the instruments. The equipment should preferably be mounted at eye level with easy access to all connections. Standard pipe mounts or vessel support members could be used to accomplish this. In the event of exposed locations, consideration should be given to suitable instrument housings; this is important in areas affected by ice, snow, and heavy rain, and also in desert locations where sand and high temperatures cause considerable difficulties.

Process connections should always be capable of being isolated with valves that are able to withstand the process fluid, cause minimum losses to the measurement, and facilitate the removal of the equipment, should this be required, without endangering personnel. To ensure this, flushing connections for a solvent (in this particular case, water) on the primary connections and control valve should be included. Always ensure the waste is disposed of safely. The material of construction for these valves should follow the process line specification.

Since the requirement is to maintain the process fluid level within the vessel, a controller and a control valve are required. The engineer will have through discussion determined the type of controller, whether electronic or pneumatic; the type of control action needed; and whether this instrument has to be located in the field, i.e., near the process vessel itself, or in the control room. If it is to be field-mounted, then the mounting requirement already discussed will also apply. At this point it is worth considering the provision of audible alarms on the measurement to give the field plant operators suitable warning of impending problems, with these alarms being repeated in the control room. Chapters 19 and 23 discuss control and alarm requirements applicable to modern distributed control systems.

The same amount of attention will be required in selecting the control valve, with particular care being taken in its sizing and characteristic—points covered in Chapter 16. It is also important to consider where the control valve is to be mounted, for it will have to be serviced, and adequate space must be provided to permit routine and breakdown maintenance in safety. Valves are usually heavy; therefore, adequate access facilities for lifting gear (if required) must be provided. If the process is continuous, then a bypass line with a second control valve should be considered. The bypass line will have to be provided with isolating valves to ensure that only one control valve operates at any time, and the control signal will have to be switched accordingly.

Now you try your hand at the following:

PROBLEMS

1. It is desired to measure the flow of air to a combustion process. The ambient air being sucked into the furnace is heavily dust-laden. Part of the conditioning demands that the air be subject to filtration to exclude air-borne solids. It is important that the air supply be maintained; hence, a blocked filter condition has to be advised for corrective action to be taken.

   (a) If a pressure drop of 50 mm (about 2 in) water gauge across the filter is indicative of an almost totally clogged element, what measuring device would be suitable for this application?

   (b) If the pressure drop across the flow-measuring device has to be a minimum, what measuring instrument or combination of instruments would you recommend?

   (c) Where would the proposed device(s) have to be installed?

   (d) Give reasons for your choice of instruments and location of the measuring devices.

2. If the ambient air of question 1 is also subject to temperature variations that are not negligible, what would you recommend as a temperature-sensing element, and what would you specify as an appropriate mechanical design of the sensor for the application that would give the necessary measurement? Give reasons for your choice, stating clearly the advantages of the proposal.

3. A volatile and taxable liquid fluid of specific gravity 0.82 has to be metered to a road tanker. Under customs and tax requirements, it is necessary to maintain accurate records for each and every delivery made. The discharge pipe is partly rigid and partly flexible and has an internal diameter of 40 mm (about 1.5 in).

   (a) How would you propose to meet the requirements for accurate measurement of the delivered quantities?

   (b) Give a diagram of the system you propose, showing clearly any specific requirements that have to be met as far as the measuring system is concerned.

4. You need to measure the flow of an oil product. In this instance the fluid conductivity is not very high. However, the quantities involved are moderately large and delivered through a metallic pipe of approximately 100 mm (4 in) internal diameter. The fluid has a viscosity approaching that of water.

   (a) What device would you use to determine the fluid flow?

   (b) Advise the user of the advantages, or limitations, if any, of the techniques you recommend.

5. A slurry formed by the mixture of a milled ore, hydrochloric acid, starch, and water is to be pumped to a processing unit called a digester. In view of the process fluid involved, it can be taken for granted that corrosion and erosion are factors that must be given due consideration. It is also appropriate to assume that the pumping pressures involved are very high. The requirement is to determine the fluid flow and pressure to the digester. Provide suggestions of alternative measuring devices or systems that will allow this to be done, and the advantages of each of the proposed devices and/or systems. State any precautions that have to be taken by the user to preserve the safety of equipment and personnel involved.

6. Chemical seals are very useful in the segregation of aggressive or toxic process fluids from the measuring sensor. Discuss this statement, making references to the criteria that must be taken into consideration when such accessories are introduced into a measuring system.

7. A cylindrical process vessel of 3 m diameter and 3.25 m between the vessel tan lines contains a hot (135°C) and aggressive fluid whose level has to be determined. To keep the fluid at conditions that are conducive to further processing, it is necessary to keep the contents agitated and maintain the temperature. The maximum level of the fluid is 2.5 m and the minimum 0.25 m, and the pressure is 1.25 bars. If the materials with which the process fluid is compatible are nickel chrome or 316 stainless steel, make recommendations for suitable measuring devices and give reasons for your selection. If there is any data you think is necessary to obtain before the instrumentation can be provided, state it clearly. Control of the liquid level and temperature should not be considered, since this will be provided by others.

8. The level of a water reservoir has to be determined. The reservoir was not built for the purpose, and it has been found that beyond a distance of 15 m from the most convenient bank, the bed slopes away very rapidly. Soundings taken previously indicate that the bed at the distance referred to is on average representative of the reservoir depth.

   If the annual variation of the water level is 1.5 m, specify the method you would employ to make the measurement. In severe drought conditions the water can fall approximately 5.25 m from a normal average level of 15 m, and the measuring system must be able to determine this so that timely precautions can be taken.

   Data are required not only at the site, but also at the control center located a considerable distance away. It is therefore necessary to provide suitable means of data transmission. Any suggested instrumentation must take account of this important requirement. Electrical power can be provided if required.

9. In question 8, it has been decided that in order to minimize inconvenience to the water authority's customers, a channel is to be built to divert some water from a nearby river. However, to avoid the deleterious effects of withdrawing excessive quantities, it is mandatory that the amount of water diverted be recorded. Suitable sluice gates will be provided and regulated by a control system provided by others to ensure that untoward effects are kept to an acceptable minimum.

   What method would you employ to determine the water flow and the transmission of the relevant information to the main control center (MCC) Provide sketches of your proposal.

10. A certain manufacturing company uses a variety of preprocessed liquid chemical fluids of specific gravity that varies between 0.72 and 1.72 in the preparation of its products. These raw materials have to be held onsite in very large storage tanks. The storage tanks vary in size from 7 to 13.4 m in height in steps of 1.6 m and 4.5 m to 8.4 m diameter with a difference between the tank diameters of 1.1 m. The contents of the tank farm have to be determined for the following purposes:

    (a) Evaluating the total inventory of each fluid

    (b) Determining when to order replenishment

    (c) Determining financial investment

To monitor the vessel contents, it is decided to provide the process operatives with a single large-level indicator and a multipoint selector switch. The indicator scale should be in percentage.

(a) Describe in detail how this requirement can be fulfilled using a specific gravity variation of 0.2 between fluids.

(b) Do you think it necessary to know the fluid specific gravity in order to provide the measurement? State clearly your reasons.

(c) Assume that the tank barreling effect is not to be accounted for, and the minimum level is to be no less than 0.5 m from the base, and the maximum level within 0.4 m of the top of the vessel. What would be the resulting effect if the level in each tank had to be displayed in units of the actual height of the liquid within it?

(d) Replenishing is to be effected when the fluid level is 0.2 m above the low limit. If the price difference between each two fluids is $1385.50 (£923.60)/m^3, calculate the financial investment when there are a total of 5 tanks and all the tanks are full. (The most expensive fluid is in the smallest tank, and the least costly in the largest represents an investment of $10,000.00 (£6666.67), an assumed rate of $1.5 to £1.00.)

11. You need to provide a local, as well as a remote, measurement of the level of a liquid in an open-top vessel of rectangular cross section that is continuously being supplied with fluid through a pipe mounted above the top. The arrangement makes the liquid fall under gravity into the vessel causing the free surface to be considerably disturbed. The specific gravity of the fluid is 1.3, and the total level to be measured is 5 m. In addition, personnel access is very restricted on all sides, so that a level-measuring instrument mounted on or near the sides of the vessel is not a desirable option. However, it is possible to use an existing walkway structure centrally located above the vessel for mounting instrumentation.

Describe with supporting sketches how this can be done and what additional facilities, if any, would be required to make the measurement. Assume that others will build any mechanical structure associated with the measurement from a design provided by you.

If there are multiple solutions, then all should be detailed.

12. A particular furnace burns waste wood and wood bark that result from a manufacturing process. It is necessary to determine the oxygen content of the exhaust gas from the furnace. What instrument would you specify to meet the requirement, and what advice would you give the user on the matter of ensuring a particle-free gas sample?

13. Describe the principle of operation of the oxygen analyzer you specified for question 12, and give a clearly labeled sketch of the instrument involved.

14. The hydrogen ion concentration of a liquid fluid is very often used in the process industry to determine its degree of acidity or alkalinity. Discuss this statement and show how this is made possible.

15. It is required to determine the mass flow of particular liquid fluid contained in a 1.5-in (approximately 38-mm) feed line. What instrument(s) would you

specify if:

  (a) The cost of purchase had to be taken into account.

  (b) The cost was of secondary importance to the amount actually delivered, and additionally it was necessary to obtain a measurement of fluid density.

  (c) Is it necessary to always consider measurement accuracy as a criterion for instrument choice? What would be a suitable alternative?

16. A particular heater is designed to operate continuously at a temperature of 1100°C with occasional excursions to 1300°C. The furnace atmosphere is very slightly reducing. The heater is to be controlled at a temperature of 1000°C.

  (a) Specify the temperature sensor to be used.

  (b) If the sensor immersion is to be 15 in (381 mm) and the mounting nozzle including the flange is 6 in (152.4 mm), what would you state the total length of the sensor to be?

  (c) If there are any points that the user should consider or on which you need clarification, state these.

  (d) Give a sketch of your proposed sensor along with the material of construction of the sensor.

17. Milk has to be metered into a processing vessel. The pump being used is not designed to produce a high discharge pressure. In view of this, it is necessary that the flow sensor be one that produces the minimum pressure loss. The system must be able to withstand sterilization and therefore due attention is to be paid to this aspect. Propose a measuring instrument and its advantages to the user. Provide a sketch of the device you would specify.

18. An application involves flow measurement of a particular fluid to a processing unit operating at 5.9 kg/cm^2 (approximately 84 lb/in^2) with a suitable device. The fluid has the following characteristics:

  (a) It is an emulsion of edible fats and water to which a colorant and an inhibitor have been added. The fluid has a viscosity almost equivalent to that of water.

  (b) The process line is 100 mm (4 in) NB (nominal bore) in 316 stainless steel, and operates at 6 kg/cm^2 (about 85 lb/in^2) and 25°C (about 170°F).

If the cost of providing the measurement reduces in magnitude according to the listed order of the instruments, specify the type of measuring instrument(s) and installation arrangement you would recommend, giving reasons for your choice (or choices, if there is an alternate). Choose among the Coriolis mass meter, magnetic flowmeter, target flowmeter, vortex flowmeter, orifice plate, and D/P cell.

PART II

PROCESS CONTROL

CHAPTER 6

Gain

This chapter represents the beginning of a fascinating field of study and work, namely, *process control* theory and practice. The initial paragraphs will set out in very broad terms the basic requirements of a controller, although not specifically of a process controller. The general description of a controller is then developed into a broad-based block diagram presentation of a typical industrial process controller. This method of presentation has been chosen to assist the reader in understanding the fundamental requirements, before actually developing the mathematics of the control functions, which is given in later chapters.

Since all control equipment and processes exhibit inherent qualities of *gain*, or the difference between the magnitude of an input resulting in the output, it is important that we understand what is really happening. The gain of a process stage or control equipment is the relationship between the input and the resultant output (or the vector ratio of output to input). The effect of joining gains in *series* (sequentially) or in *parallel* (simultaneously) will be developed systematically, with the use of some quite straightforward mathematics. This chapter will present the fundamental ideas involved and the techniques used.

Please refer to the List of Notations on page xxxvii.

BASIC ELEMENTS OF A CONTROLLER

As the demand for better, cheaper, and more varied material increases, so does the complexity of modern industrial processes to satisfy customer demands. This results in an increasing number of process parameters to be monitored and controlled to much closer tolerances than ever required previously.

Let us start by trying to understand what a controller is and finding out the basis of its functionality. A very general definition of a controller might be a device or a person that is capable of maintaining conditions or other persons within acceptable predefined limits. The foregoing statement is the widest possible definition and has been deliberately chosen so that the essential idea can be grasped. For any controller to be able to function as such, it is vital that four fundamental requirements be met, without any one of which it cannot truly be called a controller:

1. It must have an input in the form of a measurement.

2. It must have the means for generating and setting a desired value (setpoint).

3. It must have the ability of comparing the measurement with the desired value and as a result producing the difference (error) between the two.

4. It must be capable of producing an output of sufficient magnitude and direction to correct any deviation from the desired value.

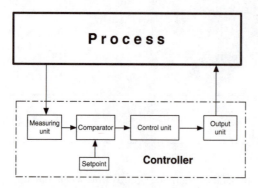

Figure 6.1: Block diagram of a controller.

Figure 6.1 is a block diagram of a typical controller depicted as a device, but is nonetheless applicable to the human controller as well, given a little imagination. If we were to expand this block diagram to include the control terms that are normally associated with automatic process control, the diagram would appear as in Figure 6.2.

BASIC ELEMENTS OF CONTROL

Let us consider a simple plant in which the control is achieved manually and also assume that the scheme has been designed to supply water at a particular temperature to another part of the process. For the purposes of our discussion we shall use direct steam heating, control it with a manual valve, and monitor the temperature of the outgoing water. Figure 6.3 is a diagram of the process. The process operator is advised of the required outlet temperature. The actual value attained by the water is continuously displayed on the temperature indicator, and the operator manipulates the valve in the steam line when any departure from the required (desired) temperature is observed, by an amount he or she estimates as being necessary to bring the outgoing water back to the required temperature.

From what has been said, it will be appreciated that the control achieved will be a rather hit-and-miss affair. For our discussion we shall assume that the plant is

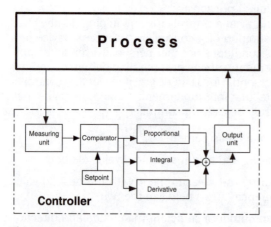

Figure 6.2: Block diagram of three-term controller.

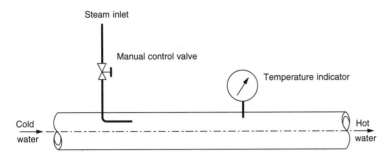

Figure 6.3: Simple hot water system.

under steady-state control—i.e., the outgoing water is at the desired temperature. For our simple system to achieve equilibrium at the required (desired) temperature, the desired value must be fixed and the following must apply:

1. The cold water must be at a constant inlet temperature and flow rate.

2. The heating steam must be of constant enthalpy (heat content) and flow rate.

3. The demand on the output of hot water must be constant.

In addition, the following applies in real-life situations, as nothing can remain constant indefinitely:

4. The operator must be ever vigilant of the temperature.

A disturbance or change, in any or all of these factors, will result in a deviation from required temperature of the outgoing hot water. The disturbance could be periodic or random or take any of the following forms:

• Step

• Ramp

• Transitory

• Cyclic (damped or undamped)

It is assumed that cyclic disturbances can be resolved into a number of purely sinusoidal components that can be treated separately. In a real plant, the most important cyclic changes that occur are those brought about under automatic control during the period of recovery after a process upset.

   The effect that a disturbance will have on a plant depends on its point of entry into the control system. For convenience, a system can roughly be divided into two parts, the *supply side* and the *demand side*, and in the small system we are just considering, the supply side is the steam while the demand side is the heated water. In general it is easier to control a supply-side disturbance. For this system under discussion, the hardware fitted to the process is the temperature sensor and the valve in the steam supply line, the operator being the link between these two items of hardware, providing the "intelligence" for the control action. Adjustments to the position of the steam supply valve are made on the basis of the temperature of the outgoing water, and in this regard the operator acts as the means of relating the measurement provided by the sensor to the corrective action provided by the valve. The operator is the means of completing the *loop* formed by the sensor, operator, and valve to make the action a cohesive entity, thus putting the control system under *closed-loop control*. With a varying output dependent on the input values of steam and water flow and temperature, it should be fairly easy to visualize that there is an overall gain aspect to the system, made up of the individual gains of the controller and the process itself. The system gain is usually referred to as the *closed-loop gain*.

There is another system of control in which the measurement does not relate directly to the process that is being controlled, the link between what is being measured and what is controlled being indirect. In this system the controller, like every other, can take action only on the information presented. As an example, let us consider the following scenario:

- The operator is human, and the system is a living room with a room heater. The heater controls available to the operator are manual only, and there is a temperature sensor with an indicator to measure the temperature of the air outside the room.

- The operator has no access to the room, but is instructed to maintain a suitable temperature for living within the room.

Under this operating scenario, it will be impossible to hold the operator responsible for unsuitable temperatures within the room, for the obvious reason that the only information available to the operator to assess the inside temperature condition is the air temperature on the outside of the room. Such a system is referred to as being under *open-loop control*. Once again it should be fairly easy to visualize that there is a gain aspect to the system, made up as before of the individual gains of the controller and the process itself. In this case the system gain is usually referred to as the *open-loop gain*. As it stands, such a temperature control loop cannot meet the objective of providing suitable temperature conditions in the living room and must therefore be modified. An easy way out of the situation, while still maintaining the manual control feature, would be to provide the operator with an additional temperature indicator arranged to sense the room inside temperature. This arrangement will enable the operator to see the effect his or her actions have on the environment within the room, the result being that alterations can be made to the heater controls to take account of changes in temperature of the air outside the room. This modified control loop illustrates in a simple way the basic concept of a *cascade* system.

RELATIONSHIP BETWEEN GAIN AND FEEDBACK

From what has been discussed so far there is one central idea coming through strongly, and this is known as feedback. What we are actually doing is monitoring the process, altering that item that most influences a change if required, observing the results of the alteration, and, dependent on the observations made, realtering, if necessary, the item that influenced the change.

Figure 6.4 depicts a simple control unit in which the output is linearly proportional to the input—i.e., with no phase shift. Assigning symbols to this diagram we have:

$$V \propto \theta$$

where θ is the input and V is the output; from which we can say:

$$V = G\theta$$

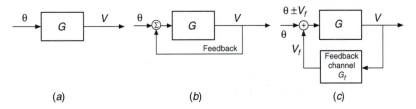

(a)   (b)   (c)

Figure 6.4: Block diagrams of gain.

where G is the constant of proportionality, in this case the gain; and therefore

$$\frac{V}{\theta} = G$$

Figure 6.4b shows the generalized format of feedback, but Figure 6.4c shows in real terms what happens in a loop. When we connect the output back to the input through a feedback unit with a gain of G_f, we then have:

$$\frac{V_f}{V} = G_f$$

where V_f is the output of the feedback unit and G_f is the gain.

The feedback can be arranged to "assist" or "oppose" the circuit input θ, resulting in conditions at the input unit of either $(\theta + V_f)$ or $(\theta - V_f)$. The first of these terms is called *positive feedback*, and the second is called *negative feedback*.

Positive feedback, when applied to the control unit, has the effect of increasing the overall gain and will preclude any possibility of stability. Control loops with positive feedback will lock at one extreme signal limit or the other and are therefore of little use in practice. (Positive feedback, however, is sometimes a design feature of some individual circuit elements of peripheral or other units such as trigger, trip, or alarm units.) For this reason, we shall concentrate on negative feedback and see what results from its inclusion in the system.

Considering the circuit of Figure 6.4c (where the summer is in the subtracting mode and V_f is really $+V_f$), the actual input to the amplifier is $(\theta - V_f)$, so that

$$V = G(\theta - V_f)$$

But

$$V_f = G_f V$$

Therefore

$$V = G(\theta - G_f V)$$

from which

$$V(1 + GG_f) = G\theta$$

Rearranging, we have:

$$\frac{V}{\theta} = \frac{G}{1 + GG_f}$$

Now, if we look at the effect of changes in the separate components of the overall gain, i.e., V/θ, we have two main cases to consider. The standard method to obtain the derivative of a quotient is:

$$y = \frac{u}{v} \qquad \frac{dy}{dx} = \frac{v\frac{du}{dx} - u\frac{dv}{dx}}{v^2}$$

1. *change of unit or amplifier gain.* Differentiating $\frac{V}{\theta} = \frac{G}{1+GG_f}$ with respect to G by the standard method of a quotient, we have:

$$\frac{dV}{dG} = \left[\frac{1}{(1+GG_f)^2}\right]\theta$$

But

$$\frac{V}{\theta} = \frac{G}{1+GG_f}$$

$$\frac{V}{\theta G} = \frac{1}{1+GG_f}$$

so that

$$\frac{dV}{dG} = \frac{V}{\theta G}\left[\frac{1}{1+GG_f}\right]\theta$$

Rearranging, we have

$$\frac{dV}{V} = \frac{dG}{G}\left[\frac{1}{1+GG_f}\right]$$

If we make G very large, then this last equation will reduce with very little error to:

$$\frac{V}{\theta} = 1/G_f$$

From this we can see that V/θ is very nearly independent of changes in G. This makes it most useful in electronic amplifiers where stability is so important and where components may affect the gain G of the amplifier.

2. *Change of feedback unit or path gain.* Having seen the effect of varying G, let us now observe what happens when G_f is allowed to vary. Once again we shall use the equation:

$$\frac{V}{\theta} = \frac{G}{1+GG_f}$$

This time differentiate it with respect to G_f and note that the numerator on the right-hand side is the derivative of the denominator to give

$$\frac{dV}{dG_f} = -\left[\frac{G^2}{(1+GG_f)^2}\right]\theta$$

by using the standard method of determining the derivative of a quotient. But

$$\frac{V}{\theta} = \frac{G}{1+GG_f}$$

Therefore

$$\theta = \left[\frac{1+GG_f}{G}\right]V$$

so that by using

$$\frac{dV}{dG_f} = -\left[\frac{G^2}{(1 + GG_f)^2}\right]\theta$$

and substituting for the value of θ just derived, we have:

$$\frac{dV}{dG_f} = -\left[\frac{G}{(1 + GG_f)}\right]V$$

Therefore

$$\frac{dV}{V} = -\left[\frac{G}{(1 + GG_f)}\right]dG_f$$

Multiplying the numerator and denominator of the right-hand side by G_f, we have:

$$\frac{dV}{V} = -\left[\frac{GG_f}{1 + GG_f}\right]\frac{dG_f}{G_f}$$

If G is still large, the product GG_f is very much larger than 1, so that the terms within the brackets tend to 1.0, and this equation can be written as:

$$\frac{dV}{V} = -\frac{dG_f}{G_f}$$

The negative sign shows that the output change is an increase if the feedback decreases, and vice versa—i.e., output V is inversely proportional to G_f.

The findings of both parts of the foregoing analysis show that, in practical cases of adequate gain and feedback, it is possible to ensure that overall gain is substantially independent of amplifier unit gain, but is determined almost independently by the feedback applied, i.e., by the amount and stability of the feedback path and/or components.

GAIN OF A COMBINATION OF UNITS

For components in a feedback system having gains G_1, G_2, G_3, etc., arranged in series as shown in the Figure 6.5, we have:

$$V_a = G_1\theta \qquad V_b = G_2V_a \qquad V_0 = G_3V_b$$

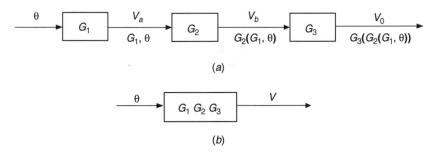

(a)

(b)

Figure 6.5: *(a)* Gains in a series arrangement. *(b)* Equivalent circuit.

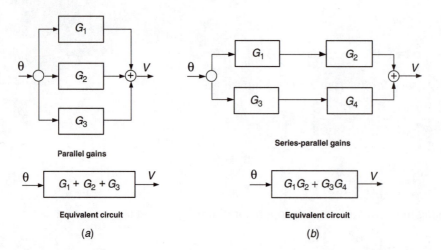

Figure 6.6: (*a*) Parallel gains and equivalent circuit. (*b*) Series-parallel gains and equivalent circuit.

But

$$G_2 V_a = G_2 G_1 \theta \qquad G_3 V_b = G_3 G_2 G_1 \theta$$

The system is therefore equivalent to a single control unit of gain equal to the product of the individual gains, or $G_1 G_2 G_3$. Therefore

$$\frac{V}{\theta} = G_1 G_2 G_3$$

If the units were arranged in parallel as shown in Figure 6.6*a*, then the input θ would be applied to each of the units, with the result that the output from each unit would be $G_1\theta$, $G_2\theta$, $G_3\theta$; and all these outputs will be added together (summed), resulting in:

$$V = G_1\theta + G_2\theta + G_3\theta$$
$$= \theta(G_1 + G_2 + G_3)$$

Therefore

$$\frac{V}{\theta} = G_1 + G_2 + G_3$$

If four units were now arranged in a series-parallel arrangement as in Figure 6.6*b*, then we can see that for the upper series the output would be $G_2(G_1\theta)$ and for the lower series the output would be $G_4(G_3\theta)$. The actual output would be the sum of the outputs of these two:

$$V = G_2(G_1\theta) + G_4(G_3\theta)$$

Therefore

$$\frac{V}{\theta} = G_1 G_2 + G_3 G_4$$

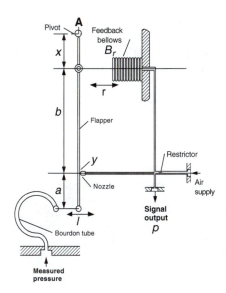

Figure 6.7: Flapper-nozzle assembly of a pneumatic instrument.

NEGATIVE FEEDBACK APPLICATION EXAMPLE

We shall now consider a simple practical example to see how negative feedback is applied to a pneumatic controller. Figure 6.7 is an illustration of this application involving real-world instruments. The sensor of the instrument is a C-type Bourdon tube that will tend to straighten out under the action of the applied pressure. The sealed end of the tube is linked to one end of a baffle plate that is pivoted at the opposite end. This arrangement will allow the flapper (baffle) to move under the action of the applied pressure; the actual movement is of the order of 0.0003 to 0.003 in. A nozzle is placed directly in front of the baffle and an air supply is connected to the nozzle, the bore of which is usually of the order of 0.02 to 0.04 in. The position of the nozzle with respect to the baffle can be adjusted within a small range in order to allow calibration of range and zero. A restrictor having a bore of 0.25 to 0.5 times the nozzle bore is placed in the air supply line to the nozzle.

Let us see how the system functions, and to do this, let us add some dimensions to those shown in Figure 6.7. Let l be the movement of the Bourdon tube at the free end and y the corresponding movement of the flapper (baffle) at the nozzle; then for a movement l of the Bourdon about the pivot point A, the movement y at the nozzle is given by:

$$\frac{l(b + x)}{(a + b + x)}$$

This is opposed by the bellows, which moves an amount given by:

$$\frac{lx}{(a + b + x)}$$

which is the negative feedback in the system. The net movement is the difference between these two movements, and the resulting output is given by:

$$y_{\text{net}} = \frac{lb(B_r\, p)}{(a + b + x)}$$

where B_r = the bellows rate (say, in/(lb/in^2)) and p = the output pressure (say, lb/in^2). The overall effect of this is to eliminate the errors that would be predominant due to the positioning of the flapper, which in a real operational instrument would be moving. The bellows provides the stabilizing force to accurately position the flapper with respect to the nozzle and stabilize the gain.

Why is a restrictor present in the air supply line to the nozzle? To find out, let us remove the restrictor and consider what the result will be. The position or spacing of the flapper or nozzle determines the output pressure; hence, with the flapper furthest away from the nozzle, the output pressure will reduce to zero, as escape of the supply air would be virtually unrestrained. Within the nozzle there will be no significant change in pressure until the flapper is moved almost up against the nozzle. If the resulting output pressure was to be recorded, the trace would be seen as a series of very sharp peaks and troughs, as the flapper moved toward or away from the nozzle. The addition of a restrictor to the bellows unit allows a delay to be built into the system (what we have here is equivalent to an electrical capacitor and resistor in series, a filter, in fact) that will allow the measurement pressure changes to be followed more clearly by the pen recorder.

SUMMARY

1. There are four essential features for any controller, animate or inanimate: it must have a measurement as input; must offer the means for generating and/or accepting a setpoint; must have the ability of comparing the setpoint with the measurement and registering the resultant error; and must be capable of producing an output of sufficient magnitude and direction to correct any deviation from the desired value.

2. Process disturbances can be periodic or random or take any of the forms categorized as: step, ramp, transitory, and cyclic (damped or undamped). It is assumed that cyclic disturbances can be resolved into a number of purely sinusoidal disturbances that can be treated separately.

3. A system can roughly be divided into a supply side and a demand side. In general, it is easier to control a supply-side disturbance.

4. Feedback control systems are closed-loops in which the measurement is altered directly by controlled output.

5. An open loop control system is one in which the control action is independent of the output. The accuracy of an open loop depends on the establishment of the correct input-output relationship.

6. Gain is output divided by input, or

$$\frac{V}{\theta} = G$$

where θ is input, V is output, and G is gain.

7. When the output V is connected back to the input through a feedback unit,

$$\frac{V_f}{V} = G_f$$

where V_f is the output of the feedback unit and G_f is the gain.

8. The feedback can be arranged to assist or oppose the input θ, resulting in conditions at the input unit of either $(\theta + V_f)$ or $(\theta - V_f)$. The first of these terms is called positive feedback and the second negative feedback.

9. Positive feedback, when applied to the control unit, has the effect of augmenting the imbalance and will preclude any condition of stability. Negative feedback, on the other hand, has the effect of stabilizing the system.

10. For components in a system of gains G_1, G_2, G_3, etc., arranged in series:

$$V_a = G_1\theta \qquad V_b = G_2 V_a \qquad V_0 = G_3 V_b$$

But

$$G_2 V_a = G_2 G_1 \theta \qquad G_3 V_b = G_3 G_2 G_1 \theta$$

Therefore

$$\frac{V}{\theta} = G_1 G_2 G_3$$

The system is therefore equivalent to a single control unit of gain equal to the product of the individual gains

11. If components in a system with gains G_1, G_2, G_3 are arranged in parallel, then the input θ will be applied to each of the units separately with the result that the respective outputs are $G_1\theta$, $G_2\theta$, $G_3\theta$. These outputs will be added (summed) together, resulting in:

$$V = G_1\theta + G_2\theta + G_3\theta$$
$$= \theta(G_1 + G_2 + G_3)$$

Therefore

$$\frac{V}{\theta} = G_1 + G_2 + G_3$$

12. If components in a system with gains G_1, G_2, G_3, and G_4 are arranged in series-parallel, then the output will be the sum of two series outputs:

$$V = G_2(G_1\theta) + G_4(G_3\theta)$$

Therefore

$$\frac{V}{\theta} = G_1 G_2 + G_3 G_4$$

CHAPTER 7

Delays or "Lags"

In this chapter we shall be continuing our study of control, but this time we shall consider the actual implications of a control strategy and how it is affected by being applied to a real process environment. We are all aware that no action in this world can be implemented at the instant we would wish it to occur. All actions are bounded by time. Suppose, for example, we add sugar to a cup of tea. Immediately after the addition it is not possible to taste the sweetness. We have to wait for a period of time to allow the sugar to dissolve and disperse through the volume of the tea before the effect of the addition can be tasted. Stirring the tea will accelerate the dispersal, and the elevated temperature of the tea itself will accelerate the process of dissolving the sugar. If we analyze the situation a little more we will also see that, apart from the time needed for the sugar to disperse and dissolve, time is also required, however short it may be, for our taste system to appreciate the sweetness.

In control terminology, all of these delays are called *lags*. We shall be directing our attention to this subject to try to understand the underlying principles, with the objective of trying to minimize their effect on the control itself.

LAGS ASSOCIATED WITH CONTROL

In the last chapter we saw how control of a simple process could be effected by manual means and how stability can be improved or achieved by feedback. Furthermore, having also defined the requirements for a controller, let us now spend some time in considering the delays that are present in real-life control systems.

MEASUREMENT LAGS

Let us start at the sensor that provides the measurement. In order for this device to produce a valid signal of the parameter it is to measure, it must be in equilibrium with the process medium. If the process medium changes from this equilibrium state to another condition, the sensor must be able to respond to the change, but once again the signal produced must represent the equilibrium state of the process medium at the new, changed condition. The time involved for the sensor to correctly produce a signal representing the measured parameter is termed a *measurement lag*.

As an example, consider a filled-system temperature sensor. Suppose the sensor is measuring the temperature of a process fluid. If we now suppose the process fluid temperature starts to rise, the fluid will have to impart its change in temperature to the fluid filling of the sensor before the change can start to be shown on the indicating mechanism. However, it will not be able to do this before it has caused the temperature of the bulb material itself to change, and therefore the overall response to the change will depend on the thermal capacity of the bulb plus its contents—in other words, the sum of the product of the mass and the specific heat for each of the

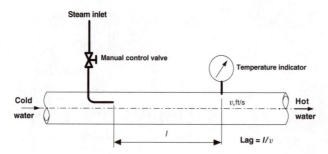

Figure 7.1: The simple hot water system of Figure 6.3, with l/v lag indicated.

two mechanical components of construction—and will be influenced by the rate(s) of heat transfer. If a thermowell is involved, then the overall measurement lag will also include a separate lag item due to it.

DISTANCE/VELOCITY LAGS

The next of these delays is the *distance/velocity* effect or lag. A distance/velocity lag is defined as the time interval between a change in the value of a signal and its appearance at a later part of the system at its new value, arising solely from the finite speed of propagation of the signal. The distance/velocity lag is one of the most difficult to overcome, as it involves the actual design and configuration of the plant itself, and therefore any steps taken at the plant design stage to reduce its effects will pay off in a better control system.

Figure 7.1 is similar to Figure 6.3, but we have added some detail to make it more realistic. For instance, there must be some physical distance between the steam inlet and the downstream point at which the temperature is sensed. If we, say, let this distance be l, in feet (meters) and if we let the water velocity be v, in ft/s (m/s), then the time interval between a change in steam supply and the indication of its effect will be l/v, i.e., distance/velocity lag. If t' is the time interval for this to be observed, then:

$$t' = l/v$$

It is obvious that no corrective action can be taken within this time interval, i.e., until some change arrives at the point of measurement. Meanwhile, temperature changes in the process fluid will continue to occur during this period and they could be increases or decreases or oscillations, and they could occur rapidly or slowly, or they could be wholly random! In any event, some process fluid will already have left the process at a temperature other than that which was desired, and there will be a further time interval before corrective action can be taken. If d (ft or m) is the bore of the pipe, then the volume (ft^3 or m^3) of fluid that has left the pipe is:

$$t'v\frac{\pi d^2}{4}$$

From the foregoing it is clear that the longer the time interval, the greater will be the departure from the desired state.

For this simple case we have not concerned ourselves with the conditions existing upstream of the steam and water inlets, i.e., before the water and steam are made to enter our small process. In a real process plant, these supply lines would be shared with other parts of the manufacturing process, in which case the conditions within any other pipelines, or the supplies themselves, could be varying all the while. Let us assume that the variations are sinusoidal and therefore of uniform period. Then the measurement will be given as:

$$\theta = \theta_{av} + A \, \sin2\pi \frac{t}{\tau_0}$$

where θ = measured variable
θ_{av} = average value of the measurement (flow, for instance)
τ_0 = period of sinusoidial variation
A = amplitude
t = time

A point worth remembering is that the amplitude of any type of disturbance is unchanged as a result of lags (delays), although it is subject to the delays. If τ_d is the effect of the delay on the signal, then the measurement will be given by:

$$\theta = \theta_{av} + A \, \sin2\pi \frac{t - \tau_d}{\tau_0}$$

The angular difference between the signals at the output and the input will give the phase shift, and if this angle is ϕ, then:

$$\phi = 2\pi \frac{t - \tau_d}{\tau_0} - 2\pi \frac{t}{\tau_0}$$

$$= -2\pi \frac{\tau_d}{\tau_0}$$

$$= -360° \frac{\tau_d}{\tau_0}$$

where the negative sign indicates a phase lag or delay at the output relative to the input.

The following are a few points worth remembering:

1. If separate distance/velocity lags occur in a process line, their effects are additive and can be considered as a single lag, provided that the behavior of the intermediate signals between the two lags is not of great importance.

2. With incompressible fluids, a change in the rate of flow or in pressure takes place simultaneously and at all points, provided the process system is completely filled. It therefore follows that, in any completely filled incompressible fluid flow system, there cannot be a distance/velocity lag. However, if the system has parts that are not completely filled, or if the fluid contains some entrained compressible fluids, then distance/velocity lags will occur. Furthermore, in the latter case a change in temperature or purity will, of course, be subject to distance/velocity lags.

3. Pure distance/velocity lags hardly ever occur as a single entity in a real plant. For example, suppose we are measuring the temperature of hot water in a pipeline. The heater, whether it be direct steam or indirect tube, will be physically located a distance away from the temperature detector, in which case some of the heat imparted to the water will be lost to the pipe itself, some to the insulation (if any), and some to the surrounding air. From this it is clear that the temperature detected

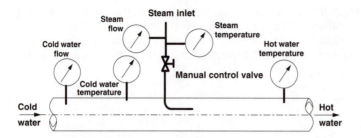

Figure 7.2: Improved hot water system.

at the point of measurement will be less than at the heat source. Increasing the separation distance between the source and the detector will increase the lag and the temperature difference, and therefore we do not have a pure distance/velocity lag, but a combination of lags.

4. A stand-alone controller, without any built-in predictive element, e.g., the Smith-predictor, cannot overcome the effects of distance/velocity lags in a control loop, for it does not have powers of intuition! It can only respond to a change in the signal presented to it from the sensor and take the appropriate action as soon as it sees the signal.

5. For a step change in the process, the action of the controller will be to eventually bring the controlled variable back to the desired value, but for a transitory change the corrective action will be too late to eliminate the disturbance; it will also introduce an equal but opposite deviation one period later.

As mentioned earlier, eliminating the effects of distance and velocity is costly in both time and money; the best alternative is to try to minimize the lag and live with the results of the effort expended. Redesigning part or all of the controls in addition to the alternative plant improvements can also help toward a more controllable process.

Let us see how we could improve the simple process control system we used in Figure 7.1 so that a much more stable outlet temperature is obtained. Figure 7.2 shows what can be done, even though the human operator is still manipulating the control valve. From the information now available to the operator, there is every possibility that the outlet water will be nearer the required temperature, for the operator is able to see whether the inlet cold water or heating steam flow and temperature have changed in any way and to compensate for these changes. However, be aware that any compensatory changes made will have to be anticipatory, based on the operator's experience of the system behavior, which is mainly in the way the final temperature measurement responds to a change in the operational setting of the steam valve.

TRANSFER LAGS

The next type of delay we shall consider is the *transfer lag*. This lag occurs when matter or energy is transferred to a *receiver* via some form of constriction. From this, it should be clear that this is analogous to the classic electrical *CR* (resistance-capacitance) network, and transfer lags are exponential in character in a similar way.

EXAMPLE 7.1   143

The causes of transfer lags are often complex, and, as a consequence, the plant response will in most cases be extremely difficult to calculate. Experimental analysis is the alternative; it can be employed in those instances when examination of the results of test measurements or trials gives a good idea of the process unit characteristics. By expressing the variations of output in terms of the input as a differential equation, a response curve can be developed.

We shall first consider two practical examples, which approximate single exponential stages. These examples are chosen to show that:

1. A single resistance associated with a single capacity does give an exponential response.

2. Exponential responses can also be obtained as a result of energy or matter transfer even if no physical resistance to the transfer exists; for example, heating a body involves an energy transfer with no resistance offered by the body involved.

3. Assumptions must be made to simplify the considerations in order to assess the characteristics.

EXAMPLE 7.1

Consider Figure 7.3. In the simple level system shown, let us assume that equilibrium has been established, with the inflow equal to the outflow. Now let there be a sudden increase in the inflow, which is shown in the inflow graph as a step change. The tank level will begin to rise, resulting in an increase in the head of the liquid. To restabilize the level, the outflow rate will increase until a new equilibrium level is established with outflow equal to inflow again. Since the rate of rise of level is proportional to the difference between the inflow and outflow rates, and this continually decreases until it reaches zero, the rate of rise of the tank level will also decrease to zero as the new equilibrium level is reached. Therefore, the relationship between the liquid level and time will be exponential and the system will consist of a single exponential stage. (The derivation and proof of the exponential form of the relationship will be developed in the course of the next example, but it follows the same basic analysis as for an electrical RC circuit.)

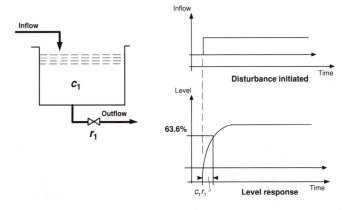

Figure 7.3: Level control system for Example 7.1.

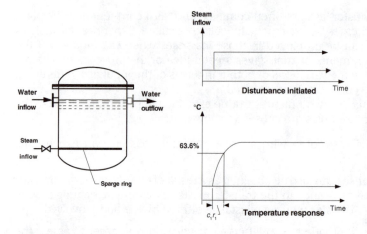

Figure 7.4: Temperature control system for Example 7.2.

EXAMPLE 7.2

In this example we shall consider the transfer of energy by direct steam injection, but to do this we must first make the following assumptions, and consider Figure 7.4.

Assumptions

1. The thermal capacity of the whole system is the sum of the thermal capacities of the water and the vessel walls.

2. Heat losses from the system are negligible.

3. Heat supplied is uniformly distributed throughout the vessel instantly.

4. A step change in the heating steam supply upstream of the valve will produce an immediate step change in steam flow into the vessel from the sparge tube nozzles.

Reasons for the Validity of the Assumptions

1. It is more than likely that assumption 1 will be attainable, as efficient circulation implies a high water speed over the vessel and hence a low boundary resistance to heat flow between water and wall.

2. Having the vessel well lagged, along with efficient circulation by the steam sparge nozzles, will help ensure that assumptions 1, 2, and 3 are valid.

3. Assumption 4 will be approximately true if the volume of the pipe between valve and sparge tube is small.

Let us now look at this example mathematically and see what the result will be. Suppose a step change occurs in the steam rate, say, due to a supply pressure change. Let the step change in heat input be W, kcal/min, and let the temperature change be T_2, in °C, at a time Δt minutes after the change, then:

$$W\Delta t = c_{\mathrm{Th,w}}\frac{dT_2}{dt}\Delta t + c_{\mathrm{Th,v}}T_2\Delta t \tag{7.1}$$

where $c_{\mathrm{Th,w}}$ and $c_{\mathrm{Th,v}}$ are the thermal capacities of the water in the vessel and the

EXAMPLE 7.2   145

water leaving the vessel. When new equilibrium is reached,

$$W = c_{Th,v} T_p$$

where T_p is the final change in water temperature. Substituting into Equation (7.1), we have:

$$c_{Th,v} T_p \Delta t = c_{Th,w} \frac{dT_2}{dt} \Delta t + c_{Th,v} T_2 \Delta t$$

$$c_{Th,v} T_p \Delta t - c_{Th,v} T_2 \Delta t = c_{Th,w} \frac{dT_2}{dt} \Delta t$$

Dividing through by Δt, we have:

$$c_{Th,v} T_p - c_{Th,v} T_2 = c_{Th,w} \frac{dT_2}{dt}$$

and therefore

$$c_{Th,w} \frac{dT_2}{dt} - c_{Th,v} \left(T_p - T_2 \right) = 0$$

from which, by integration,

$$T_2 = T_p \left(1 - e^{-(t/c_{Th,w})(1/c_{Th,v})} \right)$$

The time constant of this exponential stage is $\tau = c_{Th,w} c_{Th,v}$, although there is no resistance in this plant to the transfer of heat; and it proves that the relationship between T_2 and T_p is exponential. We can then write it in the form:

$$T_2 = kW \left(1 - e^{-(t/c_{Th,w} c_{Th,v})} \right)$$

where $T_p/W = k$ (a constant) for the given conditions, since the final change in equilibrium value of the level or temperature is proportional to the change in water flow rate or heat supply, respectively.

Suppose the outflow from the vessel of Figure 7.3 in Example 7.1 feeds a second vessel of capacity c_2 and having an inlet resistance of r_2. This will then represent two resistance-capacitance systems of time constants, say, τ_1 and τ_2, these constants being evaluated from $c_1 r_1$ and $c_2 r_2$, respectively. A knowledge of the process input θ_1 and the process output V_2 will give the response of the system, which can be found by solving the second-order differential equation that describes the system behavior, the results of which for convenience is shown below:

$$V_2 = k\theta_1 \left[1 - \frac{\tau_1 \tau_2}{\tau_2 - \tau_1} \left(\frac{1}{\tau_1} e^{-t/\tau_2} - \frac{1}{\tau_2} e^{-t/\tau_1} \right) \right]$$

So far we have discussed single- and two-stage exponential systems; let us now compare the responses of both these systems. Taking into consideration the equation

$$T_2 = kW \left(1 - e^{-t/c_w c} \right)$$

we developed for Example 7.2 and converted it to a more generalized form using θ_1 as input and V_2 as output, c as c_w, and r as c, we can write it as:

$$V_2 = k\theta_1 \left(1 - e^{-t/cr} \right) \tag{7.2}$$

In this form it is applicable to either single level or temperature systems. Differentiating Equation (7.2), with respect to t, we have:

$$\frac{dV_2}{dt} = \frac{k\theta_1}{\tau_1} e^{-t/\tau_1}$$

where $\tau_1 = cr$; and at $t = 0$,

$$\frac{dV_2}{dt} = \frac{k\theta_1}{\tau_1}$$

and if we now take the equation for the two-stage system,

$$V_2 = k\theta_1 \left[1 - \frac{\tau_1 \tau_2}{\tau_2 - \tau_1} \left(\frac{1}{\tau_1} e^{-t/\tau_2} - \frac{1}{\tau_2} e^{-t/\tau_1} \right) \right]$$

and differentiate it with respect to t, we have:

$$\frac{dV_2}{dt} = k\theta_1 \left(\frac{e^{-t/\tau_2} - e^{-t/\tau_1}}{\tau_2 - \tau_1} \right)$$

at $t = 0$, this derivative will be zero.

Hence, for a single exponential stage, the output V_2 increases immediately as the step change occurs, and at a rate proportional to the input θ_1; whereas for two exponential stages, the output V_3 has a low rate of increase initially and subsequently increases gradually. Furthermore, the response is slowed for $\tau_1 = \tau_2$ and improves as the ratio τ_1/τ_2 departs from unity for a given total value $\tau_1 + \tau_2$. If there are a number of exponential stages, then the initial response will be slower and slower; the effective or apparent time constant or response at each successive stage is slower (each individual stage is unaffected), to the point where the overall response is effectively that of the slowest.

The following are some examples of transfer lags that are encountered in instruments we have discussed in previous chapters.

1. Differential-pressure instruments using the force balance principle are extremely fast and provide very small transfer lags. Other devices, mainly motion balance instruments, are not so responsive, and the lags are significant when they are used.

2. Bourdon-type instruments have minimal lags if the elements are located near the point of measurement; increasing the separation distance increases the lag.

3. Filled thermal systems and electrothermal measuring devices have lags mainly due to their installation in thermowells. However, a correctly designed thermowell, by keeping the space between the bulb and the thermowell bore as small as possible, should keep these to a minimum by aiding thermal contact and transfer.

4. Float-operated level instruments have lags introduced by the friction in the pulleys, and on large tanks these could be significant.

5. Analytical instruments, in general, have the largest lags. These lags are inherent in the technique used to obtain the measurement, which by its nature demand time to complete. Ways are continually being explored to reduce these to a minimum, or to find other ways of making the measurement that do not involve excessive time delays.

SUMMARY

1. A measurement lag is defined as the time required for the sensor to produce a signal that represents the measured parameter.

2. A distance/velocity lag is the time interval between a change in the value of a signal and its appearance at a later part of the system at its new value, arising

from the signal's finite speed of propagation via the process medium. It is affected by the design and configuration of the plant itself.

3. A transfer lag is the delay that occurs when matter or energy is transferred to a capacity through a resistance, such as to a volume via a restriction.

4. The amplitude of any type of disturbance is unchanged as a result of lags (delays).

5. If, in a process line, the distance between the point where a change to the process medium is effected and the point where the process medium is measured is l, and if the velocity of the process medium is v, then the time interval between the change and its effect being measured will be l/v (distance/velocity). If t' is the time interval, then:

$$t' = l/v$$

6. If we assume that the variations are sinusoidal and of uniform period, then the measurement will be given as:

$$\theta = \theta_{av} + A \sin 2\pi \frac{t}{\tau_0}$$

where θ = measured variable
   θ_{av} = average value of the measurement
   τ_0 = period
   A = amplitude
   t = time

7. If τ_d is the effect of the delay on the (downstream) signal, then:

$$\theta = \theta_{av} + A \sin 2\pi \frac{t - \tau_d}{\tau_0}$$

8. The phase shift will be given by subtracting the input angle from the output angle. If this angle is ϕ, then:

$$\phi = 2\pi \frac{t - \tau_d}{\tau_0} - 2\pi \frac{t}{\tau_0}$$

$$= -2\pi \frac{\tau_d}{\tau_0}$$

$$= -360° \frac{\tau_d}{\tau_0}$$

where the negative sign indicates a phase lag.

9. A single resistance associated with a single capacity gives an exponential response.

10. Exponential responses can be obtained as a result of energy or matter transfer even if no physical resistance to the transfer exists. Assumptions must be made to simplify the considerations in order to assess the characteristics.

11. If there are a number of exponential stages, then the initial response will be slower and slower; the effective or apparent time constant or response at each successive stage is slowed (each individual stage is unaffected) to the point where the response is substantially that of the slowest.

12. Examples of transfer lags encountered in instruments:

   a. Differential-pressure instruments using the force balance principle are extremely fast and provide very small transfer lags. Motion balance instruments are not so responsive; when they are used the lags are significant.

   b. Bourdon-type instruments have minimal lags if the elements are located near the point of measurement; increasing the separation distance increases the lag.

   c. Filled thermal systems and electrothermal-measuring devices have lags mainly due to their installation in thermowells. Keeping the space between the bulb and the thermowell bore as small as possible should keep these to a minimum.

   d. Float-operated level instruments have lags introduced by the friction in the pulleys, and on large tanks these could be significant.

   e. Analytical instruments, in general, have the largest lags, inherent in the techniques used to obtain the measurement. Ways are being explored to reduce the lags, as are new techniques of making the measurement.

CHAPTER 8

Control Actions

So far we have been concerned with looking at the plant itself to ascertain the conditions that prevail there, which in turn influence the control that can be effected. We have seen the part played by lags and how these can affect the system. In this chapter we shall devote our attention to the effect that a controller has on the process. We shall examine in stages the basis upon which an industrial controller performs its task. Our gradual development of the ideas will refocus on those concepts that were propounded in the Introduction to this book. We will study the scientific reasons for the techniques used. Furthermore, the concept of the individual control terms used in a three-term controller will be explained, and we shall develop the mathematics for each of the terms involved.

A PRACTICAL THREE-TERM PROCESS CONTROLLER

When we first started on the subject of control in Chapter 6, a brief very general specification was given of what the attributes of a typical controller must be—animate or inanimate. We also produced a simple block diagram of what we could expect to find in a typical industrial controller. Let us now expand that diagram and add some more detail to make it a practical instrument and one that can be found either as hardware or forming the basis of a software algorithm that can perform the function of process control. The diagram in Figure 8.1 is based on Figures 6.1 and 6.2, with some additional features and facilities included.

From Figure 8.1, we can see that it is now possible to have a setpoint or desired value that either is capable of being set, locally, that is, at the instrument itself, or is provided from another source. The setpoint could be either the output of another (remote) controller to form a cascade loop, or the computed value of a calculation performed in a computing device (also remote); the computations are sometimes compensated values of the measured variable, but usually they are some form of feed-forward arrangement, or in the simplest case the setpoint could even be the output from a manual station remotely located, perhaps on a central control panel or console. Be aware that the feed-forward arrangement requires the characteristics of the process to be converted into a simple mathematical model and, it is vital to have the mathematical model (calculation) absolutely correct, which requires a good understanding of the process; otherwise, the control loop could be doing some very unexpected things indeed, and in the worst case will render the plant uncontrollable.

An additional Auto/Manual switch has been shown, and this will allow the operator to intervene should it be necessary. The controller would be in Auto under normal circumstances once the loop had been tuned to the process, but when operator intervention is desired, then this switch will have to be operated to connect the manual station to the output; and the control output to the plant will then be adjusted manually to achieve the desired results. The problem here is that after the

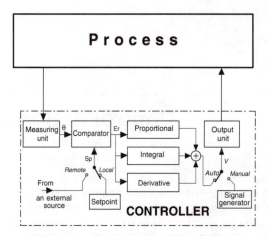

Figure 8.1: Block schematic of a three-term
controller with added functionality.

operator has intervened, the controller cannot be left in Manual, for then the loop
would *always* demand operator attention; therefore, steps have to be taken to return
the controller to Auto as soon as practicable. If the switching is done without any
thought, then certainly the process will "bump." This means that it will shift by an
amount that represents the difference between the output value before the operator
intervened and that produced after the intervention. To avoid this undesirable bump
in the process, the operator must prebalance the two differing outputs, i.e., must ensure that the output produced automatically by the control term(s) or algorithm and
the output produced manually via the manual station are of the same magnitude, before effecting the transfer of the controller back to Auto. Modern controllers perform
the balancing of the controller and manual station outputs every time a transfer from
Manual back to Auto is made, without any help from the operator, a feature referred
to as *automatic bumpless transfer.*

BASIC (SIMPLER) CONTROLLER ACTIONS

ON-OFF ACTION (THE SIMPLEST)

Let us for the moment leave this complex controller and consider the simpler
forms of controller that are adequate for many less critical situations, but we shall
return to the three-term controller later in this discussion. The simplest controller is
the On/Off device, an instrument with only two control states, On and Off. In this
respect, a common household switch is an On-Off controller. In such devices, apart
from the two states already mentioned, no other states can exist. How does such a
controller operate? Let us assume that we are examining an electrical thermostat,
say, one that is fitted on the secondary hot water cylinder or radiators in a domestic
heating system. This instrument is provided with the following:

1. A temperature sensor that has a direct mechanical link to a single-pole, single-
   throw electrical switch

2. A means of setting the trip point of the switch, to allow the user to set the desired water temperature—in other words, the means to adjust the setpoint of the
   controller

The instrument operates as follows. The temperature sensor monitors the water temperature, and for as long as the temperature is below the desired value (setpoint) allows power to be applied to the heating element, and the water temperature rises. The sensor, upon "seeing" the change, when the water temperature goes just above the desired value, trips the switch to cut off the power. When the temperature of the water falls just below the desired value, the switch allows power to be applied once more.

This switching of the power is the means of controlling the water temperature. The temperature interval between the "just above" and the "just below" values is called the *deadband*; nothing can happen within this band. In electrical parlance the deadband is called the *switching differential*. From this it is clear that no control can be initiated until the measurement crosses the setpoint. The result of this application and disconnection of power causes the temperature in the water cylinder to fluctuate about the desired value, the amount of fluctuation being a function of the switching differential of the switch and the deadband of the sensor. The process will therefore cycle continuously; and, depending on the total magnitude of the fluctuation, the average temperature will vary accordingly.

MULTISTEP ACTIONS

If the water temperature had to be held more closely, the On-Off control system would not be acceptable. To achieve this, it will be necessary to change the mode of control, and in this case we shall attempt to keep the method simple and relatively inexpensive by using what is generally known as *two-step with overlap*. Figure 8.2a shows graphically the simple On-Off control action, and Figure 8.2b illustrates the two-step with overlap. In Figure 8.2b, the band within which the control is effected is 4° (i.e., ±2° of desired value), and this spread is not unreasonable to expect from such a control system.

To further improve the control, using a valve with discrete positions provided by detents on the operating mechanism, the two-position valve used so far will have to

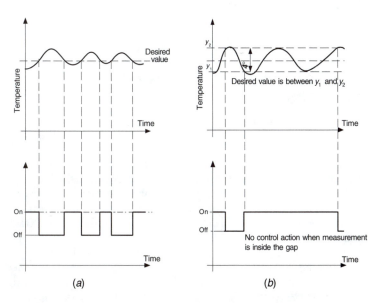

Figure 8.2: (a) On/Off control action. (b) Two-step with overlap.

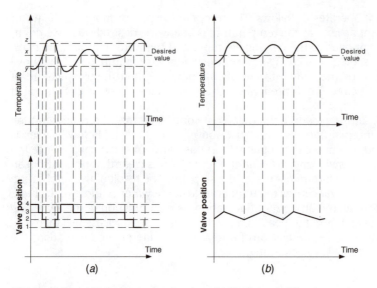

Figure 8.3: (a) Multistep control action. (b) Single-speed floating action.

be replaced by one that has more set operating positions, say, 4, as shown in Figure 8.3a, which will allow a control action called *multistep* to be used. The results are also shown on the diagram, where x is the desired value.

FLOATING ACTIONS

Alternatively, changing the control valve for a fully variable device will enable *floating action* to be used and thereby improve the quality of control even more. Figure 8.3b actually shows a single-speed floating action, but if the speed of action was to be varied to suit the speed of the process changes, we would obtain what is termed *multispeed floating action*.

MATHEMATICAL FUNCTIONS USED IN AUTOMATIC PROCESS CONTROLLERS

Figure 8.4 is once again the same three-term controller we discussed earlier, but now the mathematical functions that each term computes have been added—they are defined and derived later in this chapter.

Let us take a more detailed look at the controller starting at the input or measurement, i.e., the signal coming from the sensor or transmitter, representing the variable being measured, usually in terms of either a continuously varying electrical current or a train of pulses. In some instances the measurement could even be a pneumatic signal, changed via a converter to an electrical signal. The input signal will have to be converted into a standard form that can be manipulated by the controller and is assigned definite values (in engineering units). This exercise is usually called *scaling* and is especially important when analog instrumentation is being used, for the transmitted signal of the measuring instrument(s) must be made to correspond to the process variables they represent and to each other. For example, suppose we have to sum two

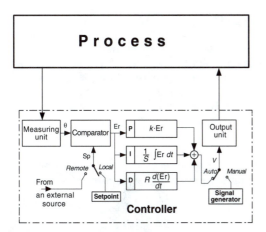

Figure 8.4: Block diagram of a three-term controller with the math functions of the control terms shown.

flows each of a range from 0 to 100 units; then if maximum values of flow are being delivered, the total flow will be 200 units. Now, suppose further that each transmitter produces a signal in the range, say, 0 to x units, therefore each flow of 0 to 100 units will be represented by its transmitter signal range of 0 to x units. It then follows that the sum of the two transmitter outputs will be $2x$ units. But, the analog unit cannot handle $2x$ units. It is therefore necessary at all times to present internally to the actual summing component, of the summing device, signals for each flow that have values proportional to half the respective transmitter signal range; so that when these are added together a signal from the summing unit in the range 0 to x units is produced. This signal will then represent at any instant a true value of totalized flow that is within the maximum overall flow range of 200 units. To achieve the scaling it must be possible to adjust the gains and bias settings of the summing unit input channel amplifiers, for it is through the manipulation of these two adjustments that correspondence is brought about between signal and process variable. In the example given, halving the signal is achieved by adjusting the gain only, because the measurement parameter ranges are zero-based, i.e., 0 to 100 units. If on the other hand, the parameters were not zero-based, say, for example, 10 to 100 units, then the bias would have to be adjusted also in order to shift the starting point of the measurement parameter range upward. With an analog controller the working signal range is matched or equated to the process variable signal range; in digital controllers the worry is taken care of by the electronics, which accomplish this automatically, for the pulse train can represent discrete numbers, and manipulation becomes an exercise in binary arithmetic and of no concern to the process operator or personnel involved.

The conditioned (scaled) signal is then applied to the *comparator*, which is really a subtracting unit that takes the difference between the setpoint and the measured variable producing the *error* signal. Note that it is the setpoint − measured variable which means that

- A measurement *above* the setpoint gives a positive error, which requires a decreasing (i.e., negative-going) output.

- A measurement *below* the setpoint gives a negative error and requires an increasing (i.e., positive-going) output.

This is the convention we will follow for assigning the correct algebraic sign to the output, but others may prefer it the other way round. There is no standardization in the terminology used for the measurement:output direction; some refer to what we have just described as *direct action* and others refer to it as *reverse action*. It is safest to spell the requirement out, as done here. At any rate, the result of the comparison (difference) is called the *error* and in the classic controller is the signal on which the control terms operate. (The classic controller is also called an *interactive controller*, as explained in more detail in Chapter 10.)

The error signal is applied to the *control unit*, which produces the appropriate controlled output to reduce the error—in other words to bring the measurement and setpoint values back to coincidence. To achieve the necessary control output, the control unit uses the following control actions, which are basically mathematical:

a. Proportional

b. Integral

c. Derivative

An automatic controller can employ all three of these together or any of the combinations (a), (b), (a + b), or (a + c). Derivative action is never applied on its own for reasons that will be explained later; and integral action on its own is not very common. When integral action (b), or (a + b) is used, the application is one that has a measurement prone to relatively slow continuous change, and since integral action produces an output for as long as there is a difference between setpoint and measurement, the controlled output will adjust the controlled device to respond accordingly.

Let us now concentrate on each of the individual control terms to see what they accomplish and how they do it.

PROPORTIONAL CONTROL ACTION

Proportional action produces an output that is proportional to the error and can be expressed as:

$$V \propto -\text{Er}$$

where V is the output and Er is the deviation (error); or

$$V = -k_1 \text{Er}$$

It is assumed that for a small range of variations about a mean working value, the component units of a system behave in a linear manner. Hence, under steady-state conditions, each position M of the correcting element corresponds to a discrete value of the controlled condition applied, which is given by:

$$M = KV$$

$$= -Kk_1 \text{Er}$$

where M is the correction being applied—e.g., by opening and closing of a control valve—K is a constant for the plant and the correcting unit motor, and k_1 is the proportional action factor of the controller.

The proportional relationship between the output and error implies that, for each value of Er, the controller will produce a correction M as long as conditions that give rise to an error remain. If another disturbance occurs to cause a deviation, the controller will apply a correction of $-Kk_1$ Er dependent on the magnitude of Er.

The effect of the disturbance will be reduced by M, and once the signals have been restored to their original values, Er and M will be zero. If k_1 is not too large, then restoration will be through a series of damped oscillations. However, if the disturbance is of the nature of a permanent change, the desired value will not be attained, but a new equilibrium value will be achieved. This new value is called the *control point*, with the difference in either direction between the desired value and the control point termed the *offset*. Why does offset occur? The reason is as follows. A correction that is proportional to the deviation is applied, so there is no correction if there is no deviation. If there is a sustained change, there will be sustained correction, and therefore, there will come a time when equilibrium will be established with the control point offset from the desired value.

The value of the offset can be found by considering the value of the deviation, which would have occurred without control. Let this value be Er_p; then $-Er_p$ would have to be the correction applied by the controller to correct completely. If the controller reduces the actual deviation to the value of the offset, which is, say, Er_x then the correction applied is

$$M = -(Er_p - Er_x)$$

But

$$M = (Kk_1 Er_x)$$

Therefore

$$(Kk_1 Er_x) = Er_p - Er_x$$

and

$$Er_x = \frac{Er_p}{Kk_1 + 1}$$

From this it will be seen that as Kk_1 increases, the offset decreases for a given value of Er_p; but since K is dependent on the plant, it will also affect the value of k_1 that can be used. In practice the values of Kk_1 are very small, so that significant offset, dependent on the value of Er_p, for load changes can be anticipated, and therefore some other means have to be found to reduce the offset. This is achieved by the use of the next type of control action.

INTEGRAL CONTROL ACTION

Integral action produces a signal component in the controller output that is proportional not only to the value of the deviation but also to the integral over the time it persists, or, expressed mathematically,

$$V \propto - \int_0^t Er\, dt$$

From this equation it can be seen that, as long as any deviation exists, the integral action signal will continue to increase, and as a result the deviation due to any load change must eventually be reduced to zero. If k_2 (constant of proportionality) is the integral action factor of the controller, then:

$$V = -k_2 \int_0^t Er\, dt$$

If the input error signal Er to a proportional + integral controller remains constant for a time, the contribution of the proportional action to the output V will remain constant at k_1 Er, but the contribution of the integral action will increase from zero to

$$-k_2 \int_0^t \text{Er}\, dt$$

At some time t, the contributions of the two actions will become equal, and which condition defines the integral action time when:

$$-k_1 \text{Er} = -k_2 \int_0^t \text{Er}\, dt$$

The integral action factor of a practical controller is the integral action time and is conventionally stated in repeats per minute, and usually denoted by the letter S. Since Er is held constant, we have:

$$-k_1 \text{Er} = -k_2 \text{Er}\, t \qquad \text{from which} \qquad t = \frac{k_1}{k_2} = S$$

DERIVATIVE CONTROL ACTION

Derivative action provides a signal to the controller output proportional to the rate at which the deviation changes. If the error input signal Er increases from zero at a constant rate after equilibrium, the derivative action will immediately provide a contribution to the output that will be:

$$-k_3 \frac{d\,\text{Er}}{dt}$$

Since derivative action is never used on its own, we combine it in a proportional + derivative controller, the proportional action will increase continually as Er increases, and at some time t the contribution of the two actions will be equal or:

$$-k_1 \text{Er}\, t = -k_3 \frac{d\,\text{Er}}{dt}$$

from which $t = k_3/k_1 = R$, the derivative action time, in a practical controller.

COMBINING THE BASIC CONTROLLER EQUATIONS

From the foregoing we can say the following:

For a proportional-only controller:

$$V = -k_1 \text{Er}$$

For a proportional + integral controller:

$$V = -k_1 \left(\text{Er} + \frac{1}{S} \int \text{Er}\, dt \right)$$

For a proportional + derivative controller:

$$V = -k_1 \left(\text{Er} + R \frac{d\,\text{Er}}{dt} \right)$$

For a proportional + integral + derivative controller:

$$V = -k_1 \left(\mathrm{Er} + \frac{1}{S} \int \mathrm{Er}\, dt + R \frac{d\mathrm{Er}}{dt} \right)$$

This last equation is the summation of all three control actions.

PROPORTIONAL BAND OF A CONTROLLER

Frequently we hear the phrase "the proportional band of the controller is . . . ." What is this proportional band? The proportional band of a controller is the percentage of the controller measurement signal range that will drive the controller output through its full operating range. Figure 8.5 explains graphically what is meant.

Changing the proportional band of the controller has the effect of changing the slope of the output/measurement curve and therefore at the same time changes the amount the control device (valve or drive) will move for the full range of measurement change. The larger the proportional band, the smaller the amount of change in the position of the control device. Alternatively, we can say that, if the proportional band is zero, we will get On/Off control action.

For those electronically minded, the proportional band is the inverse of gain:

$$\text{Proportional band} \propto \frac{1}{\text{gain}} \qquad \text{or} \qquad \text{gain} = \frac{100}{\text{proportional band}}$$

$$\text{or} \qquad \text{proportional band} = \frac{100}{\text{gain}}$$

Therefore, if the proportional band is zero, the gain will be infinity; i.e., control action will be On/Off.

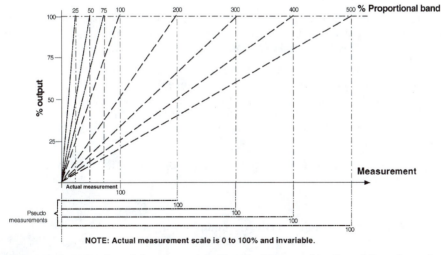

Figure 8.5: Definition of the proportional band of a controller. Sloped lines from the origin denote different proportional bands.

SUMMARY

1. A setpoint may be set either locally, i.e., at the instrument itself, or provided from another source, e.g., remote Auto/Manual station, or it could be from another controller to form a cascade loop, or from a calculation in which case it could be some form of feed-forward arrangement. In the feed-forward arrangement it is vital to have the model (calculation) absolutely correct; otherwise the control loop could perform strangely or the plant could become uncontrollable.

2. Under normal circumstances the controller would be in Auto once the loop has been tuned to the process. When operator intervention is desired, the controller will have to be switched to Manual to connect the manual station to the output and adjusted manually to drive the measurement to the desired value. It is not usual to leave the controller in Manual, but steps have to be taken when returning it to Auto to avoid bumping, that is, shifting by the amount of the difference between the output value before intervention and that after the intervention.

3. The simplest controller is the On-Off device, which has only the two control states On and Off. A switch is an On-Off controller, for it can exist in no other state apart from On or Off.

4. On-Off action would cause the measurement to fluctuate about the desired value, the amount of fluctuation being a function of the switching differential of the switch itself plus the deadband of the sensor. The process would therefore cycle continuously, and the average value would depend on the total magnitude of the fluctuation. The average value would tend to stay constant provided that no oddities or nonlinearities existed.

5. An On-Off control system would not be acceptable if the measurement had to be held more closely. To achieve this it will be necessary to change the method of control. A simple and relatively inexpensive method is a control action generally known as two-step with overlap.

6. To improve on On-Off control in flow loops, a valve that has more operating positions replaces the valve with only two discrete positions, and multistep control action is implemented.

7. A fully variable control valve will enable floating action to be used and thereby improve the quality of control even more. Multispeed floating action can be obtained by varying the speed of action to suit the speed of the process changes.

8. Requiring an operator to care for a process is a limitation. An automatic controller can look after the process continuously. To achieve the necessary control output, the control unit uses the following control actions: (a) proportional, (b) integral, and (c) derivative. An automatic controller can employ all three of these together or any of the combinations: (a), (b), (a + b), or (a + c). Derivative action is never applied on its own.

9. Proportional action produces an output that is proportional to the error and can be expressed as:

$$V \propto -\mathrm{Er}$$

where V is the output and Er is the deviation (error); or

$$V = -k_1 \mathrm{Er}$$

It is assumed that for a small range of variations about a mean working value, the component units of a system behave in a linear manner. Hence, under steady-state conditions, each position M of the correcting element corresponds to a discrete value of the controlled condition applied, which is given by:

$$M = KV$$

$$= -Kk_1 \text{Er}$$

where M is the correction being applied—e.g., by opening or closing of a control valve—K is a constant for the plant and the correcting unit motor, and k_1 is the proportional action factor of the controller. The proportional relationship between the output and error implies that for each value of Er the controller will produce a correction M as long as conditions that give rise to an error remain. If another disturbance occurs to cause a deviation, the controller will apply a correction of $-Kk_1 \text{Er}$ dependent on the magnitude of Er.

10. With proportional action there will always be offset, because if there is a sustained change, there will be sustained correction, and therefore there will come a time when equilibrium will be established with the control point offset from the desired value. The value of the offset can be found by considering the value of the deviation that would have occurred without control. Let this value be Er_p; then $-\text{Er}_p$ would have to be the correction applied by the controller to correct completely. If the controller reduces the actual deviation to the value of the offset, say, Er_x, then the correction applied is:

$$M = -(\text{Er}_p - \text{Er}_x)$$

But

$$M = Kk_1 \text{Er}_x$$

Therefore

$$Kk_1 \text{Er}_x = \text{Er}_p - \text{Er}_x$$

and

$$\text{Er}_x = \frac{\text{Er}_p}{Kk_1 + 1}$$

11. Integral action produces a signal to the controller output that is proportional to the value of the deviation and to the integral over the time it persists, or:

$$V \propto - \int_0^t \text{Er} \, dt$$

As long as any deviation exists, the integral action signal will continue to increase, and as a result the deviation due to a given load change must eventually be reduced to zero. If k_2 is the integral action factor of the controller, then:

$$V = -k_2 \int_0^t \text{Er} \, dt$$

12. The integral action factor of a practical controller is the integral action time and is usually stated in repeats per minute and is usually denoted by S. If the input error signal Er to a proportional + integral controller is maintained constant for a time, the contribution of the proportional action to the output V will remain

constant at k_1 Er, but the contribution of the integral action will increase from zero to

$$-k_2 \int_0^t \text{Er}\, dt$$

At some time the contributions of the two actions will become equal, or

$$-k_1 \text{Er} = -k_2 \int_0^t \text{Er}\, dt$$

Since Er is held constant, we have:

$$-k_1 \text{Er} = -k_2 \text{Er}\, t \qquad \text{from which} \qquad t = \frac{k_1}{k_2} = S$$

13. Derivative action provides a signal to the controller output at the rate at which the deviation changes. If the error input signal Er increases from zero at a constant rate after equilibrium, the derivative action will immediately provide a steady contribution to the output that will be:

$$-k_3 \frac{d\,\text{Er}}{dt}$$

In a proportional + derivative controller the proportional action will increase continually as Er increases, and at some time the contribution of the two actions will be equal or:

$$-k_1 \text{Er}\, t = -k_3 \frac{d\,\text{Er}}{dt}$$

from which $t = k_3/k_1 = R$, the derivative action time, in a practical controller.

14. Considering each of the permissible combinations of control terms we can say:

For a proportional-only controller:

$$V = -k_1 \text{Er}$$

For a proportional + integral controller:

$$V = -k_1 \left(\text{Er} + \frac{1}{S} \int \text{Er}\, dt \right)$$

For a proportional + derivative controller:

$$V = -k_1 \left(\text{Er} + R \frac{d\,\text{Er}}{dt} \right)$$

For a proportional + integral + derivative controller:

$$V = -k_1 \left(\text{Er} + \frac{1}{S} \int \text{Er}\, dt + R \frac{d\,\text{Er}}{dt} \right)$$

15. The proportional band of a controller is defined as the percentage of the controller measurement signal range that will drive the controller output from zero to 100 percent. The larger the proportional band, the smaller the amount of change in the position of the control device. Some may prefer to define the proportional

band as the inverse of gain:

$$\text{Proportional band} \propto \frac{1}{\text{gain}} \quad \text{or} \quad \text{gain} = \frac{100}{\text{proportional band}}$$

$$\text{or} \quad \text{proportional band} = \frac{100}{\text{gain}}$$

Therefore, if the proportional band is zero, the gain would be infinity—i.e., control action would be On/Off.

CHAPTER 9

Control Equations

In this chapter we shall focus our attention on the way an industrial controller operates to see how the mathematics of the control actions, developed in the previous chapter, are put to practical use. We shall also show, by relatively simple mathematics, some of the relationships between the control actions and the process to which they have been applied, the objective being to understand what is occurring in the process stream under the action of the controller. This will also allow us introduce the formal techniques for putting a controller into service on a process stream and to be confident that the instrument will provide the necessary adjustments for the process to return the measurement back to the desired value whenever there is a divergence between the two. Such divergences occur whenever a steady-state condition of the process is altered in some way by the demand or supply, this change in operating conditions being called a *process upset*.

We shall also explain the well-known phenomenon of an ever-increasing controller output observed when a controller is either deliberately or inadvertently disconnected from the controlled device. This phenomenon is called *integral saturation* or *reset wind-up*.

PROCESSING A SIGNAL WITHIN A CONTROLLER

We have so far discussed in basic theoretical terms the conditions that the plant presents to the control instrument and have also looked at the components forming the build up of the controller itself from a theoretical aspect. Furthermore, we have developed mathematical equations that describe the control actions necessary to bring the process to the desired condition (setpoint) whenever the process state departs or diverges from that which is required. A departure of the process from the setpoint is manifested in a divergence of the measurement from the setpoint.

The block diagrams of the controller as explained in earlier chapters showed the instrument in its most fundamental form, so let us now proceed to put a little more flesh on the bones and explain some of the components in more detail.

AT THE INPUT

We shall start at the input, or measuring unit, and work our way through. This unit is designed in two forms, one that is capable of accepting the signal from the *primary element* directly, and the other, via a transmitter that has a primary element connected directly to it to form a composite unit, or via a transmitter, mounted separately from the primary element, which produces a proportional signal that is connected to the controller input unit. Examples of primary elements that can be directly connected to the controller input unit are thermocouples and resistance bulbs. Examples of measuring devices that are not capable of direct connection, but require a transmitting mechanism in order to provide a compatible measurement signal, are orifice plates, pH electrodes, and magnetic flowmeters, but with each of these a separately mounted transmitter is a necessity, and those with integral transmitting mechanisms include Coriolis mass flowmeters, vortex flowmeters, target flowmeters, and several others. The incoming signal is conditioned (scaled) and converted to a form that can easily be manipulated by the various circuits and components within the instrument itself. The conditioned incoming signal is passed on to the comparator and could also be, and often is, applied to a pointer-type analog indicator or made to drive a digital display, both of which are graduated or scaled in engineering units to the range of the measurement. (See chapter 8 for a description of what scaling entails.)

AT THE COMPARATOR

The comparator receives the scaled signal and makes a comparison between it and the desired value. As we have said before, the comparator is, in essence, a subtracting unit, where the difference between the setpoint and measurement signal is determined, the difference being the error or deviation. In pneumatic instruments the desired value (setpoint) is usually in the form of a spring-operated diaphragm, manually adjusted to alter the pressure of an air supply, the modified pressure being indicated by a pointer. It is usual for the setpoint and the measurement pointers to be opposite each other for convenience of comparison. Manual setpoint adjustment is carried out at the instrument and is therefore local to it. Replacing the spring-operated diaphragm with a bellows unit connected to a pneumatic signal generated external to the instrument allows the setpoint to be adjusted by the remote signal. All local/remote setpoint controllers have a changeover switch that allows either local or remote adjustment of the setpoint. There are also controllers without a changeover switch that permit either a local or a remote setpoint only.

In electronic instruments (we shall start with the old and move on to the new further along in this section), the setpoint can "symbolically" be thought of as developed across a potentiometer that forms part of one arm of a Wheatstone bridge, the measurement and its associated resistance forming the other arm. Any imbalance of the bridge will then be proportional to the difference between setpoint and measurement and is thus the error or deviation. In the less recent electronic instruments, both the setpoint and measurement drive their own moving-coil meters to provide the readouts for the operator, but it is not difficult to envisage DVMs (digital volt meters) being used increasingly in place of the moving-coil instruments; in which case the readout will be digital, i.e., alphanumeric. The local/remote switch will be an electronic unit that performs similarly to the pneumatic instrument. In most modern electronic controllers and distributed control systems (DCSs), which we shall look at in Chapter 19, the generation of the error signal is not via a Wheatstone bridge. In the case of "single-box" electronic control instruments it could be via operational amplifiers, and in the case of DCSs by software algorithms. Selection of local or remote setpoint, Auto or Manual, etc., with this modern equipment can be made by external switching logic. This facility increases the versatility of the instrument when applied to advanced control schemes.

AT THE CONTROL UNIT

The error signal thus generated is passed on to the control unit. In pneumatic instruments this will consist of a number of bellows operating on a common beam, which achieves balance when the measurement is equal to the setpoint. Subunits in the form of needle valves and adjustable calibrated springs allow the control terms to be set. In electronic controllers the control unit comprises a set of circuits dedicated to performing the control functions, the circuits being provided with adjustments to facilitate the setting of the control terms. In the case of the programmable single-box controllers that are in increasingly common use today, all the electronics is contained in an application-specific integrated circuit (ASIC) and all adjustments are made by pushbuttons located on the faceplate, the complete instrument being software-driven by conveniently configurable algorithms.

AT THE OUTPUT

The output unit is provided with a means of changing the signals used within the confines of the controller itself into units that can be transmitted to and used by the receiving instrument or final control device such as a pneumatic valve, a current to pneumatic (I/P) converter driving an actuator, or an actuator directly. It is additionally provided with the means of setting any necessary bias. This output signal is normally a small electrical current that is sufficient to actuate a single control device with a specific load (electrical resistance), and, because of this, care must be exercised to ensure that the load connected to the controlled output is within the driving capability of the controller output current signal; otherwise, some form of current amplifier will be required to increase the output load capability.

In all electronic control systems using current signals, the loads must be connected in series on both the input and the output sides of the controller. This being the case, the resistances of the loads are additive, and care must therefore be exercised to ensure that the load capabilities of the output unit are not exceeded. *Parallel connection* of the loads using a current signal is possible, but not recommended at all, because predictable current sharing is not possible, and there are too many other problems associated with this method.

ILLUSTRATIONS OF TYPICAL PROCESS CONTROLLER ACTIONS

PROPORTIONAL ACTION

We shall now consider a simple process consisting of a distance velocity lag (dead time) and select a suitable controller to maintain the process conditions required. In view of its simplicity, let us look initially at a single-term proportional controller. As mentioned earlier, the output V from such an instrument is related to its input in the following way:

$$V = -k_1 \mathrm{Er}$$

where Er is the error or deviation from setpoint. We can rewrite this equation in terms of the proportional band and the output bias:

$$V = \frac{100}{P} \mathrm{Er} + b$$

where P = percentage proportional band
  Er = error, i.e. deviation (setpoint − measurement)
  b = output bias

Note: In the above and all the equations henceforth in this chapter, the 100 in respect of the proportional band is 100%.

From this equation it is clear that as P approaches zero, the gain approaches infinity, and as P approaches 100, the gain approaches unity. Furthermore, when there is no error, the output of the controller is equal to the bias. Remember, the percent proportional band of a controller is equal to the inverse of the gain multiplied by 100. As there are no dynamic elements in the controller, and the process is one of dead time, the entire phase shift of $180°$, $\phi = -180° = -\pi$, will take place in the plant itself—as shown in Chapter 7, when we saw a sine wave measurement delayed through a dead time process—and will determine the natural period. Substituting this into the equation

$$\phi = -2\pi \frac{\tau_d}{\tau_0}$$

as derived in Chapter 7, when we considered the effects a distance velocity lag (a dead time process) had on the measurement. We have

$$-\pi = -2\pi \frac{\tau_d}{\tau_0}$$

and solving for the periodic time τ_0, we have:

$$\tau_0 = 2\tau_d \tag{9.1}$$

where τ_d is the effect of the distance velocity lag (dead time). It is clear that under proportional-only control the process will cycle with a period of twice the dead time. If we set the proportional band to 100%, the process will oscillate undiminished and the loop gain will be unity, since the lag does not contribute anything to the gain.

To dampen the oscillations, we shall have to increase the proportional band, so if we now set the proportional band to 200%, the loop gain will be 0.5, and the effect will be a reduction by one-half in amplitude of each successive half cycle, and this is called *quarter-amplitude damping.* It is important to remember that on proportional controllers there is only one adjustment—the proportional band—and this also affects the damping.

As stated in Chapter 8, all proportional-only controllers will have an offset Er_x whenever there is a load change. The amount of the offset from the desired value can be written in a more general way than was shown in Chapter 8 to account for bias:

$$Er_x = \frac{P(V - b)}{100}$$

where P = percent proportional band
   V = controller output
   b = output bias

Let us now consider the proportional control algebraically, and let the controller output at any instant V_n be equal to the measurement after a dead time period; let this be denoted by τ_d.

Then

$$V_n = \theta_{n-1} \quad \text{where} \quad n = t/\tau_d$$

This represents a process with unity gain and a lag (dead time) of τ_d. Let us now connect a controller to close the loop, when this is done the result is

$$\theta_n = \frac{100}{P}(Sp - V_n)$$

where   Sp = setpoint
        θ = measured variable
        V = controlled output

At the next period, the following will prevail:

$$\theta_{n+1} = \frac{100}{P}(Sp - V_{n+1}) = \frac{100}{P}(Sp - \theta_n)$$

Let us suppose the starting conditions are: Sp = 0, V = 0, b = 0, and P = 200%; or, writing values in terms of percentage at the initial conditions, Sp = 0%, V_0 = 0%, θ_0 = 0%.

Let us now make a setpoint change of 50%; then:

Setpoint	Output	Measurement
$Sp_1 = 50\%$	$V_1 = 0\%$	$\theta_1 = 0.5(50 - 0) = 25$ = output V_2 one dead time later
	$V_2 = 25\%$	$\theta_2 = 0.5(50 - 25) = 12.5$ = output V_3 one dead time later
	$V_3 = 12.5\%$	$\theta_3 = 0.5(50 - 12.5) = 18.75$ = output V_4 one dead time later
	$V_4 = 18.75\%$	$\theta_4 = 0.5(50 - 18.75) = 15.625$ = output V_5 one dead time later
	$V_5 = 15.625\%$	
	$V_\infty = 16.667\%$	$\theta_\infty = 16.667$

The controller comes to rest at 16.667% above the bias. The offset is (Sp − V_∞) or (50 − 16.667), which is 33.333%.

From the preceding it will be seen that the output is a series of damped oscillations with a period of twice the dead time as calculated in Equation (9.1) and the amplitude of each peak is reduced by a quarter, also as stated, and the offset is equal to:

$$\frac{P}{100}(\theta - \text{bias}) = \frac{200}{100}(16.667 - 0) = 2(16.667) = 33.33\%$$

Proportional action is never used in applications where the proportional band is to be no wider than a few percent; a different type of action will be required in these cases. This control action is *integral* action.

INTEGRAL ACTION

Let us see what happens when we use the following equation for integral action in terms of integral action time:

$$V = \frac{1}{S}\int \text{Er} \, dt$$

where V = output
        S = integral action time
        Er = deviation (error)

With this type of action the output will change as long as there is an error, and the rate of change of output is proportional to the deviation, or:

$$\frac{dV}{dt} = \frac{Er}{S}$$

Before we use this controller in a closed loop, we should define its gain and phase characteristics, for we are interested in the behavior at the natural period τ_0 of the loop. If, therefore, we introduce a sinusoidal input to the instrument as we considered earlier, in other words:

$$Er = A \sin\left(\frac{2\pi t}{\tau_0}\right)$$

the output will then be:

$$V = \frac{1}{S} \int\left(A \sin\frac{2\pi t}{\tau_0}\right) dt = \frac{A\tau_0}{2\pi S}\left(-\cos\frac{2\pi t}{\tau_0}\right) + V_0$$

because

$$\int \sin\theta \, d\theta = -\cos\theta$$

where V_0 is the output at zero time. To evaluate the gain and phase, we must reduce the output to the same form as the input. Since we can write the cosine in terms of the sine, we can say:

$$-\cos x = \sin\left(-\frac{\pi}{2} + x\right)$$

and rewrite the output equation as:

$$V = \frac{A\tau_0}{2\pi S} \sin\left(-\frac{\pi}{2} + \frac{2\pi t}{\tau_0}\right) + V_0$$

The phase shift is given by the output angle minus the input angle, or:

$$\phi = \left(-\frac{\pi}{2} + \frac{2\pi t}{\tau_0}\right) - \frac{2\pi t}{\tau_0}$$

from which we see that the phase angle is $-\pi/2$ or $-90°$. An integrator will exhibit this angular phase lag regardless of the period.

The gain is given by the amplitude of the output divided by the amplitude of the input:

$$G = \frac{A\tau_0/2\pi S}{A} = \frac{\tau_0}{2\pi S}$$

When an integral-only controller is put into a loop that is then closed, the sum of the phase shift of the controller and the process must be equal to $-\pi$ at the natural period τ_0, or:

$$-\pi = -\frac{\pi}{2} - \frac{2\pi \tau_d}{\tau_0} \tag{9.2}$$

The second term of Equation (9.2) is phase shift of the measurement through a dead time as derived in Chapter 7. Or, in terms of degrees,

$$-180° = -90° - 360°\frac{\tau_d}{\tau_0}$$

Solving for τ_0, we have

$$\tau_0 = 4\tau_d$$

Notice carefully that for this controller the period is twice that for proportional control, as only 90° of the phase shift takes place in the process (the dead time), the integrator taking care of the remainder.

To sustain oscillation, the loop gain must be 1.0, but since the process gain is 1.0, the integrator gain for this condition must also be 1.0, or controller gain must be

$$G = \frac{\tau_0}{2\pi S} = 1.0$$

and solving for S we have:

$$S = \frac{\tau_0}{2\pi} = 2\frac{\tau_d}{\pi}$$

To summarize, a process lag of 1.0 min would cycle with a period of 4.0 min, sustained by a reset time of $2/\pi$ or approximately 0.65 min. Hence, quarter-amplitude damping can be obtained by halving the gain or, in other words, *doubling* the reset time, i.e., taking $2S$. An integral controller has only one adjustment, which affects the damping. It does, however, avoid offset but reduces the speed of response.

PROPORTIONAL PLUS INTEGRAL ACTION

Let us now combine the two types of controllers we have just been talking about and discuss the two-term P + I controller that is so common in most process plants. The two-term P + I controller combines the features of both the control actions we have been discussing so far. As you will recall from Chapter 7, the controller is represented by the equation below:

$$V = \frac{100}{P}\left(\text{Er} + \frac{1}{S}\int \text{Er}\,dt\right)$$

We have already determined the performance of each control action separately, and we can instinctively expect the performance of the combination, depending on the settings of the proportional and integral terms, to be somewhere in the following range:

$$4\tau_d > \tau_0 > 2\tau_d$$

where $4\tau_d$ is for the integral and $2\tau_d$ is for the proportional terms.

An infinite number of settings can be found to provide constant damping, the only requirement being that the loop gain must be achieved. We have seen that in order to have quarter-amplitude damping the following must be obtained:

$$\frac{100}{P} = 0.5 \qquad \text{proportional-only controller}$$

$$\frac{\tau_0}{2\pi S} = 0.5 \qquad \text{integral-only controller}$$

Therefore, to obtain quarter-amplitude damping with the two-term instrument, the sum of the gains must also equal 0.5, but since the proportional and integral gains are

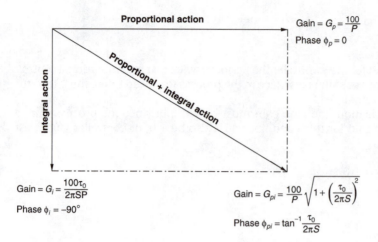

Figure 9.1: Proportional and integral gains.

not in phase, the required gain of the P + I controller will be the vector sum of the gains of the two control components, i.e., the *resultant* shown in Figure 9.1. In this figure, because there are a number of similar terms, subscripts are added to differentiate them: i.e., p indicates proportional, i indicates integral, and pi indicates proportional + integral.

This controller has two adjustments, proportional band and integral action time, both of which affect the stability of the control loop. Making the integral time infinite reduces the effect of the integral action almost completely, but never gets rid of it, and will again result in offset, as in the proportional-only case. On the other hand, by reducing the integral time, we are forced to set the proportional band very high and make the controller behave as a pure integrator. An acceptable combination would be to have a proportional band that will give good damping and an integral time to give good recovery.

PROPORTIONAL PLUS INTEGRAL PLUS DERIVATIVE ACTION

The addition of derivative action to the controller produces the three-term instrument with which we are all familiar. The most important thing to remember here is that *derivative action on its own is not a controlling action*. To benefit from this, the action must be combined with others, usually proportional, or proportional + integral. In the past, derivative action was made to act on the controller output in the form of a lag in the feedback path of the controller amplifier. By doing this, the effect was only available when there was a change in output, but the measurement could have changed within the time that it took for its effect to be felt at the output. This clearly was not acceptable in the light of experience; hence, most if not all modern controllers have derivative action take effect with the measurement. On fast-changing processes, this action is very important, since it will produce a control action as soon as a change in the measurement is sensed, and moreover this action increases with faster transient changes of the measured variable. However, do not use a three-term controller on a flow control loop.

A cautionary note: Do not as a matter of routine specify a three-term instrument for use on temperature control loops. Despite the popular belief that all temperature control loops are fast, try and analyze the system requirement first, as sometimes a less sophisticated instrument could give the required results and therefore be less expensive.

INTEGRAL SATURATION OR WINDUP

The last item to study in this chapter is *integral saturation* or *reset windup*, as it is sometimes called. We know that integral action will produce an output for as long as there is a difference between the measurement and the setpoint, or in other words, for as long as there is an error signal. The effect of this would eventually be to drive the output off scale, i.e., to either limit, a condition that is encountered any time the control loop is "opened," i.e., on plant shutdown, or when there is a very rapid and massive change in the measured variable, or when the controller is transferred to manual control. What really happens is that when the control loop is opened or the change in measurement is so massive, the integral action, because it is trying to raise the measurement upward toward setpoint, will force the proportional band up (increase)—i.e., the gain down—to try and restore the measurement and setpoint balance, but with no response from the loop. The integral action will continue to integrate the error. Eventually, because the output is going nowhere as in the case of a true open circuit, or because there is no further response from the valve, which has been driven to its limit, or because the controlled output is deliberately cut off, as in manual operation, the controller reaches a point when even 100% output has no effect in reducing the error to zero. The overall effect will eventually be that the output, if it is increasing, will be driven to the full value of its power supply (20 lb/in^2 in the case of a pneumatic instrument, or whatever the output circuit or unit operating voltage is for an electronic instrument) and will lock up at this value. The process is now out of control. If, in the case of the open circuit, the output was then to be suddenly restored, the controller would apply the full output to the controlling device; this action, however, will not restore immediate control of the process. In the case of the massive measurement change, suppose there is a gradual reduction in the amount of change after the peak value has been reached. The controller, having already reached a value greater than the 100% output, will now have to commence integrating back downward toward the 100% value. During this time, there is still no response from the control valve and there is therefore no control of the process. Transfer to manual will give results as described for the open circuit, but the severity will depend on the magnitude of the manual change made. The effect of the control action will only begin to be seen when the controller output comes below 100% of its output value. This return could be a long time, for it will only be at the reset (integral) rate.

To overcome this serious problem, it is necessary to stop the integral action from continuing to drive the output under its own dictate, but yet let the integrator assume the value of the adjusted controlled output. The latter has important implications, especially when the controller is in manual, for then the output, e.g., to the valve, can be driven or placed anywhere within the output signal range dependent on where the operator chooses to set it.

Almost all automatic controllers are provided with means of transferring the output back from Manual to Auto without causing the process to bump, and this is achieved by balancing and locking the signal produced by the control terms against the manually generated signal so that there is no output change when the controller is transferred to Auto and the control terms are allowed to take over again. All this balancing is accomplished automatically by the instrument and called *bumpless transfer.*

SUMMARY

1. An industrial process controller can be considered to consist of the following functions:

   - Input or measuring
   - Comparison
   - Controlling
   - Output or correcting

2. The input or measuring unit is designed in two forms, one that is capable of accepting the signal directly from the primary element and the other, from the primary element that is integral with a transmitter, or a primary element connected to a separate transmitter. In each case, the transmitter produces a proportional signal that is connected to the controller unit.

3. The comparator receives the signal from the input or measuring unit and makes a comparison between the input signal and the desired value. In essence, the comparator is a subtracting unit where the difference between the setpoint and measurement signal is determined, the difference being the error or deviation Er.

4. The error signal generated in the comparator is passed on to the control unit. The control unit can be any one of the following:

   - Proportional only, in which the output will be proportional to the error.

   - Proportional + integral, in which the output will contain not only a component proportional to the error, but also one that will persist and continually increase for as long as there is an error detected.

   - Proportional + derivative, in which the output will contain not only a component proportional to the error, but also an additional one that will change at the rate of change of the error.

   - Integral only, in which the output will be available and continually increase for as long as there is an error.

   - Proportional + integral + derivative, in which the output will not only contain components proportional to the error, but will change at the rate at which the error changes and persist and continually increase for as long as the error is detected.

   Each type of control action is most often called a control term, and sometimes shortened to *term*.

5. The output unit is provided with a means of changing the signals used within the controller itself to a signal that can be transmitted and used by a receiving instrument or control device. In addition it is provided with the means of setting a bias if necessary. The output signal is normally sufficient to actuate a single control device.

6. Selecting a suitable controller to maintain the process conditions required for a simple process consisting of a distance velocity lag (dead time) could, in view of the simplicity, be a single term proportional-only controller. The output from such an instrument is related to its input by

   $$V = -k_1 \mathrm{Er}$$

where Er is the error or deviation from setpoint. We can rewrite this equation in terms of the proportional band and the output bias as:

$$V = \frac{100}{P} Er + b$$

where P = percentage proportional band
Er = error, i.e., deviation (setpoint − measurement)
b = output bias

7. Under proportional-only control, a process consisting of dead time will cycle at a period of twice the dead time.

8. A process consisting of dead time and oscillating undiminished under proportional control signifies a loop gain of unity. Doubling the proportional band will halve the loop gain, and have the effect of a reduction in amplitude of the oscillation by one half in each of successive half cycle. This is called quarter-amplitude damping.

9. On proportional-only controllers, there is only one adjustment, the proportional band, and this affects the damping.

10. All proportional controllers will have their outputs settle having an offset Er_x following any load change. The amount of the offset from the desired value is given by:

$$Er_x = \frac{P(V - b)}{100}$$

where P = percent proportional band
V = controller output
b = output bias

11. Proportional action is never used in applications where the proportional band is to be no wider than a few percent.

12. Integral action is given in terms of integral action time by:

$$V = \frac{1}{S} \int Er \, dt$$

where V = output
S = integral action time
Er = deviation (error)

13. With integral action, the output will change as long as there is an error and the rate of change of output is proportional to the deviation, or:

$$\frac{dV}{dt} = \frac{Er}{S}$$

14. In an integral action controller, introducing a sinusoidal input to the instrument to determine the natural period τ_0 of the loop results in:

$$Er = A \sin\left(\frac{2\pi t}{\tau_0}\right)$$

The output will then be:

$$V = \frac{1}{S} \int \left(A \sin \frac{2\pi t}{\tau_0} \right) dt = \frac{A\tau_0}{2\pi S} \left(-\cos \frac{2\pi t}{\tau_0} \right) + V_0$$

where V_0 is the output at zero time.

15. To evaluate the gain and phase of an integral action controller, we must reduce the output to the same form as the input. Or we can say:

$$-\cos x = \sin \left(-\frac{\pi}{2} + x \right)$$

We then rewrite the output equation as:

$$V = \frac{A\tau_0}{2\pi S} \sin \left(-\frac{\pi}{2} + \frac{2\pi t}{\tau_0} \right) + V_0$$

The phase shift is given by the output angle minus the input angle:

$$\phi = \left(-\frac{\pi}{2} + \frac{2\pi t}{\tau_0} \right) - \frac{2\pi t}{\tau_0}$$

from which we see that the phase angle is $-\pi/2$ or $-90°$.

16. The gain of the integral controller is given by the amplitude of the output divided by the amplitude of the input:

$$G = \frac{A\tau_0/2\pi S}{A} = \frac{\tau_0}{2\pi S}$$

17. When the integral controller is in the loop; then the sum of the phase shift of the process and the controller must be equal to $-\pi$ at the natural period τ_0, or:

$$-\pi = -\frac{\pi}{2} - \frac{2\pi \tau_d}{\tau_0}$$

or, in degrees,

$$-180° = -90° - 360° \frac{\tau_d}{\tau_0}$$

Solving for τ_0, we have:

$$\tau_0 = 4\tau_d$$

18. For an integral-only controller, quarter-amplitude damping can be obtained by doubling the reset time. Such a controller has only one adjustment, which affects the damping. It avoids offset but reduces the speed of response.

19. The two-term P + I controller combines the features of both the control actions and is represented by

$$V = \frac{100}{P} \left(Er + \frac{1}{S} \int Er \, dt \right)$$

20. The performance of the P + I controller lies somewhere in the range:

$$4\tau_d > \tau_0 > 2\tau_d$$

depending on the settings of the proportional and integral terms. An infinite number of settings can be found to provide constant damping, the only requirement being that the loop gain must be achieved.

21. For the two-term instrument the sum of the gains must equal 0.5. The components of the proportional and integral gains are out of phase with each other. For the two-term instrument the vector sum of these two control components or resultant of these two gains will be the gain of the two-term controller.

22. This controller has two adjustments, proportional band and integral action time, both of which affect the stability of the control loop. An acceptable combination would be to have a proportional band that will give good damping and an integral time to give good recovery.

23. *Derivative action on its own is not a controlling action.* To obtain benefit the action must be combined, usually with other control actions either proportional, or proportional + integral.

24. Integral saturation, or reset windup, can occur only when a controller incorporates integral control action and under open-loop conditions. Integral action will produce an output for as long as there is a difference between the measurement and the setpoint—in other words, for as long as there is an error signal. The effect of this will be to eventually drive the output off scale, and even up to the limit value of the power supply, if the output is increasing, whether this is pneumatic or electric. The integral term will saturate when there is a massive change in measurement, or the controller output to the control device is interrupted for a time, or when the controller is transferred to Manual.

CHAPTER 10

Controller Tuning

In this part of our study we shall initially continue the investigation of the control terms of automatic process controllers that we commenced in the previous chapter and direct our attention again to the three-term controller. We shall also see how the control terms themselves react with each other, and the effect that these control-term interactions have on the process itself. The characteristics of some commonly used instruments and control devices will be given, and the combination of these interactions in a realistic application to a process will be developed. The object of this exercise is to move on to understanding how the plant will behave under the demands of an automatic process controller. With this knowledge, we shall develop procedures to assist in the installation of an automatic process controller in a control loop and tune the instrument so that it will be possible for the process to be manipulated in an optimum manner by the controlled output to produce material at—or, more correctly within—a predetermined margin of acceptability.

CONTROL TERM INTERACTION IN A THREE-TERM CONTROLLER

Note in all the control equations in this chapter, the 100 in respect of the proportional band relates the function into gain terms where 100% is unity gain.

In Chapter 9, we discussed the makeup of single- and two-term controllers and application to process loops. The single-term devices were proportional and integral, and the two-term instrument a combination of the two, i.e., P + I. We shall complete our investigation and consider the three-term device now, and go through the same sort of analysis as we did before using the control action relationships derived in Chapter 8. The equation for this type of control is:

$$V = \frac{100}{P}\left(\mathrm{Er} + \frac{1}{S}\int \mathrm{Er}\,dt + R\frac{d\,\mathrm{Er}}{dt}\right)$$

which shows the adjustments (P, S, and R) that are available on a real instrument. Furthermore, it also shows that each control term has its own gain, and Chapter 9 showed how the natural period of the plant is determined, and how the effects of the controller terms are out of phase with each other. Therefore, in order to determine the overall gain of this controller, it is necessary to perform a vector addition. This addition is shown in Figure 10.1.

Note that while the vector for the derivative is shown as vertical, this is not strictly true because of the gain limitations on the derivative action. However, the inaccuracy is not severe above a period of $2\pi R$, and from the diagram, the phase angle is:

$$\phi = \tan^{-1}\left(\frac{2\pi R}{\tau_0} - \frac{\tau_0}{2\pi S}\right)$$

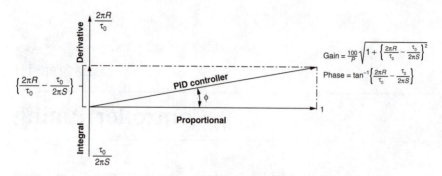

Figure 10.1: Resultant gain and phase of a three-term controller.

and the gain is:

$$G = \frac{100}{P} \sqrt{1 + \left(\frac{2\pi R}{\tau_0} - \frac{\tau_0}{2\pi S}\right)^2}$$

Controllers that have integral and derivative actions in series will always be subject to interaction between these two modes of control. While a fully noninteractive controller is theoretically (mathematically) simpler in construction, it is expensive to produce.

Figure 10.2*a* is a block diagram of a typical interactive controller, and Figure 10.2*b*, of a noninteractive controller. We can reduce the diagram for the interacting controller to the expression:

$$V = \frac{100}{P}\left(1 + \frac{R}{S}\right)\left(\text{Er} + \frac{1}{S+R}\int \text{Er}\,dt + \frac{\frac{d\,\text{Er}}{dt}}{1/R + 1/S}\right) \tag{10.1}$$

The effectiveness, or the results obtained at a practical level when an interactive controller is in operation of the three-term adjustments, illustrates the interaction. It has been found that even with a variety of different settings on the control terms the results obtained are almost virtually unchanged, which resulted in the drawing up of tuning maps. Letting the *effective value* of each control term relative to the actual

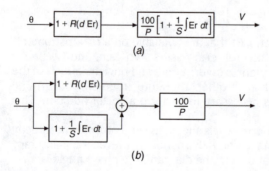

Figure 10.2: (a) Interacting controller. (b) Noninteracting controller.

instrument setting be indicated by a prime, we have:

$$P' = \frac{P}{1 + R/S}$$

$$S' = S + R$$

$$R' = \frac{1}{1/R + 1/S}$$

or, rewriting Equation (10.1)

$$V = \frac{100}{P'} \left(Er + \frac{1}{S'} \int Er\, dt + R' \frac{dEr}{dt} \right) \tag{10.2}$$

If we examine the preceding equation(s), we find that:

1. It is impossible to make the *effective* derivative action time R equal to or greater than the integral action time.

2. As the derivative time approaches the integral time, further adjustment of R will produce very little change in the *effective* derivative time.

3. If derivative time is greater than the integral time, then the derivative time is affected more by the integral setting; but if the integral time is greater than the derivative time, then the integral time is affected more by the derivative setting.

Hence, it is clear that the interaction very severely limits the range of the *effective* settings. The integrated error is a function of the *effective* values of the proportional and integral action and is given by:

$$\frac{Er}{dV} = \frac{P'S'}{100} = \frac{P(S + R)}{100\,(1 + R/S)} = \frac{PS}{100}$$

The integrated error is unaffected by the interaction.

A more severe form of interaction occurs in pneumatic controllers where integral and derivative actions are in parallel feedback about the amplifier; pneumatic controllers with an anti-windup switch are connected in this manner. The difference between the anti-windup and a conventional interacting controller is the *effective* proportional bandwidth of the instrument, which for a controller with anti-windup, is shown below:

$$P' = P \left(\frac{1 - R/S}{1 + R/S} \right)$$

Under these circumstances:

- When $R = S$, the *effective* proportional bandwidth is zero—i.e., the effective gain $= \infty$.

- When $R > S$, the *effective* proportional bandwidth is negative—i.e., of reverse phase—and in a negative-feedback controller the actual feedback becomes positive, which means great care is required when making adjustments to avoid a disaster.

METHODS OF LOOP AND CONTROLLER TUNING

We have completed our review of some of the commonly available controller configurations and will now therefore look at the methods of setting up the controllers for service in the assigned loops. These are some ways already defined and used in industry:

- Frequency response method
- Ziegler-Nichols method

THE FREQUENCY RESPONSE METHOD

The frequency response method was in fashion in the 1950s. The analysis comprised disconnecting the pneumatic signal from the controller to the control device (valve, damper, etc.) and replacing it with a fixed pressure of a magnitude that maintained the process at the desired value. The controller was then put into Manual mode, and the signal from the measuring device, i.e., the field transmitter, applied to a pen on a chart recorder. A sinusoidial signal was superimposed on the steady pressure applied to the control device (valve, damper, etc.) and also to a second pen on the chart recorder. The ratio of amplitudes gave the attenuation at the period used, and the phase lag of the output signal behind the input signal could be measured. The test was carried out at several different frequencies in the likely working range, with the phase lag and attenuation being plotted against the operating period to give the frequency response of the plant. The method suffered from two major shortcomings:

(1) The tests were very time consuming.

(2) It assumed that the process was linear and invariant.

This frequency response method of analysis is really suitable only for fast linear devices such as instruments. The difficulty in using frequency response as a method of analysis lies in the fact that the plant response will also have to be included with that of the control instrument in looking at the complete loop, and the two responses are not the same.

There are other methods for getting information from the plant quickly and easily that will suggest corrective measures for most cases. Before proceeding with the ways of conducting the tests, let us consider two points that will avoid running into serious problems later:

1. *Do not test for steady-state gain.* Steady-state gain for a single capacity *is not constant*; it will vary. For example, the steady state-gain with flow and level is not the same. Dynamic gain, however, *is constant*. With a control loop in a process, because we are concerned only with the loop or dynamic gain, the natural period of the loop τ_0 is all that matters to us.

2. *Do not test for time constants.* Once again considering the single-capacity liquid level process, the time constant may vary with flow without affecting the dynamic gain. A nonlinear element in a process is highly possible. *To be meaningful, the test requires the process to come to rest after a disturbance. A process that is not self-regulating will never come to rest.* Furthermore, the test requires the control loop to be open until a new steady state is reached, and this could mean a very long time.

ALTERNATIVE TESTS TO OBTAIN PLANT DATA

The simplified testing consists of one open- and one closed-loop test. For the tests, the proportional action of the controller is all that is required, adjust the integral action to maximum, and the derivative action to minimum values if these settings are available. The procedures are as follows:

- *The open-loop test.* With the controller in Manual, step the control valve sufficiently to observe an effect on the measurement at the controller input. Note the time interval between the disturbance and the first indication of a response. This is the distance/velocity lag (dead time), τ_d.

- *The closed-loop test.* Transfer the controller to Automatic. Adjust the proportional band to give nearly undamped oscillations. Note the period of the oscillations τ_0 and the proportional band that achieves it. Two complete cycles are sufficient to measure this.

With the data obtained; a great deal of information on the dynamic elements in the process can be obtained. Let τ_0 be the natural period of the loop and τ_d the periodic time. Then:

1. If $\tau_0/\tau_d = 2$, the process is pure distance/velocity lag (dead time) as shown in Chapter 9.

2. If $2 < \tau_0/\tau_d < 4$, dead time is dominant.

3. If $\tau_0/\tau_d = 4$, there is a single dominant capacity.

4. If $\tau_0/\tau_d > 4$, there is more than one capacity.

5. The value of the proportional band setting that caused the loop to go into oscillation is equal to the product of the gains of the other elements in the loop at τ_0.

If we combine the information obtained from this analysis with the characteristics of the known elements, we can arrive at a very accurate idea of the process. For example:

- If a process is known to contain one principal capacity and $\tau_0/\tau_d = 4$, then there is no need to look for any other constants.

- If the time constant of this capacity is known, then process dynamic gain G_{Dyn} at τ_0 can be calculated. **Note:** The dynamic gain G_1 of the controller is defined as the change in output brought about by a change in the error signal, i.e., deviation Er.

- Now, if we combine this calculated value with the gains of the transmitter, and control valve and the proportional band of the controller, we can determine the process gain, $G_{process}$:

$$G_{process} = \frac{P}{100\, G_1\, G_{transmtr}\, G_{valve}}$$

The meaning of the terms are given later in this chapter.
Remember: since these tests are made at only one operating point, they will not show any nonlinearity.

- Closed-loop response should also be observed at other flow conditions to find any change in damping.

- If the period varies with flow, a variable dynamic element is present.

- A pH loop is normally very nonlinear and easily identified by the distorted waveform produced, but less severe nonlinear measurements may not be detected unless a change of setpoint is made.

For a thorough analysis, the closed-loop test should be repeated for several values of measurement and setpoint to identify the situation correctly.

Some terms shown in the previous equation have not been defined previously. First the transmitter gain $G_{transmtr}$, is equal to 100%/span and is derived directly from the initial definition of gain, which is output/input. Therefore, in this case the 100% is the output produced for a full-span (input) change. It follows, then, that "$G_{transmtr}$" is not a pure number, for it has the dimensions and units of the measurement, and therefore must be multiplied by other gains around the loop in order to make the loop gain dimensionless.

If the transmitter is nonlinear, then $G_{transmtr}$ will not be a constant, and this will show a marked effect on the stability of the control loop. The most common nonlinear transmitting instrument is the differential-pressure transmitter when used, e.g., with an orifice plate, in a flow-measuring loop. In this situation the output will vary as the square of the flow. Each D/P instrument in the flow transmitting mode will have its own calibrated span, and in addition to this the nonlinearity is applied as a coefficient.

The following is a mathematical explanation. Let h = the dimensionless differential (fraction of head at full flow) and f = dimensionless flow (fraction of full flow). Then $h = f^2$ and differentiating this with respect to f, we have:

$$\frac{dh}{df} = 2f \qquad \text{where} \qquad \frac{dh}{df} = \text{dimensionless gain}$$

Therefore

$$G_{transmtr} = 2f \frac{100\%}{\text{span}}$$

where span is the measurement span.

We shall now consider an example. Suppose the differential flowmeter has a scale (span) of 0 to 500 gal/min. Then applying the preceding definition, we have:

$$G_{transmtr} = \begin{cases} (2f \times 0.2\%)/(\text{gal/min}) & \text{with units applied to span} \\ 0.4\%/(\text{gal/min}) & \text{at full flow } (f = 1.0 \text{ or } 100\% \text{ span}) \\ 0.2\%/(\text{gal/min}) & \text{at 50\% flow } (f = 0.5) \\ 0.0\%/(\text{gal/min}) & \text{at zero flow} \end{cases}$$

From this it is clear that, because of the nonlinearity, the control loop will not perform consistently at different flow rates. If we did not take account of this and adjusted the proportional band of the controller for acceptable damping at 50% flow, then at 100% flow the loop would be undamped, and would be excessively slow at near zero flow. We can put it another way and say that the loop will be underdamped (unstable) at 100% because of the higher $G_{transmtr}$, but overdamped at 0% because the total loop gain approaches zero. Control will improve if we take the square root of the transmitter output signal to make the output linear with respect to flow.

Valve gain can be defined as the change in delivered flow versus the percentage change in valve stem position. For a linear valve, the gain is the rated flow under normal process conditions at full stroke:

$$G_{valve} = \frac{\text{maximum flow}}{100\%}$$

For example, if a linear valve is capable of delivering 500 gal/min when fully open at process conditions, then:

$$G_{valve} = \frac{500}{100\%} = \frac{5\,\text{gal/min}}{\%}$$

This relationship is constant for a given valve type. This gain also has dimensions in units of flow and when it is compared with $G_{transmtr}$, the % is in the denominator.

It is not possible to manufacture a perfectly linear valve, but this really poses no problem, since perfect linearity is not demanded by the control loop. Some valves are deliberately made nonlinear in order to accomplish the required control function. The most common nonlinear valve is the *equal-percentage* type. What does equal percentage mean? It means that for a given change of valve stem position, the flow rate will change by a certain percentage of the present flow regardless of the value of that flow, or, mathematically:

$$\frac{dF}{F} = k\,dl$$

where l is the fractional stem position and k is a constant A typical value of k for equal percentage valves is 4. This gives:

$$\frac{dF}{dl} = 4F$$

Therefore

$$G_{valve} = 4F\,\frac{\text{maximum flow}}{100\%}$$

An interesting feature of equal-percentage valves is that the loop gain is not affected by the valve size, since the fractional flow multiplied by the maximum flow equals the actual flow being delivered. Hence, the valve gain is a function of the actual flow and not of the valve size and varies directly with the controlled flow, which makes the valve sizing not a critical requirement. Hence, valves with an equal-percentage characteristic are the most popular selection.

Variable dynamic gain occurs in processes where the values of the secondary elements, principally dead time, vary with flow. These variations cause proportionate changes in the period of the control loop, which in turn affect the dynamic gain of the principal capacity—e.g., two vessels (exponential stages) with one filling the other as described in Chapter 7.

A *heat exchanger* can be considered a single-capacity plus dead time process, the dynamic gain of which is expressed as:

$$G_{Dyn} = \frac{\tau_0}{2\pi V/F}$$

where V is the total volume and F the flow. Let the dead time vary with flow through the tubing, which has a volume V' then:

$$\tau_d = \frac{V'}{fF}$$

where f is the fractional flow and F is the maximum flow. The period of oscillation varies with τ_d and since there is a single dominant capacity, then, $\tau_0/\tau_d = 4$, as stated earlier. Hence:

$$\tau_0 = 4\tau_d = \frac{4V'}{fF}$$

and the dynamic gain is given by:

$$G_{Dyn} = \frac{4V'/fF}{2\pi V/F} = \frac{2V'}{\pi V f}$$

The dynamic gain is thus inversely proportional to product flow, which means that as flow tends to zero, the gain tends to infinity. If this variation is not compensated for, severe problems can be encountered, especially at startup when flows are low.

ZIEGLER-NICHOLS METHOD

We shall now discuss the Ziegler-Nichols method, but to do this we shall have to define the term *subsidence ratio*, which is used in this procedure. (**Caution**! Please be aware that the notation used in this section is that usually associated with this specific method and will differ from that used elsewhere in this book.) In practice the degree of damping is not calculated from plant records, but is usually estimated directly from a chart recording of the process measurement. The subsidence ratio is the ratio of the peak values of the first two consecutive oscillations preceding the final control point. Figure 10.3 shows the method. In this figure the control response curve illustrates a step change made in the process variable by some requirement of the process; the other two curves are the controller responses to the change produced by two separate controllers (for illustration purposes only) as they bring the measurement and setpoint to coincidence. The first of the controller responses is associated with a P + I instrument and the second associated with a PID instrument.

The formulas given by Ziegler and Nichols for the proportional band and the integral and derivative actions used a subsidence ratio of 4:1, as they felt this would give the best recovery. The formulas were given in terms of a *proportional action factor* K_u, which, together with proportional action alone, gives continuous oscillations of period T_u in the process, and of K_1, the controller gain.

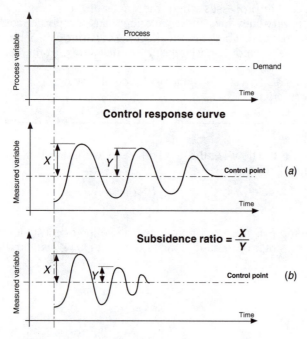

Figure 10.3: Response curves for (a) PI and (b) PID controllers.

They are read as follows:

Proportional action: $K_1 = 0.5K_u$

Proportional + integral: $K_1 = 0.45K_u$ $\qquad S = \dfrac{T_u}{1.2}$

Proportional + integral + derivative: $K_1 = 0.6K_u$ $\qquad S = \dfrac{T_u}{2}$ $\qquad R = \dfrac{T_u}{8}$

In order for the method to be applied, it is necessary to produce a curve that illustrates the response of the process to a change in correcting signal. This curve is called the *process reaction curve* and is obtained as follows:

A chart recorder is assigned to the measurement signal of the control instrument in order to produce a hard copy of the measurement being fed to the controller.

The control signal to the correcting device (control valve) is disconnected from the controller but is then instead connected to a manual loading station.

A step change is made to the signal that is applied to the control device (the valve actuator), which will result in a change ΔP in the air pressure of the pneumatic signal that actually drives the control device.

In this way the lag due to the whole system, excluding the controller, is obtained and therefore represents the response of the process only. The information thus generated is plotted on the recorder and results in a graph as shown in Figure 10.4:

From experience it is nearly always found possible to measure the slope of the response curve in the vicinity at the point of transition shown in Figure 10.4. This slope, which is shown as F in the figure, varies with the magnitude of the step signal, although the intercept, denoted by L in the figure, always remains the same, or, in other words, L is a constant.

The value of K_1 is calculated from this figure. Ziegler and Nichols stated that the best proportional action factor K_1 for proportional control of the process was

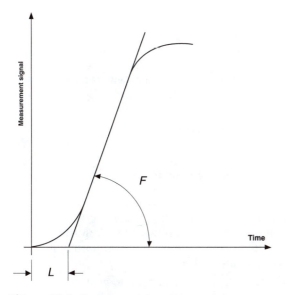

Figure 10.4: Process reaction curve.

proportional to $1/FL$, which, together with a subsidence ratio of 4:1, would allow good recovery:

$$K_1 = \frac{\Delta P}{FL} = \frac{1 \Delta Q}{kFL}$$

where $\Delta_Q = k\Delta P$ is the potential correction, and k is the plant constant for the given condition.

With the aid of the process reaction curve, the previous equations can be rewritten in terms of F and L in Figure 10.4:

Proportional action: $K_1 = 0.5K_u = \dfrac{1}{F'L}$

Proportional + integral action: $K_1 = 0.45K_u = \dfrac{0.9}{F'L} \qquad S = \dfrac{T_u}{1.2} = \dfrac{4L}{1.2}$

Proportional + integral + derivative:

$$K_1 = 0.6K_u = \frac{1.2}{F'L} \qquad S = \frac{T_u}{2} = 2L \qquad R = \frac{T_u}{8} = 0.5L$$

where

$$\frac{1}{F'} = \frac{1 \Delta Q}{KF}$$

A PRACTICAL METHODOLOGY FOR CONTROL LOOP TUNING

To conclude this discussion, a practical method will now be described. The procedure is one that has been proved successful on most occasions encountered in real plants. For the purpose of evaluating the required controller settings, we shall leave these in terms of τ_d and τ_0 for the simple reason that the natural period of the loop varies with the controller settings. We use the two vector diagrams given earlier in Figure 9.1 (for P + I) and Figure 10.1 (for PID) to accurately compute the phase and the gain. For quarter-amplitude damping, equate the gain product of process and controller to 0.5, and proceed as follows:

Initializing
1. Set integral action time to maximum.
2. Set derivative action time to minimum.
3. Adjust (reduce) the proportional band to a value that will just put the closed loop into oscillation. (This adjustment will usually result in a proportional band of a few percent.)

Primary settings
1. Measure the period of oscillation τ_0 and set both, the integral and derivative action times to $\tau_0/2\pi$.
2. For P + I controllers, set the integral action time to $\tau_0/2.4$.
3. If the controller is a P + I + D with anti-windup feature, always keep the integral action time greater than twice the derivative action time. This will ensure the proportional action remains stable.

Damping
1. Readjust the proportional band to give the desired degree of damping.

Final settings   1. If τ_0 is higher than before, increase both the integral and derivative settings.

2. If τ_0 is lower than before, decrease both the integral and derivative settings. **Note:** With a two-term controller, τ_0 will increase by about 50%.

Interaction will always result if we attempt to tune the controller for maximum performance.

SUMMARY

Note: In all the control equations, the 100 in respect of the proportional band is 100%.

1. In a three-term (PID) controller, the phase angle is given by

$$\phi = \tan^{-1}\left(\frac{2\pi R}{\tau_0} - \frac{\tau_0}{2\pi S}\right)$$

and gain is given by

$$G = \frac{100}{P}\sqrt{1 + \left(\frac{2\pi R}{\tau_0} - \frac{\tau_0}{2\pi S}\right)^2}$$

2. Controllers that have integral and derivative actions in series will always be subject to interaction between these two modes of control.

3. The interacting controller can be reduced to the mathematical expression:

$$V = \frac{100}{P}\left(1 + \frac{R}{S}\right)\left(Er + \frac{1}{S+R}\int Er\,dt + \frac{\frac{d\,Er}{dt}}{1/R + 1/S}\right)$$

4. If the *effective* value of each control term relative to the actual instrument setting is indicated by a prime, then:

$$P' = \frac{P}{1 + R/S}$$

$$S' = S + R$$

$$R' = \frac{1}{1/R + 1/S}$$

These equations show that:

(a) It is impossible to make the *effective* derivative action time R equal to or greater than the integral action time.

(b) As the derivative time approaches the integral time, further adjustment of R will produce very little change in the *effective* derivative time.

(c) If derivative time were greater than the integral time, then the derivative time is affected more by the integral setting; but if the integral time were greater than the derivative time, then the integral time is affected more by the derivative setting.

5. Interaction very severely limits the range of the *effective* settings on an actual instrument.

6. The integrated error is a function of the *effective* values of the proportional and integral action:

$$\frac{\text{Er}}{dV} = \frac{P'S'}{100} = \frac{P(S + R)}{100(1 + R/S)} = \frac{PS}{100}$$

7. The frequency response method for tuning a controller suffered from two major shortcomings: the tests were very time-consuming and it assumed that the process was linear and invariant. Frequency response as a method of analysis is really suitable only for fast linear devices such as instruments. When applied to a control loop the frequency response of the plant will also have to be included with that of the control instrument, but the two responses are not the same.

8. As an alternative to the frequency response method, a simplified test consisting of one open- and one closed-loop test, will get information from the plant quickly and easily. To perform the open loop test, set the controller in Manual, and step the control valve sufficiently to observe an effect. Measure the time interval between the disturbance and the first indication of a response. This is the distance/velocity lag (dead time), τ_d. To perform the closed-loop test, transfer the controller to Automatic with maximum integral action time, and minimum derivative time, so that it behaves as a proportional-only controller. Adjust the proportional band to give nearly undamped oscillations. Note the period of the oscillations τ_0 and the proportional band that achieve it. Two complete cycles are sufficient to measure this.

9. In the open- and closed-loop methods of conducting tests, avoid running into serious problems by first not testing for steady-state gain, for it is not constant; it will vary even for a single capacity, but dynamic gain is constant, and the natural period of the loop τ_0 is all that matters. Second, not testing for time constants, as the time constant even for a single capacity may vary, and a nonlinear element in a process is highly possible. The tests require the process to come to rest after a disturbance; a process that is not self-regulating will never come to rest. Furthermore, the control loop is to be open until a new steady state is reached, and this could be a very long time.

10. With the data obtained from the open- and closed-loop tests, a great deal of information on the dynamic elements in the process can be obtained. Let τ_0 be the natural period and τ_d the periodic time. Then:

    • If $\tau_0/\tau_d = 2$, the process is pure distance/velocity lag (dead time).

    • If $2 < \tau_0/\tau_d < 4$, dead time is dominant.

    • If $\tau_0/\tau_d = 4$, there is a single dominant capacity.

    • If $\tau_0/\tau_d > 4$, there is more than one capacity.

    • The value of the proportional band setting that caused the loop to go into oscillation is equal to the gain product of the other elements in the loop at τ_0.

11. The transmitter gain G_{transmtr} is equal to 100%/span and is derived directly from the initial definition of gain. The 100% is the output produced for a full-span change. G_{transmtr} is not a pure number, but has the dimensions of the measurement

and must be multiplied by other gains around the loop to make the loop gain dimensionless.

12. Valve gain can be defined as the change in delivered flow versus the percentage change in valve stem position. For a linear valve the gain is the rated flow under normal process conditions at full stroke:

$$G_{valve} = \frac{\text{maximum flow}}{100\%}$$

This gain also has dimensions, and when it is compared with $G_{transmtr}$, the % is in the denominator.

13. If we combine the information obtained from the analysis conducted in item 10 with the characteristics of the known elements, we obtain a very accurate idea of the process:

(a) If a process is known to contain one principal capacity and $\tau_0/\tau_d = 4$, then there is no need to look for any other constants.

(b) If the time constant of this capacity is known, then its dynamic gain G_{Dyn} at τ_0 can be calculated.

(c) Combining this calculated value with the gains of the transmitter, control valve, and the proportional band of the controller, we can determine the process gain, $G_{process}$:

$$G_{process} = \frac{P}{100 G_1 G_{transmtr} G_{valve}}$$

Since these tests are made at only one operating point, they will not show any nonlinearity.

(d) Closed-loop response should be observed at other flow conditions to find any change in damping.

(e) If the period varies with flow, a variable dynamic element is present.

(f) A very nonlinear loop, e.g., pH, is easily identified by the distorted wave-form produced; less severe nonlinear measurements may not be detected unless a change of setpoint is made.

14. It is not possible to manufacture a perfectly linear valve. Perfect linearity is not demanded by the control loop. Some valves are deliberately made nonlinear.

15. The most common nonlinear valve is the equal-percentage type: for a given change of valve stem position, the flow rate will change by a certain percentage of the present flow regardless of the value of the present flow, or:

$$\frac{dF}{F} = k \, dl$$

where l is the fractional stem position and k is a constant. A typical value of k for equal percentage valves is 4. so that:

$$\frac{dF}{dl} = 4F$$

Therefore

$$G_{valve} = 4F \frac{\text{maximum flow}}{100\%}$$

In equal-percentage valves the loop gain is not affected by the valve size, since the fractional flow multiplied by the maximum flow equals the actual flow being delivered. The valve gain is a function of, and varies directly with, the actual flow. Hence, valves with an equal-percentage characteristic are the most popular selection.

16. The Ziegler-Nichols method of controller tuning uses the subsidence ratio in computing the tuning constants. The subsidence ratio is the ratio of the peak values of the first two consecutive oscillations preceding the final control point.

17. The Ziegler-Nichols formulas for the proportional band and the integral and derivative actions used a subsidence ratio of 4:1, as it was felt this would give the best recovery. The formulas given in terms of a proportional action factor K_u, which, together with proportional action alone, gives continuous oscillations of period T_u in the process, are as follows:.

$$\text{Proportional action: } K_1 = 0.5 K_u = \frac{1}{F'L}$$

$$\text{Proportional + integral action: } K_1 = 0.45 K_u = \frac{0.9}{F'L} \qquad S = \frac{T_u}{1.2} = \frac{4L}{1.2}$$

Proportional + integral + derivative:

$$K_1 = 0.6 K_u = \frac{1.2}{F'L} \qquad S = \frac{T_u}{2} = 2L \qquad R = \frac{T_u}{8} = 0.5L$$

where

$$\frac{1}{F'} = \frac{1 \Delta Q}{KF}$$

18. To apply the Ziegler-Nichols method, it is necessary to produce a curve called the process reaction curve that illustrates the response of the process to a change in correcting signal.

PART III

APPLICATIONS

CHAPTER 11

Flow, Pressure, and Level Control

Having now studied some available controllers, we shall devote our time to applying these instruments to some typical processes found in industry. We shall not be concerned initially with the more difficult applications, but let us be aware from the outset that many of these exist and will give us all the opportunity and challenge of devising solutions. Unknown scenarios are part and parcel of the excitement and stimulation of process control, even though developing the solutions is a demanding phase that has to be endured. Let us not be beguiled by the impression that process control constitutes a never-ending "high," as it certainly does not. We shall find the majority of the time is devoted to the more mundane aspects of choosing and specifying sensors, instruments, and final actuators to replace failed or worn-out equipment. But even this exercise of carrying out the day-to-day activities of a control engineer can be a means of widening our knowledge of this, our chosen field of employment, for new methods of measurement and control equipment are being brought to the market continually.

We shall find that, in a few process plants some control loops function because of the ingenuity of the process operators and not because they were designed well. The process operator can be, and usually is, a very valuable source of information. Remember, the operator has to live with the process all the time and therefore has an incredible amount of information and experience to offer regarding process and plant behavior. You should use the operator's expertise to develop and improve the control strategies, especially in difficult applications, as it will pay dividends both in human relations and financial terms.

Having now introduced the ideas and methods of control, let us now consider how these can be applied in the real world of process control.

FLOW

The first control loop we consider is flow. Apart from being one of the most important parameters in any process, it is also one in which the measured and controlled variables are the same. Opening a valve in a flow line starts a flow, although the response is not quite instantaneous. If the fluid is a gas, then it will expand due to a change in pressure; it is therefore obvious that the delivered amount will vary with the change in pressure and hence with the flow. Inertia is very important in liquid systems. Flow cannot be started or stopped without acceleration or deceleration of the fluid. In steady-state conditions the flow velocity varies with pressure drop:

$$v^2 = k_{\text{Flow}}^2 \, 2g \frac{\Delta p}{\rho}$$

where v is the velocity, k_{Flow}^2 is the flow coefficient, g is the acceleration due to gravity;

Δp is the pressure drop, and ρ is the density. But, velocity is proportional to flow, or:

$$v = \frac{Q}{\alpha}$$

where Q is the flow and α is the cross-sectional area. If we combine the two preceding equations, we have:

$$\Delta p = \frac{u^2 \rho}{2gk_{Flow}^2} = \frac{Q^2 \rho}{2g\alpha^2 k_{Flow}^2} \tag{11.0}$$

However, if the applied force $\alpha \, \Delta p$ exceeds the resistance to the flow, then acceleration occurs.

Now, force = mass × acceleration, and therefore, we can write an equation for the unsteady state as follows:

$$\alpha \, \Delta p - \frac{\alpha Q^2 \rho}{2g\alpha^2 k_{Flow}^2} = m\frac{dv}{dt}$$

where m is the mass and t is time in seconds. But, $v = Q/\alpha$. Therefore

$$\alpha \, \Delta p - \frac{\alpha Q^2 \rho}{2g\alpha^2 k_{Flow}^2} = \frac{m}{\alpha}\frac{dQ}{dt} \tag{11.1}$$

The mass of fluid in the pipe is given by:

$$m = \frac{l\alpha\rho}{g} \tag{11.2}$$

where l is the length. Therefore

$$\alpha\Delta p - \frac{\alpha Q^2 \rho}{2g\alpha^2 k_{Flow}^2} = \frac{l\alpha\rho}{g\alpha}\frac{dQ}{dt}$$

by substituting Equation (11.2) into Equation (11.1). Simplifying, we have:

$$\frac{2g\alpha^2 k_{Flow}^2 \Delta p - Q^2 \rho}{2g\alpha k_{Flow}^2} = \frac{l\rho}{g}\frac{dQ}{dt}$$

Simplifying by canceling g on the right-hand side, we have:

$$2g\alpha^2 k_{Flow}^2 \Delta p - Q^2 \rho = 2\alpha k_{Flow}^2 l\rho\frac{dQ}{dt} \tag{11.3}$$

Dividing through by ρQ and transposing, we have:

$$\frac{2g\alpha^2 k_{Flow}^2 \Delta p}{\rho Q} - Q = \frac{2\alpha k_{Flow}^2 l}{Q}\frac{dQ}{dt}$$

$$\frac{2g\alpha^2 k_{Flow}^2 \Delta p}{\rho Q} = Q + \frac{2\alpha k_{Flow}^2 l}{Q}\frac{dQ}{dt}$$

From this equation, which is the standard form of the differential equation (11.3), we find the time constant is given by the coefficient of dQ/dt, or:

$$\tau = \frac{2\alpha k_{Flow}^2 l}{Q} \tag{11.4}$$

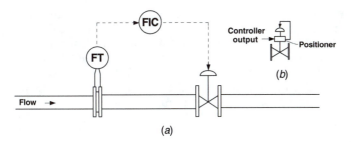

Figure 11.1: (a) A typical flow control loop. (b) Alternative arrangement for valve.

but

$$\Delta p = \frac{Q^2 \rho}{2g\alpha^2 k_{\text{Flow}}^2}$$ from (11.0)

at steady state as previously noted. Therefore

$$k_{\text{Flow}}^2 = \frac{Q^2 \rho}{(\Delta p)2g\alpha^2}$$

If we replace k_{Flow}^2 with the steady-state equivalent, we can write Equation (11.4) as:

$$\tau = \frac{2\alpha Q^2 \rho l}{2g\alpha^2 \Delta p Q}$$

$$= \frac{lQ\rho}{g\alpha \, \Delta p}$$

It is thus clear that *the time constant varies with both flow and pressure drop, because of the square law relationship between the two, and is not zero except at zero flow.* In a flow control loop, the only fundamental dynamic element is the time constant.

Let us analyze a typical flow control loop using pneumatic instrumentation, for by this we are more capable of seeing what is taking place, since the system is more mechanical. Figure 11.1 shows the arrangement, which consists of:

1. A flow transmitter (FT)

2. A flow indicating controller (FIC)

3. A flow control valve

The transmitter amplifier in FT has dynamic properties, and because of it the lag of the flowing fluid is isolated from that of the pneumatic signal transmission line from transmitter to controller. However, there is no such isolation between the controller and the control valve, and interaction can occur.

In this system there will be transmission lags, the values of which depend on the lengths of the transmission line and the bore diameter of the pipe, and can be represented as dead time plus first-order lags.

The control valve is a different story, because the valve actuator/motor is not a constant-volume device, as every change in pressure results in a corresponding change in volume of the valve motor. Therefore, this device does not behave as a first-order lag. The valve motor operates at a limited velocity, based on the maximum airflow that can be delivered into an expanding volume. The result of this is that *the time constant is small for small changes, and large for large changes*. Hence, a valve cannot be represented by a single time constant. The valve gain is higher at low flows, whereas the transmitter gain is low (see Chapter 10). This results in a gain product that tends more to uniformity than either of the two separate gains. We can improve the situation by adding a booster in the form of a *positioner* to the control valve as shown in Figure 11.1*b* and installing it in place of the normal in-line valve shown, but this will not reduce the proportional band of the controller. Much greater improvement can be effected if we mount the controller on the control valve itself, as this will eliminate the pneumatic transmission line between controller and valve.

Suppose we now use electronic instead of pneumatic instrumentation. Then, the first noticeable thing will be the absence of pneumatic transmission lines, resulting in a lower natural period for the loop. This advantage is soon lost, for high-frequency disturbances, which may be either periodic or random, will take on prominence. The frequencies at which these disturbances occur are too high for the control action to eliminate and are usually due to pump vibration and stream turbulence. These disturbances are called *noise*, and *the presence* of *noise in a system is sufficient cause to prohibit the use of derivative control action in flow control loops*.

This, then, is a brief look at flow control. Flow controllers seldom have proportional bands below 100%. Integral action is mandatory, and derivative action is never used.

PRESSURE

Now let us take a look at the parameter pressure:

- If the process fluid is a gas, the gas law $pV^\gamma = K$ will apply and cause the pressure and volume to be inversely proportional, with enthalpy (a thermodynamic function of a system equivalent to internal energy + pressure × volume, or heat, from the Greek *enthalpein* "to heat in") playing only a small role.

- On the other hand, if the fluid is a vapor in equilibrium with its liquid, enthalpy change will produce very marked changes in pressure, while volumetric changes will have less of an effect.

- However, if the fluid is a liquid, which means that it is considered to be incompressible, then the volume is more or less unaffected by pressure or enthalpy changes.

It is clear that pressure exerts great influence on the thermodynamic properties of a compressible fluid and is a function of the state of the fluid under consideration, but it is a poor measure of the heat or mass of the fluid.

If we consider a "perfect" gas, then:

$$pV = m\Re T$$

where p is the pressure, V is the volume, m is the mass, $\Re$ is the universal gas constant, and T is the absolute temperature. Then:

$$\frac{dp}{dt} = \frac{dm}{dt}\frac{\Re T}{V}$$

If T is held constant—and with $\Re$ being a known constant—then the rate of change of mass is the difference between mass inflow and mass outflow, or:

$$\frac{dp}{dt} = \frac{Q_m}{V}(q_{\text{in}} - q_{\text{our}})$$

where Q_m is the nominal mass flow, q_{in} is the fractional inflow, and q_{out} is the fractional outflow. Integrating, puts pressure in terms of flow, we have:

$$p = \frac{1}{(V/Q_m)} \int (q_{in} - q_{out})dt$$

Normally, gas pressure processes are self-regulating except at zero flow, for the obvious reason that pressure always influences inflow and outflow. The process is basically a single-capacity system with the transmitter and valve adding small secondary lags. Gas pressure is normally easy to control, with controllers for these systems usually having narrow proportional bands.

In systems where liquid and vapor are in equilibrium, the difference between the inflow and outflow of vapor will change the pressure, because of the change in material balance:

$$V\frac{dp}{dt} = Q_{vap,in} - Q_{vap,out}$$

where Q_{vap} is the vapor flow. If the enthalpies of inflow and outflow differ, resulting in changes between the liquid and vapor phases (material change), the system pressure will be affected, and from an energy balance point of view:

$$Q_{vap,in} W_{in} - Q_{vap,out} W_{out} + Wq_{in} - Wq_{out} = VW_{vap}\frac{dp}{dt}$$

where W is the enthalpy, Wq is the amount of heat transferred, and W_{vap} is the heat of vaporization.

Mass and heat flow both affect pressure. If the net change of enthalpy across a process is zero, then mass flow alone is a suitable parameter to control. For example, in pressure reduction of saturated steam there is no enthalpy change across the reducing valve. If the process is one of heat transfer—e.g., distillation columns, boilers, evaporators—then pressure control can be used to close or regulate the heat balance.

Pressure control of a liquid is the same as flow control, for pressure is related to the flow in the pipeline and vice versa, the only dynamic contribution being inertia. In a flow loop the process gain is by definition 1.0; in a pressure loop there must be a conversion from flow to pressure units. For liquids in a pipeline with a resistance of r to the flow inserted, the pressure upstream of the resistance will vary as a function of (flow)^2, or

$$p = p_0 + \frac{Q^2}{r^2}$$

where p_0 is the static pressure at no flow. Differentiating pressure with respect to flow, we have:

$$\frac{dp}{dQ} = \frac{2Q}{r^2}$$

which is the process gain.

Usually, the indicated pressure changes less than the change in valve position, resulting in a lower proportional band for the controller. Problems with noise are similar to those encountered with flow, but, in this case, pressure snubbers can be used to give stable outputs.

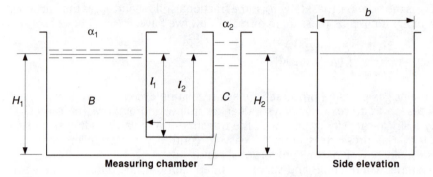

Figure 11.2: Interconnected tank level control.

LEVEL

Level control of a liquid, though appearing to be simple, is really quite difficult, and we shall now give it some consideration. As we all know, a liquid has no inherent shape, but takes up the contours of the vessel containing it. Visualize a cup of water. The sides of the cup restrain the liquid, but there is one surface that is unrestrained. The unrestrained open surface can be disturbed and caused to ripple; that is, it can be made to oscillate. Like every other material, once driven into oscillation it can be made to stay that way as long as the initiating action can be maintained. We shall now try to understand this resonance and its effects.

To start the investigation, let us look at Figure 11.2, and let the level in $C(H_2)$ in limb C momentarily exceed the level in $B(H_1)$ in limb B starting from a position of rest.

Then, because of the difference in forces, there will be a downward acceleration in $C(H_2)$. **Note:** In the following the suffixes 1 and 2 refer to chambers B and C, respectively. If ρ is the fluid density, H is the liquid head, α is the cross-sectional area for the respective U arms or legs, m is the mass, and v is the velocity, then:

$$\rho H_2\alpha_2 - \rho H_1\alpha_1 = -m_2\frac{dv_2}{dt} - m_1\frac{dv_1}{dt}$$

since

$$v_1 = \frac{\alpha_2}{\alpha_1}v_2 \qquad m_1 = \frac{\rho l_1\alpha_1}{g} \qquad m_2 = \frac{\rho l_2\alpha_2}{g}$$

where l is the length of the column surface and g the acceleration due to gravity. Then, by substituting, dividing through by ρ, and rewriting, we have:

$$H_2\alpha_2 - H_1\alpha_1 = -\frac{l_2\alpha_2}{g}\frac{dv_2}{dt} - \frac{l_1\alpha_2}{g}\frac{dv_2}{dt} \qquad (11.5)$$

Because we have caused an acceleration in one leg of the system we have initiated an oscillation. Hence, let the average level be H_{avg}. The level H_2 in C is related to the average level H_{avg} by:

$$H_{avg} - H_1 = (H_2 - H_{avg})\frac{\alpha_2}{\alpha_1} \qquad (11.6)$$

or, rewriting, Equation (11.6) we can say:

$$-H_1\alpha_1 = H_2\alpha_2 - H_{avg}\alpha_2 - H_{avg}\alpha_1$$

By substitution into Equation (11.5) and simplifying the right-hand side, we have:

$$H_2\alpha_2 + H_2\alpha_2 - H_{avg}\alpha_2 - H_{avg}\alpha_1 = -\left(\frac{l_2+l_1}{g}\right)\alpha_2 \frac{dv_2}{dt}$$

$$2H_2\alpha_2 - H_{avg}\alpha_2 - H_{avg}\alpha_1 = -\left(\frac{l_2+l_1}{g}\right)\alpha_2 \frac{dv_2}{dt}$$

Dividing through by α_2 and simplifying, we have:

$$2H_2 = H_{avg}\left(1+\frac{\alpha_1}{\alpha_2}\right) - \left(\frac{l_1+l_2}{g}\right)\frac{dv_2}{dt}$$

Transposing and dividing through by the coefficient of H_2, we have:

$$H_2 + \left(\frac{l_1+l_2}{2g}\right)\frac{dv_2}{dt} = \frac{H_{avg}}{2}\left(1+\frac{\alpha_1}{\alpha_2}\right) \tag{11.7}$$

But dv_2/dt in Equation (11.7) is really the rate of change of level, in limb C, or

$$\frac{dH_2}{dt}$$

or acceleration, so we can rewrite Equation (11.7) as:

$$H_2 + \left(\frac{l_1+l_2}{2g}\right)\frac{d^2H_2}{dt^2} = \frac{H_{avg}}{2}\left(1+\frac{\alpha_1}{\alpha_2}\right)$$

which is the acceleration in B and an undamped second order differential.

The U tube formed by B and C will vibrate at a natural period given by:

$$\tau_0 = 2\pi\sqrt{\frac{l_1+l_2}{2g}}$$

which is unaffected by any property of the fluid—e.g., density, area—except the total length of the bounding surface $(l_1 + l_2)$. It must be pointed out that *liquids do not require U-shaped containers to oscillate; they will oscillate in any vessel.* If the containing vessel were a circular tank of diameter D, the period of oscillation would be given by:

$$\tau_0 = 2\pi\sqrt{\frac{D}{2g}}$$

Liquids in rectangular tanks can oscillate at two different periods corresponding to the two transverse dimensions. The phase shift of a resonant element is exactly $-90°$ at its natural period, irrespective of the damping applied. Since the integration of flow into average level already represents a phase shift of $-90°$; the total phase shift in the process from flow to measured level will be $-180°$ at the natural period of the vessel.

In real process plants the vessel dimensions commonly fall between 2 and 200 ft, with the liquids oscillating most often with a period between 1 and 10 s. This is of serious consequence in vessels having time constants less than 1 min, for then the stability of the control loop will be affected. Levels in vessels are most often accompanied by turbulence and splashing, thus making the measurements noisy; and sometimes, dependent on the vessel connection configuration, the fluctuations can be of the order of 20% to 30%, which is not uncommon in boiling liquids.

Controllers used on liquid-level applications usually have quite narrow proportional bands. However, if there are large fluctuations on the free surface of the liquid, a narrow proportional band will be totally unsuitable, for the surface peaks and troughs appearing in the measurement could force the control valve to both its limits. To overcome this, the proportional band is widened and integral action is provided on the controller and relied on to maintain stable control.

SUMMARY

1. Flow, apart from being one of the most important parameters in any process, is also one in which the measured and controlled variable is the same. Flow cannot be started or stopped without acceleration or deceleration of the fluid.

2. Under steady-state conditions the flow velocity varies with pressure drop:

$$v^2 = k_{\text{Flow}}^2 2g \frac{\Delta p}{\rho}$$

where v is the velocity, k_{Flow}^2 is flow coefficient, g is the acceleration due to gravity, Δp is pressure drop, and ρ is density.

3. Velocity is proportional to flow, or

$$v = \frac{Q}{\alpha}$$

where Q is flow and α is cross-sectional area. Combining this with the equations of item 2 and rewriting, we have:

$$\Delta p = \frac{u^2 \rho}{2gk_{\text{Flow}}^2} = \frac{Q^2 \rho}{2g\alpha^2 k_{\text{Flow}}^2}$$

4. If the applied force $\alpha \, \Delta p$ exceeds the resistance to the flow, then acceleration occurs. Now force = mass × acceleration, and therefore, we can write an equation for the unsteady state as follows:

$$\alpha \, \Delta p - \frac{\alpha Q^2 \rho}{2g\alpha^2 k_{\text{Flow}}^2} = m \frac{dv}{dt}$$

where m is the mass and t is time in seconds.

5. Since $v = F/\alpha$

$$\alpha \, \Delta p - \frac{\alpha Q^2 \rho}{2g\alpha^2 k_{\text{Flow}}^2} = \frac{m}{\alpha} \frac{dQ}{dt}$$

by rewriting the equation of item 4, and the mass of fluid in the pipe is given by:

$$m = \frac{l \alpha \rho}{g}$$

where l is the length. Therefore,

$$\alpha \, \Delta p - \frac{\alpha Q^2 \rho}{2g\alpha^2 k_{\text{Flow}}^2} = \frac{l \alpha \rho}{g \alpha} \frac{dQ}{dt}$$

by substituting Equation (11.5.2) into Equation (11.5.1)

6. The time constant derived from the equation of item 5 is given by the coefficient of dQ/dt, or:

$$\tau = \frac{2\alpha k_{\text{Flow}}^2 l}{Q}$$

If we replace k_{Flow}^2 with the steady-state equivalent, we can write this as:

$$\tau = \frac{2\alpha Q^2 \rho l}{2g\alpha^2 \, \Delta p Q}$$

$$= \frac{l Q \rho}{g\alpha \, \Delta p}$$

7. In a flow control loop the only fundamental dynamic element is the time constant. The time constant varies with both flow and pressure drop because of the square law relationship between the two, and is not zero except at zero flow.

8. Pressure is a function of the state of the fluid under consideration. It is a poor measure of the heat or mass of the fluid.

9. For a "perfect" gas:

$$pV = m\mathfrak{R}T$$

where p is the pressure, V is the volume, m is the mass, $\mathfrak{R}$ is the universal gas constant, and T is the absolute temperature; or:

$$\frac{dp}{dt} = \frac{dm}{dt}\frac{\mathfrak{R}T}{V}$$

10. In the equation of item 9, if T is held constant—and with $\mathfrak{R}$ being a known constant—then the rate of change of mass is the difference between mass inflow and mass outflow:

$$\frac{dp}{dt} = \frac{Q_m}{V}(q_{\text{in}} - q_{\text{out}})$$

where Q_m is the normal flow, q_{in} is fractional inflow, and q_{out} is fractional outflow. Integrating, we have:

$$p = \frac{1}{V/Q_m} \int (q_{\text{in}} - q_{\text{out}})dt$$

11. In systems where liquid and vapor are in equilibrium, the difference between inflow and outflow of vapor will change the pressure, due to the change in material balance:

$$V\frac{dp}{dt} = Q_{\text{vap,in}} - Q_{\text{vap,out}}$$

where Q_{vap} is the vapor flow.

12. If the enthalpies of inflow and outflow differ, resulting in changes between the liquid and vapor phases (material change), the system pressure is affected, and from an energy balance point of view:

$$Q_{\text{vap,in}} W_{\text{in}} - Q_{\text{vap,out}} W_{\text{out}} + Wq_{\text{in}} - Wq_{\text{out}} = VW_{\text{vap}}\frac{dp}{dt}$$

where W is enthalpy, W_q is heat transfer, and W_{vap} is heat of vaporization.

13. Pressure and flow control of a liquid are the same, for pressure is related to the flow in the pipeline; the only dynamic contribution is inertia.

14. For liquids in a pipeline with a resistance of r to the flow inserted, the pressure upstream of the resistance will vary as a function of (flow)^2, or

$$p = p_0 + \frac{Q^2}{r^2}$$

where p_0 is the static pressure at no flow. Differentiating pressure with respect to flow, we have:

$$\frac{dp}{dQ} = \frac{2Q}{r^2}$$

which is the process gain.

15. The unrestrained open surface of a liquid in a vessel can be caused to ripple, that is, oscillate. In common with every other material, once driven into oscillation, liquids can be made to stay that way for as long as the initiating action can be maintained.

16. If two vessels of unequal cross-section containing a liquid are joined via the base to form a U-tube, and the level of liquid in the larger vessel subjected to a cyclic change, the liquid in both vessels will vibrate at a natural period given by:

$$\tau_0 = 2\pi \sqrt{\frac{l_1 + l_2}{2g}}$$

where l is the length of the interconnecting column in each vessel and subscripts 1 and 2 refer to the large and small vessels, respectively. The period of oscillation is unaffected by any property of the fluid—e.g., density, area—except the total length of the interconnecting liquid column surface ($l_1 + l_2$).

17. Liquids do not require U-shaped containers to oscillate; they will oscillate in any vessel. If the containing vessel were a circular tank of diameter D, the period of oscillation will be given by:

$$\tau_0 = 2\pi \sqrt{\frac{D}{2g}}$$

18. Liquids in rectangular tanks can oscillate at two different periods corresponding to the two transverse dimensions. The phase shift of a vibrating element is exactly $-90°$ at its natural period, irrespective of the damping applied.

19. The integration of flow into average level represents a phase shift of $-90°$. At the natural period of the vessel, the phase shift in the process from flow to measured level will be $-180°$.

CHAPTER 12

Control of a Stirred Reactor and Product Composition

A REACTION PROCESS

Examples of plant equipment often encountered in process plant in the pharmaceutical or industrial chemical industry are the reaction and mixing vessels; the latter for the treatment of by-product from the process, as will be described in this chapter. The reaction vessel, usually abbreviated to *reactor*, provides us with many examples of control function, of which temperature is an important and excellent platform to illustrate some of the approaches to them used in industry. In these manufacturing processes, raw materials are combined to form the final product. Measured quantities of ingredients, very often in liquid form, are mixed together, and dependent on the materials used, the process could give rise to heat being evolved, or the materials could require heat to be added as the combination occurs; these effects are called *reactions*. Reaction processes that evolve heat are *exothermic*, and those that need heat to be supplied *endothermic*. To assist in the uniform dispersal of the ingredients and obtain an even temperature distribution through the contents, a motor-driven stirrer is used; with this arrangement, the reactor is called a *stirred reactor*. Temperature is important, for all things are affected by temperature and in different ways. It is one of the parameters that will cause a material to change state (in the physical sense). The important thing we must remember is that temperature is not a measure of the heat of a body, but it is a means of determining the amount of heat that a body has for a given condition. There is a very close relationship between the parameters temperature and pressure. This dependence of one on the other is demonstrated most vividly in compressible fluids, of which the gases are the most common.

Given sufficient time, all bodies will attain the same temperature, provided no other heat or cooling source is involved. The uniform temperature will not be that of the hottest or the coolest, but one that is determined by the materials themselves, depending on the mass and heat capacity of each body involved. It is a well-known fact that one result of all our actions, whether biological or mechanical, is the generation of heat.

Temperature is difficult to both measure and control, for we are really involved with the transfer of energy. In the real world this energy transfer takes place by three methods: (1) conduction, (2) convection, and (3) radiation. When the mechanism by which the transfer is effected is flow, the transfer can vary continuously dependent on the prevailing conditions.

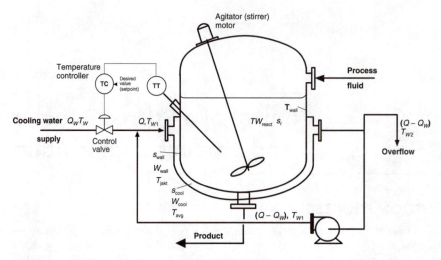

Figure 12.1: Schematic diagram of a stirred tank reactor. Only terms used in the computations are shown.

CONTROL OF TEMPERATURE

ILLUSTRATION: CONTROL OF A STIRRED REACTOR

Let us look at a stirred tank reactor, cooled by a constant controlled flow of cold water, and assume a constant evolution of heat by the material within the reactor. The arrangement described is shown in Figure 12.1. In this system there are five important elements, all subject to change:

1. Heat (thermal) capacity of the reactor contents

2. Heat (thermal) capacity of the jacket contents

3. Heat (thermal) capacity of the walls

4. Dead time of circulation

5. Lag in the temperature sensor

Since all heat transfer takes place through the walls, there will be interaction between items 1, 2, and 3. The temperature sensor has a very small thermal capacity and therefore can be neglected, as it will produce no significant interaction with the other elements. From this we find the process to be essentially one of four capacities and a dead time. We must compute the time constants involved with each of the capacities, and to do this we shall have to write a heat balance for each heat transfer surface in the steady and unsteady states.

The heat balance in simple terms is:

Total heat in = total heat out + (capacity × rate of change of temperature)

For the reactor wall, we have:

$$W_{\text{heat}} = k_{\text{wall}} W q_\alpha (T - T_{\text{wall}}) + W_{\text{react}} s_r \frac{dT}{dt}$$

where W_{heat} = rate of heat evolution
        k_{wall} = heat transfer coefficient

$$Wq_\alpha = \text{heat transfer area}$$
$$T = \text{reactor temperature (i.e., of the contents)}$$
$$T_{\text{wall}} = \text{wall temperature on contents side}$$
$$W_{\text{react}} = \text{weight of the reactants}$$
$$s_r = \text{specific heat of reactants}$$

We can rewrite the preceding equation as:

$$k_{\text{wall}} Wq_\alpha T_{\text{wall}} + W_{\text{heat}} = k_{\text{wall}} Wq_\alpha T + W_{\text{react}} s_r \frac{dT}{dt}$$

Dividing through by $k_{\text{wall}} Wq_\alpha$ we have:

$$T_{\text{wall}} + \frac{W_{\text{heat}}}{k_{\text{wall}} Wq_\alpha} = T + \frac{W_{\text{react}} s_r}{k_{\text{wall}} Wq_\alpha} \frac{dT}{dt}$$

The reactor temperature responds to the temperature of the wall with a thermal time constant given by

$$\frac{W_{\text{react}} s_r}{k_{\text{wall}} Wq_\alpha}$$

that is

$$\tau_{\text{wall}} = \frac{W_{\text{react}} s_r}{k_{\text{wall}} Wq_\alpha}$$

and with a steady-state gain of 1.0. Often, $k_{\text{wall}} Wq_\alpha$ is unknown, and we can then substitute

$$\frac{W_{\text{heat}}}{T - T_{\text{wall}}}$$

to give

$$\tau_{\text{wall}} = \frac{W_{\text{react}} s_r}{W_{\text{heat}}} (T - T_{\text{wall}})$$

Reasoning the same way, the reactor outside wall (jacket side) responds to the changes in the inner wall (contents side), the thermal time constant for the reactor jacket-wall being given by:

$$\tau_{\text{rejakt}} = \frac{W_{\text{wall}} s_{\text{wall}} x}{k_{\text{rewall}} Wq_\alpha}$$

where $W_{\text{wall}} = \text{weight of wall}$
$s_{\text{wall}} = \text{specific heat of wall}$
$k_{\text{rewall}} = \text{thermal conductivity (since only the construction material of the reactor is considered)}$
$x = \text{wall thickness}$

If, however,

$$\frac{x}{k_{\text{rewall}} Wq_\alpha}$$

is unobtainable then use the following:

$$\tau_{\text{rejakt}} = \frac{W_{\text{wall}} s_{\text{wall}}}{W_{\text{heat}}} (T_{\text{wall}} - T_{\text{jakt}})$$

where T_{jakt} is temperature of the wall on the jacket side.

The temperature of the outside wall of the reactor will respond to the temperature of the coolant, the thermal time constant for which is given by:

$$\tau_{cool} = \frac{W_{cool}s_{cool}}{k_{cool}Wq_\alpha}$$

where W_{cool} = weight of the coolant in the jacket
s_{cool} = specific heat of the jacket
k_{cool} = heat transfer coefficient

If $k_{cool}Wq_\alpha$ is unknown, then use the following:

$$\tau_{cool} = \frac{W_{cool}s_{cool}}{W_{heat}}(T_{wall} - T_{avgc})$$

T_{avgc} = average coolant temperature
The time constant for the temperature bulb is given by:

$$\tau_{bulb} = \frac{W_{bulb}s_{bulb}}{k_{bulb}Wq_{bulb}}$$

where W_{bulb} = weight of the bulb
s_{bulb} = specific heat of the bulb
Wq_{bulb} = surface area of the bulb

Data on the response of bulbs and on heat transfer conditions for most types of thermal systems are available in published form.

Each of the foregoing time constants will have an effect on the average and measured temperature of the coolant, and since we are using water flow to achieve this we must include the effect on the coolant in our considerations. Let us add a stream of water at a flow of Q_w and a temperature of T_w to the coolant recycle operating at $Q - Q_w$ and a temperature of T_{w2}, then the mixture returning to the reactor is Q at a temperature T_{w1}. The heat balance (in terms of flow) is:

$$QT_{w1} = Q_wT_w + (Q - Q_w)T_{w2}$$
$$= Q_wT_w + QT_{w2} - Q_wT_{w2}$$

rearranging the terms, we have:

$$Q_wT_{w2} - Q_wT_w = QT_{w2} - QT_{w1}$$
$$Q_w(T_{w2} - T_w) = Q(T_{w2} - T_{w1})$$
$$T_{w2} - T_{w1} = \frac{Q_w}{Q}(T_{w2} - T_w) \tag{12.1}$$

Relating this to heat load, we have:

$$T_{w2} - T_{w1} = \frac{W_{heat}}{Qs_{cool}} \tag{12.2}$$

We require the average coolant temperature; hence, we use:

$$T_{w2} = T_{avg} + \frac{T_{w2} - T_{w1}}{2}$$
$$= T_{avg} + \frac{W_{heat}}{2Qs_{cool}} \tag{12.3}$$

Substituting Equation (12.3) into Equation (12.1) and noting that $T_{w2} - T_{w1}$ is in fact W_{heat}/Qs_{cool} from Equation (12.2), we have:

$$\frac{W_{heat}}{Qs_{cool}} = \frac{Q_w}{Q}\left[\left(T_{avg} + \frac{W_{heat}}{2Qs_{cool}}\right) - T_w\right]$$

$$= \frac{Q_w T_{avg}}{Q} + \frac{Q_w W_{heat}}{2Q^2 s_{cool}} - \frac{Q_w T_w}{Q}$$

Transposing we have:

$$\frac{Q_w T_{avg}}{Q} = \frac{W_{heat}}{Qs_{cool}} - \frac{Q_w W_{heat}}{2Q^2 s_{cool}} - \frac{Q_w T_w}{Q}$$

Multiplying by F/F_w and solving for T_{avg}, we have:

$$T_{avg} = \frac{W_{heat}}{Q_w s_{cool}} - \frac{W_{heat}}{2Qs_{cool}} + T_w$$

$$= T_w + \frac{W_{heat}}{s_{cool}}\left(\frac{1}{Q_w} - \frac{1}{2Q}\right)$$

To obtain the reaction process gain we differentiate with respect to F_w (cooling water inflow) to give:

$$\frac{dT_{avg}}{dQ_w} = -\frac{W_{heat}}{s_{cool}Q_w^2}$$

This shows that the relationship between coolant temperature and water flow is non-linear, and an equal-percentage valve may be used to partially correct the situation.

Most temperature control loops are difficult for the following reasons:

1. Time constants are interactive and difficult to identify.

2. Distributed lags make performance hard to predict.

3. All heat transfer processes are nonlinear.

One can conclude from the entire foregoing analysis that there is no such thing as a typical temperature control system. Each application must be considered individually and solutions based on the specific requirements provided.

CONTROL OF A MIXING PROCESS

We shall now consider a *composition* control loop to see what we can learn from it. Composition or make-up is a property of a material and, as such, it is part of the flowing stream of that material. Dead time is always in the loop, and incomplete mixing and the difficulty of obtaining samples further complicate matters. To a very large extent the composition of a material is a function of the machinery producing it.

Let us consider the simple blending system shown in Figure 12.2. In this system we are producing a mixture only—there is no chemical reaction—and therefore the object is to control the percentage of a single component by adding the required amount of concentrate to the inflow of effluent from the process before it is disposed of by other means. If the agitator could mix the contents completely in zero time, the problem would be reduced to one of a single capacity, and it would be easy to control. Unfortunately, it is not possible to have complete mixing instantaneously

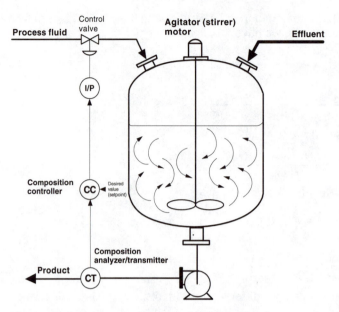

Figure 12.2: Schematic diagram of a simple blending system.

and as a result we are therefore burdened with process lag or dead time, which in turn places limitations on the controller and the speed of response.

There is a level control loop that maintains a fairly constant volume in the vessel, but in the interest of simplicity and clarity this loop has not been shown. If V was the volume of the concentrate and Q the flow rate, and if no mixing at all took place, then there would be a plug of material leaving the inlet and reaching the outlet after a time interval of V/Q. For the case we have just chosen to describe, the dead time $\tau_d = V/Q$, and the lag (delay) $\tau_1 = 0$. If, on the other hand, there was perfect mixing, then $\tau_d = 0$ and $\tau_1 = V/Q$. It is within these two limits that all real processes fall. Most mechanical agitators of the type illustrated behave and are treated as pumps that circulate material at a uniform rate from the bottom to the top.

Let the circulation rate be Q_{circ}; then the time taken for a particle to travel from the bottom to the top will be given by:

$$\tau_d = \frac{V}{Q_{circ} + Q} \tag{12.4}$$

How well the mixing is done can be defined as the ratio of upflow to downflow:

$$\frac{Q_{circ}}{Q_{circ} + Q}$$

Because the whole content of the vessel is not fully mixed instantly, the time it takes to produce a completely mixed product in part of the vessel can be considered the delay or time constant for the vessel:

$$\tau_1 = \frac{V}{Q} \frac{Q_{circ}}{Q_{circ} + Q} \tag{12.5}$$

If we write Equation (12.4) in the same format as τ_1, we have:

$$\tau_d = \frac{V}{Q} \frac{Q}{Q_{circ} + Q} \tag{12.6}$$

that provides a basis upon which to make a comparison between the two.

If we add Equations (12.5) and (12.6), we have:

$$\tau_1 + \tau_d = \frac{V}{Q}$$

This shows that:

- the average particle cannot be held in the vessel longer than the retention time V/Q, irrespective of whether it is mixed or not, confirming that no mixing can be perfect.

- The ratio τ_d/τ_1 or Q/Q_{circ} can be used as a means of determining how difficult it is to control the production of a mixture.

Please remember that the example used is a very oversimplified one and has been chosen only to illustrate the following:

1. The actual mechanism of mixing is not discrete.

2. Flow generated by the agitator is not unidirectional, as it is turbulent and almost certainly multidirectional.

3. Since there is flow, there is bound to be some mixing due to the small amount of turbulence in producing the flow.

4. Some mixing will occur, even without an agitator, due to diffusion taking place.

SUMMARY

1. The relationship between coolant temperature and water flow is nonlinear, and an equal-percentage valve will partially correct the situation. Most temperature control loops are difficult for the following reasons:

    (1) Time constants are interactive and difficult to identify.

    (2) Distributed lags make performance hard to predict.

    (3) All heat transfer processes are nonlinear.

2. There is no such thing as a typical temperature control system. Each application must be considered individually and solutions provided based on the specific requirements.

3. The composition of a material is a property of the material itself, and when considering a liquid the composition is part of the flowing stream of that material.

4. Dead time is always part of a composition loop. Incomplete mixing and difficulty of obtaining samples further complicate matters. To a very large extent, the composition of a material is a function of the machinery producing it.

5. For a single-process fluid and a dilutant being mixed in a vessel without chemical reactions, if the agitator could mix the contents completely in zero time, then the

problem in producing a mixture of the two components would be reduced to one of a single capacity and controlling it would be easy.

6. A particle in a fluid mixture cannot be held in the production vessel for a period longer than the retention time, which is defined as the ratio of the volume of the fluid and its flow rate (V/Q).

7. The ratio of retention time to process lag (τ_d/τ_1) can be used to determine how difficult it is to produce a mixture.

CHAPTER 13

Surge Control of Centrifugal Compressors

INTRODUCTION TO CENTRIFUGAL COMPRESSORS

A compressor is a machine that is used to compress a gas. This means that it changes the volume of the fluid it takes in at the suction end to a much smaller volume at the discharge end. Compressors are used in process industries as a means to pump up gases to suitable conditions for further processing in, say, reformer reactors or furnaces, or for transportation by pipelines for either direct use or storage. The physical act of compression, by reducing the fluid volume, changes its internal energy; in turn affecting its thermodynamic properties of pressure and temperature, both of which are increased. The temperature rise due to compression is not of great significance, but indicative of the upward change in heat energy that requires removal by coolers in the discharge line. The pressure rise, on the other hand, in the form of a much higher pressure at the discharge end of the machine, is the most important and sought-after effect. This increase in pressure energy can be made visible, quite easily, by pressure gauges fitted to the suction and discharge pipes of the machine involved. The *pressure ratio*, a dimensionless number derived from the value of the discharge pressure divided by the value of the suction pressure, is fixed by the design of the machine. Sometimes the term turbo-compressor is used; this really means that the prime mover of the compressor is a turbine (steam or gas driven).

SURGE IN CENTRIFUGAL COMPRESSORS

Since the process uses compressed gas, the gas demand will vary dependent on the formulation of the product as processing proceeds. This changing requirement by the process has an effect on the compressor itself, most especially when the gas demand suddenly drops low and a condition called *surging* occurs. Surge is a phenomenon that is particular to the combined use of gases with centrifugal compressors. The sudden drop in the amount of fluid required on the *demand side* (the discharge end of the compressor equipment) means that the *supply side* (the suction end of the compressor equipment) has more fluid than is required by the situation, thus causing the imbalance. Surge is an unstable condition liable to propagate very swiftly, which, if left unattended, would seriously damage the machine itself and very likely the plant around it, also.

The objective of any control system used in such an application is to monitor the process continuously, detect the beginnings of the potentially dangerous conditions just described, and automatically prevent the machine from entering the surge regime. The avoidance of surge conditions is accomplished by maintaining the flow of the fluid through the compressor, and the technique by which avoidance is accomplished is by diverting some of the fluid from the discharge of the machine back to the suction side. In other words, we arrange to "fool" the machine into believing that there has been no change. We really have not altered the fact that demand has fallen, but what we have accomplished is the prevention of extremely costly repair or replacement of the compressor and plant. The remainder of the control system outside the surge prevention should be designed to ensure that the supply is sufficient for the demand. In this chapter we shall first analyze the older, linear type of controls, and later we shall show the newer techniques available with modern microprocessor-based controllers. We shall be considering only the centrifugal compressor, as it is this type that is most widely used in the process industry.

The following outlines the principles of centrifugal compressor control only. It is vital that all national and local codes of practice for safety, electrical wiring, and operating procedures be strictly adhered to.

OPERATING RANGE OF A CENTRIFUGAL COMPRESSOR

Centrifugal compressors are normally driven by prime movers (steam or gas turbines, electric motors, etc.) of high rotational speed. The performance of the machine is normally depicted by characteristic curves that relate the rotational speed, thermodynamic head (pressure rise), and flow rate through the machine. Figure 13.1 is a typical example of the characteristic curves for a variable-speed compressor. The speed curves, which have a drooping character, show that increasing the flow rate reduces the generated head. When we move along a particular speed curve in the direction of increased flow rate, we reach a choked condition known as the *stonewall*. The stonewall represents the limit of the machine throughput, which is accompanied

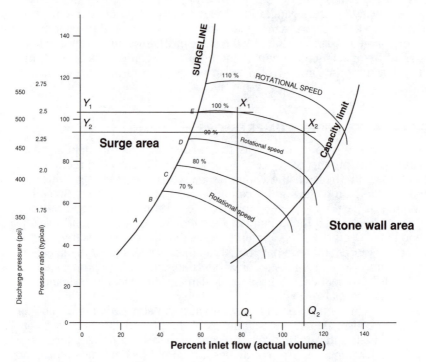

Figure 13.1: Characteristic curves for a typical variable-speed compressor.

by a loss of the generated head and establishes the capacity limit for that particular rotational speed. If we now move along the speed curve in the direction of reduced flow rate, we will approach a point of instability called *surge*. At the point of instability the generated head collapses very suddenly, and very rapidly, causing flow reversals through the machine accompanied by severe pulsation and damaging vibration. In view of what has been said, the normal operating range of the compressor must lie within these two limiting constraints of surge and stonewall.

THEORY OF COMPRESSION

In any continuous compression process the relationship between the absolute pressure and the Volume is expressed as:

$$pV^n = C = \text{constant}$$

If these were to be plotted for each value of n a family of polytropic curves would be derived, and work would be performed when proceeding along any one of the curves from p_1 to p_2, where p_2 is the higher pressure, or we can say:

$$W = \int p \, dV$$

where W is the work done. Therefore, the amount of work done is dependent on the polytropic curve involved and increases with an increase in the value of n. For adiabatic (no heat added or subtracted) compression, the work involved is the least and is signified by $n = \gamma$, where γ is the ratio of specific heats (c_p/c_v). Most compressors work on a polytropic curve that approaches the adiabatic, hence, the calculations are based on the adiabatic:

$$\frac{p_2}{p_1} = \left(\frac{V_1}{V_2}\right)^{\gamma}$$

$$\frac{T_2}{T_1} = \left(\frac{V_1}{V_2}\right)^{\gamma-1}$$

from which we can say:

$$\frac{p_1}{p_2} = \left(\frac{T_2}{T_1}\right)^{\frac{\gamma}{\gamma-1}}$$

A variable-speed machine of a single compression stage normally exhibits a parabolic surge line. Multiple-stage variable-speed machines normally exhibit a surge line that breaks away from the parabolic form and bends forward to the right as shown in Figure 13.1. Similar surge line characteristics apply to constant-speed machines that have inlet guide vane manipulation capability.

With the foregoing limitations imposed, the objective for this type of compressor is to keep the machine operating at the highest performance for the chosen speed and within the operating range. In all single-stage centrifugal pumps and compressors, the flow through the machine varies directly with the rotational speed, and the head (pressure) produced varies as the square of the rotational speed. It is possible to write these two relationships as simple math equations:

$$Q = kN \tag{13.1}$$

$$H_{ad} = k_1 N^2 \tag{13.2}$$

where Q is the flow rate, H_{ad} is the adiabatic head, and N is the rotational speed. If we combine Equations (13.1) and (13.2), we have

$$H_{ad} = k_2 Q^2 \tag{13.3}$$

Plotting Equation (13.3) will give a parabola. It is this expression that compressor manufacturers use to produce the surge curves, such as the one shown in Figure 13.1, that they provide with their machines. The adiabatic head is not measurable directly, but it is related to other measurable parameters such as compression ratio, gas molecular weight, gas specific heat ratio, and gas inlet temperature.

INHERENT INSTABILITY OF CENTRIFUGAL COMPRESSORS

Since we are now aware that the compressor can be driven into a condition of instability at any of the chosen speeds within the operating range, let us discuss how this unwanted and potentially dangerous situation arises and the controls that can be designed to avoid it. Using Figure 13.1, suppose the machine were running at 100% speed and operating at X_2; for this condition, the actual inlet flow is Q_2 and the head is Y_2. If now there were to be a change in the load, i.e., decrease:

1. This will cause the operating point to move to the left, say, to X_1, for the speed remains the same.

2. The actual inlet volume flow decreases, say, to Q_1, which will result in an increase in head to Y_1.

3. If the demand continues to fall, the result will be to further push the operating point leftward until it reaches the surge line.

4. At this point the compressor is unable to increase the head for the given rotational speed, since the speed curve is practically flat at this point.

5. What follows is a rapid pulsation in both the flow and discharge pressure, which produces high-frequency reversals in the axial thrust on the compressor shaft. In some cases the axial thrust can become so violent that mechanical damage is caused to the machine.

THE SEARCH FOR A SOLUTION—FLUID MECHANICS OF COMPRESSORS

In order to provide a control system as a solution, we must first understand the phenomenon from the viewpoint of the machine and the compressor system variables. To appreciate what occurs in a compressor, it is suggested we start by observing what occurs when a fluid is made to flow across a body. In a compressor the "body" is the vanes that make up the rotor and stator of an axial machine or the impeller of a centrifugal machine. All bodies have a surface, and when a fluid passes over this surface, a boundary layer develops. The boundary layer on a solid surface has been shown to depend predominantly on the shear stress of the surface on the fluid, and to be nearly independent of the properties of the main stream. If pressure increases in the direction of motion, the boundary layer may affect the main stream by producing *breakaway*, sometimes also called *separation*. This is where problems with the machine start as we shall see.

Figure 13.2 will help us visualize the situation that occurs. In the diagram, we consider a flat surface *OABCD*, along which there is a flow of fluid forming a boundary layer on the leading edge *OAB*. Following the diagram from left to right, one can

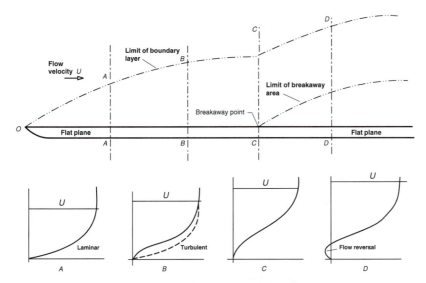

Figure 13.2: Boundary layer on a flat surface. Each of the plots shown at bottom refers to the conditions at the corresponding section above. In the plots, the x axis is the velocity U, and the y axis is pressure.

observe that a separation region builds up the further we move from the leading edge. C is the point of separation, and at position D the opposing pressure force is of such a magnitude that it actually causes the forward movement of the fluid near the surface to cease and reverse direction. The point at which the reversal is initiated is termed the *breakaway point*. This flow reversal is the cause of the severe problems encountered in rotary compressors.

How we avoid flow reversal is best seen by understanding how the fluid flows through the compressor vanes, and how we can maintain the velocity relationships on the suction and discharge sides of the machine. An axial compressor is always thought of as a row of rotor blades followed by a row of stator blades. For simplicity and the purposes of this description, we shall initially consider a single rotor/stator pair in detail. The velocity triangles are depicted as shown in Figure 13.3 and obtained as follows. Considering the first rotor section of the compressor, suppose the fluid approaches the rotor with an absolute velocity C_0 and at an angle α_0. With the blade speed equal to u, the relative velocity u_{r0} is the vector difference between the fluid (absolute) and blade velocities. If the angle formed by the joining of these vectors is β_0, the rotor blade inlet angle must also be approximately β_0. This is necessary to ensure that there is a smooth transition in the flow of the fluid along the blade. The path between successive blades is made divergent in order to diffuse the fluid and to reduce the relative forward velocity of the fluid. If the relative exit velocity is u_{r1} and the exit angle is β_1, then by vector addition of the relative exit velocity and the blade speed we will obtain the absolute velocity C_1 and its direction α_1 at the outlet of the rotor. C_1 is always greater than C_0 on account of the work done by the fluid on the rotor. Once again the fluid is diffused into the stator, where the inlet angle of the blades is made approximately β_1, the intent as before being to have minimum obstruction to the forward flow of the fluid.

If the fluid is to enter another similar stage, then $C_2\alpha_2$ must be made equal to $C_0\alpha_0$. The work done in the stage is given by:

$$W = mu(C_{1w} - C_{0w}) \tag{13.4}$$

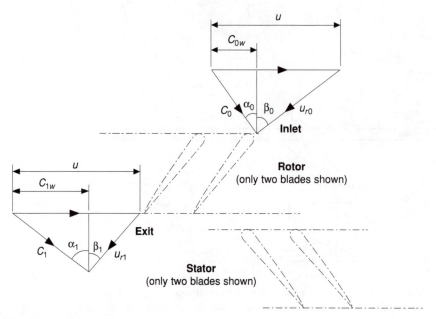

Figure 13.3: Turbine blade velocity diagram. u = blade velocity; C_0 = absolute fluid velocity.

where W is the rate of work transfer, m is the rate of mass flow, and u is the blade speed.

In the stages just described, there is an increase in enthalpy and pressure of the process fluid in both the rotor and the stator. It has been found that the losses associated with fluid friction and blade tip clearance are minimized when the pressure rise is equally divided between rotor and stator. For this reason, a 50% reaction is normally used, a result of this being that the blade shape is the same for both. In a compressor, the design is such that the flow is always diffusing, whereas in a turbine it is always accelerating. The natural tendency of a diffusing flow is to break away from the walls of the diverging passage, reverse its direction, and flow back in the direction of the falling pressure gradient.

Equation (13.4) is a one-dimensional expression, which is suitable for turbines but does not adequately cover the need for determining the work done in a compressor stage. In order to obtain the average work done in a compressor, there are two important points to consider. First, the result obtained from the equation has to be multiplied by an empirical factor representing the limit to the work that can be put into the stage, which for the first few stages is nearer unity, but typically 0.86. The other point is the limitation imposed by the rate of divergence of the passage between successive blades, which is a function of the angle through which the fluid is deflected, which for the rotor is $(\beta_0 - \beta_1)$ and for the stator $(\alpha_0 - \alpha)$. The term $(\beta_0 - \beta_1)$ reflects the limitation on the pressure ratio that can be obtained which, when combined with the stressing of the rotor blades, determines the permissible relative gas velocities (which must be subsonic) in order to avoid the formation of shock waves.

It is possible to increase the delivery pressure of a positive displacement compressor running at constant speed by throttling the discharge and reducing the mass flow. The mass reduction/pressure increase process can continue, theoretically, until the mass flow is zero. However, the axial compressor operates on a restricted mass flow range for any given speed; if the mass flow rate is reduced below a particular point in this range, then the directions of the velocities relative to the blades are so different from the actual angle of the blades that they cause the flow to break down completely, causing the compressor to go into the surge state. This is a condition that does not persist continuously, and the phenomenon exhibits itself as a series of flow breakdowns and recoveries. Accompanying each surge and recovery is induced mechanical vibration of the rotating parts of the machine. These vibrations unbalance the rotor and cause physical damage to the blades of the rotor and stator bearings, shaft, and sometimes even the casing of the machine itself. The repair or replacement costs are in most instances high, and for very large machines can be counted in millions.

LINEAR METHOD OF COMPRESSOR CONTROL

It is important to remember that there is no direct method of providing a measurement of the adiabatic head; however, the compression ratio is an available parameter that is readily calculated and is related to the adiabatic head. Let R_v be the compression ratio then:

$$R_v = \frac{p_2}{p_1}$$

where p_1 is the suction pressure and p_2 is the discharge pressure. The following equation can be found in *Perry's Chemical Engineers Handbook*. This equation, relevant to the FPS (Foot Pound Second) system of measurement, and differs from the equation for the SI (Système International) system of measurement, relates the compression ratio to the adiabatic head.

$$H_{ad} = \frac{\gamma}{\gamma - 1} \Re T \left[\left(\frac{p_2}{p_1} \right)^{(\gamma-1)/\gamma} - 1 \right] \tag{13.5}$$

where γ = specific heat ratio = c_p/c_v
$\Re$ = universal gas constant = 1545 ft/lbmol·°R
T = absolute temperature, °R
p_1 = suction pressure
p_2 = discharge pressure

from which we can say:

$$\left(\frac{p_2}{p_1} \right)^{(\gamma-1)/\gamma} - 1 = \frac{H_{ad}}{\frac{\gamma}{\gamma-1} \Re T} \tag{13.6}$$

If we write ϕ instead of $(\gamma - 1)/\gamma$ we have

$$\left(\frac{p_2}{p_1} \right)^{\phi} = 1 + \left(H_{ad}\phi \frac{1}{\Re T} \right)$$

$$\frac{p_2}{p_1} = \left(1 + \frac{H_{ad}\phi}{\Re T} \right)^{1/\phi} \tag{13.7}$$

If W_m is the molecular weight of the gas and z the supercompressibility factor, we can write Equation (13.7) as:

$$\frac{p_2}{p_1} = \left(1 + \frac{H_{ad}W_m\phi}{\Re Tz}\right)^{1/\phi}$$

$$= \left(1 + \frac{H_{ad}W_m\phi}{1545Tz}\right)^{1/\phi} \tag{13.8}$$

But since p_2/p_1 is the compression ratio $= R_v$, we can write Equation (13.8) as:

$$R_v = \left(1 + \frac{H_{ad}W_m\phi}{1545Tz}\right)^{1/\phi}$$

For ease of reference, we also show the SI version of the same compressor formula, also to be found in *Perry's Chemical Engineers Handbook*, and for the purpose of showing the slight difference in form and the units involved,

$$H_{ad} = \frac{\gamma}{\gamma - 1}\frac{\Re T_i}{9.806}\left[\left(\frac{p_2}{p_1}\right)^{(\gamma-1)/\gamma} - 1\right]$$

where $\Re =$ universal gas constant, J/kg·°K $= 8314/$molecular weight
          $T_i =$ inlet gas temperature, K
          $p_1 =$ inlet absolute pressure, kPa
          $p_2 =$ discharge absolute pressure, kPa

If the compressor handles only one type of gas at near constant inlet temperature, molecular weight, specific heat, and supercompressibility, then in Equation (13.8) the terms T, W_m, ϕ, and z will be constants, and therefore can be replaced by a single constant K_1. This will result in an even simpler equation:

$$R_v = (1 + K_1 H_{ad})^{1/\phi} \tag{13.9}$$

The equation just derived is nonlinear, but in the cases where the gas is air, natural gas, or another of the more common gases and the temperature is around ambient, then $(R_v - 1)$ can be substituted for H_{ad} without distorting the curves too greatly.

    M. H. White of the Foxboro Company produced the graph in Figure 13.4 to illustrate the small variation that exists between a trace of the adiabatic head versus actual inlet volume, on the one hand, and the trace of $(R_v - 1)$ versus actual inlet volume, on the other, for the same range of heads. The gas involved is natural gas and shows the error involved in substituting $(R_v - 1)$ for the adiabatic head H_{ad}. Please note that at greater heads the error can increase and may require adjustments to be made to correct it. However, in the majority of cases the error is small enough to be neglected.

    For all centrifugal machines, pumps, and compressors, the adiabatic head is proportional to the square of the speed:

$$H_{ad} \propto N^2$$

or we can say

$$H_{ad} = K_2 N^2$$

and the flow rate is directly proportional to the speed,

$$Q \propto N$$

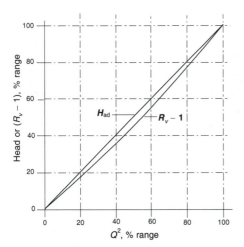

Figure 13.4: Graph showing flow versus adiabatic head and pressure ratio factor (not to scale). (From M. H. White, "Surge Control of Centrifugal Compressors," *Chemical Engineering*, July 18, 1988, pp. 39–42. Reproduced with permission of the publisher.)

or we can say

$$Q = K_3 N$$

in view of the last two equations, we can combine, substitute for N, and rewrite the relationship as:

$$H_{ad} = K_4 Q^2 \qquad (13.10)$$

But as we have seen, $R_v - 1$ can be substituted for H_{ad} with only very small error; thus, we can write

$$R_v - 1 = K_4 Q^2 \qquad (13.11)$$

The differential pressure across the compressor is:

$$\Delta p = p_2 - p_1$$

but

$$p_2 = p_1 R_v$$

therefore

$$\Delta p = p_1 R_v - p_1$$
$$= p_1 (R_v - 1)$$

therefore

$$R_v - 1 = \frac{\Delta p}{p_1}$$

Hence, we can say:

$$\frac{\Delta p}{p_1} = K_2 Q^2 \qquad (13.12)$$

To measure the actual inlet flow we must have a primary device in the suction line. If this is a venturi meter or an orifice plate, then we will have the flow related to the differential head across the device, but this will be in volumetric terms, which must be converted to mass. The equation below gives the relationship.

Let k_3 be the meter constant, V is the specific volume, m is the mass flow, T is the temperature, and p is the pressure with subscripts 1 to indicate inlet. Then

$$V = k_3 \frac{T_1}{p_1}$$

$$m = k_4 \sqrt{\frac{H_d p_1}{T_1}}$$

Because

$$Q = mV$$

$$= k_5 \sqrt{\frac{H_d p_1}{T_1}} \left(\frac{T_1}{p_1} \right)$$

$$= k_5 \sqrt{\frac{H_d T_1}{p_1}} \qquad (13.13)$$

Assuming the inlet temperature T_1 is constant, then:

$$Q = k_6 \sqrt{\frac{H_d}{p_1}}$$

$$Q^2 = k_7 \frac{H_d}{p_1} \qquad (13.14)$$

Substituting this into:

$$\frac{\Delta p}{p_1} = K_2 Q^2$$

we have:

$$\frac{\Delta p}{p_1} = k_8 \left(\frac{H_d}{p_1} \right)$$

$$\Delta p = k_8 H_d \qquad (13.15)$$

This last equation is the one that will be used to design the control system; however, there are two important conclusions that can be drawn:

1. $\Delta p \propto H_d$, i.e., the relationship is linear.

2. The surge is not affected by the variations in the suction pressure p_1.

From Figure 13.5, it is clear that if we move the control line toward the left, then we shall be approaching the surge regime. On the other hand, moving the control

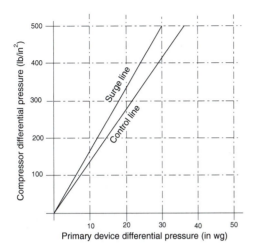

Figure 13.5: Graph of surge line with the control line displaced approximately 20% of compressor differential pressure (not to scale). (From M. H. White, "Surge Control of Centrifugal Compressors," *Chemical Engineering*, July 18, 1988, pp. 39–42. Reproduced with permission of the publisher.)

line toward the right and therefore away from the surge line will result in a waste of compressor motive power. It follows, therefore, that there must be a position for the control line that is optimal, both in avoiding surge and conserving motive power; this position has generally been found to be placed about 10% to 20% to the right of the surge line. It must be pointed out that as a matter for an individual application, there are instances when this control margin is reduced, and the compressor operated at margins less than 10%.

While first developing the theory, we made the assumption that the temperature of the inlet gas was constant. This is not what happens in reality, for the inlet temperature can vary, and we must be able to act accordingly. Let us start by looking once again at the following equation, derived from Equation (13.8), which relates the compression ratio to the inlet temperature:

$$R_v = \left(1 + \frac{H_{ad} W_m \phi}{1545 Tz}\right)^{1/\phi}$$

In this instance we shall assume that H_{ad}, W_m, and ϕ are constant. However, it must be pointed out that z will be affected by changes in temperature; but, for the purposes of the analysis, we shall assume that it is constant also. From these assumptions, we shall reduce the equation to:

$$R_v = \left(\frac{1 + K_3}{T}\right)^{1/\phi} \tag{13.16}$$

Figure 13.6, also taken from M. H. White's work on compressor controls, shows the plot of Equation (13.16) for two gases, air and natural gas whose molecular weights and specific heat ratios differ widely. The curves show the changes over a range of 120°F and adiabatic head of 40,000 ft·lb/lb. We have already shown that

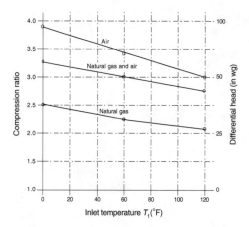

Figure 13.6: Graph of inlet temperature versus compression ratio and differential head for air and natural gas (not to scale). (From M. H. White, "Surge Control of Centrifugal Compressors," *Chemical Engineering*, July 18, 1988, pp. 39–42. Reproduced with permission of the publisher.)

$\Delta P = p_1(R_v - 1)$, from which it will be seen that ΔP will vary directly with $(R_v - 1)$ for any p_1; and the shapes of the curves for R_v versus T_1 and ΔP versus T_1 are approximately the same.

Let us now consider the effect of changes in the differential pressure across the orifice, i.e., changes in the suction flow due to the temperature variations. For this we shall need to reconsider Equation (13.13):

$$Q = k_5 \sqrt{\frac{H_d T_1}{p_1}}$$

$$Q^2 = k_9 \left(\frac{H_d T_1}{p_1}\right)$$

Therefore

$$H_d = k_9 \left(\frac{Q^2 p_1}{T_1}\right)$$

Hence, for specific values of Q and p_1, we can say:

$$H_d = \frac{k_{10}}{T_1} \tag{13.17}$$

where k_{10} is the combined constant.

Figure 13.6 shows that both Δp and H_d vary inversely with T_1; or if the temperature effects on both were identical, the system would always compensate. This is really not the case. The actual effect is a change in the slope of the surge curve due to the differing magnitudes of the effect that the temperature has on each parameter. If the temperature effects are too great, then this also must be taken care of; however, temperature compensation must always be provided in applications where the compressor is operated near surge for most of the time.

Changes in molecular weight of the gas will affect the control of the compressor and also have to be accounted for. In most instances a compressor will be handling a single-composition gas all of the time, but there are times when this is not the case. Changing the gas composition will in most instances result in a change in molecular weight, which in turn will affect the surge curve, which is essentially a plot of adiabatic head versus inlet volume, and as such we shall have to go back to the Equation (13.8) for R_v to determine the effect:

$$R_v = \left(1 + \frac{H_{ad} W_m \phi}{1545 Tz}\right)^{1/\phi}$$

For fixed values of H_d, T, and z, this results in:

$$R_v = (1 + K_4 W_m \phi)^{1/\phi} \tag{13.18}$$

where K_4 is the combined constants.

Equation (13.18) shows that there is a nonlinear relationship between the molecular weight and the pressure drop across the compressor. Furthermore, a molecular weight change will be reflected in a new value for the specific heat ratio and hence ϕ. For some of the lighter hydrocarbons there are known definite relationships between the molecular weight and ϕ, and an accurate calculation can be made; but for those cases where there is no correlation between the specific heat ratio and the molecular weight, individual calculations must be made to determine the magnitude and the direction of the change. In these cases we shall have to use Equation (13.13) and modify it to account for the molecular weight:

$$Q = k_5 \sqrt{\frac{H_d T_1}{P_1}}$$

$$= k_5 \sqrt{\frac{H_d T_1}{P_1 W_m}} \tag{13.19}$$

Hence

$$H_d = \frac{k_5 Q^2 p_1 W_m}{T_1}$$

Using specific values for Q, p_1, and T_1, we can say:

$$H_d = k_{11} W_m \tag{13.20}$$

Comparing Equations (13.18) and (13.20) will show that the system cannot be self-compensating for changes in molecular weight, for any change shifts the surge line as a result. The reason for the shift is that the change in molecular weight will alter the effect of the specific heat ratio. If the same compressor is to be used for different, unrelated gases, then the slope of the surge line must be calculated for each gas that has to be processed. In the event that, because of the difference in the gases, it is impossible to effect control by the assignment of a single control line, then adjustments to the control line must be made accordingly and, this is usually done manually.

As stated earlier, and repeated here for convenience, Equation (13.15) is the basic equation used to control a compressor:

$$\Delta p = k_8 H_d$$

The instrumentation to implement is shown in Figure 13.7. From the figure it can easily be seen that the measurement (MV in the figure) for the controller is produced

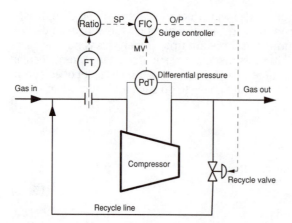

Figure 13.7: Basic control scheme for a compressor
(linear method).

by the differential-pressure transmitter (PdT), which must have a high range, since
it is measuring the pressure difference across the compressor and the constant of
proportionality provided by the ratio station. Furthermore, the flow indicating controller (FIC) must have some means of preventing the integral term saturating, since
for the most part the compressor will be made to operate some distance to the right
of the control line, and therefore under "safe" conditions for the machine itself.

    If a standard P + I controller is used with its setpoint (SP in the figure) at a
value on the control line, there is going to be a difference between the measurement
(produced by FT the flow transmitter) and the setpoint and, because of this the
controller will drive the output (O/P in the figure) in a direction to attempt to reduce
the difference. The inevitable result is that the controller output will be driven to its
limit and beyond, possibly up to the controller circuit power supply level. If, now, the
measurement starts to approach the control line (setpoint, set manually by the ratio
unit in the figure) rapidly, then it will for sure overshoot before the controller will
respond, for the simple reason that the integral term is saturated and will unwind at
the rate set on the integral term. During this interval it is certain that the compressor
will be in surge, a condition that we must avoid.

THE BATCH ACTION CONTROLLER

    The surge controller must be provided with a means of avoiding the shortcomings
we have noted, which will involve a device to shift the proportional band to the same
side of the control line as the measurement. In this way, if the measurement enters the
proportional band, the controller will respond immediately. Figure 13.8 is a simple
block diagram of a typical arrangement for such a device. The batch unit shown is for
high limiting and operates as follows. The controlled output is applied to a summing
junction where it is subtracted from the signal set on the high limit module (Hi Lim);
the difference is the input to a high-gain amplifier, the output of which is applied to
a high signal selector (Hi Sel), the second input of which is obtained from a preload
module (Preload) that is in fact a low limiter. The selected signal forms a feedback
path to the integral control action component to limit its operation when the output
V goes beyond that set on the high limit module. As long as the output V is lower
than that produced by the high limit module, the amplifier output is high and will be
selected by the high signal selector (Hi Sel). This high signal is applied to one input

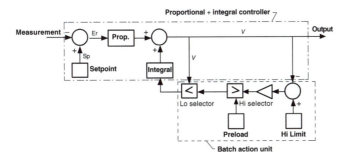

Figure 13.8: Schematic diagram of a batch action controller.

of a low signal selector (Lo Sel), the other input of which is the controlled output V; since the controlled output is lower than that produced by the high gain amplifier it will be selected and applied to the reset feedback input of the integral control action component and will allow it to process the signal in the normal way to produce the integral action of the controller.

When the controlled output is equal to that set on the high limit module, the input to the high gain amplifier goes to zero and the feedback signal is driven downward. If the error Er exceeds the proportional band set on the controller, the high gain amplifier may drive the feedback signal below the low saturation limit, making it unable to keep the controller output V and the high limiter output in balance. If now the error Er decreases then the controller output V could remain low for some time before control of the process can be regained. This extended period to regain control, the opposite of integral saturation, is minimized by the preload module, which prevents the high gain amplifier output going below that set in the preload module. When the error Er is reduced to zero then the output V would be equal to that obtained by the preload module. The adjustment provided by the preload module is invaluable to the operator when starting up the process, as it allows the amount of over- or undershoot to be trimmed to an optimum. Batch action can be configured in DCS systems.

ENHANCING THE BASIC LINEAR CONTROL SCHEME FOR GAS TEMPERATURE VARIATION

We can modify the control scheme given earlier to take care of temperature variations of the inlet gas, and while our method will not be strictly accurate theoretically, the amount of inaccuracy introduced is negligible. The important thing is that it keeps the control scheme simple and easy to understand.

In Figure 13.9 the ratio module shown in Figure 13.7 has been replaced with a divider module ($\div$ in the figure), the temperature measurement (from TT the temperature transmitter) being in absolute units. The equation being solved is:

$$\Delta P = k_{12}\frac{H_d}{T_1} \tag{13.21}$$

where k_{12} is a scaling constant. This sort of temperature correction is very often used on air compressors, and manual adjustments are made to the slope of the control line to account for seasonal temperature changes.

Sometimes because there is insufficient straight run of pipe to allow the flow measurement to be made on the suction side of the compressor, it is possible to install the flow primary device in the discharge pipeline. This arrangement is not

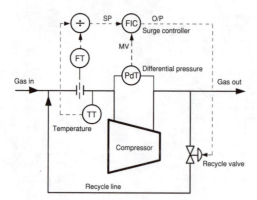

Figure 13.9: Basic control scheme for a compressor (linear method) modified for temperature correction.

ideal, but it must be remembered that while the flow is measured at the discharge side, the calculations must be referred to the suction side, so that a corrected value of the head will have to be calculated for use in the basic equation:

$$\Delta p = k_8 H_d$$

Under steady-state conditions, the amount of material flowing is the same at both the suction and discharge ends, and using weight then:

$$W_1 = W_2$$

where the subscripts 1 and 2 are inlet and outlet, respectively. If there is a primary device in the suction line, then the weight flow will be given by:

$$W = k_{13}\sqrt{H_d \frac{p_1}{T_1}} \qquad (13.22)$$

where W is the weight. Now, using the fact that the weight of material coming in is equal to the weight of material going out, we can say:

$$k_{13}\sqrt{H_{d1}\frac{p_1}{T_1}} = k_{13}\sqrt{H_{d2}\frac{p_2}{T_2}} \qquad (13.23)$$

We shall have to design the primary devices for the same maximum flow, and therefore the constants will be the same:

$$\sqrt{H_{d1}\frac{p_1}{T_1}} = \sqrt{H_{d2}\frac{p_2}{T_2}}$$

Therefore

$$H_{d1} = H_{d2}\left(\frac{p_2 T_1}{p_1 T_2}\right) \qquad (13.24)$$

We can see that we have referred the discharge flow to the suction flow even though we do not actually have a flow primary element in the suction line. This relationship will remain true for all maximum values on the scale of flow. The surge and control lines can be drawn in the manner previously described and the ratio calculated as before. The system can be detailed as shown in Figure 13.10.

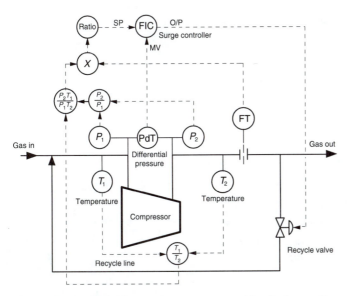

Figure 13.10: Variable-speed compressor with discharge flow measurement and pressure/temperature compensation. For clarity the computational details have been expanded individually.

It is quite normal for the suction temperature to vary considerably, but in some machines the temperature ratio T_2/T_1 is designed to be uniform; in such cases Equation (13.24) will be reduced to:

$$H_{d1} = k_{14}H_{d2}\frac{p_2}{p_1} \tag{13.25}$$

The effect of this is to simplify the system diagram, for it will eliminate the temperature transmitters and the associated computations, resulting in Figure 13.11.

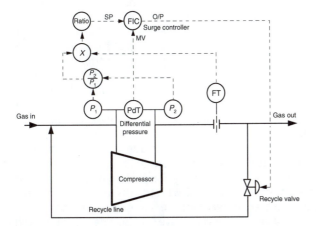

Figure 13.11: Variable-speed compressor with discharge flow measurement and uniform temperature ratio. For clarity the computational details have been expanded individually.

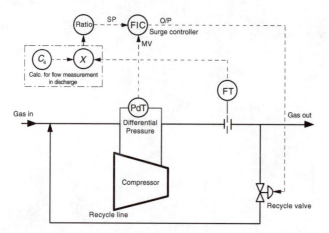

Figure 13.12: Single-speed compressor with discharge flow measurement. For clarity the computational details have been expanded individually.

Now if the compressor runs at one speed only, then the system can be simplified even further, for the simple reason that the compression ratio is constant, and the operating equation will be:

$$H_{d1} = k_{15}H_{d2} \qquad\qquad (13.26)$$

The system will then be as shown in Figure 13.12, where there is only one difference from the basic system, in that the flow primary element is in the discharge line.

NONLINEAR METHODS OF CONTROL—COMPRESSORS WITH INLET GUIDE VANES

Some compressors are fitted with guide vanes on the inlet, or have adjustable blades on the stator. These modifications are usually to be found on constant-speed machines and are used to control the discharge flow. As will be appreciated, a change in the position of the blades or the vanes will result in a change of slope of the surge line; therefore, for each position of the vane there will be a new control line. In most instances there is a nonlinear relationship between the slope and the vane position, which varies from machine to machine. The surge lines exhibit a nonparabolic form, as shown in Figure 13.13; hence, it is very difficult to determine a mathematical relationship for this. It has been found that opening up the inlet vanes has the effect of reducing the slope of the surge line—in other words, it tends to move the surge line toward the right—and closing the vanes tends to move the surge line to the left. Since the control line is on the right of the surge line and the surge line is determined for the maximum movement of the vanes, it follows that by using a linear approximation for the control line, the compressor is protected, but with the disadvantage that the safety margin—i.e., the separation distance between the surge line and the control line—is made too wide for economy, as the recycle valve will be open for a greater proportion of the time. The modern microprocessor-based controllers are able to replicate any shape curve using multipoint (x, y) coordinate characterization. This feature permits the implementation of the actual compressor characteristic without resorting to a linear approximation.

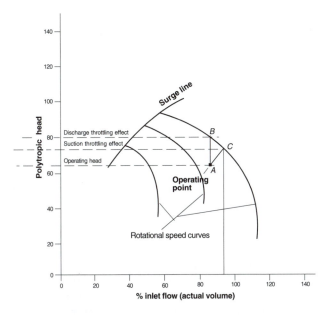

Figure 13.13: Effect of inlet guide vanes on shape of surge line.

THE ASYMMETRIC CONTROLLER

The asymmetric controller is shown schematically in Figure 13.14; where the input signal-characterizing feature is also included. The controllers used previously were standard instruments having P + I control terms and fitted with a batch action switch. A batch switch was required to prevent integral saturation (windup), since during normal operation the operating point of the machine at full flow conditions is well away from the setpoint of the controller. Without an anti-integral windup feature, the controller has to see a deviation of flow below setpoint before acting, which is too late, for the compressor has already entered the surge region. With a batch action switch, the controller will act as soon as the flow changes and before

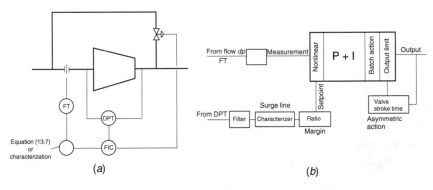

Figure 13.14: (*a*) Basic compressor control system. (*b*) Block schematic of asymmetric controller.

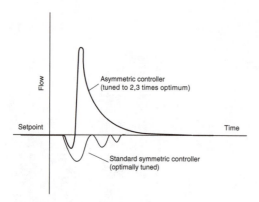

Figure 13.15: Closed-loop response curves.

crossing the setpoint into the control margin region. Batch action enables the upper range value of the proportional band to be placed on the control line. As standard PID controllers are symmetrical in action, any oscillation can cause hunting with equal amplitudes above and below the setpoint, leading to incursions into the surge region. What is required for surge protection is a control algorithm with asymmetric action and nonlinear gain control that makes the gain adaptive and causes it to increase as the measurement goes further into the margin region. This feature will make the corrective action stronger for cases of more severe decreases in flow rate. For a comparison between the two, symmetric and asymmetric control actions, see the closed loop responses in Figure 13.15.

A unique, microprocessor-based controller is now available that has the dual advantages of being able to operate a standard instrument suitable for any PID control application, while also incorporating all the features for asymmetrical action, setpoint characterization, flow compensation, digital logic capability to respond to plant shut-down signals, noise compensation, and many more, making it very suitable for advanced compressor control.

MULTIPLE COMPRESSOR CONFIGURATIONS

Compressors can be arranged in series or parallel configurations. We shall now consider each arrangement in turn.

COMPRESSORS IN SERIES

When two machines are connected in series, one has to look at the best way to control them, and the following considerations have to be taken into account.

1. Is the mechanical drive to the shaft of the compressor obtained from a common source—i.e., a common motor or drive turbine?

2. Is the mechanical drive to the shaft of each machine independent of the other?

3. How many recycle lines are there?

A solution has to be determined on the basis of the answers to these questions. If the answer to question 2 is yes and the answer to question 3 is two, i.e., one to each

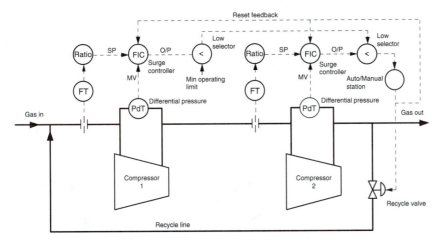

Figure 13.16: Control scheme for two compressors in series.

machine, then the solution would be to treat each compressor as an individual entity and use the information given so far to design the controls. If, on the other hand, the answer to question 2 is no, then it follows that the answer to question 1 will have to be yes, and under these circumstances the answer to question 3 is most often a single recirculation line with a single surge curve provided by the compressor maker that applies to both machines. The control system will then have to be considered further. If the answer to question 2 is yes and the answer to question 3 is one, then in this case, too, further consideration will have to be given to the control system.

We shall now consider series compressors driven independently but having a single recycle valve. The system is shown in Figure 13.16. In this system each compressor can be driven at speeds that are independently variable, but since there is only one recycle control valve, this valve must always be accessible to whichever controller that demands it. To achieve this each controller output is fitted with a low signal selector (< in the figure). One of the low signal selectors has a maximum signal as one of the inputs, thereby forcing the selector to choose the controller output. The second selector chooses the lower of the two controller outputs and directs it to the control valve. From this arrangement it will be appreciated that since only one controller is in charge, steps must be taken to prevent the unselected controller from saturating. This is achieved by feeding the driving output, i.e., the selected (lower) controlled output to the recycle valve, back to both controllers as the reset feedback signal. The arrangement of selecting the signal automatically is generally referred to as an *auto selector*. The system shown applies specifically to compressors that will always work together. It can be modified quite easily to permit a chosen machine to work on its own, which is a facility that can prove useful during periods of low loads and during maintenance. While only two machines are shown in Figure 13.16, the scheme is nonetheless fundamentally applicable to any number of compressors in series.

COMPRESSORS IN PARALLEL

Machines arranged in parallel formation are a great deal more difficult to control than those arranged in series. When arranged in parallel, each compressor discharges into a common header; because of this the compressors will interact with each other.

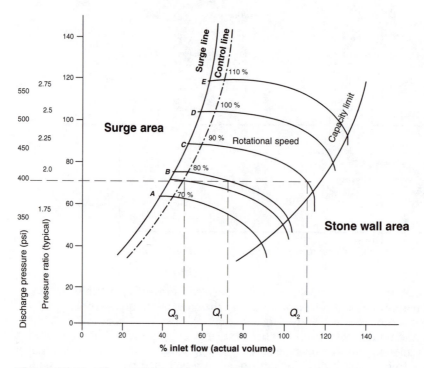

Figure 13.17: Effect on capacity for compressors running in parallel.

Each compressor is provided with its own surge controller. The load is controlled either by variable speed, suction throttle, or Inlet Guide vanes. It is important that the units be balanced for appropriate load sharing and preventing one machine from being starved into surge conditions. Running the compressors at the same speed may not guarantee uniform load sharing to prevent one machine from working harder. It is also important that the controllers bring the operating points to the same distance from their respective surge control lines. Arrangements of parallel compressors are very likely to have wide changes in the load, which could require single or multiple-compressor units to satisfy the demand. The reasons can be clearly seen from Figure 13.17.

Note: In the discussion that follows the matter of compressor load balancing will not be discussed, for the simple reason that it is a topic on its own. However, suffice it to say that in situations where compressors are arranged in parallel, the balancing of the load is most important and great care must be paid to this aspect of the control system.

From Figure 13.17, it can be seen that if both machines were operating at, say, 80% speed and they were balanced, then the volume would be Q_1. Now, suppose that one of the compressors had its speed reduced to, say, 95% while maintaining the same head; then in this case its delivered volume would be Q_3. Clearly, there is a difference, which could result in one of the machines going into surge due to the fact that there is a reduction in the flow through the machine. The situation gets worse as the speed difference increases. For example in the case shown in the diagram, if one machine were run at 90% and the other at 105%, then the speed difference is very large and surging will almost certainly ensue in the slower machine. In these circumstances, what sort of control scheme can be suggested? It will be appreciated

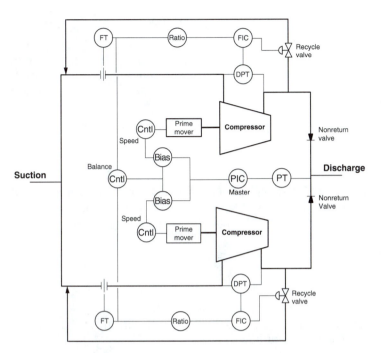

Figure 13.18: Control scheme for two compressors running in parallel.

that any scheme put forward must be suitable under all conditions. An arrangement as shown in Figure 13.18 illustrates a scheme that could meet the requirements for load sharing control of parallel machines, of surge protection, and even allow for one compressor to be started under automatic control while the other is running. The scheme will be valid for any number of compressors in parallel.

SUMMARY

1. A compressor is a machine that is used to compress a gas, by changing the volume of the fluid it takes in at the suction end to a much smaller volume at the discharge end. Compression therefore increases the energy of the fluid.

2. The increase in energy of a compressed fluid creates a much higher pressure at the discharge end of a compressor. The large difference between suction and discharge pressures results in a much higher velocity of the fluid at the suction end, and thus an increase in the kinetic energy of the fluid at the suction end.

3. The pressure ratio between the discharge and suction ends is fixed by the design of the machine. It is a dimensionless number.

4. When demand for the fluid decreases, the discharge pressure will rise. The pressure energy of the fluid at the suction end is less than the discharge pressure; hence, the gas will start to flow backward through the machine, because fluids always flow from a region of high to a region of low pressure.

5. If a reverse flow situation is allowed to continue, it will cause the gas from the discharge header to also be pulled through toward the suction side of the

compressor. This will, in turn, result in a reduced discharge pressure and cause the gas to go forward toward the discharge side again.

6. The back-and-forth flow of the fluid is called, surging. Surge is a phenomenon that is particular to the combined use of gases with centrifugal compressors. It is an unstable condition which if left without attention would seriously damage the compressor itself.

7. In any continuous compression process the relationship between the absolute pressure and the volume is expressed as:

$$pV^n = C = \text{constant}$$

If these were to be plotted for each value of n, a family of polytropic curves would be derived and work would be performed when proceeding along any one of the curves from P_1 to P_2, where P_2 is the higher pressure, or we can say:

$$W = \int P \, dV$$

where W is the work done.

8. In all centrifugal machines, the head (adiabatic head) produced is proportional to the square of the rotational speed, and the flow rate is directly proportional to the rotational speed, in other words:

$$Q = kN$$
$$H_{\text{ad}} = k_1 N^2$$

where Q is the flow rate, H_{ad} is the adiabatic head, and N is the rotational speed. If we combine these two equations, we have

$$H_{\text{ad}} = k_2 Q^2$$

9. There is no direct method of providing a measurement of the adiabatic head. The compression ratio is an available parameter that is readily calculated and is related to the adiabatic head. If R_v is the compression ratio, then:

$$R_v = \frac{p_2}{p_1}$$

where p_1 is the suction pressure and p_2 is the discharge pressure.

10. The following equations found in *Perry's Chemical Engineers Handbook*, relate the compression ratio to the adiabatic head. For the FPS (foot-pound-second) system of measurement

$$H_{\text{ad}} = \frac{\gamma}{\gamma - 1} \Re T \left[\left(\frac{p_2}{p_1} \right)^{(\gamma-1)/\gamma} - 1 \right]$$

where γ = specific heat ratio = c_p/c_v
$\Re$ = universal gas constant = 1545 ft/lbmol·°R
T = absolute temperature, °R
p_1 = suction pressure
p_2 = discharge pressure

For the SI (Système International) system of measurement:

$$H_{ad} = \frac{\gamma}{\gamma - 1} \frac{\Re T_i}{9.806} \left[\left(\frac{p_2}{p_1} \right)^{(\gamma-1)/\gamma} - 1 \right]$$

where $\Re$ = universal gas constant J/kg·K = 8314/molecular weight
        T_i = inlet gas temperature, K
        p_1 = inlet absolute pressure, kPa
        p_2 = discharge absolute pressure, kPa

11. Writing ϕ instead of $(\gamma - 1)/\gamma$ in the FPS equation of item 10, we have:

$$\left(\frac{p_2}{p_1} \right)^{\phi} = 1 + \left(H_{ad} \, \phi \frac{1}{\Re T} \right)$$

$$\frac{p_2}{p_1} = \left(1 + \frac{H_{ad} \, \phi}{\Re T} \right)^{1/\phi}$$

If W_m is the molecular weight of the gas and z the supercompressibility factor, we can write the preceding equation as:

$$\frac{p_2}{p_1} = \left(1 + \frac{H_{ad} \, W_m \, \phi}{\Re T z} \right)^{1/\phi}$$

$$= \left(1 + \frac{H_{ad} \, W_m \, \phi}{1545 T z} \right)^{1/\phi}$$

But since p_2/p_1 is the compression ratio = R_v, we can write this equation as:

$$R_v = \left(1 + \frac{H_{ad} \, W_m \, \phi}{1545 T z} \right)^{1/\phi}$$

If the compressor handles only one type of gas at near constant inlet temperature, molecular weight, specific heat, and supercompressibility, then in the foregoing equation the terms T, W_m, ϕ, and z will be constants and therefore can be replaced by a single constant K_1. This will result in an even simpler equation:

$$R_v = (1 + K_1 H_{ad})^{1/\phi}$$

This equation is nonlinear, but where the gas is air, natural gas, or another of the more common gases and the temperature is around ambient, then $(R_v - 1)$ can be substituted for H_{ad} without distorting the curves too greatly.

12. The differential pressure across the compressor is:

$$\Delta p = p_2 - p_1$$

But

$$p_2 = p_1 R_v$$

Therefore

$$\Delta p = p_1 R_v - p_1$$
$$= p_1 (R_v - 1)$$

Therefore

$$R_v - 1 = \frac{\Delta p}{p_1}$$

Hence we can say:

$$\frac{\Delta p}{p_1} = K_2 Q^2$$

13. To measure the actual inlet flow, we must have a primary device in the suction line. If this is a venturi meter or an orifice plate, then we will have the flow related to the differential head across the device, but this will be in volumetric terms that must be converted to mass. If k_3 is the meter constant, V is the specific volume, m is the mass flow, T is the temperature, and p is the pressure, with subscript 1 to indicate inlet. Then

$$V = k_3 \left(\frac{T_1}{p_1} \right)$$

$$m = k_4 \sqrt{\frac{H_d p_1}{T_1}}$$

Since

$$Q = mV$$

$$= k_5 \sqrt{\frac{H_d p_1}{T_1}} \left(\frac{T_1}{p_1} \right)$$

$$= k_5 \sqrt{\frac{H_d T_1}{p_1}}$$

Assuming the inlet temperature T_1 is constant, then:

$$Q = k_6 \sqrt{\frac{H_d}{p_1}}$$

$$Q^2 = k_7 \frac{H_d}{p_1}$$

Substituting this into:

$$\frac{\Delta p}{p_1} = K_2 Q^2$$

we have:

$$\frac{\Delta p}{p_1} = k_8 \left(\frac{H_d}{p_1} \right)$$

$$\Delta p = k_8 H_d$$

This last equation is the one used to design the control system.

14. Temperature variations at the inlet affect the suction flow and as a result the differential pressure across the orifice plate or venturi. To visualize the effects,

we shall need to consider:

$$Q = k_5 \sqrt{\frac{H_d T_1}{p_1}}$$

$$Q^2 = k_9 \sqrt{\frac{H_d T_1}{p_1}}$$

Therefore

$$H_d = k_9 \left(\frac{Q^2 p_1}{T_1} \right)$$

For specific values of Q and p_1 we can say:

$$H_d = \frac{k_{10}}{T_1} \qquad\qquad (13.27)$$

where k_{10} is the combined constant.

15. Changes in molecular weight of the gas—changes the gas composition—will also affect the control of a compressor, affecting the surge curve. The surge curve is a plot of adiabatic head versus inlet volume. We use the following for R_v to determine the effect, assuming imperial units.

$$R_v = \left(1 + \frac{H_{ad} W_m \phi}{1545 Tz} \right)^{1/\phi}$$

If we keep H_{ad}, T, and z constant, i.e., as fixed values, we can rewrite this equation as:

$$R_v = (1 + K_4 W_m \phi)^{1/\phi}$$

where K_4 is the combined constant.

16. A changed molecular weight will be reflected in a new value for the specific heat ratio and hence ϕ. For some of the lighter hydrocarbons there are known definite relationships between the molecular weight and ϕ, and as a result an accurate calculation can be made. However, for those cases where there is no correlation between the specific heat ratio and the molecular weight, individual calculations must be made to determine the magnitude and the direction of the change. In these cases we use:

$$Q = k_5 \sqrt{\frac{H_d T_1}{P_1}}$$

$$= k_5 \sqrt{\frac{H_d T_1}{p_1 W_m}}$$

Hence

$$H_d = \frac{k_5 Q^2 p_1 W_m}{T_1}$$

using specific values for Q, p_1, and T_1 we can say:

$$H_d = k_{11} W_m$$

CHAPTER 14

Control of Steam Generation Plant

INTRODUCTION TO STEAM GENERATORS (BOILERS)

We shall now consider the control of one of the most common items found in any process plant—the steam generator. The reason for its widespread use is that almost all manufacturing processes require steam to manufacture their product(s), and because it is available—albeit at a slightly extra, but nonetheless economical, cost—as a heat source, or as an energy supply in some part of the process of manufacturing the materials for products. From what has just been stated, do not conclude that steam is the only means of motive power; electricity provides the bulk of energy for driving machinery.

Steam generators, or boilers, as some call them, all operate in fundamentally the same manner. The process of steam generation can be described simply as applying heat to water to make it boil, and, once the water is boiling, to maintain it that way to collect the steam. This simplicity in concept really belies the very difficult problems involved, for the steam generator is among those processes that are the most interactive on any plant. The other important thing to realize is that there are in reality a surprisingly small number of control loops involved.

The following outlines the principles of steam generator control only. It is vital that all national and local codes of practice for safety, electrical wiring, and operating procedures be strictly adhered to.

PRINCIPLES OF STEAM GENERATION—A REVIEW

Before going further, let us recapitulate some fundamentals of steam generation. The first is that steam is the result of applying heat to water, and vapor is evolved long before the water begins to boil. In boiling therefore, there are definitely two phases present: namely liquid and vapor. The vapor or steam can best be described as minute droplets of water held in suspension in the air above the surface of the liquid.

The next fundamental point is that the temperature at which the liquid begins to boil is highly dependent on the pressure at the liquid surface. In general it can be said that, as the pressure on the surface increases, the point at which the liquid boils (the *boiling point*) also increases, and if the pressure decreases, then the boiling point decreases. To permit universal discussion of ideas and obtain repeatable data, the procedure is standardized and a datum fixed; the boiling point of water is determined by maintaining the pressure on the surface of the liquid at one bar—or 14.696 lb/in^2 g, the equivalent in imperial units—and recording the temperature at which the water boiled, usually taken as 100°C (or 212°F in imperial units).

It will be appreciated that all materials are subject to the influence of heat, the transfer occurring all the time even though the material bodies are not in physical contact. There are three methods of heat transfer: conduction, convection, and radiation. *Conduction* is a method most often associated with heat transfer in solid materials. To understand the mechanism of the method, one has to go down to the atomic level and consider the effect that heat addition has on the electrons surrounding the nucleus. Without going into quantum mechanics, it can be very simply envisaged that adding heat energy will cause the electrons to accelerate and will increase their energy, but although the electrons are moving very rapidly, there is no physical movement in the heated material to indicate the increase in energy; instead, the change in energy level is manifested in a temperature rise of the material. We can conduct an experiment to show this; if we take a reasonable length of, a metal rod and hold one end in the flame of a Bunsen burner, after a short time, the end we are holding will get hotter. Nothing has moved, but the heat from the flame has traveled toward our hand. If the added heat is increasing, then the temperature will continue to rise also until such time as we let go; the heat energy will begin giving off rays in the visible band of the spectrum. In ferrous material, this is indicated by the familiar initial dull red, which progresses right through to incandescent white. The color changes for increasing temperature in nonferrous metals are different to that described.

Convection is a method associated only with liquids and gases. Heat is transferred by causing particles nearest the heat source to move away from the heat source and be replaced by cooler ones within the fluid. This phenomenon can be made visible by a very simple experiment in which a quantity of water in a conical glass flask is subjected to a heat source and allowed to gain a little heat; then a crystal of potassium permanganate ($KMnO_4$) is dropped into the fluid. The salt, being soluble in water and more dense, sinks to the bottom of the flask, where it is nearest the heat source. The dissolving salt gives off streams of purple-colored solution of $KMnO_4$ rising from the crystal. The colored streams eventually bend back toward the bottom of the flask as they are replaced by the hotter rising upstream, tracing the complete path of the heated fluid. Lowering a thermometer into the water will show that it is hotter at the bottom than at the top, confirming that heat is being transferred. If the heating is continued, then eventually the water will boil. It will be appreciated that the same thing occurs when gases are heated even though an experiment such as conducted for water cannot be repeated for a gas. To prove that this is true, consider a hot air balloon; it rises only because the air within the envelope is hotter and therefore less dense than the surrounding outside air. This is another effect of the parameter temperature: it affects the density of material. Gases show quite dramatically the effect of density change.

Radiation is the transfer of heat by invisible waves. The only practical way to observe this phenomenon is to place a hot body of known temperature near to, but not touching, another cooler body also of known temperature and wait a while. At the end of the wait period, take a temperature reading of both the bodies. In every case the hot body would have become cooler and the cool body would have become hotter even though the two bodies were not in physical contact. The change in temperature can only be explained by the fact that there must have been a transfer of heat by some sort of wave motion from the hotter to the cooler body and in addition some heat transfer to the surrounding atmosphere and container by the same method. The quantity transferred can be approximately calculated by taking into account the masses of the bodies, their specific heats, and the change in temperatures. The heat transfer to the surrounding is a lot more difficult and perhaps impossible to determine with any degree of accuracy unless very elaborate precautions are taken.

There is another concept that we should come to terms with before we leave this review of physics, and that is what happens when water is changed into steam. Reverting to the conical flask with water being heated up, for this part of the experiment let us start with a measured mass of water. In determining data on steam generation, the conventional reference amount of water is 1 kilogram, or 1 pound mass in imperial units. The reason for choosing mass is to more easily understand the conversion and relate it to a measurable parameter. We shall assume that the water is at 0°C (32°F), for it has been internationally agreed that at this temperature the heat content (enthalpy) is zero. Two thermometers and a pressure gauge are inserted through the stopper of the flask, and the thermometers adjusted so that one measures the temperature of the water only and the other the temperature of the space above the water only. The pressure is adjusted if necessary to read 1 bar (14.69 lb/in^2) and sealed off. Heat is applied, and a note made of the pressure, the temperatures, and the liquid level. With this data it is possible to construct a very rudimentary steam table.

The important thing to remember is that *as long as there is water in the flask, there will be two phases present, liquid and vapor*; therefore, under these conditions *the steam produced is considered to be wet*. When the last drop of water has just turned to steam, then there is only a single phase present, and the steam is considered to be *dry and saturated*. If any more heat is added after the last drop of water has just turned to steam, then this addition will go toward increasing the *internal energy* of the steam and the result will be to *superheat* the dry steam. The range of data gathered from successive conditions of water to wet steam to dry saturated to superheated steam has been completely documented in what are called "steam tables" or, more correctly, the table of the thermodynamic properties of water.

Regarding the matter of dryness of the steam, the amount of water contained is "qualified" and given a scale value in the range 0 to 1.0, where zero represents total wetness, or entirely liquid, and 1.0 represents dry saturated steam; the amount of water held is in the form of vapor, and when the steam reaches the condition of dry saturated there is no more water vapor present.

THE STEAM GENERATOR OR BOILER

It may be useful at this stage to outline the general parts and functionality of an industrial steam generator, or, more commonly, a boiler—this latter name, it is felt, does not really define the equipment and its function correctly. The raw materials required by the steam generating plant are water, fuel, and air. The "art" in steam raising is the manipulation of these three raw materials to produce the required amount of steam, of the correct quality, at the most economical conditions both in financial and raw material terms.

Let us take these in turn, and consider the water circuit first:

1. The water required is stored in the water/steam drum (in the industry, normally referred to as the steam drum only). It has to be replenished as more steam is generated. Therefore, to meet the demands on the drum for steam, there must be a sufficient water supply, capable of being regulated, available to it at all times. The amount of regulation applied to the water supply will depend on the quantity of steam demanded by the process.

2. In water tube-type generators, the water drum is in communication via a number of interconnecting tubes with another drum, the *mud drum* deep within the

heated area of the unit. The purpose of this arrangement is to:

a. Reduce the volume of the water in contact with the heat source.

b. Increase the surface area in contact with the heat source, thereby gaining rapid heat transfer to the water from hot gases evolved during the combustion of the fuel. The furnace internals have specially designed baffles to allow the hot gases to make several passes over the water tubes to effect maximum heat transfer.

Water-tube boilers are the ones generally found today in most plants, whereas the Lancashire-type generators are more commonly found in the older land-based plants and classic antique locomotives.

Now we consider the fuel and air circuit:

3. The fuel used could be solid, liquid, or gas. Solid and liquid fuels have to be held in the plant itself; therefore, an inventory and replenishment scheme has to be in place to avoid the problems that would be inevitable if they ran out. Fuel gas, on the other hand, is usually delivered by pipeline and drawn upon as required in metered quantities, with usage recorded for heat balancing and financial accounting purposes.

To operate efficiently the fuel has to be used in amounts that will ensure its complete combustion in the appropriate amount of air. Any fuel that is not fully burned will result in unnecessarily high operating cost and air pollution. Inefficient combustion is, therefore, a financial and material waste. Lean-fuel (air-rich) combustion always results in higher flame temperatures, detrimental to the construction material of the generating plant itself, and impinges directly on the running cost of the process, as it will involve more frequent replacement of the affected parts. Thus, control of the fuel used in the process will be necessary for obtaining the most all-around benefit.

4. Air, or, more correctly oxygen, is vital to support the combustion of the fuel. The air required is made available to the furnace by the *forced draft fan* located in the air suction duct. As already implied, the air supply requires regulation, which is carried out by either a damper on the downstream side of the fan, or by a series of parallel vanes built into the suction inlet of the fan itself to ensure the correct amount at the most suitable condition is available for burning the fuel. The fan could be steam or electrically driven. Lean-fuel combustion (too much air), as stated earlier, will result in higher flame temperatures at the point of combustion, while fuel-rich (too little air) mixtures on the other hand, result in incomplete combustion, and hence, lower flame temperatures, lower heat transfer, and less water/steam conversion. Insufficient air is therefore a waste of fuel and translates directly into financial loss.

5. As we have just seen, the fuel and air are interdependent, and brought together for burning in the *combustion chamber*, or furnace, in the accepted parlance. There is a very definite relationship between the air and the fuel, which depends on the amount of carbon (C) contained in the fuel. It will be appreciated that what we are really doing when we burn a material is conducting a chemical change by combining the element carbon, in the fuel, with the oxygen (O_2), in the air, to form the compound carbon dioxide (CO_2), which can be formed only when there is complete combustion, i.e., when the correct amounts of the two elements are available. It is therefore of the utmost importance to maintain the correct relationship between the fuel and the air, or as it is usually called, *fuel/air ratio*.

6. It is also advisable, especially in industrial steam generators, to monitor and manipulate the exhaust gases to maintain suitable conditions within the furnace. To maximize the benefits, it will be necessary to monitor the contents of the exhaust

gas for either oxygen or combustibles content (be aware that incomplete combustion results in the formation of carbon monoxide (CO), a fuel, since there is unburned carbon remaining) and to use this information to trim the fuel/air ratio as necessary.

With this brief outline of the principles and the control requirements, one can see that a steam generator, though simple in concept, rapidly becomes a highly interactive process, where small changes in one parameter affect many others throughout the plant.

GENERAL DESCRIPTION OF THE PROCESS PLANT

Figure 14.1 depicts a typical water-tube steam generator:
With this figure we shall try to see how maximum use of the heat input can bring about economies in running the plant. To see how this is achieved and to understand why the components are located where they are, let us consider the following:

1. The combustion air, is drawn into the combustion chamber from the surroundings where it (air) is subject to the seasonal temperature variations that normally occur at the location. Generally, it is expected that it will be cold in winter and spring, and not so cold in the summer and fall. However, in a country like the UK, changes in temperature can be quite rapid and often unpredictable. To avoid excessive temperature fluctuations in the furnace, in situations of rapid changes in ambient air conditions, the incoming air is preheated to bring stability. In the majority of cases, the cost of running a separate air heater is minimized, or avoided, by using the hot furnace exhaust gases to perform this function.

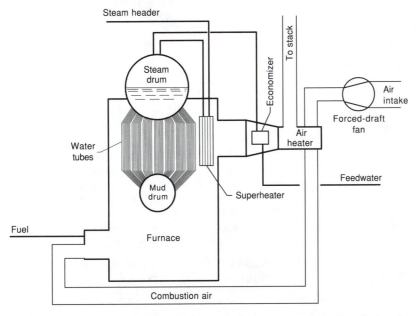

Figure 14.1: Schematic diagram of a typical steam generator. For clarity, the superheater has been shown nearer the gas exhaust. It should be on the opposite side nearer the burners.

2. The feedwater for the steam generator is also in most cases drawn from a water treatment plant that conditions the water by chemical additions to eliminate as far as possible any algae and, calcium carbonate, the latter being the cause of lime scale, a chemical compound often found fouling a common stovetop or electrically heated kettle. In the domestic scenario water treatment for the complete eradication of calcium carbonate would be too expensive, and if done could pose a health hazard. However, the important thing to remember is that the water, too, will in most instances be at ambient temperature and will have to be raised to boiling point to be useful. Raising the feedwater temperature will therefore result in lowering the amount of fuel used, thus reducing the running cost. The greater the amount of heat we can add to the "raw" incoming water, the less heat we shall have to add to cause it to boil. The heat added to the feedwater is carried out in the *economizer*, an equipment unit placed in the path of the hot exhaust gases going to the stack.

At this point it is worth remembering that, attractive as this may appear, it is not possible to recover all the heat from the exhaust gases, for the reason that there will come a time when the increased cost (worsening economics) of heat recovery combined with the increased thermodynamic inefficiency of the plant will far outweigh the gains for doing so.

3. It is important that the drum itself always contains water, for if it were ever to run dry, then certainly there would be an explosion, because the constrained steam would increase in pressure and try to expand. The repair or, in the worst case, replacement cost after such events will be very high. The steam generated in the steam drum will be wet, since it is always in contact with water. If dry saturated steam (DSS) is required from the steam generator, then the wet steam will have to be superheated—raised up to a much higher temperature—to some extent and thereafter either allowed to cool to the required condition through heat losses or *desuperheated* by the addition of cooler water. In desuperheating, the cooler water is called *spray water*, because it is added in the form of a spray in order to reduce the mass of the water and ensure almost instant change of the liquid to steam. In Figure 14.1, the desuperheater is not shown; however, one can visualize that it will be located somewhere after the superheater. Remember that desuperheating also causes an increase in mass, because we are adding a mass of liquid to the mass of the generated steam.

PLANT OPERATION

Let us suppose that a steam generator is up and running and supplying steam to some part of a collective manufacturing process. We call this manufacturing process a *load*, for that is what it appears like to the steam generator; suppose also for the present that the load is *steady*, i.e., the amount of steam generated and supplied is sufficient for the needs of the load. (**Note:** To enable an understanding of the general principles involved, the descriptions that follow for the various states are of necessity simplified). Under these *steady-state conditions*, the following will apply:

1. The amount of steam taken by the load will be made up by a similar amount of feedwater being added to the steam drum.

2. In view of this steam/water balance, the level in the steam drum will remain constant.

3. The amount of fuel and air used will also be steady, for the heat released by combustion is sufficient to meet the steam demand.

4. These steady-state conditions, where all the (otherwise variable) parameters are in equilibrium, is brought about by the instrumentation and control systems involved.

Now let there be a disturbance in the steady-state balanced system, say, an *increase* in steam demand by the load. Under these circumstances the system will become upset and the following will now apply:

1. The amount of steam taken by the load will increase.

2. The water level in the drum will change in response to the load increase.

3. Since the drum level has changed, then it follows that the feedwater flow will also have to change (increase) in order to maintain the water level.

4. It will be necessary to evaporate more water to meet the increased demand for steam, and to accomplish this it will be necessary to increase the heat supply. Therefore, an increase in fuel will be required, and as a direct result an increase in the amount of air to support the combustion will be necessary.

If the disturbance from the steady-state condition is a *decrease* in the steam demand by the load, the outcome under these circumstances will be different, and the following will now apply:

1. The amount of steam taken by the load will decrease.

2. The level in the drum will change in response to the load decrease.

3. Since the drum level has changed, then it follows that the feedwater flow will also change (decrease) in order to maintain the drum level.

4. To meet the decreased demand, it is necessary to reduce the amount of steam being generated, and to accomplish this it will be necessary to decrease the heat supply. Therefore, a decrease in fuel will be required, along with a corresponding decrease in the amount of air, to support the combustion of a reduced fuel input.

Having now defined, in general terms, the conditions prevailing under steady state, increased and decreased demand, let us now see how we can proceed with the design of a control strategy to respond to these several needs.

THE CONTROL SYSTEM

One of the most important requirements for the control system is that it be able to recognize changes in the demand of the load. There are two process parameters that can serve as a basis for this determination; the first is steam flow and the second is steam pressure. Since there is a choice, which shall we use? These two parameters, we find, are really concerned with the same process fluid, namely the steam to the load. The choice is now reduced to determining the parameter that is most responsive to the changes, which leaves us with the steam pressure as the only parameter that fulfills the requirement, as it rises very rapidly for a load reduction, and falls equally fast for a load increase. In the accepted terminology, the steam header pressure is called the *master pressure* for it sets the whole control strategy moving in the correct way.

The next objective is to determine the amount of material that has to be provided to meet the demand. To determine this we shall have to know how much steam the load demands and balance this demand by an equivalent amount of water. This requirement therefore stipulates that the flow of both steam and feedwater must be measured and compared and any difference compensated for by supplying or inhibiting feedwater to the steam drum.

The general outline of the steam generation plant shown in Figure 14.1 will be used as the base on which to build the system, adding more detail to it at each stage of development, so that eventually a very graphic idea of the instrumentation and control requirements of such a plant will be obtained, and hopefully understood. It is very important to appreciate from the outset that what we will discuss are the basic techniques used; these can be and have been enhanced in several different ways, but the fundamental principles have not changed. The very advanced techniques employed today are beyond the scope of this book.

WATER AND STEAM CIRCUIT

Since we started to discuss the way in which the feedwater is converted into steam, perhaps it is now a good point at which to present the water and steam circuit of the steam generator. In Figure 14.2 we have shown only the water and steam process lines. On the steam side we have not shown any desuperheating, for we wish to keep the system simple at the moment. We shall discuss this a little later on so that a complete picture can gradually be built up.

Turndown

On the flow measurements being discussed, orifice plates and differential-pressure transmitters have been shown, for the simple reason that these "old faithful"

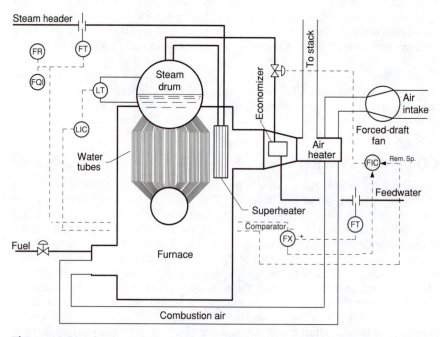

Figure 14.2: Schematic diagram of a typical steam generator with three-element feedwater controls. For clarity, the superheater is shown nearer the gas exit; it should be on the opposite side nearest the burner end.

combinations are still to be found in many plants, even though they do not really give the turndown necessary for such applications. What does turndown mean? This is the same as defined much earlier in Chapter 1 and is the ratio of the maximum expected flow to the minimum expected flow:

$$\text{Turndown} = \frac{\text{maximum flow}}{\text{minimum flow}}$$

The minimum is normally referred to as the base condition. If say the maximum flow was 1000 units and the minimum flow was 100 units, then from the foregoing we have a 10 to 1 turndown.

Normally one can expect a turndown of 3 to 1 from an orifice plate and differential-pressure transmitter combination, but most often the turndown is restricted to even less than this in order to give the necessary accuracy. The in-line devices, such as vortex meters, electromagnetic flowmeters, or Coriolis mass flowmeters, could be used, and they will provide much larger turndown capability. All the latter-mentioned instruments have high turndowns, in the region of approximately 10 times larger than the orifice plate/differential-pressure transmitter combination, and in some cases even more—e.g., the Coriolis mass flowmeter. The choice of any of these instruments will depend on the application and on the operating conditions in particular. Please be aware of the temperature limitations of all these instruments when used on steam, and in particular the use of the Coriolis meter especially when two-phase flow is a possibility; in such cases do not use the Coriolis meter. The vortex meter will in most cases be the wisest choice, as this instrument can now be made for use on high-temperature and pressure steam services.

Referring to Figure 14.2, we see that the measurements of both the steam flow and the feedwater flow are taken to a device named FX, which is a comparator, where a subtraction of steam from feedwater is made to determine the amount of makeup feedwater required. The comparator is calibrated so that, at steady state, the output is exactly half the output span, which for scaling purposes is called zero; in other words, the instrument has a center-based zero scale to allow the output to be driven in the correct direction depending on whether the steam or the feedwater is the larger. The comparator output provides the measurement for the feedwater controller (flow indicating controller FIC in the figure). If desired, the center-zero scale can be avoided by using a suitable bias that lowers the zero to the position of minimum value on the scale.

Steam Drum Level—"Drum Level Control"

The steam drum level is measured using a differential-pressure instrument. This is necessary, as the space above the water in the drum contains steam whose pressure will be varying dependent on load, certainly above ambient pressure, and therefore must be allowed for. The differential-pressure instrument automatically takes care of this and produces a signal that is directly proportional to the liquid head or drum level. Since it is necessary for the drum to contain water at all times, boilermakers set the allowable working range of the drum level; all that is necessary for users then is to measure the changes in level about an optimum value set at mid-working range and called the *normal working level* (NWL) and to use this as the measurement for a level controller (level indicating controller LIC in the figure). The setpoint for LIC is an operator-determined value set close to the NWL, as the optimal value can only be determined after the generator has been commissioned and all the equipment comprising the load is known. The controller output is used in cascade as the setpoint for the feedwater flow controller (FIC). There are important considerations to be taken into account:

- The level controller (LIC) *must have an anti-reset windup facility* to avoid saturating the integral control term if the process operator changes the flow controller (FIC) from remote to local setpoint drive, which will be the situation when starting up the steam generator.

- The flow controller (FIC) *must not have a derivative control term* as the derivative term would react to the noise content of the measurement signal, for it is not possible to have noiseless flow. If derivative control action was fitted, then for certain the system would continuously oscillate and never settle to a steady value.

- There is a very interesting phenomenon that occurs in the steam/water drum when the system is responding to a change. To visualize the situation, let us assume that there is an *increase in load* demand; under normal circumstances one would naturally be inclined to think that since more material is being withdrawn from the drum then the drum level will fall. This is in reality incorrect, for the true situation when the load increases is that the pressure is reduced; this pressure reduction, coupled with the density change of the feedwater, will in fact cause the drum level to *rise*. The reason for the density change is that when the pressure is reduced, the boiling point of the water is lowered and more steam bubbles are generated and are found to be in the water. In the same way, with a *decrease* in load, due to the increased pressure the boiling point and accompanying density of the water will tend to rise as more bubbles of steam are forced out, and the steam drum level will *fall*. In the parlance of the boiler specialists, this unusual phenomenon is termed "shrink-and-swell." Therefore, when the steam drum level falls, the system is really demanding more feedwater to be put into the drum; under no circumstances in this situation should the control device be allowed to close off the feedwater supply.

Three-Element Feedwater Control

The system shown in Figure 14.2 is considered to be the most efficient method of controlling the water circuit in the steam generator. The modus operandi of the system is as follows. Let us assume the system to be in a steady-state condition. Then

- The level in the steam drum is steady, the level transmitter (LT in the figure) is calibrated with a center-zero range to correspond to the normal liquid level in the steam drum.

- The level controller (LIC in the figure) is a two-term (P + I) instrument with a center-zero scale; the output calibrated under these conditions to be at 50% and the output signal from the controller such that an *increasing measurement* gives *increasing output* to compensate for the shrink-and-swell effect.

- The flow transmitters (FTs in the figure) are calibrated so that half the output span represents the steady-state flows of both the steam and the feedwater; hence, the output from each of the transmitters is 50%.

- The flow controller (FIC in the figure) is arranged to have a signal that is a *decreasing output* for an *increasing measurement*.

Under this state the output from FX (the comparator) is adjusted to be at 50%, which is the zero for the system, and hence the feedwater control valve will be driven to a position that balances the steam outflow with the feedwater inflow.

With a change in the demand for steam, both the steam flow transmitter and the drum level transmitter will change their outputs. Let the change in demand be for more steam; then:

1. The steam flow transmitter output will increase, this will cause the output from comparator FX to fall, since it is arranged to subtract the steam flow from the water flow.

2. A falling measurement on the feedwater flow controller FIC will cause its output to increase, and will make the feedwater valve open even larger to admit more feedwater to the steam drum, to ensure that the amount of steam taken out is replaced by more feedwater and maintain the status quo.

When the change in the demand is for less steam, then it is easy to see that the control action of the system will be the reverse of that just described.

Real-System Behavior

Some words gained from experience.

3. A real system can never remain balanced (under steady-state conditions) over indefinite or prolonged periods, for there will always be some changing steam demand by equipment somewhere on the process. The control loops will therefore have to be dynamically balanced, and facilities will have to be provided for boiler room personnel to intervene if necessary.

4. The range of feedwater flow should exactly equal the range of steam flow, as this will ensure balanced control loops; and the quantity of feedwater supplied against each demand change has to equal that withdrawn as steam.

5. It will be found that, with the two controllers arranged as shown in Figure 14.2, flow control is a very fast loop, whereas level control is much slower.

Before we leave this section it is worth mentioning that there are other methods of controlling the steam and water circuit that are financially less costly, but suitable only for some applications. These alternatives will now be discussed.

Single-Element Feedwater Control

In this scheme, drum level is the measured parameter and the feedwater flow the manipulated parameter. From this it will be understood that there is only a single drum level controller involved, and as before the control valve is installed in the feedwater line. From our discussion of three-element feedwater control, it will be appreciated that the scheme now being considered will be suitable for processes that can have no sudden and dramatic changes in steam demand, and to ensure smooth operation the steam generator for such a system is fitted with a large feedwater capacity.

Two-Element Feedwater Control

In a two-element control scheme, there are two measurements made, steam drum level and feedwater flow, but only the feedwater flow is manipulated. From this it can be surmised that, because the objective is to maintain water level in the drum, there is only one controller, and it is associated with the drum level. The controller output manipulates the feedwater valve as before. However, the flow measurement is used in a feed-forward computation that adjusts the setpoint of the level controller. (**Note:** The feed-forward model has to be accurate for stability to be obtained in the control system.)

Before we proceed further it is very important to understand that even though we are considering the control system as being made up of a number of small, individual systems that appear to be self-sustaining, this is really not so. Each system depends on the others, and because of this, a steam generator is a very fine example of a highly interactive control system that can, and does, require a good understanding of the underlying principles to control effectively.

FUEL AND AIR CIRCUIT

We shall now consider the fuel and air circuit shown in Figure 14.3. To avoid confusion, the water and steam circuit we have already discussed has not been shown on the diagram, for it is hoped that by doing this, it will make for a clearer picture of the control concepts to be obtained.

Fuel Combustion

From Figure 14.3 it will be seen that the following measurements are made: air flow, fuel flow, steam pressure, furnace pressure, and stack pressure. Every parameter mentioned has an important role to play in obtaining the heat energy required to generate the steam. As will be appreciated, there is a definite relationship between the amount of fuel burned and the quantity of air used to support the combustion. When "normal" fuels (coal, oil, and gas) are used, the most significant combustible element is carbon (C), which is combined with the oxygen (O_2) in the air to release heat in the process.

How does one calculate the amounts of each of these elements so that an appropriate size for the fuel pipeline and the sizes of the forced- and induced-draft fans can be determined? A simple and quite effective way is to use the "Atomic Weight" of the two elements involved, and reason as follows. Carbon will combine fully with oxygen to produce carbon dioxide, or in symbolic terms:

$$C + O_2 = CO_2$$

Putting in the atomic weights for each element, we have:

$$12 + 16 + 16 = 44$$

where 12 is the atomic weight of carbon, and 16 is the atomic weight of oxygen. We can therefore say that, if the fuel contains 12 lb of carbon, then it will require 32 lb of oxygen to combine with and will produce 44 lb of carbon dioxide as a result.

The air we breathe does not contain pure oxygen, but a number of additional gases as well. However, we can simplify the actual composition by assuming that only two main gases make up the air we breathe—nitrogen (N_2) and oxygen, and for the purposes of this exercise we can consider their percentages to be 77% nitrogen and 23% oxygen. If now we consider "normal conditions," i.e., atmospheric pressure 14.69 lb/in^2 and temperature 60°F, or the equivalent in metric terms, we can say that for one pound weight of air there is 0.23 lb of O_2 and 0.77 lb of N_2.

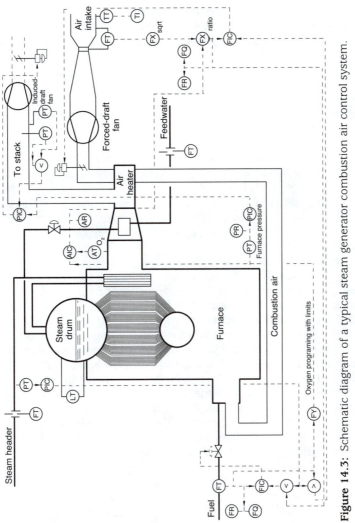

Figure 14.3: Schematic diagram of a typical steam generator combustion air control system.

From this data we can calculate the amount of air required to convert 12 lb of carbon to CO_2 as follows. We have shown that 12 lb of carbon requires 32 lb of oxygen to produce 44 lb of CO_2. We have also shown that each pound weight of air contains 0.23 lb of oxygen, hence we shall require:

$$\frac{32}{0.23} = 139.13043 \, \text{lb air}$$

This gives the theoretical exact amount of air required to make a complete conversion of the carbon to carbon dioxide. However, for reasons of overcoming any variation in the chemical content of either the fuel or the air, or some error in the measurements, the theoretical amount of air is never only provided: there is always a greater amount, referred to in the trade as *excess air*; the actual quantity is usually of the order of 10% more than that theoretically determined. It follows, therefore, that if in a given furnace 10% excess air is provided, then under ideal conditions where all the carbon in the fuel is converted to carbon dioxide then there will be 23% of the 10% excess air, or 2.3% oxygen content in the flue gas, and the combustion of the carbon will be complete.

Under actual operating conditions, the quantity of fuel will vary to meet the demands of the steam required by the load; therefore, it is necessary to be able to provide the appropriate amount of air for each change in demand and still maintain the amount of excess air. This need is satisfied by ratioing the fuel flow to the airflow and treating the fuel flow as the controlled, and the airflow as the uncontrolled, or "wild," variable. Figure 14.4 is a simplified schematic diagram of the control loop only. From this diagram it will be appreciated that the signal from the airflow measurement is modified in the ratio unit so that the setpoint of the fuel flow controller will be driven to a position that will maintain the fixed relationship between the amount of fuel and the air, depending on the actual amount of fuel flowing in the supply line. This is the principle upon which the basic control scheme shown in Figure 14.3 is designed, but in the scheme shown in Figure 14.3 the basic control scheme has been slightly modified to account for the oxygen or combustibles content of the exhaust gases passing to the stack. By adjusting the ratio unit setting by an amount dependent on the actual O_2 or CO content to bring it into line with the requirements

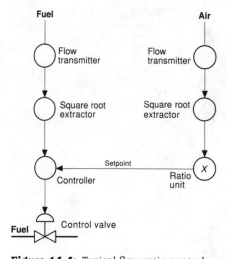

Figure 14.4: Typical flow ratio control.

of the allowable excess air for which the system has been designed, the effect of the modification is to make sure there is always complete combustion and the air/fuel ratio (A/F ratio) remains intact. As stated earlier, the system shown is one in which the flows are measured by differential-pressure instruments, which have a square law relationship to flow and are therefore linearized in the square root extractors shown before being manipulated.

We shall now consider how the system responds to a change in steam demand, for this will not only affect the water and steam circuit, but will also impose changes in the fuel and air circuit at the same time.

Master Pressure Control

The steam header pressure, a very important measurement, is monitored and used to manipulate the heat supply to the steam generating plant, and if a number of steam generators are involved the *common header pressure* (the single combined header formed by joining the individual headers from each steam generating plant) is used to bring other steam generators online to meet the demand. As mentioned earlier, the steam header pressure is the most responsive parameter to load changes and therefore crucially important, for it dictates the direction the whole control system should take when a demand is made. In order to ensure that this is accomplished, it is necessary to have a controller called the *master pressure controller* assigned to the task. This controller needs to be a very responsive device, and therefore, a three-term instrument is most often used, the action of the controlled output from this controller being an increasing output signal for a decreasing measurement.

The principle of operation is shown through the simple schematic diagram in Figure 14.5. On this diagram we have included the master pressure controller, which orchestrates the correct direction for the control system. Let us see how the controls work, but before we proceed further, let us consider the way we want the fuel and air controllers to function. To do this we must first consider, how the fuel control valve and the air supply damper, which are the two controlled devices involved and are linked also to personnel and plant safety, are to be set in the failed state. What

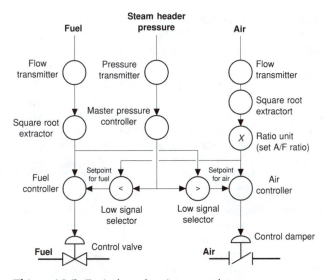

Figure 14.5: Typical combustion control system.

is demanded of the failed state is a definite and safe final position to which each of these controlled devices is driven in the event of a failure of the motive power supply (major failure). In the language of the valve and damper actuator manufacturers who actually implement this failed state requirement, it is referred to as the "failure mode" of the valve or damper. With these considerations in mind, we can generally say:

1. We shall require the *fuel control valve* to be *fully shut* to prevent fuel from being injected into a furnace that could after use be very hot, and therefore very dangerous from a safety aspect.

2. We shall require the *combustion air supply damper* to be *fully open*, to ensure that in the event of a failure the furnace is purged of all combustible material and therefore safe from a safety point of view.

Based on these requirements:

1. We require the *fuel controller* to respond to an *increasing measurement* change with a *decreasing controlled output*.

2. We require the *combustion air controller* to respond to an *increasing measurement* change with an *increasing controlled output*.

Now, to start on the control loop operation, let us assume for simplicity that the process is in a steady state.

1. In this condition, the system is *in balance*, with the master pressure controller measurement and setpoint in coincidence. The feedwater, fuel, and air control devices are driven to positions that allow such conditions to prevail.

Now we introduce a process disturbance:

2. Let us make a change in the steam demand—for more steam, which will be indicated by an increasing steam flow into the load. The increased flow makes the pressure in the steam header fall as soon as the demand is sensed. The setpoint and measurement of the master pressure controller will diverge.

3. The master pressure controller now has a measurement that is lower compared with the setpoint. As a result, the controller will begin to increase its output to bring the measurement and setpoint back together again (coincidence).

4. The output from the master pressure controller is applied to both the low and the high signal selectors, where the second input to the low selector is taken from the measurement of the combustion airflow, and the second input to the high selector is taken from the measurement of the fuel flow. (**Note:** Signal selectors, by the nature of the function they fulfill, require a minimum of two signal inputs. In modern distributed control systems, selector algorithms are usually capable of selecting one from several inputs.) The low selector output is the setpoint of the fuel flow controller, and the high selector output is the setpoint of the airflow controller.

5. Under the disturbance we have introduced, the pressure controller output will increase, and in the case of the low selector, will be higher than the signal from the airflow measurement, causing the air measurement signal to be selected. But for combustion reasons, the airflow is always greater than fuel flow; therefore, the result will be an increase in the setpoint of the fuel controller. This will cause more fuel to be directed to the burners.

6. At the same time the high selector, to which the pressure controller output and the measurement of the fuel flow are also applied, selects the pressure controller output, since it is greater than the fuel flow signal, which results in an increase in the setpoint of the airflow controller. This will change the air controller output and cause more air to be admitted to the furnace to support the increased fuel supplied.

7. The increased fuel and air supplied to the furnace will inevitably result in an increase in the heat applied to the water and, as a consequence, will increase the quantity of water evaporated. The increased supply of steam will therefore meet the increased load demand.

8. These new conditions will cause the pressure and flows to steady out at new values that satisfy the new load demand.

For a decrease in demand, the reverse will occur, thus ensuring that in all cases the control loops will act in parallel (i.e., behave as two independent control loops) when the system is under steady-state conditions and will act in series (i.e., behave as interdependent control loops) for any system disturbance. This is commonly referred to in the trade as *cross limiting*. The advantages of the cross limiting, which is sometimes also called *lead-lag* are:

- On an increase in steam demand, the fuel demand cannot increase until an actual increase in airflow measurement is sensed.

- On a decrease in steam demand, the air demand is not decreased until an actual decrease in fuel flow measurement is sensed

- If there is an inadvertent decrease in the measured airflow, the fuel demand is immediately reduced by an equal amount.

- If there is an inadvertent increase in the measured fuel flow, the air demand is immediately increased by a similar amount (within the design calculated air/fuel ratio).

Multiple-Fuel Firing Systems

In some countries the change of fuels described in this section must be carried out manually, but the principles involved and discussed here are valid. The individual country's national and local regulations must be adhered to at all times.

In some plants, there is a requirement to use more than a single type of fuel. This could be dictated by reasons of cost or availability. Under these circumstances the firing control must be designed to take care of these requirements. Figure 14.6 shows an arrangement for permitting the use of two fuels in the burners.

The system in Figure 14.6 operates as follows. As before, the firing demand is generated by the master pressure controller and applied as the setpoint to both the air and fuel controllers via the cross-limiting circuits, which operate as described earlier. However, the scheme demands that the fuels, though different, be made to resemble each other. The easiest way to do this is to consider the calorific values of each and apply a factor to one of the fuels so that it can resemble the other in calorific value. The factor used will take into account that the amount of heat being given up by the fuel will be a function of the flow rate. The fuel controller will then be, strictly speaking, a total heat controller. The summator is used to calculate the

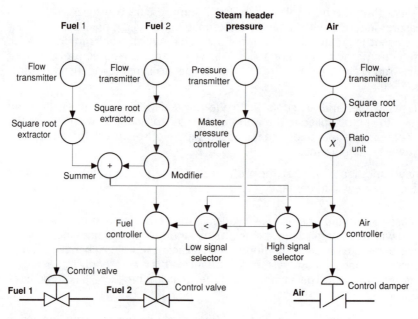

Figure 14.6: Typical dual-fuel combustion control system.

total quantity of heat being supplied. The functioning of the system henceforward is exactly as described earlier, in that under steady-state conditions, each controller will be acting in parallel with the other, but under a process upset, the controllers will be acting in series with one another to maintain the status quo under the changed circumstances.

Preferential Fuel Firing

Operating the system may once again be dictated by the need to fire two fuels, but as an alternative there should in addition be the facility for

• Burning one fuel in preference to the other

• Varying the amounts of one fuel over the other.

Figure 14.7 shows an arrangement that will provide the means of accomplishing these variations. The system operates as follows. Once again the firing demand is generated by the master pressure controller, and applied as the setpoint to both the air and fuel controllers via the cross-limiting circuits. In this control scheme there are two independent fuel controllers, one for each fuel, and for the purposes of this description let us call one of the fuels 1 and the other fuel 2.

As before, we have to make one of the fuels resemble the other by equating the calorific value. Let the modification be carried out, say, on Fuel 2.

The firing rate demand after passing through the low selector in the fuel circuit is applied as an input to a multiplying and a subtracting unit, before it is made to appear as the setpoint of each of the fuel controllers. The multiplier unit is associated with fuel 1 and the subtractor unit with fuel 2. In the case of the fuel 1 multiplying unit, the second input is the multiplier term provided by a manually adjustable signal

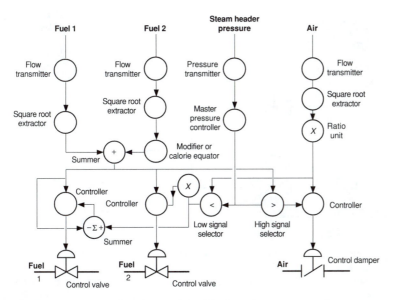

Figure 14.7: Typical preferential dual-fuel combustion control system.

generator, which can be set at any value within a range 0 to 1.0. However, be aware that setting the multiplying unit to zero with the fuel 1 controller in the Auto mode will make the controller go out of control, as this will drive the setpoint of the controller to the minimum value and cause the controlled output to be driven to the other extreme. If the intention is to disregard fuel 1, then it is necessary to *shut off the fuel* line itself *before* setting the multiplying unit to *zero*; this method will make the measurement of the fuel 1 controller drop to the minimum value, and when zero is set there is coincidence between measurement and setpoint, making the output of the fuel 1 controller stay where it is. To use fuel 1 on its own, all that is necessary is to adjust the signal generator unit to make the fuel 1 multiplying unit output equal to 1.0.

The subtractor unit associated with fuel 2 takes the difference between the low selector output and the common measurement of both controllers and applies the resulting output as the setpoint of the fuel 2 controller; this has the effect of making fuel 2 provide the balance of the fuel to fulfill the total fuel requirement demanded by the system. Adjusting the multiplying unit to 0.5 (50%) will result in equal amounts of fuel being provided, and by appropriately setting the multiplying unit to any other value the amounts of each fuel used can be varied; in this way preferential quantities of each fuel can be provided to meet the firing demand.

Control Valve Rangeability

On steam generator systems, because the fuel supply to the burners has to be regulated down to very low values, the control valves associated with fuel can cause a few problems. What we are concerned with relates to the turndown of the valve itself; all valve manufacturers define the *rangeability* of their equipment as the ratio of maximum to minimum flow, and if we take this as read, then we could argue with some conviction that, if the valve is a tight shutoff device, then the valve has infinite rangeability, which is clearly not the case. This calls for a slight modification of the

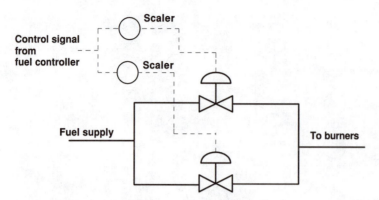

Figure 14.8: Split controller output for two valves.

definition of rangeability, which should read as the *ratio of maximum controllable flow to minimum controllable flow.*

The wide rangeability requirement for the control valves on fuel lines can be tackled in several ways. The first is to have two or more valves, and to arrange for these to be piped in parallel. If this is done with a two-valve arrangement, then the controller output must be split. This means that one of the valves will have to operate for that part of the flow that corresponds to the lower portion of the output range, and the other, over that part of the flow that corresponds to the remaining upper portion of the output range. The actual point at which the split occurs will have to be decided carefully, and sufficient flexibility retained in the arrangement to permit changes of the point of split to be made if necessary. The arrangement is shown in Figure 14.8.

Systems such as described in Figure 14.8 have been used with limited success, for in practice the rangeability required was most often even greater than that achievable with this arrangement. To achieve greater rangeability, alternatives had to be found, and the arrangement shown in Figure 14.9 is a method that can provide, and has, provided, rangeability of the order of 20:1.

In this system, the fuel controller output is used to regulate the main fuel supply line and provides the setpoint for a pressure control loop, which regulates a

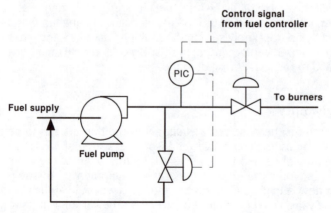

Figure 14.9: Split controller output for high rangeability.

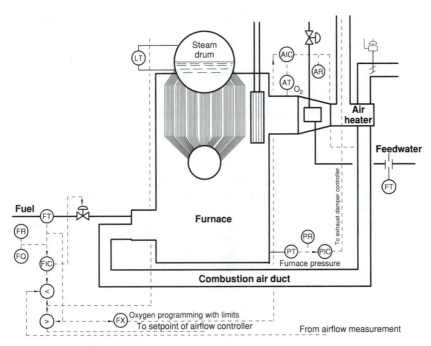

Figure 14.10: Expanded schematic diagram of combustion air control system.

recirculation path around the pump. With a reducing control signal, the pressure indicating controller (PIC in the figure) output diverts more fuel to the pump's suction side to allow the fuel supply control valve to adhere more closely to the actual requirement.

Fuel and Air Cross Limiting and Oxygen Trim of Combustion Controls

Since the operation of fuel and air control schemes have now been discussed in some detail, it will be a suitable point for returning to the original system drawing to see how this idea of cross limiting is included. The scheme shown previously in Figure 14.5 is in reality an enlarged view of part of the relevant portion excluding the oxygen trim of the complete fuel/air control scheme shown in Figure 14.10, which shows the positioning of the measuring and controlling point flue gas oxygen of the fuel/air system relative to the plant, and how the cross limiting and air flow control interconnections are incorporated.

The Master Pressure Controller dictates not only the manner in which the steam/water system will respond to a change in load, but also how it is involved with the fuel/air system as well. By including an oxygen programming device, with appropriate limits, on the fuel measurement, it can also be used to alter the signal that drives the setpoint of the oxygen controller. The output from the oxygen controller (AIC in the figure) is used to modify the measurement of the airflow so that it reflects the requirements of the actual amount of O_2 required by the fuel. The effect of the slight modification of having oxygen programming is to automatically alter the amount of oxygen required to burn the fuel in response to varying fuel flows. The output from the oxygen controller forms one of the inputs to the ratio unit on the airflow measurement and adjusts the amount of air required; this is called *oxygen trimming*. Once set, the ratio need not be altered until a different fuel is used; therefore the boiler operator will need to be advised of the fuel analysis to be able to adjust the ratio if necessary.

Be aware that while the system is shown with an oxygen trim, there is no reason for it not to have a CO measurement instead, however, using combustibles as the trim would make evaluating the percentage excess air more difficult.

Combustion of Additional Elements in the Fuel

It may seem out of place considering combustion once again, but the preceding discussion outlined principles to deal with only a single combustible component in a fuel. We shall now give attention to principles for dealing with a more realistic fuel containing other combustibles and its combustion air requirements.

The method outlined earlier for determining the quantity of air required was based on a fuel containing carbon only. However, in reality other combustible elements are also contained in a fuel, most often hydrogen (H_2) and sulfur (S), and we will have to take this into account and provide the necessary amount of air to burn these latter elements also. The technique used is similar to that used before (recall $C + O_2 = CO_2$); as a memory aid the procedure is repeated.

The *combustion of hydrogen* can be represented by the equation

$$2H_2 + O_2 = 2H_2O$$

using atomic weights this gives:

$$2(1 + 1) + (16 + 16) \qquad \text{or} \qquad 4 + 32$$

This means that 1 lb of hydrogen will require 8 lb of oxygen. The *combustion of sulfur* can be represented by the equation

$$S + O_2 = SO_2$$

Using atomic weights, this gives:

$$32 + (16 + 16) \qquad \text{or} \qquad 32 + 32$$

This means that 1 lb of sulfur will require 1 lb of oxygen.

Hence, we can say that, if a certain fuel contains C pounds of carbon, H pounds of hydrogen, and S pounds of sulfur, then each pound of the fuel will require

$$\frac{32}{12}(C) + 8(H) + 1(s) \quad \text{pounds wt } O_2$$

and this will give a theoretical value for the air of:

$$\frac{1}{0.23}\left[\frac{32}{12}(C) + 8(H) + 1(S)\right] \quad \text{pounds wt}$$

We can use the following equation to calculate the volume that this weight of air represents:

$$pV = m\Re T$$

$$V = \frac{m\Re T}{p}$$

where m is the mass, $\Re$ is the universal gas constant, T is the absolute temperature, p is the pressure, and V is the volume. If we were using imperial units, then for the mass we could use the calculated weight in pounds directly and 53.3 for $\Re$, where

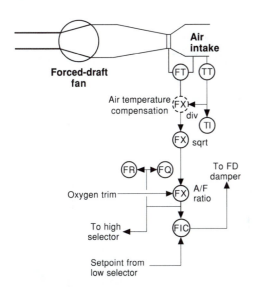

Figure 14.11: Schematic diagram of air temperature measurement.

the units for $\Re$ is ft·lb/lb·°F, T is in °R, and P in lb/in^2 Abs. This will give a result in cubic feet for the air. (With suitable adjustments to the units, and the constants involved, the foregoing can be adapted for metric units. The value of $\Re$ in SI units is 8.3J·mol^{-1}·K^{-1}.)

Combustion Air Temperature Control

Let us now look at the method used to compensate the combustion air for temperature variations. For this we shall use the diagram shown in Figure 14.11. Note that this diagram is an enlargement of the relevant part given in Figure 14.12, which we shall look at shortly.

From Figure 14.11 we can see that both the differential pressure created by the venturi section of the intake and the temperature of the combustion air are measured. As shown, the differential pressure is divided by the temperature. The ambient pressure generally does not fluctuate very rapidly and can be treated as constant; but, in some places in the world where the ambient temperatures fall within the design limits of the primary flow-measuring element, it may not even be necessary to compensate for ambient temperature variations, and in these circumstances only the differential head and ambient pressure compensation need be used. Why do we need to compensate for ambient temperature and pressure? The answer lies in the gas laws, which state that:

$$\frac{p_1 V_1}{T_1} = \frac{p_2 V_2}{T_2}$$

where p is the pressure, V is the volume, and T is the temperature; and with which air, even though not an ideal gas, has to conform. The differential pressure across the primary element (venturi tube in this particular instance), per Bernoulli, is proportional to the volumetric flow as shown in Chapter 1; and in the loop shown, we are really correcting the differential head for the variations in ambient temperature. Therefore,

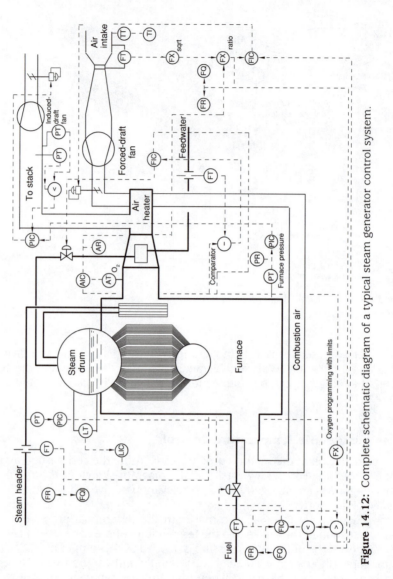

Figure 14.12: Complete schematic diagram of a typical steam generator control system.

taking the square root after the head compensation will give the actual volumetric airflow corrected for temperature. If we took the square root of the differential head before performing the division, we would really not be compensating the flow for temperature; for the differential head, although proportional to the volumetric flow, would still be uncorrected.

Furnace Pressure Regulation

The last item we consider is the pressure within the furnace, which intuition says should be low in order to let the combustion air and fuel be drawn in and ignited, the maximum allowable pressure being only 1 in or approximately 25 mm wg (water gauge/column) or thereabout. To ensure that there is a draft through the furnace, an induced draft (ID)—i.e., suction—fan is installed. In large steam generating plants, the ID fan is a necessity, because of the very large gaseous volumes involved. The complete loop for a stack that is some distance away from the furnace is a simple cascade loop of a furnace pressure (primary) controller (PIC in Figures 14.3 and 14.10) whose setpoint is determined by the process operator, on to a stack pressure (secondary) controller (PIC in Figure 14.3) that manipulates the stack damper according to the dictates of the furnace pressure. The point to remember here is that the measurements to both the primary and secondary controllers is the lowest pressure measured in the furnace and stack, and is the lowest of a number of measurements made in these areas and obtained by selection, using low signal selectors (< in Figure 14.3 and 14.12). The number of signal selectors used will depend on the number of measuring points and the type of instrumentation in the control system involved, for example, a DCS signal selector has a capability of accepting more than two signals, with the Foxboro I/A Series capable of selecting any one from eight inputs. (**Note:** In Figures 14.3 and 14.12, for clarity the low signal selectors on the furnace pressure measurement are not shown, and only two measurements are depicted on the stack.)

THE COMPLETE CONTROL SYSTEM

The complete control system up to this point will look like that shown in Figure 14.12. The figure covers all the topics that we have discussed so far. There are more loops that can be considered; for instance, some consider reliance on one type of fuel to be important and should be avoided; there are control schemes to address this valid point, of which we have shown a small selection, but to give an idea of what the possible design criteria might involve we shall only state some of the questions that the system designer might ask. The following queries are not exhaustive and are divided into two main categories—fuels and general topics.

Fuels

Multiple Fuels

Note: When multiple fuels are involved, there could be restrictions imposed by local safety and environmental bodies that must be followed, but whose implications and requirements are beyond the scope of this book.

1. Is the alternative fuel a liquid or a gas? For example, is it possible to have natural gas as the primary fuel and fuel oil as the alternate, or vice versa.

2. What type of burners are fitted to the steam generator? It is possible to have burners that can handle oil and gas separately, but only one fuel at a time without dismantling the fuel connections. One assumes that the lines will have to be purged before a changeover is carried out, or even that the fuel connections will have to be dismantled and reassembled when a change is made. On the other hand are the burners suitable for handling the two fuel simultaneously?

3. What are the calorific values of the two fuels? Combustion control will have to be handled on the basis of the total heat available.

4. If fuel oil is one of the fuels used, what is the distribution system to the burners? This question needs to be answered when oil is the sole means of heat. There are two types of distribution possible:

   a. The fuel oil is supplied to the burners in quantities that are totally combusted.

   b. The fuel oil is supplied to the burners in quantities that are in excess of that which can be totally combusted, and the unused oil is returned to the supply tank. This practice avoids fuel stagnation in the lines at low firing rates and the type of burner used in these instances is called a *spill burner*. To determine the amount of fuel used, one has to measure both the amount supplied and the amount returned and subtract one from the other.

There are other questions on the topic of fuels alone, but for the present we do not intend to consider these.

General Topics

The other important points that have to be considered are:

1. Burner management systems (BMS). These systems take care of the normal running of the burners through sequenced firing of each burner, monitoring the condition of the flame, and providing alarms and shutting down the generator in the event of adverse conditions. **Safety regulations in most countries today insist that a recognized BMS be provided on every installation.**

2. Deaeration and treatment of the feedwater itself.

3. Steam distribution systems.

4. Desuperheater controls.

5. The economic package boilers (fire-tube type) are now much in use for the smaller applications and sometimes more than one is used to provide the necessary amount of steam.

These topics should be covered as further reading to obtain a fuller understanding of this subject.

SUMMARY

1. It has been internationally agreed that the heat content (enthalpy) of water is at 0°C or 32°F is zero.

2. In determining data on steam generation, the conventional reference amount of water is 1 kilogram mass, or 1 pound mass in imperial units.

3. When water is being boiled, there will be two phases, liquid and vapor, present for as long as there is water in the containing vessel, and under these conditions the steam produced is considered to be wet.

4. When the last drop of water has just turned to steam, there is only a single phase present, and the steam is considered to be dry and saturated.

5. Any heat added after the last drop of water has just turned to steam will go toward increasing the internal energy of the steam, and the result will be to superheat the dry saturated steam (DSS) and convert it to superheated steam.

6. There are two main types of steam generation plant: water-tube boilers and the economic package boilers (fire-tube type), which are now much in use for the smaller applications where more than one is used to provide the necessary steam.

7. The main parts of an industrial water-tube steam generator are:

   a. The steam drum, which provides the water capacity for the plant and the source of wet steam.

   b. The tubes connecting the steam drum to the mud drum. Both of these are in contact with the hot combusted gases and cause the water contained within them to be rapidly converted to steam.

   c. The feedwater valve, which regulates the amount of feedwater supplied to the boiler.

   d. The superheater, which is provided as the means of increasing the steam energy. It is always in the steam line in series with the steam drum.

   e. The desuperheater, which is provided as the primary means of controlling the temperature of the superheated steam when necessary, and is always in series with, but after, the superheater. As a secondary effect it provides a steam mass increase without application of additional heat.

   f. The economizer, which raises the temperature, thereby increasing the enthalpy, of the incoming feedwater, and reduces the heat required to convert the water to steam. It is located in the exit path of the exhaust gases from the furnace to the stack.

   g. The air heater, which heats the incoming combustion air with heat extracted from the hot exhaust gas, is located in the exit path of the exhaust gases from the furnace to the stack.

   h. The furnace, or space where the combustion air and fuel are combined and allowed to expand. Ignition of the fuel occurs at the burners. The furnace internals are specially designed to allow the hot gases to make several passes over the water tubes to effect maximum heat transfer.

   i. The stack, which provides the path for the exhaust gases to be released to the atmosphere.

   j. The forced-draft (FD) fan forces air into the combustion chamber and the amount of air regulated by a damper on the downstream side of the fan, or by a series of parallel vanes built into the suction inlet of the fan itself. The fan could be steam or electrically driven.

k. The induced-draft (ID) fan, which draws out the products of combustion from the furnace and ensures balanced pressure conditions within the combustion chamber. To regulate the amount of pressure, a damper is fitted, and manipulated by the control system.

8. Under steady-state conditions of the load,

   a. The amount of steam taken by the load will be made up by a similar supply of feedwater to the steam drum.

   b. Because of this steam/water balance, the level in the steam drum will remain reasonably steady.

   c. The amounts of fuel and air providing the heat to generate steam to meet the demand will also be steady.

9. For an increase in steam demand by the load, the balanced system will be upset; and

   a. The amount of steam taken by the load will increase.

   b. The water level in the drum will change in response to the load increase.

   c. Since the drum level has changed, the feedwater flow will also have to change (increase) to maintain the water level.

   d. More water will have to be evaporated to produce the extra steam; therefore, the heat supply will be increased, resulting in an increase in the amounts of both fuel and air to support the combustion required.

10. A decrease in steam demand by the load will also upset a steady-state condition, and:

   a. The amount of steam taken by the load will decrease.

   b. The level in the drum will change in response to the load decrease.

   c. Since the drum level has changed, the feedwater flow will also change in order to maintain the drum level.

   d. To meet the decreased demand for steam, the amount generated under the steady-state conditions will have to be reduced, making it necessary to decrease the heat supply. Therefore, the amount of fuel and combustion air will decrease in proper proportion.

11. Three-element feedwater control ensures sufficient water in the drum at all times to meet demands. It is a feedforward method of control that uses the load (demand) to balance the feedwater input (supply) to the drum; at steady-state the two must be equal. To determine the magnitude of the demand or imbalance, the steam flow is subtracted from the feedwater flow. The demand is the measurement of the feedwater controller that regulates a control valve in the feedwater supply line. The setpoint of the feedwater controller is provided by the output from the drum level controller. In this control system the most important parameter is the water level in the drum, and the process operator makes the only adjustment required, the setpoint of the drum level controller.

12. Shrink-and-swell is a phenomenon observed in the steam/water drum. The water level rises when steam demand is increased (water is withdrawn from the drum) and falls when the demand is reduced. This is the converse of what would normally be expected.

13. The air required for fuel combustion depends on the quantity of carbon and other combustibles in the fuel and is calculated by using the atomic weights of the elements. Carbon will combine fully with oxygen to produce carbon dioxide:

$$C + O_2 = CO_2$$

Putting in the atomic weights, we have

$$12 + 16 + 16 = 44$$

where 12 is the atomic weight of carbon and 16 is the atomic weight of oxygen. Fuel containing, say, 12 lb of carbon will require 32 lb of oxygen to combine with to produce 44 lb of carbon dioxide. Air contains roughly 77% nitrogen and 23% oxygen at normal conditions (pressure 14.69 lb/in^2 and temperature 60°F). For 1 pound mass of air there is 0.23 lb of O_2 and 0.77 lb of N_2. Thus, the amount of air required to convert 12 lb of carbon to CO_2 is:

$$\frac{32}{0.23} = 139.13043 \text{ lb air}$$

This gives the theoretical amount of air required to make a complete conversion of the carbon to carbon dioxide.

14. The other combustible elements most often also contained in a fuel are hydrogen and sulfur. The air requirements for burning these are the same as before. Combustion of hydrogen can be represented by the equation:

$$2H_2 + O_2 = 2H_2O$$

Using atomic weights, this gives:

$$2(1 + 1) + (16 + 16)$$

or

$$4 + 32$$

This means that 1 pound of H_2 will require 8 pounds of O_2. Combustion of sulfur can be represented by the equation:

$$S + O_2 = SO_2$$

Using atomic weights, this gives:

$$32 + (16 + 16)$$

or

$$32 + 32$$

This means that 1 pound of S will require 1 pound of O_2. Hence, if a certain fuel contains C pounds of carbon, H pounds of hydrogen, and S pounds of sulfur, then each pound of the fuel will require:

$$\frac{32}{12}(C) + 8(H) + 1(S) \text{ pounds wt } O_2$$

to give a theoretical value for the air of:

$$\frac{1}{0.23}\left[\frac{32}{12}(C) + 8(H) + 1(S)\right] \text{ pounds wt}$$

We can now calculate the volume that this weight of air represents:

$$pV = m\Re T$$

$$V = \frac{m\Re T}{p}$$

where m is the mass, $\Re$ is the universal gas constant, T is the absolute temperature, p is the pressure, and V is the volume.

15. The theoretical amount of combustion air is never provided exactly, for if there are any variations in the components of the fuel or some error in the measurements, then the quantity of air provided needs to be sufficient to allow for these. Hence, an extra amount of air, usually of the order of 10%, called excess air, is provided.

16. The quantity of fuel burned will vary to meet the steam demand of the load. Therefore, the appropriate amount of air must be available for each change in demand. This need is satisfied by ratioing the fuel flow to the airflow and treating the fuel flow as the controlled variable and the airflow as the "wild" or uncontrolled variable.

17. The parameter most responsive to load changes is the steam header pressure; it is used to dictate the direction the whole combustion control system should take. To do this a master pressure controller is assigned to this measurement. The controller needs to be very responsive. Most often a three-term device is assigned. The controlled output has to respond so that a decreasing measurement results in an increasing output.

18. The fuel controller must respond to changes so that an increasing measurement results in a decreasing output. The combustion air controller must respond to changes so that an increasing measurement results in an increasing output.

19. The air and fuel control system operates in series while the system is in a steady state, but in parallel when the system is unbalanced. The words series and parallel are used as in the electrical circuit sense. The parallel operation is brought about by the action of cross limiting the air with the fuel and the fuel with the air.

CHAPTER 15

Process Dryers

INTRODUCTION TO PROCESS DRYING

Some manufacturers use drying as an intrinsic part of their process for manufacturing the product they offer to the marketplace. Others, motivated by the universal desire to maximize on their outlay and thereby increase profits, turn to the drying process to recover material that would otherwise be considered waste. In the harsh economic world in which we live today, the reality of viability coupled to profits is the motivation for all modern businesses. Without this motivation the prospect of being forced out is very real indeed. Manufacturers, in the food and drug industries in particular, are sensitive to the rapid changes in fortune that so easily can be their lot if they do not exercise care in maintaining their market share. It is for this very reason that companies in this sector of manufacturing process industries seek to squeeze as much financial benefit as possible out of both plant and material.

In most instances the raw material used in the food sector is in fact "natural," and this can be said almost without fear of contradiction. The material obtained originates in the very natural process of self-reproduction and is provided by those who either intentionally grow it or harvest it from an environment where it grows on its own. However, in all instances the producers of the raw material are absolutely dependent for its quantity and quality on the cooperation of Nature, who, as we all know, can be a very fickle partner indeed.

In this chapter we shall not be discussing the methods the raw material *producers* use to ensure either quantity or quality, but how raw material *processors* can make full use of the raw products obtained and thereby ensure that they, the processors, receive the maximum return on their investment. We shall be concentrating on a process that is particularly important in the food industry, and shall be giving attention to this sector of the manufacturing process industry. The principles discussed apply equally to other industries where product drying is involved.

The following outlines the system operating principles for drying only. It is vital that all national and local codes of practice for safety, electrical wiring, and operating procedures be strictly adhered to.

PRINCIPLES OF SOLIDS DRYING

In the food and drug industries, it is vital for health reasons, as well as essential for preserving the integrity of the finished product, to ensure that both the product and any mediums with which it comes in contact are free from bacteria harmful to humans. Achieving this bacteria-free environment is a separate issue that will not be expanded on in this chapter on product drying. Suffice it to say that within the food and drug industries absolute cleanliness, in the widest sense, is of paramount importance and nothing short of the utmost attention to it is acceptable. It is even necessary to absolutely avoid all contact with any materials that cannot be sterilized, or any that can impart poisonous residue in whatever form.

SOME BASIC DEFINITIONS

Humidity is defined as the amount of water vapor contained in a gas. Since we are dealing with air, this definition can be rephrased as the amount of water vapor contained in the air. *Absolute humidity* is defined as the humidity of a gas expressed as the mass of water in grams per cubic meter of the gas. When dealing with air, this is expressed as the mass of water in grams per cubic meter of air. *Relative humidity* is defined as the ratio of the amount of water vapor in a gas at a specific temperature to the maximum capacity of the gas at that temperature. With air, it is the ratio of the amount of water vapor in the air at a specific temperature to the maximum capacity of the air at that temperature.

THE WET- AND DRY-BULB HYGROMETER

In industry it is quite normal to determine relative humidity as a process measurement since it, rather than humidity or absolute humidity, is likely to be relevant at a practical level. Absolute humidity, on the other hand, is likely to be a valuable measurement in the laboratory in determining the allowable moisture content of the product as delivered to the user. It is a difficult measurement to make. However, because of the relative ease of making the measurement, some may wish to use the *dew point* instead as a measure of absolute humidity, but remember that this is valid only if the pressure is constant. The simplest-to-understand instrument for measuring relative humidity is the wet-and-dry bulb hygrometer. In this type instrument there are two thermometers, both of which are exposed to the atmosphere being measured, but one of which has its bulb directly and the other indirectly exposed. The indirect one has its bulb enclosed in a sleeve or wick that is constantly water impregnated. The equipment is described here mainly for the purpose of illustrating the principles involved.

Figure 15.1 is a schematic of a typical hygrometer using the principle of wet and dry bulbs. In operation the ambient air enters the instrument as shown and in so doing passes directly over the dry-bulb resistance thermometer (RTD—resistance temperature detector—in the figure) where its temperature is measured. The air continues its passage through the U-shaped pathway on to the wet-bulb resistance thermometer (RTD). This latter temperature sensor is so named for the fact that the sensitive bulb is covered by a porous sleeve that terminates in a wick that dips

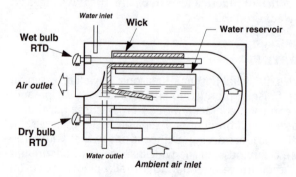

Figure 15.1: Schematic arrangement of wet and dry-bulb hygrometer.

constantly into the water contained in a reservoir, the level of which is maintained by a standpipe weir of which the water outlet is a part. There is a constant, very slow flow of replenishing water into the reservoir through the water inlet connection. The ambient air has to pass through the porous wick arrangement, and in doing so causes some of the moisture to evaporate, the amount evaporated being dependent on the temperature of the incoming air. It is easy, therefore, to see that the wet bulb will measure a lower temperature for the same incoming ambient air and as a result there will be a temperature difference between the dry- and wet-bulb readings. This difference in temperature is called the *wet bulb depression*. It is this value that is related to the relative humidity at the dry-bulb temperature, the relationship giving rise to a family of curves. The characteristic of this family of curves is nonlinear and while the inferred relative humidity will approach zero it will never actually attain a zero value. This is due to the method of determining the wet bulb depression, which requires water to be always present at the wet bulb. If at any time the wet bulb had no moisture surrounding it, then its reading would be the same as that of the dry bulb and there would be no difference between the values of the two RTDs. Hence, this technique of determining relative humidity precludes ever measuring a zero value for it. It must be pointed out that instruments of this type require thorough and regular maintenance, as they are prone to blockage of the wick with solids contained in the water supply.

(Of course, there are other hygrometers using the same principle, which are of a much simpler design, or some that use other techniques. If the measurement needed is to determine the general conditions prevailing in, say, a warehouse the instrument used is usually a sling hygrometer. This instrument consists of a rotating part that contains the wet and dry bulbs—which are two mercury-in-glass thermometers—encased in a rectangular light metal housing and a handle about which the metal case rotates. A water-saturated soft cotton swab retained in a thin metal crucible with a wide perforation except for the side near the dry bulb, which is solid to ensure bulb dryness, surrounds the wet bulb. In use, the whole instrument is whirled round on its handle, hence its name. After a short period the temperature readings are taken and recorded. From this data the relative humidity is obtained from tables.)

VISUALIZING THE DRYING PROCESS

Drying is entirely a surface phenomenon, and for most solids the drying process consists of driving off liquid from the solid, which itself may, or may not, absorb the liquid. When the solid to be dried is capable of absorbing a liquid it is said to be *hygroscopic*. For a solid to "feel" wet a film of liquid must surround it. In the case of process drying in the food industry the absorbed moisture is in almost all cases water, and the drying agent is usually hot dry air.

Figure 15.2 shows diagrammatically what is involved when we consider product drying in general as a process.

The diagram concentrates on only a single hygroscopic particle of the wet product. The depth of the liquid penetrating the solid will depend on the absorbency of the solid material. It is therefore clear that the acceptable dryness will be the resultant depth of residual moisture after the excess moisture has been driven off. While we are considering the excess that is driven off, it is important to realize that a certain amount will remain within the solid whether we want it or not. Remember, the air that surrounds us contains moisture at all times, and if a product is hygroscopic, even if it started off absolutely bone dry, as soon as we exposed it to the atmosphere it would start to absorb some moisture. Once again, the amount taken up by the product will depend on the amount of atmospheric moisture, the product's absorbency, and the duration of the exposure.

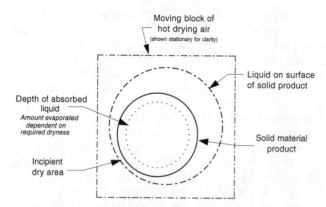

Figure 15.2: Schematic diagram of drying a single solid particle of product.

To visualize what is taking place, and for simplicity, the diagram shows a single particle of the product as a uniform sphere. It must be emphasized that this spherical shape may not really be the case at all; particle shape is entirely dependent on the product itself and varies from one material to the next. Since the solid particle can be irregular in shape, it follows that there will be a film of water of nonuniform cross-sectional thickness that surrounds it. In the figure, the dotted line within the solid represents the depth of moisture penetration prior to the start of the drying process. During drying, the hot air is forced to move around the solid product. At the individual particle level, when this water-enclosed particle of material is exposed to hot, dry air, the film of water will start to evaporate first. The result of the nonuniform thickness of the film of water will give rise to locations on the surface of the solid that will tend toward dryness in a shorter time than others, because a small mass of liquid when heated will vaporize much earlier than a larger mass of liquid.

While the surface moisture is being evaporated the temperature at the surface of the solid remains constant and because there is moisture present will approximate the wet bulb temperature. If dry bulb temperature, as represented by the temperature of the hot drying air, is held constant (i.e., no change is made to the hot air temperature), then the rate of evaporation from the surface will remain constant until a dry area is obtained on the surface of the solid. The evaporation rate when dry areas appear is not dependent on the total moisture content of the particle, the time for complete vaporization being dependent only on the mass of liquid.

On the particle itself, areas of small liquid mass—therefore, of shorter drying time periods—are referred to as areas of *incipient dryness*. Our simplified pictorial representation shows the solid sphere displaced to one side to permit a thin area of moisture to be depicted as an area of incipient dryness.

The evaporation rate decreases continuously as soon as the first dry area appears on the solid. None of the penetrated moisture will begin to evaporate until all the surface moisture has completely vaporized, because the hygroscopic nature of the material the moisture film will tend to "feed" the penetrated moisture.

The evaporation rate decreases further when the absorbed moisture within the solid has to be dried, the process being a logical one that starts with the surface moisture vaporizing first, followed by that which is embedded. This gives rise to well-defined drying zones as shown in Figure 15.3.

The smaller the solid particle size, the greater will be the amount of moisture that it will retain, because of the larger surface area of solid per unit volume of material available, and as a result, the higher will be the evaporation rate. From a heat transfer

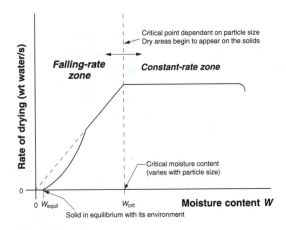

Figure 15.3: Ideal drying curve for uniform size particles.

point of view, the rate of evaporation from a solid is proportional to the temperature difference between the air and the solid. Therefore, the wet-bulb temperature plays a very important part in the solids drying process. In Figure 15.3, note:

- The drying process shown is progressive with time.

- The characteristic of the initial portion of the curve is due to the fact that the material has to change from its normal equilibrium state to one approaching vaporization with very little moisture evaporating in the early stages as each moisture particle absorbs heat until vaporization temperature is attained. The amount evaporated in these early stages is liquid mass-dependent. As the liquid mass reduces, uniformity results.

- The size of the solid particle will affect the position, the point of *critical moisture content*, but it will not change the slope of the curve in the falling-rate zone. This is shown in Figure 15.4.

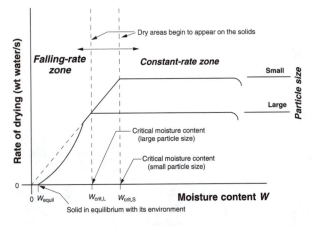

Figure 15.4: Ideal drying curve for particles of varying size.

PSYCHROMETRY

Since in process drying we are involved in driving off water from material, under-standing the relationships between the different parameters connected with wetness is a vital part of the exercise. These relationships are given visibility in diagrams called *psychrometric charts*. Several of these charts are available, each for a different liquid material, but the one we are most interested in is the one concerned with water and what occurs to this medium when subjected to changes in ambient pressure and temperature. Since actual atmospheric pressure is so variable, the traditional psychrometric chart for water gives data for water at the accepted standard atmo-spheric pressure. The Carrier Corporation, maker of air conditioning systems, has produced one of the most comprehensive charts for water (*Psychrometric Chart*, Carrier Corporation, copyright 1946), as it is vital for their field of work and exper-tise. It is recommended that the reader take the opportunity of viewing this important document. In Figure 15.5 the curves and description are based on the Carrier docu-ment and will detail how the information contained in the chart should be read and interpreted. Note that in Figure 15.5 the enthalpy (total heat) curves of the original document have been omitted for simplicity and clarity, since we do not require this information for the control system design. This diagram is not to scale, since it is only intended to convey the orientation and meaning of the curves of the real document.

A most important point to remember is that moisture will not exist as a single phase to the left of the saturation line; it will condense as dew, frost, or fog. In use the starting point for the diagram can be the junction of any two values that are available. As an example, suppose the wet- and dry-bulb temperatures are known. Then:

- The point of intersection of these two values obtained by using the vertical ordinates of the dry-bulb temperature and the sloped curves of the wet-bulb temperature locates the position on the chart.

- From this located point, tracing a horizontal abscissa (distance parallel to the x axis) and to its left all the way to the saturation line will give the dew point for the air at the dry-bulb temperature, which is read from the temperature values marked on the saturation line.

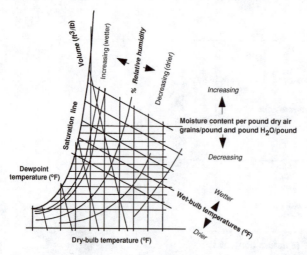

Figure 15.5: Psychrometric chart for water. This chart is drawn for a pressure of 29.92 in Hg.

- Tracing a horizontal abscissa from the same point all the way to the extreme right-hand ordinate (distance parallel to the y axis) will give the moisture content per pound of dry air at the dry-bulb temperature, which is read from the intersection with this extreme right-hand ordinate.

- The volume of the air in cubic feet per pound of air is obtained by reading the value of one of the second set of more acutely sloped curves (with respect to the curves of the omitted enthalpy values) that pass through the point of intersection of the wet- and dry-bulb temperatures.

- The parabolic curve passing through the located point will give the percent relative humidity for the known wet- and dry-bulb temperatures.

Since the chart can be constructed for only one value of ambient pressure, readings taken from it will have to be corrected for any other value of ambient pressure. This correction takes the form of the equation given in *Perry's Chemical Engineers' Handbook* and shown below:

$$H_s = H_0 + 0.622 p_w \left(\frac{1}{P - p_w} - \frac{1}{760 - p_w} \right)$$

where H_s = humidity of air at pressure P, kg/kg dry air
        H_0 = humidity read from chart based on 760 mm Hg
       p_w = vapor pressure of water at observed wet-bulb temperature, mm Hg
        P = pressure at which the wet- and dry-bulb temperatures were taken

Similar corrections will also have to be derived for the other parameters in the chart.

PULP PRODUCT DRYING IN THE FOOD-PROCESSING INDUSTRY

When we consider some industries that manufacture food for human consumption, we will find that after the usable portion has been extracted, the waste is very often used for animal feed. For example, in the processing of sugar beet for granular sugar, the pulp remaining after juice extraction by pressing is used as cattle feed and represents a significant amount of revenue. The waste from fruit juice extraction is also a potential revenue source, as animal feed. Retention of excessive moisture in the animal feed product has to be avoided, for it will inevitably accelerate its deterioration and shorten its useful life span, in turn affecting its viability in the marketplace. As a result, product drying can be an important treatment process for this waste. Drying is necessary to ensure that the money invested up front can be recouped with some profit within a reasonable time.

In practice, manufacture in the fruit and sugar industry is always periodic, dependent on the growth of raw material (sugar beet, sugar cane, oranges, grapefruit, etc.) before processing can begin. The period of processing in the sugar industry is always referred to as a *campaign*. It is a period of intense activity, the plant working flat out continuously 24 hours a day. This is necessary because, being at one time a living entity, the raw material has a very short life span before decomposition occurs. During the campaign, failure of any plant machinery can be catastrophic. It is therefore normal for the quiet periods of the year to be spent in maintenance, refurbishment, or extension of plant and equipment. But do not assume that in all sweetener (sugars) manufacturing the processes are periodic; they are not. If maize grain (sweet corn) for example, is used for sweetener syrup making, then manufacture can be continuous throughout the year, for it is possible to store this grain in good condition in suitable silos over relatively long periods.

The waste from sugar cane in sugar production very seldom, if ever, requires drying after juice extraction when used as cattle feed. However, the cane pressings, called *bagasse* are more often used for making paper or as a fuel to fire steam generators (boilers) that provide power and heat for the manufacturing process. In such cases product drying is kept to a minimum, for further use of the waste is made within very short periods.

Drying the pulp has to be accomplished with minimal additional outlay, at the same time as ensuring the shelf life of the dried product matches its ultimate use. These objectives demand that the fuels used in the heaters produce the maximum heat per unit incinerated and are the most economic in use and price, and that the amount of moisture contained in the pulp after processing is such that, for example, when the pulp is used as animal feed the dried pulp must have the longest life expectancy, whereas the moisture content is not so critical when the pulp is used for incineration. For any product, there is an optimum specification, which is determined by the manufacturer based on market research. This specification details the various components of the product and the acceptable variation of each constituent that will ensure its viability in the marketplace. For product dryness, our sole focus for the moment, there is obviously a band of acceptability, with an upper and lower limit, to the amount of moisture the product should contain. This is absolutely true of the beet and fruit pulp product we spoke of.

We shall now consider the financial implication of product drying to make a case for good control. If, suppose, we dry the pulp to the extent that the amount of water contained is tending toward the lower of the two limiting specification values, then, very briefly, we will be spending:

• A certain sum of money to provide the fuel to dry the pulp

• A sum of money per pound of product to finance the labor to monitor the process

• An amount on electricity per pound of product to drive the machinery

for which we shall obtain:

• A good product well within the product specification

• An amount of product that represents an amount of recouped money

• A satisfied user of the product

If, now, we dry the pulp to the extent that the amount of moisture contained is shifted toward the upper of the two limiting specification values, then we will be spending:

• Less money than that previously to provide the fuel to dry the pulp, for we do not need to dry the pulp as much; as a direct result, the retention time within the dryer is reduced and the pulp throughput can be increased

• Less money per pound of product than previously to finance the labor to monitor the process, for more has been produced within the same time

• Less money per pound of product than that previously spent on electricity to drive the machinery, for more has been produced within the same time for which we shall:

- Still obtain a good product well within the product specification

- Have a greater total mass produced than before, an added benefit, because the product contains more moisture and this represents additional recouped money for the benefit of the business

- Still have a satisfied user of the product. This last is also a very important point in any business.

Since it is impossible to have zero moisture in any product and there is an acceptable amount of moisture that does no serious damage to the product's shelf storage life or quality, it is possible to exploit this small amount of acceptable moisture content to advantage for increased profitability.

PROCESS DRYING EQUIPMENT

The process of drying can be divided into two main categories: adiabatic and non-adiabatic. In *adiabatic* processes, changes in the condition of a body occur without gain or loss of heat. When a dryer uses heated air and does not apply heat *directly* but indirectly through a medium, to the material being processed, the dryer is said to be an *adiabatic dryer*. When a dryer applies the heat directly to the material being processed, the dryer is said to be *nonadiabatic*. The most common types of dryers found in the food and pharmaceutical industries are adiabatic and usually use hot air as the drying medium. These industrial dryers are broadly described in the following (nonexhaustive) listing.

BATCH-TYPE DRYERS

Batch-type dryers are designed for drying only small quantities of product that is introduced to the drying process in batches. The hot air is made to pass all around the product, that is usually in special containers designed to permit the flow of air around the contents that are loaded on a trolley fitted with wheels. This type is used in the preparation of food, drugs and in the development of new products within these industries. There is a patent on the control system of the batch dryer.

FLUIDIZED-BED DRYERS

The fluidized-bed dryer is designed for use as a continuous machine and will be covered more fully later in this chapter. It is used very often in the pharmaceutical industry and sometimes in the drying of specialty foods. In this machine the product is introduced to the drying chamber, where it is made to encounter a counterflowing stream of hot air. The airflow is such that the product is held in suspension in the airstream and, when the product is dry, because of the resulting weight loss the drier solids are forced to move out of the chamber and into another, from where they can be collected or processed further. Fluidized-bed dryers can be connected in series to gradually reduce the moisture content of the product and thus preserve its quality.

AIR-LIFT DRYERS

In operation, air-lift dryers are much like the fluidized-bed dryer, but they are designed to dry fibrous material. The material is introduced to the machine, where the air stream lifts the product and causes it to dry. The floating dry product is carried to a cyclone separator, where product and air are separated, the product being collected from the bottom of the cyclone.

SPRAY DRYERS

In spray dryers, product and drying air are in counterflow and hence operate in a way that is directly opposite to that of the air-lift type. In this type of machine the product, a concentrated fine particulate slurry, is sprayed into the drying chamber from the top of the vessel against a flowing stream of hot air. Since the product is a concentrate of fine solid particulates in a relatively small amount of liquid, moisture is removed very quickly. This allows the solids to fall as a powder to the bottom of the drying chamber. This type of machine is used quite extensively in the preparation of powdered products in the food and pharmaceutical industries.

LONGITUDINAL-TYPE DRYERS

In longitudinal dryers, which are cylindrical machines as diagrammed in Figure 15.6, the hot drying air and the product are made to run either concurrent with or current opposite to each other. When operational, these dryers rotate slowly, thus causing the material to move along the axis of the machine from feed inlet to product outlet. This type of machine is used often in the beet sugar-making industry to dry the pulp obtained after the juice has been extracted from the raw material. It is almost exclusively the type of dryer used in the cement-making industry, where the drying chamber has a very large diameter and the cylinder can reach some considerable length.

THE STEAM-TUBE DRYER

The classic example of a nonadiabatic dryer is the steam-tube dryer. In this equipment, diagrammed in Figure 15.7, the wet product is dried by being brought into contact with a hot surface. The equipment comprises a set of tubes running the whole length and fitted internally around the periphery of a hollow metal cylinder, with suitable space between each to permit the passage of material and ambient air along the length of the individual tube. The ambient air is drawn into the dryer near the steam entry. The tubes, normally referred to as a tube bundle, are connected, a few at a time, in parallel (as it would be defined for an electrical circuit) through

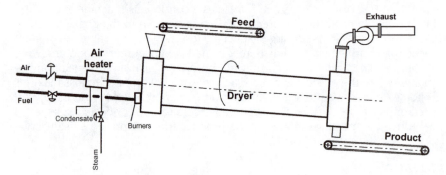

Figure 15.6: Longitudinal dryer.

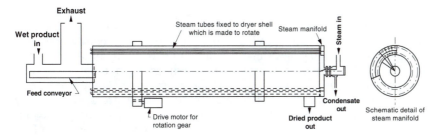

Figure 15.7: Schematic diagram of a rotating steam-tube dryer.

a fixed end cap with appropriate seals at the feed end of the dryer and a fixed steam/condensate manifold that is made to appear as a rotary valve because the whole cylindrical dryer body is made to rotate about it. A physical barrier within the manifold separates the live steam from the condensate. One end of the manifold is fixed to the source of hot steam, and at the other end the condensed steam is led away to be appropriately exhausted as condensate to drain. It is normal for the product material to be pushed through the dryer by the action of a helical feed screw in a direction contrary to the steam flow, as this is found to give the best results for drying the material. The moist air is allowed to exhaust from a point near the product entry.

DRYER OPERATION

There are many designs of pulp dryer and Figure 15.8 is just one of them. As will be seen from the diagram, the pulp and drying airflow are concurrent (in the same direction). The different types of directional arrows shown indicate the flow direction. Wet pulp brought from the juice presses on a series of roller conveyors enters the machine just after the air heater and is made to move slowly along the length of the drying chamber by the slow rotation of the dryer along its longitudinal axis. A series of short-length internally fitted baffle plates are arranged in a helical curve. The length of the baffles is such that they leave a longitudinal central area free of obstruction, thus allowing the product to fall freely during its traversal and to be exposed to the hot drying air. The helical form of the baffles is the means by which product movement is

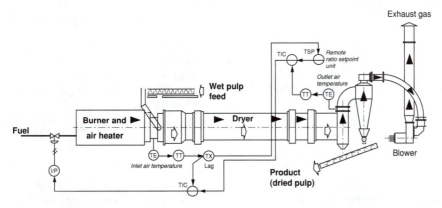

Figure 15.8: A sugar beet pulp dryer.

achieved. Hot air is forced to circulate around the wet pulp for the entire length of its traversal. At the end of its passage through the drying chamber, the bulk of the dried product falls under gravity onto the product conveyor. The product conveyor delivers the dried pulp to the packaging and storage facility. To extract any of the remaining dried pulp that could have been carried over in the exhaust airstream, the last piece of plant equipment the product encounters is the precipitator. The precipitator is effectively a large plenum chamber that allows any solids to fall free (under gravity) from the still relatively hot transport air. The collected quantity in the precipitator is periodically discharged onto the product conveyor. A blower at the base of the exhaust stack accelerates the discharge of the gas by providing secondary air at high velocity to drag the exhaust gas through and up the stack, where it emerges as large clouds of steam visible in the atmosphere.

THE CONTROL SYSTEM

Experience tells us that, for any success in devising a dryer control scheme, it is prudent that as much data as possible on machine performance first be obtained and studied. Because of the interactive nature of the parameters on these machines, most systems will use one or more cascade loops for control. It will usually be found necessary to make minor adjustments to optimize the control settings after installation and commissioning. For the drying control scheme shown in Figure 15.8, there are in reality two cascade control loops. One of these loops is directly under the control of the process plant operator and thus requires careful planning. The trend of the measured process parameters will show the plant operator the direction the control setting should move, but the amount and flexibility permitted will have to be provided for in the design of the system itself.

The heat input is manipulated to achieve drying of the product. When this is done, it has been found that the temperature of the outlet and inlet air are related to each other. This relationship, if plotted directly, is nonlinear (logarithmic); but after regression techniques have been applied to the readings obtained, a relationship is found that is almost linear. In general, for control purposes, continuous measurements of the variable of interest are required and vital to achieving success. However, in the system we have shown in Figure 15.8 we contravene this fundamental requirement. We make no direct measurement of product moisture content, the parameter of interest, but are inferentially determining the moisture in the product by measuring the exhaust gas temperature and controlling the product moisture by manipulating the temperature of the inlet air. This is necessary because product moisture is very difficult, if not impossible, to measure by direct means on a continuous basis. In order to regulate the moisture content to a desired value, we use the underlying linear relationship we have just shown to exist between inlet and outlet temperatures and assign the task of setting the required moisture content to the process plant operator. The process plant operator then has the facility of not only determining the amount of dryness, but also intervening should the need arise.

Since the relationship is linear, it will follow the general form of a straight line given by:

$$y = mx + c$$

In the case we are considering, y is the *output* from the manual setpoint generator (TSP—temperature setpoint in the figure), which is in fact the outlet air temperature that we want in order to achieve the product moisture. The slope of the curve, given by m, is the *gain* that we shall have to provide to obtain coincidence with the theoretical form. The gain will have to be calculated from the linearized curve drawn (graphed) using the actual measurements of the inlet and outlet air temperatures that we must obtain initially from the dryer itself. The constant c is the *bias* that we should provide that will allow the operator to shift the whole curve very slightly to the correct position with respect to the inlet and outlet temperatures if required. This positioning of the curve is to allow for a small change in the dewpoint of the air. Rewriting the preceding pure mathematical equation in symbols that are meaningful in process control terms, we can say:

$$T_{outlet}^{(setpoint)} = aT_{calc} + b$$

where a is the gain, b is the bias, and T_{calc} is the measured Inlet air temperature. The bias is related to the ambient dewpoint and also follows a linear relationship given by:

$$b = b_0 + f(T_{dew})$$

where b_0 is the bias at zero dewpoint and $f(T_{dew})$ has been defined by Shinskey (*Energy Conservation through Control*, Academic Press, 1978) for adiabatic dryers as a constant related to the inlet, outlet, and wet-bulb temperatures and given by

$$f(T_{dew}) = k = \frac{T_{outlet} - T_{wet\,bulb}}{T_{inlet} - T_{wet\,bulb}}$$

where k is the ratio range, which is calculated from actual values obtained from the relevant dryer. It has been found that the larger the constant k is made, the less moisture the product will contain.

A ratio station (TSP) with output bias facility that receives its measurement from a transmitter (TT temperature transmitter in the figure) connected to the inlet air temperature sensor (TE temperature element—RTD resistance thermometer detector) allows the process plant operator to set the operating moisture content by adjusting the ratio and bias for the control system. The measurement upon which the required product moisture content is determined is intentionally lagged, i.e., delayed in a temperature modifier (TX). This delay is to allow for the period of time the product is retained within the drying chamber and provided to ensure that the setpoint will respond more slowly than the outlet temperature. It is a clear case of deliberately introducing a measurement lag, normally considered to be a disadvantage, and turning it into a means of enhancing the performance of the control system.

The two controllers (TIC—temperature indicating controllers—in the figure) needed are three-term units and both must be provided with a remote setpoint facility, since they are arranged in cascade. Three control terms are necessary to permit the control action to be sensitive to the rapid changes that can occur when the system is in operation. The outlet temperature controller (TIC) receives its measurement from a sensor element (TE) located in the exhaust flue feeding a temperature transmitter (TT), and its setpoint taken from the operator-set ratio station (TSP) and its output directed to the setpoint of the inlet temperature controller (TIC) that manipulates the fuel control valve via a current to pneumatic converter (I/P in the figure) to change the heat supply to the drier. The inlet temperature controller receives its measurement from a temperature transmitter (TT) connected to a sensor (TE) located in the vicinity of the product feed inlet sufficiently removed from the direct influence of the very hot gases generated by the heater burners, but yet giving a true representation of the inlet air temperature conditions prevailing at the feed itself.

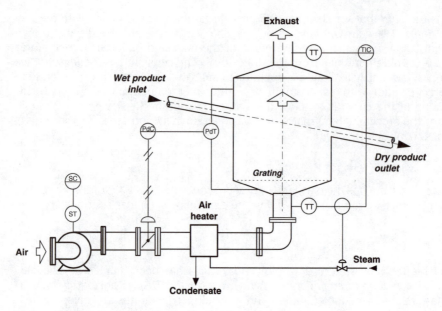

Figure 15.9: Schematic diagram of a fluidized-bed dryer.

ANALYSIS OF THE FLUIDIZED BED DRYER

OPERATION

The fluidized-bed dryer, diagrammed in Figure 15.9, is very often used in the food and pharmaceutical industries to produce specialty products in continuous quantities. The wet product is fed to the dryer toward the top end of the drying chamber and falls toward the fine mesh grating that separates the solids from the hot drying air. The drying air is fed to the chamber from beneath the mesh, and thus, acts as an inhibitor to the further fall of the solid product. From *Perry's Chemical Engineers' Handbook*, the force involved corresponds to that given by Stokes' law, usually stated as:

$$F_d = 3\pi\mu u \frac{D_p}{g_c}$$

where F_d is the drag force, D_p is the diameter of the particle, g_c is the gravitational constant, μ is the fluid (in this case air) viscosity, and u is the relative velocity between fluid and particle.

A constant flow of drying air is an absolute requirement, for it is this that holds the particles of the product suspended in the airstream. It follows that the equipment designers should size the blower to meet this requirement. However, it must be remembered that the airflow, in addition to it being able to fluidize the bed, has to be such that there is minimum solids carryover to the exhaust. The design ensures that there is an enormous amount of back mixing of solids and drying air, very unlike that found in longitudinal dryers, where back mixing is minimal or virtually nonexistent.

THE CONTROL SYSTEM

As shown in Figure 15.9, there are three loops in the fluidized-bed dryer system. Although the fan motor speed control is not often encountered, it is useful if large motor speed fluctuations exist in practice. To provide speed control, ST the speed transmitter measures the motor speed and SC the speed controller responds with an output to regulate the speed to a value determined by the process operator. The product material held in suspension in the air stream within the dryer simulates a restriction with a differential pressure across it, very similar to an orifice plate in a process flow line. Recall the relationship for the orifice plate:

$$F^2 \propto h$$

where F is the flow through the orifice bore and h is the pressure difference across the orifice plate. Therefore, in order to have constant flow through the restriction thus formed in the drying chamber, the differential pressure across it must be held constant. It is for this reason that the differential pressure across the fluid bed is measured by PdT a differential pressure transmitter in the figure and the inlet airflow manipulated by PdC the differential pressure controller acting on this measurement to hold the pressure difference constant.

On the dryer temperature control loop, we have two temperature indicating controllers TIC, one having its measurement provided by TT a temperature transmitter measuring the exhaust gas temperature, while the other receives its measurement from a temperature transmitter TT that measures the inlet air temperature. The exhaust gas temperature controller output is cascaded as the setpoint of the inlet temperature controller, the output of which manipulates the heating steam to the air heater. This arrangement produces a very stable and responsive temperature loop that has in effect a feed-forward loop on the inlet temperature and a feedback temperature loop provided by the exhaust temperature controller. The sensitivity of the feedforward loop, which by itself would make it unstable, is moderated by the stabilizing effect of the feedback loop. This control system can produce a measure of control over the product moisture range. The difficulty in providing the setpoint in the manner shown for the longitudinal dryer described earlier lies in the fact that it is not possible to measure the wet bulb temperature under the conditions of the drying atmosphere prevalent in the dryer.

Within the context of this book it is not possible to cover the control of all the dryer types used in industry. To adequately treat even those listed here will be a formidable task. However, it is hoped that what has been done provides the reader with the basic concepts of and insight to the controls associated with drying machines.

SUMMARY

1. For some manufacturing procedures, drying is an intrinsic part of manufacturing the product offered to the marketplace. In others it helps to increase output and profits.

2. Humidity is defined as the amount of water vapor contained in a gas. When dealing with air, this is defined as the amount of water vapor contained in the air. Absolute humidity is defined as the humidity of a gas expressed as the mass of water in grams per cubic meter of the gas. When dealing with air, this is expressed as the mass of water in grams per cubic meter of air. Relative humidity is defined as the ratio of the amount of water vapor in the gas at a specific temperature to the maximum capacity of the gas at that temperature. With air it is the ratio of the amount of water vapor in the air at a specific temperature to the maximum capacity of the air at that temperature.

3. Relative humidity is quite normally taken as a process measurement in industry, since it is the more likely to be relevant at a practical level.

4. Absolute humidity is a valuable measurement in determining the allowable moisture content of a product as delivered to the user, but is encountered more often in the process control laboratory. It is a difficult measurement to make. However, because of the relative ease of making the measurement some may wish to use the dewpoint instead as a measure of absolute humidity, but this is valid only if the pressure is constant.

5. The easiest relative humidity instrument to understand and use is the wet- and dry-bulb hygrometer. In this type instrument there are two thermometers, both of which are exposed to the atmosphere being measured, but one of which has its bulb directly and the other indirectly exposed. The indirect one has its bulb enclosed in a sleeve that is constantly water impregnated. The difference in measured temperature between the two is related to the relative humidity.

6. To determine the relative humidity prevailing in, say, a warehouse, the instrument used is usually a sling hygrometer, which gives the measurements of wet- and dry-bulb temperatures. The temperature sensors are mercury-in-glass thermometers. The temperature readings are taken, manually recorded, and used to determine the relative humidity from tables.

7. Drying is entirely a surface phenomenon. For most solids, the drying process consists of driving off water from the solid, whether or not the solid can absorb water.

8. For a solid to feel wet, it must be surrounded by a film of water. The solid to be dried could be hygroscopic, that is, it capable of absorbing a liquid. The depth of the liquid penetrating a hygroscopic solid will depend on the absorbency of the solid material involved.

9. In the drying of solid particles, the evaporation rate when dry areas appear is not dependent on the total moisture content of the particle. The time period for complete vaporization depends only on the mass of liquid.

10. The evaporation rate decreases continuously as soon as the first dry area appears on the solid. None of the penetrated moisture will begin to evaporate until all the surface moisture has completely vaporized, because the hygroscopic nature the material the moisture film will tend to "feed" the penetrated moisture.

11. While the surface moisture is being evaporated, the temperature at the surface of the solid remains constant and because there is moisture present will approximate the wet bulb temperature. If dry bulb temperature, as represented by the temperature of the hot drying air, is held constant, i.e., no change is made to the hot air temperature, then the rate of evaporation from the surface will remain constant until a dry area is obtained on the surface of the solid.

12. From a heat transfer point of view the rate of evaporation from a solid is proportional to the temperature difference between the air and the solid. Therefore, the wet-bulb temperature plays a very important part in the solids drying process.

13. On psychrometric charts, moisture will not exist as a single phase to the left of the saturation line but will condense as dew, frost, or fog. Since the chart can only be constructed for one value of ambient pressure, readings taken from it will have to be corrected for any other value of ambient pressure. The correction for another pressure takes the form of an equation given in *Perry's Chemical*

Engineers' Handbook

$$H_s = H_0 + 0.622\, p_w \left(\frac{1}{P - p_w} - \frac{1}{760 - p_w} \right)$$

where H_s = humidity of air at pressure P, kg/kg dry air
H_0 = humidity read from chart based on 760 mm Hg
p_w = vapor pressure of water at observed wet-bulb temperature, mm Hg
P = pressure at which the wet- and dry-bulb temperatures were taken

Similar corrections will also have to be derived for the other parameters in the chart.

14. The starting point for using a psychrometric chart can be the junction of any two values that are available. For example, if the wet- and dry-bulb temperatures are available then:

   • The point of intersection obtained by using the vertical ordinates of the dry-bulb temperature and the sloped curves of the wet-bulb temperature locates the position on the chart.

   • From this located point, tracing a horizontal abscissa and to its left all the way to the saturation line will give the dewpoint for the air at the dry-bulb temperature, which is read from the temperature values marked on the saturation line.

   • Tracing a horizontal abscissa from the same point all the way to the extreme right-hand ordinate will give the moisture content per pound of dry air at the dry-bulb temperature, read from the intersection with this extreme right-hand ordinate.

   • The volume of the air in cubic feet per pound of air is obtained by reading the value of one of the second set of more acutely sloped curves (with respect to the curves of the omitted enthalpy values) that pass through the point of intersection of the wet- and dry-bulb temperatures.

   • The parabolic curve passing through the located point will give the percent relative humidity for the known wet- and dry-bulb temperatures.

15. In *adiabatic* processes, changes in the condition of a body occur without gain or loss of heat. A dryer that uses heated air but does not apply heat *directly* but indirectly through a medium, to the material being processed, is an *adiabatic dryer*. When a dryer applies the heat directly to the material being processed, the dryer is *nonadiabatic*. The most common types of dryers found in the food and pharmaceutical industries are adiabatic and usually use hot air as the drying medium.

16. Conditions around us prohibit the possibility of any product maintaining zero amount of moisture continuously. However, there is an acceptable amount of moisture that does no serious damage to a product's shelf storage life or quality. This small amount of acceptable moisture can be and is exploited to advantage. The process of drying can be divided into to main categories: adiabatic and nonadiabatic.

17. Industrial adiabatic dryers include the following, among others:

   • The batch-type dryer, designed for drying only small quantities that are introduced to the drying process in batches. There is a patent on the control of the batch dryer.

- The fluidized-bed dryer, designed for use as a continuous machine, used very often in the pharmaceutical industry and sometimes in the drying of specialty foods. The product is introduced to the drying chamber, where it encounters an opposing stream of hot air. The product is suspended in the air stream to dry.

- The air-lift type dryers, which operate much like the fluidized bed but are designed to dry fibrous material. The material is introduced to the machine, where the air stream lifts and dries it. In the process the product is carried to a cyclone separator where product and air are separated.

- Spray dryers, operate in a manner opposite to the air-lift type. The product as a fine concentrated slurry is sprayed into the drying chamber contrary to a stream of hot air. The product dries very quickly, allowing the solid to fall as a powder to the bottom of the drying chamber.

- Longitudinal-type dryers, in which the hot drying air and product are made to run either concurrent with or in opposition to each other. These dryers rotate slowly, causing the material to move along the axis of the machine from feed inlet to product outlet.

The classic example of a nonadiabatic dryer is the steam tube dryer, which comprises a set of tube bundles running in parallel through the whole length and fitted internally around the periphery of the metal cylinder case. The tubes are spaced to permit material and ambient air to pass along its length. Air is drawn into the dryer near the steam entry, where the tube bundles are connected a few at a time through a fixed end cap with appropriate seals that is made to appear as a rotary valve. The product is pushed through the dryer in a direction contrary to the steam flow by a helical feed screw, and the moist air is exhausted from a point near the product entry.

FINAL CONTROL DEVICES

CHAPTER 16

Control Valves

In this chapter we shall be considering the way in which a process fluid is regulated so that the amount flowing at any given time meets the demand made upon it. The control of a process fluid is effected by making the fluid velocity change with respect to the demand made. Since we are manipulating velocities, we are in effect changing energy levels, reflected in the pressure that forces the fluid along a pipe, duct, or any other kind of channel. In some instances, the position of the pipe or containing duct with respect to the surface of the earth provides the motive power by potential energy.

Pipes or containing ducts will usually be of constant internal dimension(s), and therefore the volume passing through will be a direct function of the fluid velocity and the cross-sectional area. Other parameters, such as density and compressibility, also play a part in determining the quantity that flows, and these too will have to be given attention when the design and size of the valve are to be determined.

THE NEED FOR GOOD PROCESS PLANT DESIGN

So far we have discussed how different process parameters can be measured, and the basic theory governing the control of a process stream that contains a fluid subject to variation in these parameters. We shall now turn our attention to the means that translate the controller output into actual manipulation of the process fluid itself—as is the case in flow control, or the regulation of some other process parameter or energy source—ultimately resulting in correction of the measured variable so that it equals to its desired value or setpoint.

Before we go into the details of the regulating device, sometimes referred to as the *final actuator* or *control actuator*, let us be convinced of one very important fact: *a control valve dissipates energy, and energy dissipation is the sole purpose of the design.* Therefore, it should be important to consider the cost of energy in evaluating the effect of a control scheme on the overall profitability and running of the process; specifically:

1. It costs money to transfer material from one point in a process to another part of the process or to increase pressure or temperature.

2. This transfer usually involves an increase in energy also; hence, there is no advantage in providing the additional energy input only to dissipate it at the control valve.

These should be prime considerations by the process designers and the plant constructors, because in heeding these points we shall be approaching an inherently controllable process with energy dissipation at a minimum and, furthermore, we shall be contributing toward both our economic development and preservation of the environment.

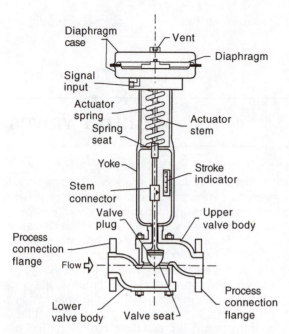

Figure 16.1: Control valve with diaphragm actuator. The valve shown requires a signal to make it open. The air failure mode is closed under the spring action.

DESCRIPTION OF A TYPICAL INDUSTRIAL CONTROL VALVE

There are several different types of control valves, the differences being mainly confined to the shape of the plug and the manner in which the device is operated. Since the word plug, in this context, is relevant to a specific part of a valve, it will be advisable to commence by defining the major parts of a control valve Figure 16.1 shows a typical control valve and will assist in understanding the way it functions. The control valve shown is one that has a *split-body* construction that was pioneered by the well-known manufacturer Fisher-Rosemount. Four through bolts, which make maintenance and seat replacement very easy, holds the lower and upper parts of the body together. Having started at the bottom of the valve, we shall therefore work upward.

THE BODY

The body is the part of the valve through which the process fluid passes, and it is here where the fluid is manipulated; for the device shown the flow is from left to right. Since the process fluid is in intimate contact with the body, the body will be subject to the same process conditions as the fluid, and the choice of material used in its construction will depend on the process fluid material and conditions with which it must be compatible.

CONNECTIONS

The means by which the valve is attached to the process line are usually called "connections"; they form an integral part of the body. For the valve in Figure 16.1, these are flanges, but alternatively they could be internal threads, butt or socket weld stubs, or fitted with sanitary connections. However, the type selected will depend on the service to which the valve will be put. Flanges could be any of the standard types available—e.g., RF (raised face), RTJ (ring-type joint), or flat face, conforming to BSI and/or DIN standards in Europe, and ANSI standards in the United States. The main criteria in determining the flange are its rating, which is governed by the process pressure and temperature; and the material, which not only has to be compatible with the process fluid, but must also align with or be superior (if it is the manufacturer's standard) to the material of the process line. Threaded (screwed) connections, like their flange counterparts, also have to conform to standards published by the standards organizations named before. Sanitary connections are very specialized, and so named because, they have no internal crevices where the process fluid can be trapped and allow bacterial growth. They also permit rapid disconnection of the valve for cleaning purposes; sanitary process connections are mandatory in the food and drug industries.

PLUG

The plug is the contoured end of the stem, by means of which the flow of the process fluid is regulated; it can be of different shapes, such as wedge, cylindrical, circular conical, or spherical. When the plug is of cylindrical or circular conical cross section, the diameter of the plug runs concentric with the valve seat, with the plane of its travel perpendicular to the seat. The contour of the plug determines the characteristic of the flow, and gives rise to the linear, equal percentage, quick opening, throttling, and other curves. The plug shape when manufactured can give rise to only one of the characteristics stated.

Even though a spherical plug was mentioned earlier, strictly speaking, spherical-plug valves (ball-type valves) do *not* have a spherical plug that fits a single seat as do the other plug valves mentioned previously. In these valves, as will be seen later in Figure 16.4, the ball is a hollow sphere with a shaped aperture to allow the process fluid to pass through from one side to the other, with the amount allowed through dependent on the degree to which the aperture is exposed to the fluid. The sphere (ball) is constrained to rotate about the stem line with its center on the line of the flow, such that the direction of fluid flow directly impacts the shaped aperture. Sealing between the body and the ball is provided by means of an accurately machined inner surface and shaped soft seals.

STEM

The stem is really a part of the plug and has a linear sliding movement for the valve illustrated. Ball-type valves, too, have a stem, but a movement that is rotational. In the device shown in Figure 16.1, the stem and plug are machined from the same bar stock, with the stem also used to guide the travel of the plug—i.e., the stem performs the same function as the sliding shaft of the power cylinder on a steam locomotive.

SEAT

For the valve in Figure 16.1, the seat is in the form of a ring having its inner face closely following the contour of the plug and into which the plug itself fits. From this it should be clear that, the nearer the plug is toward the seat, the smaller will be the flow; and the further away the plug is from the seat, the greater will be the amount of fluid that will pass. Hence, a definite and predictable relationship, depending on the shape of the plug, exists between the flow and the plug position, this relationship being graphed as one of the valve characteristics mentioned earlier. Ball valves, too, have characteristics arising from the relationship between aperture shape, ball position, and the flow passing through.

YOKE

The yoke is the support member for the top works, or drive mechanism, of the control valve. For most valves it is a metal casting, roughly the shape of a tall box without the back and the cover, and when assembled unites the top works and the body. The yoke assembly, comprising the actuator and stem return spring, is an entity, which is assembled separately from the plug-stem assembly and the body during valve manufacture—the reason for this being that yoke assemblies can be standardized, since the only considerations for them are:

1. The stem travel or stroke, which is the physical distance the plug has to move to permit the range between maximum and no flow to occur

2. The force required to drive the plug through the entire length of travel against the flow of material in the pipeline

3. Use of normal engineering materials, as problems with fluid compatibility are minimal, since the parts rarely come into contact with the process fluid.

ACTUATOR STEM

The actuator stem is a short metal shaft, of which one end is fixed to the valve drive mechanism and the other end to the valve stem. It is the means by which the movement of the valve motor (for further detail, see the subsequent description of the actuator later in this chapter) is transferred to the valve stem and plug. The term *valve motor* has been deliberately used for referring to the device, as this is the name by which it is known in the industry. The valve and actuator stems are joined via a turnbuckle assembly, which provides the means of adjusting the stem travel. None of the parts in this assembly come into contact with the process fluid, and therefore no material compatibility problems arise.

ACTUATOR SPRING

The actuator spring is a compression mechanical spring as illustrated in Figure 16.1 (or it could be an extension spring, depending on the failure mode required) of a calculated force and spring rate against which the valve motor works to position the plug. Since the majority of valves are pneumatically operated, the physical location of the spring with respect to the actuator diaphragm and the pneumatic inlet connection will determine the direction of stem travel when either the control signal is removed or the valve is taken out of service. Because this spring when relaxed determines the position the valve plug takes, this out-of-service position—or valve *failure mode*, to use the common control industry phrase—is an important factor in the design of every automatic control system. For example, in furnace control systems, it is vital that the fuel valve "fails shut" and the air damper "fails open" to protect personnel and equipment in the event of either power or signal failure. These are not the only precautions to be taken, as there are many others that a good system design should include to cover the many dangerous situations that could arise.

ACTUATOR

The actuator is usually referred to as the *valve motor* as mentioned earlier, and is the means whereby the controlled signal is made to act on the process to reduce the error. Actuators could be a diaphragm motor as shown in Figure 16.1; a cylinder actuator, which uses either a pneumatic or a hydraulic power cylinder; a squirrel cage electric motor; or a disk-type linear electric motor acting through a light efficient gearbox that converts the rotary to linear motion of the stem. The majority of actuators in a plant are pneumatically operated, mainly for financial reasons, but the other great advantage of pneumatics is that the very fine changes in movement obtained by small changes of the pneumatic signal are the most difficult to replicate electrically. Electric actuators are available, but they are expensive, heavy on power usage, and therefore of limited suitability in hazardous industries such as petroleum refining and mining, though when used on these applications they are usually built as flameproof or explosion-proof equipment.

The pneumatic diaphragm actuator shown in Figure 16.1 consists of a two-part metallic housing separated by a neoprene diaphragm, with each part airtight and sealed from the other. Depending on the valve failure mode required, the controlled signal is applied to either one side or the other of the diaphragm, with the actuator spring, etc., fitted to the unselected half. Admitting the controlled pneumatic signal into the appropriate side of the motor will cause the diaphragm to move, analogous to an inflating balloon. If, say, for the valve shown in Figure 16.1, the action is to close the valve then the upper chamber is selected; on the other hand, if the action was to open the valve, then the lower chamber would have to be selected. The spring and diaphragm are always in the relaxed state when no signal is applied to the selected chamber and, as mentioned before, the valve is in its failure mode.

Apart from the fail open (for air) and fail closed (for fuel) modes discussed earlier, there is yet another failure mode, fail fixed, that is sometimes required; what this means is that the system must ensure the valve is held at its last operated position, wherever that may be. It should be apparent that the way to accomplish this would be to "lock up" the air in the operating chamber of the diaphragm motor using a tight-shutoff electrically operated solenoid valve (tight-shutoff refers to the sealing capability of any valve, and defines the leakage across the valve seat). The solenoid valve is placed in the actuating signal line to the valve motor, and using this technique it is not difficult to see how a control valve can be driven to any of its failure modes, for all that is necessary is to insert a small three-way solenoid-operated valve in the pneumatic signal line to the control valve, and to ensure that when the solenoid is energized to motivate the plug of this small three-way valve, the control signal itself is dead-ended or vented, and the operating chamber of the actuator vented or sealed, as the case may be.

When using an electronic controller, there are two ways of using its signal to drive a pneumatic actuator:

1. Allow the electronic (current) signal from the controller to be applied to an electro-pneumatic converter, usually referred to as an I/P (where I is the symbol for electric current and P is an abbreviation for pneumatic); the unit is most often referred to as either a current/pneumatic or current/pressure converter, whose function is to change the electrical current to a proportional pneumatic signal and apply it to the actuator directly.

2. To use a *positioner* to drive the valve. A positioner is an instrument that in principle is no more than a pneumatic amplifier, for it takes in a low-range pneumatic signal from either a controller or an I/P converter and produces a proportional

pneumatic signal of a much higher range, usually 0 to 35 lb/in^2 or the metric equivalent, and a higher air throughput than the incoming pneumatic signal. When positioners are used, the diaphragm of the valve motor is of a much stronger grade and the actuator spring more substantial to withstand the greater forces involved. Positioners are explained further at the end of this chapter.

STEM CONNECTOR

The stem connector links the two separately assembled parts of the control valve—top works and valve body, including the stem—to make a complete functional unit. The connector—which must fulfill two functions, namely to be firmly fixed, but yet easily disassembled when necessary—is usually a screw-threaded turnbuckle with an additional locking pin, the turnbuckle providing some adjustment capability and the locking pin ensuring secure positioning.

VALVE TYPES

ROTARY-PLUG VALVE

The descriptions in the previous section generally covered a typical linear motion plug-type process control valve; we shall now see how circular motion of the valve plug is used to perform the task. To overcome some of what they perceived as shortcomings of the valve shown in Figure 16.1, Masoneilan, another well-known manufacturer, redesigned the instrument as shown in Figure 16.2. In this design the instrument can be used without any major change, for flow in either direction. In this type of valve the fluid normally assists the air failure mode—i.e., for fail open the flow direction is to the face of the plug, and for fail close it is reversed. The basic

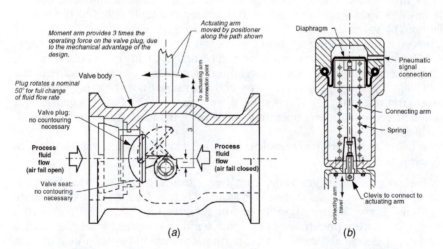

Figure 16.2: (a) Masoneilan Camflex control valve. For clarity, neither the positioner nor the Smart Valve Interface has been shown. These instruments normally bolt onto a mounting flange that is part of the valve body casting. The actuating arm is prevented from slip and secured to plug shaft by splines. (b) Rolling diaphragm positioner.

design makes use of the flexing of the plug arm to ensure firm closure. No profiling of the plug is required; the basic design provides for a linear or an equal-percentage characteristic. The plug travel is rotary and turns through a nominal 50° of arc, the applied force having a 3:1 magnification, which is obtained by judiciously placing the center of rotation a third of the length of the actuating arm below the center line of the body. If required, the valve can be manually operated by a handwheel. The positioner, being directly coupled to the valve plug, does not require a feedback linkage for stabilization. A rolling diaphragm made from Buna N, a strong, flexible material, makes the positioner with minimal moving parts a very simple and reliable device. A more recent innovation is the addition of a digital positioner, the Smart Valve Interface (SVI®); this unit has a three-term position controller and makes use of five additional parameters and an algorithm that automatically tunes the valve position as it iteratively optimizes the initial set values to reconcile response time and control stability to give setpoint/measured variable coincidence. Power to drive the instrument is from the 4- to 20-mA control signal. There are diagnostics to check the valve performance and to determine when the valve needs maintenance.

This, however, is not the end, for now we shall look at some other variations to the control valve, which are in most instances necessitated by the process fluids that are to be handled, but sometimes by the control application requirements. We shall also look at the "business end," i.e., the parts involved with the actual manipulation of the process fluid itself, of other common valves. The variations to be found here are:

- Butterfly

- Ball

- Diaphragm

BUTTERFLY VALVE

The butterfly valve shown in Figure 16.3 is mostly used on low-pressure-drop applications, and where large flow rates are involved. Three examples of the applications

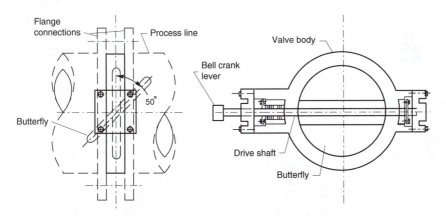

Figure 16.3: Typical wafer-type butterfly valve. The valve actuator (cylinder or diaphragm motor) is mounted such that the bell crank is operated under the action of the control signal. The butterfly should not be operated through an angle greater than 50° for full control of the process.

of this type are:

1. Control of combustion air to a furnace

2. Flow of a water supply through mains

3. For those among us who do not drive a fuel-injected automobile, the valve associated with the accelerator pedal

To describe the device, one should visualize a flat circular metallic disk with a beveled edge (shown rounded in Figure 16.3 for convenience) on the opposite faces of the disk, with the beveled surfaces confined to only half the circumference on each side. The diameter of the disk is almost the same size as the main portion of the internal bore of the valve body, and is fixed to a driveshaft capable of rotating about a diametrical axis through the thickness of the disk. The part of the internal bore through which the driveshaft is fitted is designed to form a tapered seat against which the disk comes to rest in the fully closed position. It is possible to obtain reasonably tight shut off on these valves by using "soft seats," that is seats made of some plastic material such as neoprene, although the choice of the material is governed by the process fluid and the operating conditions.

The design of the valve makes the disk or "butterfly," when operating, behave (dependent on the fluid) like an aero- or hydrofoil; therefore, the disk has the same sort of forces applied to it, and when used for control purposes the butterfly is normally allowed to rotate through 45° to 50° maximum, or less, if possible, to produce the full range of control required. The reason for this is that there is little or no control possible when the butterfly is at 90°, i.e., along or parallel to the flowstream.

Large butterfly valves normally require large forces to operate them, although the top and bottom forces on the disk are balanced moments about the shaft, shear-force due to total flow plus bearing friction is probably high; hence, the actuators are in many instances a "double acting" (needing pressure to drive in both directions, i.e., extend or contract) pneumatic control cylinder with two positioners fitted to produce the motive force necessary; the unit for splitting the control signal to drive each positioner is located either on the valve assembly, nearby, or in the control room. The body of the valve is generally of "wafer-style" construction; i.e., the valve body along with the butterfly and actuator is held between the two flanges of the pipeline, thus it is the butterfly only that obstructs the otherwise continuous bore of the pipeline. These valves are usually installed so that the butterfly driveshaft is either horizontal or vertical with respect to the earth to avoid additional forces acting on the mechanism.

BALL-TYPE VALVE

The type of valve shown in Figure 16.4 is made by Invensys Flow Control (Worcester Controls). A three-part ball assembly comprising two spherically contoured pieces and the ball, which rotates between them, regulates the process fluid. The contoured pieces are made of flexible plastic material compatible with the process fluid and are held permanently in their location on either side of the ball. The ball and one of the contoured pieces each carry a circular through hole while the other has a shaped aperture cut in it instead. The shaped aperture is presented to the flow, and the valve characteristics are determined by the shape of the aperture. The process fluid passes through, and because only the ball is able to rotate, the amount of through opening in the ball relative to the position it takes with respect to the aperture results in the pressure drop and the regulation of fluid flow through the

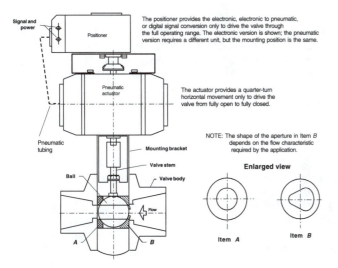

Figure 16.4: Ball valve (Invensys, Worcester, UK).

valve body. Basically the ball valve has an equal percentage characteristic. (We shall discuss "Valve Characteristics" in a little more detail later.) This type of valve is used most often to control the flow of process slurries, the reason being the through flow design with its minimal possibility of getting "hang-ups" of the process fluids. Typical exponents of this design are the Fisher Vee Ball and the Worcester ball valve.

DIAPHRAGM-TYPE VALVE

The diaphragm-type of valve, shown in Figure 16.5 has a patented design originally held by the Saunders Valve Company, now owned by Alfa Laval Ltd. It does not

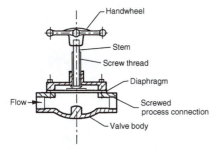

Figure 16.5: The Saunders valve (typical). The valve shown is a hand-operated type. Versions for automatic operation are also available. The principle of operation and construction of the automatic control valve remains the same. This diagram is not meant to show the actual mechanical construction of the valve, only the principle.

have a plug or ball or any other device of that nature; instead, the valve stem acts on a flexible diaphragm of a suitably stout material that seals off the process fluid from the valve stem and at the same time provides the means of manipulating the process fluid to regulate the flow. Because of the design and the resulting construction, such a valve is unable to handle high pressures; but in this type of valve, flow can be in either direction, since the internals are symmetrical about the vertical center line. Its construction makes it most suitable for corrosive fluids and slurries, for there are only small areas where the fluid can build up.

Before we proceed further, let us consider how control valves are modified for use in extreme operating conditions, such as either cryogenic or very hot fluid applications. When used in these applications, the body materials have to be chosen very carefully, for the process operating conditions are liable to change the normal characteristics of the materials used. For example, ordinary carbon steel, when subjected to very low temperatures, becomes brittle and therefore very prone to cracking. This makes it quite unsuitable as a material for the valve body. Similarly, there are other parts of the valve where the materials used must also be given due consideration. The most striking feature of the valves associated with these services is the *extended bonnet*. What is an extended bonnet? The simplest way to describe it is as an extension piece that fits between the valve body and the valve yoke, which contains seals for the sliding valve stem to enable it to fulfill the dual functions of effectively preventing the process fluid from escaping and of keeping the external conditions from affecting the process fluid. A further use for the extended bonnet is to provide the means of maintaining the degree of insulation of the process line involved, without interference to the stem itself.

VALVE CHARACTERISTICS

Valve characteristics were mentioned earlier on, and we shall now consider this important aspect of the instrument. The *characteristic* (or shape of the curve resulting from a plot of flow through the device versus valve travel) of a valve can be defined as the relationship of the flow through the valve at a constant pressure differential for a given plug position; and in order to achieve the required characteristic, it is necessary to shape the plug by contouring, or fluting, the plug.

When a valve exhibits a *linear characteristic*, the flow is *directly proportional* to the travel of the valve stem and is expressed mathematically as:

$$F \propto l$$
$$= kl$$

where F = flow
l = travel

and where k is a constant of proportionality, defined by

$$\frac{dF}{dl} = k$$

However, when a valve exhibits an *equal-percentage characteristic*, the flow through the valve is *exponentially related* to the travel of the valve stem. What this means is that, for equal changes of valve stem position, the change in flow with respect to stem travel is a *constant percentage* of the flow at the time of change; i.e., the percentage change in flow for stem travel between, say, 20% and 30% is the same

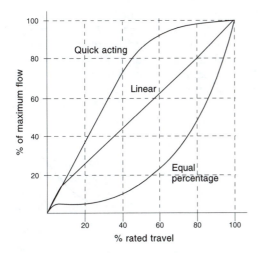

Figure 16.6: Control valve plug characteristics.

percentage change in flow for a stem travel between 40% and 50%. Mathematically, the characteristic is expressed as:

$$F = F_0 e^{nl}$$

where F = flow
F_0 = minimum controllable flow
e = 2.718
n = constant
l = valve travel

and where

$$\frac{dF}{dl} = nF_0 e^{nl}$$

When a valve exhibits a *quick-opening characteristic*, it allows maximum flow with respect to valve stem travel at low valve positions, and at high valve positions (above 70%) not very marked increases in the flow already obtained. The high flows are related to valve travel from near 0% up to 70% of the stroke.

There are other forms of control valve characteristics, but these are in essence modifications of the three main ones just defined. Figure 16.6 shows graphically the shapes of the curves one could expect for these characteristics.

VALVE PORTING

Let us now consider the valve body, and the *porting*. This latter name refers to the number of apertures through which the process fluid is made to pass and through which it is regulated by the interaction of the plug and seat as described earlier. The body of the valve is the main "retainer" of the fluid and must be robust enough to withstand the effects of erosion and corrosion, and to handle the line pressures and volumes involved.

So far we have confined our attention to single-port devices, where the forces acting on the plug are designed to be in balance. There are double-ported valves also.

Why is it necessary to have two ports? The answer is, to reduce the required force exerted by the actuator in response to a controlled signal change. This reduction is achieved by splitting equally the inlet flow stream into two, and two plugs are used to achieve the regulation (both plugs machined on the same stem and all forces acting on the two plugs designed to be balanced, i.e., areas presented to the flow are the same, for all stem positions). The plugs act in unison, one on each of the divided streams, with one plug tending to lift and the other to lower (with respect to the split flows) as the stem assembly changes position. The two streams are recombined within the valve body before being made to exit as a single flow stream.

Sometimes a larger body than necessary is chosen, but in these circumstances the ports must be reduced to compensate for the smaller flow volumes involved, with appropriate characteristic. These last statements beg the question, Why is it necessary to have large bodies with "reduced trims"? (*Trim* is the name given to the plug and seat combination, and very common in the valve manufacturing industry.) The answers to this question are:

1. To reduce the inlet and outlet flow velocities

2. To provide for increased flow volumes in the future, but good control in the interim

3. To ensure the body can withstand the pressures involved and yet have a correctly sized valve

4. To avoid the use of line size reducers, as the porting takes care of correctly manipulating the flow

There are in general three types of control valve bodies that are available for the plug-type valves:

1. Single-inlet single-outlet, straight-through construction.

2. Single-inlet single-outlet, right-angle construction. This valve is usually referred to as an *angle valve*.

3. Single-inlet double-outlet, or double-inlet single-outlet construction. This type of valve is usually referred to as a *three-way* valve.

For the second and third of these valve body types, care must be exercised to avoid unbalanced forces. For the angle valve, the inlet flow is usually through the port in the same plane as the line of action of the valve stem. This type of valve is usually installed vertically with the flow entering under the plug so that it tends to lift it. Three-way valves are used for either diverting service or proportioning service. *Flow control cannot be effected with a three-way valve.*

CONTROL VALVE SIZING

For the control strategy to be effective, the size of the valve must be correct; otherwise satisfactory control of the process will not be obtained. Proper selection of the valve is the joint responsibility of the process engineer and the control valve manufacturer. The sharing of the responsibility is as follows:

• The process engineer defines the operating conditions for the valve.

• The actual determination of the valve size is the responsibility of the valve manufacturer.

- The valve maker suggests the material of construction for the device.

- Final approval of the material selection is the prerogative of the process engineer.

    All valve sizing is based on the well-known Bernoulli relationships, with modifying factors to take care of the internal valve geometry and fluid viscosity. In order to make the procedures universal, all sizing is carried out on the basis of converting the *actual flow of the process fluid to an equivalent flow of a reference fluid*, and choosing a valve that will pass the equivalent flow at the operating pressure. In the case of incompressible fluids the reference fluid is water, and for compressible fluids the reference is air at standard temperature and pressure.

    With the need to make all sizing internationally usable, it has been agreed that this procedure be based on a sizing coefficient C_v, which has been defined, in terms of liquid flow, as the quantity in *US gallons* per minute of water that will pass through the wide-open valve with a 1.0 lb/in^2 pressure drop across it.

SIZING EQUATIONS OFFERED BY VARIOUS VALVE MANUFACTURERS

    A collection of sizing equations from various manufacturers is shown here to demonstrate the similarity of their formulas. The symbols shown in this section are defined *individually* after the formulas and are those used by the various individual manufacturers, and *they are not the same as the ones used in this book*.

Fisher-Rosemount

For the Fisher-Rosemount equations, let

$$G = \text{specific gravity of gas or vapor referred to air}$$

$$G_1 = \text{specific gravity of the process liquid}$$

$$Q_g = \text{gas or vapor flow, standard ft}^3/\text{h}$$

$$Q_1 = \text{process liquid flow, US gal/min}$$

$$Q_{sh} = \text{superheated steam flow, lb/h}$$

$$Q_x = \text{wet steam flow, lb/h}$$

$$P_1 = \text{inlet pressure, lb/in}^2 \text{ Abs}$$

$$\Delta P = \text{pressure drop across the valve, lb/in}^2$$

$$T = \text{temperature, }^\circ\text{R}$$

$$t = \text{degrees of superheat, }^\circ\text{F}$$

$$X = \text{steam quality or \% vapor expressed as a decimal}$$
$$\text{(e.g., } 90\% = 0.9, 50\% = 0.5, \text{ etc.)}$$

Then, for liquids,

$$C_v = Q_1 \sqrt{\frac{G_1}{\Delta P}}$$

For gases (critical flow, i.e., $\Delta P > \frac{1}{2} P_1$),

$$C_g = \frac{Q_g}{P_1} \sqrt{\frac{GT}{520}}$$

For gases (subcritical flow, i.e., $\Delta P > \frac{1}{2} P_1$),

$$C_g = \frac{Q_g}{(2.32 \, \Delta P)^{0.4425} P_1^{0.5575}} \sqrt{\frac{GT}{520}}$$

For steam:

• Saturated (critical flow, i.e., $\Delta P > \frac{1}{2} P_1$),

$$C_s = \frac{Q_{sat}}{P_1}$$

• Saturated (subcritical flow, i.e., $\Delta P < \frac{1}{2} P_1$),

$$C_s = \frac{Q_{sat}}{(2.32 \, \Delta P)^{0.4425} P_1^{0.5575}}$$

• Wet (critical flow, i.e., $\Delta P > \frac{1}{2} P_1$),

$$C_s = \frac{Q_x \sqrt{X}}{P_1}$$

• Wet (subcritical flow, i.e., $\Delta P < \frac{1}{2} P_1$),

$$C_s = \frac{Q_x \sqrt{X}}{(2.32 \, \Delta P)^{0.4425} P_1^{0.5575}}$$

• Superheated (critical flow, i.e., $\Delta P > \frac{1}{2} P_1$),

$$C_s = \frac{Q_{sh}(1 + 0.00065t)}{P_1}$$

• Superheated (subcritical flow, i.e., $\Delta P < \frac{1}{2} P_1$),

$$C_s = \frac{Q_{sh}(1 + 0.00065t)}{(2.32 \, \Delta P)^{0.4425} P_1^{0.5575}}$$

Fisher-Rosemount recommends that the coefficient C_v be used only for liquids. The coefficient C_g should be used for gases. The equations derived are based on extensive air tests carried out by Fisher; the results obtained were first presented to the ISA symposium in 1952. Fisher-Rosemount also recommend that the coefficient C_s be used for steam. This steam coefficient is obtained by dividing the gas sizing coefficient by 22 to convert standard cubic feet per hour to pounds per hour.

The Foxboro Company

For the Foxboro equations, let

Q = flow rate liquids, in US gal/min; gases, in standard ft^3/h

W = flow rate of vapors and steam, lb/h

T_f = flowing temperature, °R

G = specific gravity

ΔP = pressure drop, lb/in^2

P_1 = upstream pressure at valve inlet, lb/in^2 Abs

P_2 = downstream pressure at valve discharge, lb/in^2 Abs

v = downstream specific volume, in ft^3/lb

Then, for liquids,

$$C_v = Q\sqrt{\frac{G}{\Delta P}}$$

For gases,

$$C_v = \frac{Q}{1360}\sqrt{\frac{T_f G}{\Delta P(P_2)}}$$

For steam and vapors,

$$C_v = \frac{W}{63.3}\sqrt{\frac{v}{\Delta P}}$$

There are limitations on the value of ΔP that is used for gases and vapors. In these cases it may never exceed one half the absolute inlet pressure. If the pressure drop is greater than $\frac{1}{2}P_1$, then use $\frac{1}{2}P_1$ for both ΔP and the downstream pressure P_2. Under these circumstances if the process fluid is steam or vapor the specific volume will be determined for $\frac{1}{2}P_1$. If the test for cavitation and/or flashing (discussed later in this chapter) proves that this will occur, then the following equation must be used instead:

$$C_v = Q\sqrt{\frac{G}{\Delta P\,(\text{allowable})}}$$

Worcester Controls (Ball-Type Control Valves)

In all of Worcester Controls equations, the appropriate units—imperial or metric—must be used at all times. In all pairs shown, the equation to be used with imperial units is on the left, and the equation to be used with metric units is on the right. Let

C_v = valve flow coefficient

F_L = critical flow factor

G = gas specific gravity (air = 1.0)

G_f = specific gravity of liquid at flowing temperature (water = 1.0 at 60°F or 15°C)

P_1 = upstream pressure, lb/in^2 Abs, or bars Abs

P_2 = downstream pressure, lb/in^2 Abs, or bars Abs

P_c = pressure at thermodynamic critical point, lb/in^2 Abs, or bars Abs

P_v = vapor pressure of liquid at flowing temperature, lb/in^2 Abs, or bars Abs

ΔP = actual pressure drop $(P_1 - P_2)$, psi or bars

Q = gas flow rate, standard ft³/h at 14.7 psia and 60°F or m³/h at 15°C and 1013 mbar Abs

q = liquid flow rate, US gal/min or m³/h

T = flowing temperature, °R, or K

T_{sh} = steam degrees superheat °F or °C

W = flow rate, lb/h, or 1000 kg/hour

For liquids:

• Subcritical volumetric flow, $\Delta P < F_L^2(\Delta P_s)$:

$$C_v = q\sqrt{\frac{G_f}{\Delta P}} \qquad\qquad C_v = 1.16q\sqrt{\frac{G_f}{\Delta P}}$$

• Critical cavitation or flashing volumetric flow, $\Delta P \geq F_L^2(\Delta P_s)$:

$$C_v = \frac{q}{F_L}\sqrt{\frac{G_f}{\Delta P_s}} \qquad\qquad C_v = \frac{1.16q}{F_L}\sqrt{\frac{G_f}{\Delta P_s}}$$

In the equations for liquids, the sizing pressure drop is given as:

$$\Delta P_s = P_1 - \left[0.96 - 0.28\sqrt{\frac{P_v}{P_c}}\right]P_v$$

For gases and vapors:

• Subcritical volumetric flow, $\Delta P < 0.5F_L^2 P_1$:

$$C_v = \frac{Q}{963}\sqrt{\frac{GT}{P_1^2 + P_2^2}} \qquad\qquad C_v = \frac{Q}{295}\sqrt{\frac{GT}{P_1^2 + P_2^2}}$$

• Critical volumetric flow, $\Delta P \geq 0.5F_L^2 P_1$:

$$C_v = \frac{Q}{834}\sqrt{\frac{GT}{F_L P_1}} \qquad\qquad C_v = \frac{Q}{257}\sqrt{\frac{GT}{F_L P_1}}$$

For saturated steam:

• Subcritical weight flow, $\Delta P < 0.5F_L^2 P_1$:

$$C_v = \frac{W}{2.1\sqrt{P_1^2 + P_2^2}} \qquad\qquad C_v = \frac{72.4W}{\sqrt{P_1^2 + P_2^2}}$$

• Critical weight flow, $\Delta P \geq 0.5F_L^2 P_1$:

$$C_v = \frac{W}{1.83 F_L P_1} \qquad\qquad C_v = \frac{83.7W}{F_L P_1}$$

For superheated steam:

- Subcritical weight flow, $\Delta P < 0.5F_L^2 P_1$:

$$C_v = \frac{W(1 + 0.0007T_{sh})}{2.1\sqrt{P_1^2 + P_2^2}} \qquad C_v = \frac{72.4(1 + 0.00126T_{sh})W}{\sqrt{P_1^2 + P_2^2}}$$

- Critical flow, $\Delta P \geq 0.5F_L^2 P_1$:

$$C_v = \frac{W(1 + 0.0007T_{sh})}{1.83F_L P_1} \qquad C_v = \frac{83.7(1 + 0.00126T_{sh})W}{F_L P_1}$$

Masoneilan

Masoneilan's equations are sizing equations from ISA Standard S75.01 and IEC Standard 534-2. The following is only a partial extract; see the Masoneilan *Handbook for Control Valve Sizing* for full details. All equations shown are those from the ISA standard and are identical to their IEC counterparts, except that in those ISA equations marked with an asterisk, k (ISA) corresponds to γ (IEC) and γ_1 (ISA) corresponds to ρ_1 (IEC). Let

C_v = valve flow coefficient

F_F = liquid critical pressure factor = $0.96 - 0.28\sqrt{\dfrac{p_v}{p_c}}$

F_L = liquid pressure recovery factor for a valve

F_p = piping geometry factor (reducer correction)

F_{LP} = combined pressure recovery and piping geometry factor for a valve with attached fittings

p_1 = upstream pressure

p_2 = downstream pressure

p_c = pressure at thermodynamic critical point

p_v = vapor pressure of liquid at flowing temperature

q = volumetric flow rate

W = weight (mass) flow rate

N = numerical constants based on units used

γ_1 = specific weight (mass density) upstream conditions

k_1 = velocity head factors for an inlet fitting, dimensionless

Then, for nonvaporizing liquids:

Volumetric flow: $C_v = \dfrac{q}{N_1 F_p}\sqrt{\dfrac{G_f}{p_1 - p_2}}$

Mass flow: $C_v = \dfrac{W}{N_6 F_p \sqrt{(p_1 - p_2)\gamma_1}}$

For choked flow of vaporizing liquids: Liquid flow is choked if

$$\Delta p \geq F_L^2 (p_1 - F_F p_v)$$

Use the following equations:

$$\text{Volumetric flow: } C_v = \frac{q}{N_1 F_{LP}} \sqrt{\frac{G_f}{p_1 - F_F p_v}}$$

$$\text{Mass flow: } C_v = \frac{W}{N_6 F_{LP} \sqrt{(p_1 - F_F p_v)\gamma_1}}$$

For gas and vapor flow, use the following factors:

$$\text{Gas expansion factor: } Y = 1 - \frac{x}{3 F_k x_t}$$

$$\text{Pressure drop ratio: } x = \frac{\Delta p}{p_1}$$

$$\text{Ratio of specific heats factor: } F_k = \frac{k}{1.40}$$

Then, for volumetric flow,

$$C_v = \frac{q}{N_7 F_P p_1 Y} \sqrt{\frac{G_g T_1 Z}{x}}$$

or

$$C_v = \frac{q}{N_9 F_p p_1 Y} \sqrt{\frac{M T_1 Z}{x}} \qquad (*)$$

For mass flow,

$$C_v = \frac{W}{N_6 F_P Y \sqrt{x p_1 \gamma_1}} \qquad (*)$$

or

$$C_v = \frac{W}{N_8 F_p p_1 Y} \sqrt{\frac{T_1 Z}{x M}} \qquad (*)$$

CAVITATION, FLASHING, AND RANGEABILITY OF CONTROL VALVES

Most, if not all, control valve manufacturers today have their valve sizing coded in software routines. It is nevertheless very important that instrumentation applications engineers know how these devices operate, and the serious problems that errors could bring about. We shall first consider the sizing of the valve; any errors made could result in a valve that is either oversized, or undersized, but what are the results from these errors?

Let us first consider the undersized valve, and the answer here is quite obvious: it will never be able to pass the maximum flow and the corresponding pressure drop excessively high. The valve will work for low flow rates, but it seriously narrows the range over which control will be exerted. The effect that this will have on the process can be very serious indeed, especially if maximum flow is demanded to bring the product within the required product specification limits. The only solution is a replacement with a correctly sized valve—at additional cost, not only in its replacement, but also the labor, redesign, and ensuing loss of profit for the user.

Now consider the oversized valve, where the answer is again fairly easy to visualize and appreciate. Unlike the previous condition, this valve will for sure be capable of passing the maximum flow, but depending on the degree of oversize we could have a worst-case condition where, at the maximum flow, the plug is a very small distance away from the seat. Let us for the moment stay with this scenario and see what happens. With the gap between the plug and seat small, the pressure drop across it is very large, and therefore the flow velocity is going to be very high; and because the velocity is high, erosion of the plug and seat is a distinct possibility, along with the noise it will for sure generate. The foregoing leads correctly to the next points for consideration, which are *cavitation* and *flashing*. These phenomena can cause severe difficulties if correct measures to prevent them are not taken when the choice of valve is being made. To start let us first define the two words:

1. Cavitation comes from the Latin word *cavitas* meaning a "hollow" and therefore cavitation is the process of forming hollows.

2. Flashing, in the sense used here, describes a very rapid change in the phase of the fluid from liquid to vapor.

The phenomenon of cavitation occurs in liquid flow only. According to physical principles of fluid flow, the formation of hollows can only be brought about by a sudden change of pressure of the fluid from a high value to a very low value, and back up to a reasonably high value. The actual physics of how it occurs is not fully understood, but the dramatic effect of its occurrence is plain for all to see. The theory behind the thinking is that when the pressure of the fluid drops to values below the vapor pressure of the liquid, it gives rise to the formation of very minute bubbles of gas that are held in the liquid flow stream; if the pressure then rises, these bubbles collapse—implode, actually—causing very large forces to be exerted on the material nearest to them. The result of the very numerous detonations is the literal tearing away of the metal of the valve plug and seat.

Like cavitation, the phenomenon of flashing also occurs in liquid flow only, and for this to occur there must also be a change of pressure from a high value to a much lower value, below the vapor pressure of the liquid. In the case of the valve, the fluid enters as a liquid and, because of the pressure reduction, leaves the valve as a vapor.

Figure 16.7 shows graphically the ideas behind cavitation and flashing.

There is a method of predicting whether cavitation and/or flashing will occur in a control valve, and this prediction is based on the solution of the equation below.

$$\Delta p \text{ (allowable)} = F_L^2 (p_1 - p_v)$$

where F_L^2 = valve recovery coefficient
p_1 = inlet pressure
p_v = fluid vapour pressure, lb/in^2 at inlet temperature

If the actual ΔP of the valve is greater than the calculated allowable ΔP, cavitation or flashing will occur.

The valve recovery coefficient F_L, which takes into account the internal geometry of the valve body, is determined by each valve manufacturer from tests made on the range of control valves produced and is published as part of the relevant sizing data. The symbol F_L shown is the accepted ANSI symbol. Some manufacturers choose other symbols; e.g., Fisher-Rosemount uses K_c for this.

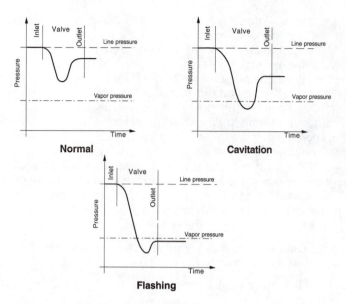

Figure 16.7: Pressure changes within a valve typical of normal flow, cavitation, and flashing.

Cavitation is always accompanied by strong audible sound, and because it is unwanted it is termed "noise." This condition must be eradicated, for, apart from the nuisance that it entails, replacing valves and rejected product are costs that must not be allowed to occur.

If the process fluid is viscous, then it is necessary to adjust the valve size to allow for this, and the adjustments are usually in the form of a multiplying factor, based on the Reynolds number that serves generally to increase the valve size. The Reynolds number is a dimensionless number characterizing the type of flow in a fluid, especially in the effects of viscosity and velocity control in fluid systems. The relationships between the fluid viscosity and the factors are nonlinear; therefore, in cases of viscous process fluids, the valve manufacturers' recommendations should be followed.

The concept of *rangeability* is also applicable to control valves, we discussed this topic initially in Chapter 14, dealing with steam generation, when we considered *turndown* with respect to instrument measurement range. The concept is still the same, but in the case of control valves the definition of rangeability is *the ratio of maximum controllable flow to minimum controllable flow*. We need to state that the limiting values of flows must be those that can be controlled effectively, because if we just made the statement read maximum to minimum values of flow, then in the case of tight-shutoff valves we could get a situation of infinite rangeability. This would theoretically be correct, but for any practical situation definitely not possible, because it is not possible to characterize a valve at no leakage or zero flow.

PRESSURE DROP ACROSS A CONTROL VALVE

One of the most important aspects of control valve sizing is the determination of the pressure drop across the device. When we first started considering control valves, we

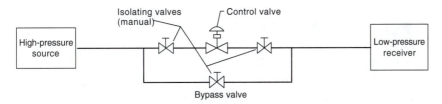

Figure 16.8: Typical control valve arrangement in a process line.

saw that the valve is a dissipater of energy; it is therefore not difficult to appreciate that when such a device is inserted in a process line, it will have to dissipate the surplus energy between what is available and what is required.

To understand what has just been said, let us consider the system shown in Figure 16.8. Let us assume that the high pressure source has an output characteristic that approximates that of a pump, in that for maximum flow the outlet pressure is lower than that for minimum flow. This is so because at low flow the velocity is lower, hence, so are the losses, and as a result the outlet pressure will be higher.

We shall now start to draw the hydraulic gradients across the whole system, shown in graphical form in Figure 16.9. The data to carry this out should be available from the following sources:

1. The discharge curve of the high pressure source, as it will give the outlet pressure of the source for maximum and minimum flow rates

2. The process piping diagrams or piping fabrication drawings, as these will give the piping run lengths and bores—normally the same throughout

3. Pipe manufacturers' published data, or a piping designer's handbook, e.g., *Flow of Fluids through Valves, Fittings, and Pipe*, Crane Company, Technical paper 410,

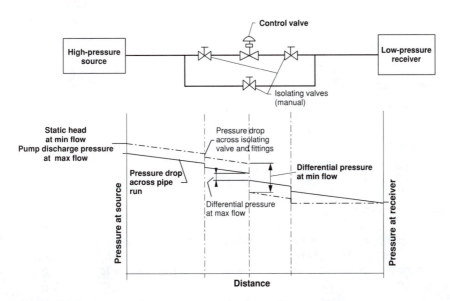

Figure 16.9: Pressure drop across control valve and process line.

Engineering Division Crane, 1969, Illinois, USA, as these will give the coefficient of piping losses (pressure drop) per unit length of pipe

4. Valve manufacturers' published data, or from a piping designer's handbook, as these will give the head loss across the isolating valves.

5. Manufacturers' published data, or from a piping designer's handbook, as these will give the head loss across the usual pipe fittings—e.g., elbows, nipples, etc.

6. The piping diagrams or vessel data sheets, as these will give the inlet pressures required by the low pressure receiver for the upper and lower flow limits.

Since all we are concerned with are the values of the pressure losses involved at each component in the process line, then pressure losses over each pipe run, the isolating valve, and the associated pipe fittings will have to be computed, and these computations used as necessary. Draw a dimensioned and scaled arrangement of Figure 16.8, and plot to scale the computed data as shown in Figure 16.9.

When plotting the computed data for the actual system, always start from either end and work toward the control valve; the reason for doing so is that the data at the ends of the system is always known, and working backwards allows the pressure drop across the valve to be determined. Also remember:

• The valve is always sized for the *maximum flow* and the *minimum pressure drop*.

• To take into account all the other requirements and cautions or caveats for the valve that were stated earlier, include maximum permitted pressure drop at minimum flow.

SOME REASONS FOR CONTROL VALVE FAILURE AND METHODS TO OVERCOME THEM

Control valves are on the whole very reliable and perform the assigned task without much intervention, apart from those periods when routine maintenance is being carried out. However, like any other instrumentation, they do occasionally fail, but when the failure is premature, then the cause must be investigated and corrected. In most instances premature valve failure is attributable to:

• Cavitation damage

• Wire drawing of the trim i.e., actual cutting away of a sliver of material like swarf when machining

• Erosion of the trim

• Resonating plug

• Corrosion

• Excessive stem packing leakage

To minimize the possibilities of a premature failure, it is suggested that preventative measures be implemented when the installation is being designed. The following tabulates, in direct relationship to the foregoing listing, methods for avoiding the stated problems:

1. Select a valve with a low-pressure recovery trim, or install a secondary flow restrictor, or relocate the valve to a more favorable position.

2. Wet steam, but most often hot steam condensate, is the cause of eating away of the carbon steel of the valve body beneath the stainless steel ring of the seat when there is no gasket or weld seal. In these applications, use or specify stainless steel or chromium-molybdenum alloy valve bodies.

3. Use hardened trims or angle valves. Angle valves are suggested for flashing condensate at the inlet of a condenser, and for erosive fluids. In erosive fluid service, use a reduced trim and a long straight oversized pipe run on the downstream side of the valve. Furthermore, in this case the fluid flow should be in the direction that does not cause the valve plug to lift.

4. Reversing the flow direction sometimes eliminates this problem; alternatively, a larger stem diameter and reduced weight of the plug can be a solution.

5. Make a better selection of valve construction materials with the aid of corrosion tables.

6. There are numerous causes for excessive stem packing leakage, some of which are:

   - Valve has a loose packing flange nut when shipped from the manufacturer.

   - Hot fluid service. A possible solution is to retighten packing after valve has been warmed up.

   - Rotating the actuator before or without loosening the stem packing, or unstable control over extended periods causing wear resulting in leakage.

   - Higher coefficient of expansion of the packing compared with the other materials used.

   - Total effective packing length less than the full length of stem travel; it must be greater.

   - Stem misalignment.

USING VARIABLE-SPEED PUMPS IN PLACE OF CONTROL VALVES

It has sometimes been suggested that variable speed pumps could be a replacement for control valves. It is argued this would avoid using two devices, i.e., one to provide the motive power to the fluid (the pump), and the other to regulate the total power actually required to accomplish the task (the control valve). However, attractive as it may appear, the suggestion should be investigated to determine whether or not it is realistic, and the following comparative points should help:

- Speed control of pumps is possible *only* when the static (suction) head pressure of the pump is relatively low.

- Rotary-type valves capable of handling high flow at lower head pressure, and therefore smaller pressure drop across the valve, are now available.

- Valve sizing is much improved, making it possible to correctly calculate low-loss valves.

- On the same service, a variable-speed pump has a lower in-line performance than a control valve.

- A variable speed pump is approximately twice as costly as a valve.

- The maintenance costs on a variable-speed pump are much higher than those of a control valve, when maintenance and downtime are included.

Despite all these issues, the variable-speed pump does have advantages and will therefore continue to be used in specific applications where greater benefit can be obtained. The following are some of the advantages of using pumps:

- They avoid the deadband experienced with a control valve, e.g., on furnace pressure control.

- A combination of variable-speed pump and control valve can be used to match process requirements with the pumped head.

- When highly corrosive fluids are involved, a single pump with special alloys can avoid the double expense of providing a suitable control valve and the pump.

- Pumps offer a possible financial advantage if ample amounts of steam are available and electrical energy is not used for the motive power—e.g., in very large electrical power generating systems where the steam is used to drive a turbine on the pumps that supply oil for lubricating turbine and feedwater machinery bearings.

ARE POSITIONERS NECESSARY WITH EVERY CONTROL VALVE?

Many process plant construction engineers specify control valve positioners with almost every valve required. From the viewpoint of instrument manufacturers, it is a sure means of increasing their revenue; but from a control engineering point of view, the real question is, Does the control loop really need the positioner? A sincere answer to this important question is, Do not use a positioner unless it is absolutely necessary. The reason is that a positioner is a controller in its own right, and because it works within a loop, we have "nested" control loops (loop within a loop); moreover, each loop will have its own gain, deadtime, and time constant that will almost certainly not coincide. The overall result will be a degradation in the quality of the control provided. Positioners are useful when:

- The control valve actuator requires an air pressure greater than 1.0 kg/cm^2 (15 lb/in^2), i.e., the normal controller or I/P converter output. In this case the positioner really acts as a signal amplifier and also provides a higher volume and rate than the controller or I/P converter.

- A single controller output is to be split between two control valves, which is called "split ranging." A much better arrangement is to have two scaling modules (input range selection, gain, and bias adjustments) between the controller output and the final actuators, which effectively appears as having two outputs from the same controller. Each of these outputs will then have the full signal range of, say, 4 to 20 mA, but will be operating from only a part of the single controller output. Some modern microprocessor-based controllers provide this split ranging facility as standard, but need the actual point of split to be configured by the user.

- Two valves have to operate in opposition to each other—i.e., one opens while the other closes.

- The deadband of the combination of actuator and valve exceeds 5% of the signal span, i.e., 0.6 lb/in^2 (pneumatic) or 0.8 mA (electronic)

ELECTRICAL CONTROL VALVE ACTUATORS

Electrical actuators have some features that are not found in pneumatic actuators. Some of these features are required in specific control applications, because of the level of the performance that they offer. Electrical actuators are needed:

- In applications where severe dynamic forces can act on the valve stem

- When high stem forces and good low-frequency responses are necessary

- Operating in locations where no instrument air is available

The main disadvantage, as stated before, is the high cost of the actuator, which can amount to approximately 5% of the total installed cost attributed to this type of actuator.

SUMMARY

1. A control valve is sometimes referred to as the final actuator or control actuator and has one very important function: to dissipate energy.

2. The main parts of a control valve are: the body, connections, plug, stem, seat, yoke, actuator stem, actuator spring, actuator, and stem connector.

3. The "business end" of the valve, i.e., the part that is involved with manipulation of the process fluid, is the plug. All forces acting on the plug are designed to be balanced.

4. The main types of control valve are: plug, butterfly, ball, and diaphragm

5. Generally there are three types of valve characteristic. The *linear characteristic* when the flow is *directly proportional* to the travel of the valve stem. Mathematically expressed as:

$$F \propto l$$

$$= kl$$

where $F =$ flow
$l =$ travel

and where k is a constant of proportionality, defined by

$$\frac{dF}{dl} = k$$

The equal-percentage characteristic allows the flow through the valve to be exponentially related to the valve stem travel. For equal changes of valve stem position, the change in flow with respect to stem travel is a constant percentage of the flow at the time of change. Mathematically, the characteristic is expressed as:

$$F = F_0 e^{nl}$$

where $F =$ flow
$F_0 =$ minimum controllable flow
$e = 2.718$
$n =$ constant
$l =$ valve travel

and where

$$\frac{dF}{dl} = nF_0e^{nl}$$

The quick-opening characteristic allows maximum flow with respect to valve stem travel at low valve positions, and very reduced flows at high valve positions. The high flows are related to valve travel from near 0% up to 70% of the stroke.

6. The body of the valve must be robust enough to withstand the effects of erosion and corrosion and to handle the line pressures and volumes involved.

7. Porting refers to the number of apertures through which the process fluid is allowed to pass and be manipulated by the action of the plug. Single-port valves have one entry and one exit, with the fluid passing through as a single flow stream that is regulated by the plug. Double-ported valves also have one entry and one exit, but the fluid passing through the body is split into two streams, which are recombined within the valve body and exit as a single flow stream. The result is to reduce the force exerted by the actuator in response to a controlled signal change. Control is achieved by having two plugs machined on the same stem acting on each of the divided streams.

8. All valve sizing is based on the well-known Bernoulli relationships, with modifying factors to take care of the internal valve geometry and fluid viscosity. To make the procedures universal, all sizing is carried out on the basis of converting the actual flow of the process fluid to an equivalent flow of a reference fluid, and choosing a valve that will pass the equivalent flow at the operating pressure. In the case of incompressible fluids the reference fluid is water, and for compressible fluids the reference is air at standard temperature and pressure.

9. To make all sizing internationally usable, a sizing coefficient C_v has been adopted and defined in terms of liquid flow as the quantity in *US gallons* per minute of water that will pass through the wide-open valve with a 1.0 lb/in^2 pressure drop across it.

10. An undersized valve will never be able to pass the maximum flow. The valve will work for low flow rates but will seriously narrow the range over which control will be exerted. An oversized valve will be capable of passing the maximum flow, but depending on the degree of oversize a worst case condition could result in a very small distance between the plug and the seat, and a very large pressure drop across the valve and a very high flow velocity. Erosion of the plug and seat is a distinct possibility.

11. Cavitation is the process of forming hollows, and the phenomenon occurs in liquid flow only. Generally, formation of hollows requires a sudden change of pressure of the fluid from a high value to a very low one and then back up to a reasonably high value. The theory is that the dropping of the fluid pressure below the vapor pressure gives rise to very minute bubbles of gas that are held in the liquid flow stream, which the rising pressure causes to collapse. The collapsing bubbles cause very large forces to be exerted on the material nearest to it, with the detonations literally tearing away the metal of the valve plug and seat.

12. Flashing also occurs in liquid flow only. It also requires a change of pressure from a high value to a value that is below the vapor pressure of the liquid. In this case the fluid enters the valve as a liquid, and because of the pressure reduction some

of the liquid "flashes," or changes to a vapor, and the fluid leaves the valve as a combination of liquid and vapor.

13. Predicting whether cavitation and/or flashing will occur in a control valve depends on the equation

$$\Delta p \text{ (allowable)} = F_L^2(p_1 - p_v)$$

where $F_L^2 =$ valve recovery coefficient
$p_1 =$ inlet pressure
$p_v =$ fluid vapor pressure, lb/in^2 at inlet temperature

If the actual ΔP of the valve is greater than the calculated allowable ΔP, cavitation or flashing will occur.

14. Control valves exhibit a concept similar to turndown when used with respect to instrument measurement range and steam generation. In the case of control valves, it is called rangeability and defined as the ratio of maximum controllable flow to minimum controllable flow.

15. Determining the pressure drop across the control valve is one of the most important aspects of control valve sizing. A graphical method of analysis of the system is by far the easiest, as all we are concerned with is plotting the scaled values of the pressure losses at each component in the process line, the plotting made to start from opposite ends (since both these values are known) and worked toward the control valve itself.

16. The following points are useful for considering variable-speed pumps in place of control valves:

    • Speed control of pumps is possible only when the static (suction) head pressure of the pump is relatively low.

    • On the same service, a variable speed pump has a lower in line performance than a control valve.

    • A variable speed pump is approximately twice as costly as a valve.

    • The maintenance cost on a variable speed pump is much higher than that of a control valve.

17. Some advantages to be gained from using pumps are:

    • They avoid the deadband experienced with a control valve, e.g., on furnace pressure control.

    • When highly corrosive fluids are involved, the use of a pump avoids expensive special alloys for a suitable control valve and pump.

18. A positioner should not be used with a control valve unless absolutely necessary. A positioner is a controller, and because it works within a loop, there will be a nested control loop. Each of these loops will have its own gain, deadtime, and time constant, which will almost certainly not coincide, the result being a degradation in the quality of the process control.

19. Positioners are useful when:

    • The control valve actuator requires an air pressure greater than 1.0 kg/cm^2 (15 lb./in^2), the positioner really acting as a signal amplifier.

- Two valves have to operate in opposition to each other—i.e., one opens while the other closes.

- High stroking speeds are required.

20. Electrical control valve actuators are required in some control applications where their level of performance is an advantage. They are desireable:

- In applications where severe dynamic forces can act on the valve stem

- When high stem forces and low-frequency responses are necessary

- Operating in locations where no instrument air is available

CHAPTER 17

Power Cylinders and Control Drives

In this chapter we shall consider some of the other equipment to drive the control device used to manipulate the process variable to bring it into line with the desired value. These in-line final *actuators* are mostly made to provide the motive power for valves, dampers, and other regulating devices which, because of their sheer size and weight, make the usual valve motors discussed in Chapter 16 inadequate. Many of these more powerful devices use a fluid as the prime mover, which can be either compressible (air) or incompressible (water or hydraulic oil); and some others use electrical power.

On what criteria does one decide whether a diaphragm, a pneumatic cylinder, or a hydraulic cylinder actuator is the most suitable? The answer to this important question in every instance rests with the manufacturer of the final control element, but it is nevertheless vital that the control engineer has a reasonably good idea of what to expect, so that any answers received from the control actuator manufacturer do not come as a surprise. It is also important for the control systems engineer to know what provisions have to be made for mounting the device, for accessing and servicing the equipment once it is installed, and, in the case of pneumatics and hydraulics, for the specification of motive medium—the air, water, oil, or electrical supply required.

Every control valve and final control element manufacturer will provide data for the selection of an actuator suitable for its products; the data supplied will include the operating pressure and the force available at the chosen operating pressure to drive the control element. It should not be difficult to appreciate that with the diaphragm actuator, the load-moving capability will largely depend on the strength of the diaphragm material when subjected to the pressure of the drive fluid.

From what has just been said it will be obvious that the devices we are about to study provide a potentially much greater force than can be obtained from the standard diaphragm-operated valve motors.

THE POWER CYLINDER

PNEUMATIC ACTUATION

We shall commence by looking at the power cylinder. As its name suggests, one can confidently visualize the movement being provided by the motion of a piston rod and its piston that slides along the internal bore and longitudinal axis of the cylinder. Because much higher pressures can be applied to the solid moving parts by the actuating fluid, a greater driving force can be achieved. Since all power cylinders have circular bores, then the force available is given by the simple relationship:

$$\text{Force} = \frac{\pi d^2}{4}(\text{operating pressure}) \qquad (17.1)$$

where d is the bore diameter. The stroke of the piston depends upon the amount of movement required by the control device; with control valves the movement is usually relatively small; but when large dampers are involved, the strokes required could be quite long. Most power cylinders are manufactured by specialist companies such as Parker Hannifin, Baldwin Fluid Power, and RGS, to name but a few.

We shall first discuss the pneumatic version of the power cylinder as we can use air and the pneumatic positioner we introduced in Chapter 16 as a means of driving and positioning the piston and rod assembly. A feedback signal, which in this particular case is a mechanical one derived from the piston rod, or "ram," as it is sometimes called, is absolutely essential for accurate positioning. It is derived from a mechanical link-and-lever mechanism in which the link is firmly fixed to the piston rod and the movable lever acts on a pneumatic relay that, dependent on the controller output, makes small adjustments to the magnitude of the position-stabilized pneumatic drive signal before it is actually applied to the device being driven. (Do not get the impression that feedback signals from cylinders always have to be purely mechanical, for it is relatively easy to translate linear movement into a proportional electrical signal, e.g., by using a potentiometer or other adjustable electrical component.) The effect of long travel on the relatively small movement at the pneumatic relay presents a particular challenge, for the mechanics of deriving the feedback have to be carefully designed to maintain its overall motion sensitivity, and any backlash in multiple links or drive gearing would have to be minimized where these are used. For control valves driven by positioners, the design constraints on the position feedback arrangement are exactly the same, the only difference being the length of stroke, which is much smaller on control valves.

Since pneumatic positioners receive control signals of 3 to 15 lb/in^2 (0.2 to 1.0 kg/cm^2) and are usually able to provide output pressures only up to approximately 35 to 40 lb/in^2 (2.5 to 2.8 kg/cm^2), a large driving force from the cylinder is obtained by increasing the cylinder bore. However, there are limits on the extent to which the pneumatic cylinder bore can be increased, and other means have to be found for those instances when the driving force required is so large that pneumatically actuated cylinders become inappropriate.

HYDRAULIC ACTUATION

Hydraulics are used to increase the driving force available with a smaller cylinder bore, and outwardly there is very little difference in design between the pneumatic and hydraulic cylinders; the differences, however, lie in the wall thickness, piston construction, and piston rod diameters and in the mounting arrangements of these two types of cylinders. In the hydraulic version these are very much more robust, as the operating pressures are much higher to enable much higher forces to be exerted on the equipment to which they are attached. In addition, there is no cushioning effect with the motive fluid used, as all hydraulic fluids are considered to be incompressible over the range of working pressure at which they are used. (Nevertheless, and for the record, hydraulic fluids will compress, but only very slightly at exceedingly high pressures, typically in the region of tens of thousands of pounds per square inch.)

ELECTROHYDRAULIC SYSTEMS

One method of driving and positioning hydraulic cylinders or directly coupled hydraulic actuators on quarter-turn valves, such as butterfly valves or dampers, or on linear-stroke control devices such as plug-type globe valves, is by using a system designed and developed by Skil Controls Ltd. This company specializes in the

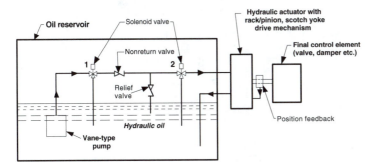

Figure 17.1: Schematic diagram of SKIL Type 416/7 electrohydraulic actuator.

manufacture and design of fail-safe final actuators, which are vital on gas and oil platforms and in "on-shore" plants where the requirement is for fast and accurate positioning of the final element. The actuator design is flexible and can also be applied in nonhazardous plants where accurate positioning and fail-safe operation are still very important.

The manner in which the final control device is positioned is shown in Figure 17.1 using the Skil 416/7 Actuator as an example; other models have systems that differ in minor details from that shown. The system is a hybrid one, in that the signals to it and the power to drive the pump are electrical, while the positioning of the control device is hydraulic. This combination of the two methods has advantages, the main being:

- It is cheaper than a fully electronic and electrical actuating system

- The precision of electronic measurement is coupled to the precise means of transferring it to the control device itself, so that the precision of control is enviably good.

- A compression spring mechanically drives the actuator, under emergency conditions and moves the actuator to a predefined safe condition that is specifically designed into the actuator.

Method of Operation—The Skil 416/7 Actuator

Before we discuss the operation of the equipment, it would be advantageous to define some of the terms that are applied to it. The important ones concern the position that the final control element is to take in the event of:

- Achieving a measurement limiting condition

- Failure either in the measurement/control signal or the loss of motive power to the actuator itself

Fail-safe operation will require driving the final control element to a position determined by the process designers to ensure personnel and plant safety. This safe position could mean that the valve or damper is forced to its fully open or fully closed position, or remains at its last operating position. The last of these three positions mentioned is termed fail-fixed.

The control system, on entering any one of these prespecified states, may require the actuator to be driven to either fully open, fully closed, to stay fixed, or to a

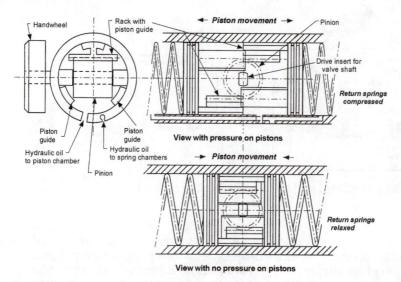

Figure 17.2: Skil Controls electrohydraulic rotary valve actuator. The valve shaft is made to rotate 90° for full piston travel. The shaft drive is shown as a track and pinion; this could be a Scotch yoke if required for heavier applications.

predetermined position set anywhere within the operating signal range. Enabling the last stated condition requires the control system designers to set the position and provide the necessary logic externally to the actuator itself.

Note: Skil Controls defines *direct action* as actuator piston retraction on spring return, and *reverse action* as actuator piston extension on spring return.

Referring to Figure 17.1, it will be appreciated that the heart of the system is the pump-and-solenoid valve arrangement, which necessarily has to be fast-acting and, if possible, should contain the minimum number of moving parts. We shall initially discuss the system up to the inlet of the actuator and then look at the different actuators involved, the reason being that there is more than one type of actuator available. The pump used on the equipment shown in Figure 17.1 is a vane-operated unit, and the actuator of a type as shown in Figure 17.2, which produces a rotary action to drive the final element (valve).

The electronics board in the equipment contains positioner circuitry, which accepts the incoming control signal, and a signal representing the valve stem position derived from the position feedback unit and compares the two. A difference detected initiates the pump and solenoid valves. Solenoids 1 and 2 are three-way units that have one input and two discharge ports; the normally closed discharge port on solenoid 1 is connected to the input of solenoid 2. When power is applied, the coils of both solenoids are energized to close the normally open discharge port to the reservoir on solenoid 1 and to open the normally closed discharge port; and to cause solenoid 2 to shut the port draining to the reservoir and open the port connected to the actuator inlet. The path now established allows the oil to be pumped to the inlet of the actuator. The timing is such that there is a smooth transition, although a relief valve is provided to relieve any pressure build up.

The actuator pistons on Figure 17.2 are driven apart by the pumped hydraulic oil to positions dependent on the value of the incoming control signal, and at the same time, as a result, this also changes the position of the feedback unit to produce a new value dependent on the different stem position. The action of driving the pistons apart forces the oil contained in the spring chambers back into the reservoir and at the same time causes the feedback unit to alter its output. This newly generated feedback output is compared with the incoming control signal, and the hydraulic pump is allowed to drive until the two signals to the positioner electronics are aligned with each other. When signal balance is achieved, solenoid 1 is deenergized, with the port connected to solenoid 2 closed and the discharge port connected to the reservoir opened to permit the still-turning pump to safely discharge excess fluid. Solenoid 2 is maintained energized with the port connected to the actuator open and the discharge port of solenoid 2 closed, thus locking up the pressure in the actuator via the nonreturn valve and the relief valve. The pump is then also stopped, enabling the whole pump pressure to drain into the reservoir in readiness for the next control signal change to be acted on. In the reverse direction, or on power failure, solenoid 2 is deenergized to open the discharge port and vent the actuator pressure to the reservoir.

Pumps

Some of these hydraulic systems—e.g., Skil Model 301/2 or 405/6/7—use either a single or a pair of patented nonrotating oscillating positive-displacement pumps instead of the vane type of Figure 17.1. The oscillating pumps have only a piston that is made to slide back and forth within a cylindrical housing that forms the pump body, the sliding action being achieved by the piston acting as a spring-loaded solenoid, the coil of which is energized by either the positive- or the negative-going half wave of the ac power supply applied to the drive circuit of the coil. Since the pumps can be supplied as a pair, one is made to respond to the positive-going half wave and the other to the negative-going half wave with the power to the drive-coil cut off automatically when the piston (solenoid core) has reached its maximum travel; the energy built up in compressing the spring while under the influence of the excited coil then takes over to ensure that the piston is returned to the starting position. The use of mechanical springs ensures that the piston always returns to the correct position and that a fail-safe condition can be achieved without the use of externally supplied power, be it from alternative or standby sources. A relatively smooth delivery at the pump discharge end is obtained by making the solenoids oscillate at the frequency of the ac supply, which could be either 50 or 60 Hz. The design of the pump internals calls for components arranged to ensure that fluid flow reversals do not occur and excessive pressure buildup is relieved. Since there are no rotating parts, there is no bearing friction, rotational inertia effects are nonexistent, and full oil delivery is achieved in the shortest possible time.

DETAILS OF THE HYDRAULIC ACTUATORS

There are three main types of actuators manufactured: (1) the rack and pinion, (2) the Scotch yoke, and (3) the rolling diaphragm, each differentiated by the means used to transmit the drive to the control device. Figure 17.2 shows the rack-and-pinion actuator, which is used in equipment where large driving forces are not necessary for moving the control device; refer to the figure while reading the following descriptions. The actuator comprises two identical die-cast pistons, which on one side has a rack and two piston supports that permit the unit to run coaxially within a cylindrical casing; the opposite face of the piston carries a recess to allow a compression spring to be carried concentrically with the piston. The pistons are assembled so that the drive racks are diametrically opposite each other with the pinion between them;

from this it can be visualized that any linear movement of the pistons will result in a rotation of the pinion. The pinion is keyed to a central shaft, the free ends of which project through the cylinder wall; on one end of the shaft a handwheel for manual operation is fitted, while the other end terminates in a suitable drive insert arranged to mate with the control device. The shaft and pinion assembly retains the pistons in the correct position with piston rings and seals fitted to make the assembly oil-tight. The two return compression springs are placed, one on each piston back face, with retaining and sealing end caps.

The unit operates as follows. Initially, the spaces that house the return springs are completely filled with the hydraulic fluid and the return line is connected to the reservoir. When the hydraulic oil from the pump is admitted to the space between the two pistons as required by the control signal, the result is to push the pistons apart against the force of the compression springs located on the opposite face of the pistons. In the process the pinion, and with it the control device, is forced to rotate an amount dependent on the magnitude of the control signal. Piston movement stops when the control signal and the actuator position signals are aligned. As the springs are being compressed, a comparable volume of oil from the spring enclosure side of the assembly is forced into the reservoir, raising its level, the amount forced back being dependent on the piston displacement. As the control signal changes and the springs relax, oil from the reservoir is forced back into the spring enclosure section of the actuator through the return pipeline due to an equalizing of pressure between the reservoir and the spring enclosures, changing the position of the control device as well.

Figure 17.3 shows the Scotch yoke actuator. The difference between this actuator and the rack-and-pinion unit lies in the design of the drive mechanism to move the control device, as in this arrangement the forces catered for are much larger, which means that the driven devices can be much larger physically and therefore heavier. The method of operation and the basic design of the pistons are the same as that shown before, but the forces to be catered for demand a larger, more substantial actuator housing and piston assembly.

The rolling diaphragm actuator is virtually identical to that shown in Figure 16.2 for linear motion with suitable modifications to allow for the fact that hydraulic fluid is involved, which demands a much more robust diaphragm material and thicker walls for the hydraulic enclosures to retain the much higher operating pressure.

Table 17.1 shows data for a range of Skil Controls actuators and is of necessity limited; the product specifications must be referred to for full information. See

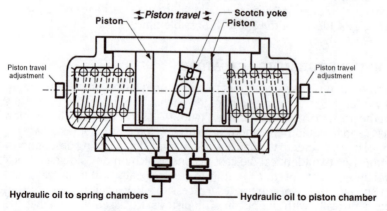

Figure 17.3: Skil Controls electrohydraulic Scotch yoke drive rotary valve actuator. Valve shaft travel is 90° rotation.

TABLE 17.1
Specification for Skil Controls Actuators with rotary movement

Model	Torque (Nm)	Torque (in lb)	Housing	Hazardous area classification	Actuator output
301*	24 to 52	212 to 460	General purpose	Not applicable	Quarter turn
302**	117 to 510	1035 to 4513	General purpose	Not applicable	Quarter turn
405/406/407*	24 to 510	212 to 4513	Weatherproof IP65	Not applicable	Quarter turn
408/409*	52 to 510	460 to 4513	Weatherproof IP65	Zone 2 Ex Nd IIC T4 FM approved	Quarter turn
413QT*	52 to 510	460 to 4510	Weatherproof IP65	E Ex dme IIB T3 Ta	Quarter turn
416QT-RC†	650 to 1320		Weatherproof IP65	Not applicable	Quarter turn
417QT-RC†			Weatherproof IP67	E Exd IIB T3 and FM	
416QT-BT†	1400 to 8530		Weatherproof IP67		Quarter turn
417QT-BT†			Weatherproof IP65	E Exd IIB T3	

Specification for Skil Controls Actuators with linear movement

Model	Thrust (kN)	Thrust (lbf)	Housing	Hazardous area classification	Actuator output
481*	1.32	300	General purpose	Not applicable	Linear 42 mm
411L*	1.76 to 8.6	400 to 2000	Weatherproof IP65	Not applicable	Linear 65 mm
413L*	1.76 to 8.6	400 to 2000	Flameproof	E Ex dme IIB T3 Ta	
416L-SR†	8.8 to 70	2000 to 16,000	Weatherproof IP67	Not applicable	Linear 89 mm
417L-SR†	8.8 to 70	2000 to 16,000	Weatherproof IP65	E Exd IIB T3 and FM	
416L-DA†	13.2 to 78.8	3000 to 17,900	Weatherproof IP67	Not applicable	Linear 89 mm
417L-DA†			Weatherproof IP65	E Exd IIB T3 and FM	

*Oscillating pump (single). **Oscillating pump (dual).
†Vane pump.

323

Chapter 22 for additional information on the table heading "Hazardous area classification"—safety issues come under this heading and are indicated typically by the code E Exd IIB T3 (European) or FM (United States). Linear output is obtained using a hydraulic cylinder.

The Skilmatic actuator is fundamentally an electrically operated system with internal hydraulic positioning, which has been specifically designed for continuous control duty under the dictates of the process controller output signal of 4 to 20 mA. The unit duty-cycle rating is therefore 100%, which is in contrast to the normal 30% to 80% rating assigned to electric motor actuators. The lower duty rating of the electric motor actuators is brought about by the number of motor starts per hour allowed for these devices.

When these actuators are used in hazardous plants, only the "certified versions" are permitted; the certification for Skilmatic actuators is for *flameproof* equipment. For the meanings of, and differences between, the terms *flameproof*, *explosionproof* and *intrinsically safe*, see Chapter 22, which deals with plant safety.

PNEUMATIC SYSTEMS

LINKS AND LEVERS

In most instances where controlled drives are specified and used for large equipment, there is also a requirement to specify and design the connecting links and levers to produce the required travel of the process correcting device, be it a large damper, control valve, or mechanical lever. The following three recommendations (by no means exhaustive) on links and levers give some of the more fundamental considerations, others that are application dependent also have to be allowed for:

1. The design should ensure that all angles formed by the pin-jointed mating components are at 90° when the system is at rest; this will allow all succeeding movements to be directly related to each other and power transmission to be smooth. Plot the locus of moving parts to ensure that adequate clearances are available, and also plot the force versus. stroke relationship, which will not be linear as toggle-action enhancement can occur hence, speed is maximum at the start and force is maximum at the end.

2. The "line of action" for all members delivering mechanical power should avoid any twisting or bending (due to excessive length) of the components, which must be of sufficient strength to deliver the power or torque required.

3. The mechanism should be as simple as possible, with as few working parts, to achieve the result required.

CONTROL DRIVES

The name *control drive* is usually given to an assembly of a pneumatic power cylinder with positioner and linkage mechanism in a protective housing supplied as a complete piece of equipment. Control drive systems are designed for the specific application; the only connections to be made by the user or installer are a link to couple the output shaft of the control drive to the control element located in the process line, the signal, and pneumatic supply to the drive. These drives are commonly used in the manipulation of the large dampers in, for example, steam generating plants and glass furnaces.

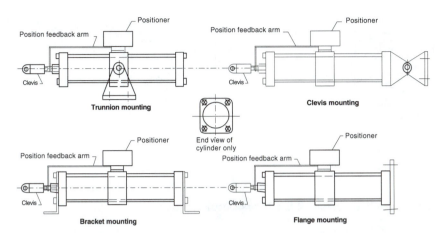

Figure 17.4: Typical pneumatic power cylinder mountings. In all cases, the actuator ram moves coplanar with the horizontal axis of the cylinder.

Sizing of the control drive requires the following:

- The total force (including friction) required to move the control element.

- The angular distance through which the control element is to be moved.

- The separation distance between the drive and the control element. This is usually the center-to-center distance of the two drive levers involved. It is normal for the link to be made adjustable, and usually achieved by a turnbuckle.

MOUNTING PNEUMATIC POWER CYLINDERS

Figure 17.4 shows a number of the standard ways in which pneumatic power cylinders can be mounted. Selection of the mounting arrangement depends on the result to be accomplished.

HEAVY ELECTRICAL MOTOR-DRIVEN ACTUATORS

In this section we shall be considering those final actuators that are electrically driven using single-phase or three-phase ac 50/60 Hz or dc electrical power to provide the motive force. It must be understood that the motors that drive the mechanism are not the standard versions found in other applications, but they are especially designed so that the torque delivered is derived easily from low-inertia components of the motor. For this reason the motors are longer and slimmer than their equivalent industrial counterparts.

EXAMPLE: THE ROTORK ACTUATOR

We shall be describing in particular an electrical drive that is manufactured by Rotork. This company produces a variety of actuators that are used throughout the world for driving large control valves, penstocks (sluice gates to control water flow), and other such devices. Figure 17.5 depicts the "A" range of the Rotork product line, comprising the Syncroset and Syncropak actuators. The Syncroset

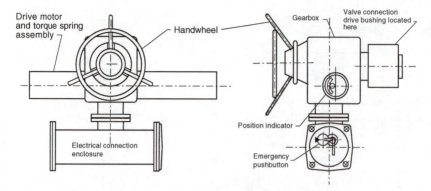

Figure 17.5: Schematic diagram of the Rotork electrical actuator.

actuator is suitable for applications where minimal equipment is required at the driven unit location; it normally has the standard three-phase motor and reduction gearbox, but no reversing contactors, which must be separately provided and installed. The Syncropak actuator, on the other hand, provides the user with the full capability of local or remote control, position indication, and motor reversing. The equipment requires that only a three-phase supply be connected to make it operational.

As mentioned before, control technology, like every other field of science, is continually advancing, and the field of actuator drives is no exception, the most recent offering being the Rotork "IQ" range. This is an "intelligent device" that does away completely with the need to perform a number of fine mechanical adjustments when setting the equipment up, or routinely maintaining it. Figure 17.6 shows the

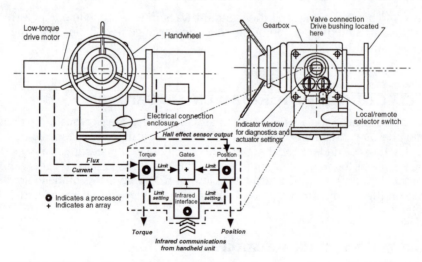

Figure 17.6: The Rotork IQ intelligent electrical actuator. The Intelligent Drive Interface has been superimposed to show how the electronics are incorporated with the mechanical detail of the actuator. Covers do not have to be removed to reconfigure the actuator; it can be done (not always) remotely from the control room by a Modbus or other data link, or handheld infrared communicator.

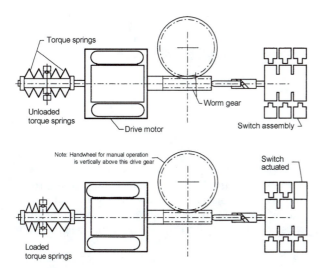

Figure 17.7: Rotork actuator drive assembly.

latest offering of the IQ actuator. It is instructive to compare the two actuators—the IQ versus the "standard" (A range)—to see the effect of advancement.

With the inclusion of microprocessors in the actuator design, much closer positioning and therefore greatly refined control are a distinctly achievable possibility. Furthermore, integrating these new devices into the overall control and MIS (Management Information Systems) strategy is much more readily accomplished. This development removes all the compatibility and reliability problems that used to exist, for it will no longer be necessary to ensure that the various signals produced by the device are available at acceptable measurement ranges and to provide numerous interface units between the actuator and the "system." The availability of the actuator position and delivered torque signals for transmission to the control room will enable these parameters to be monitored or included where required in a control scheme.

Figure 17.7 is a diagrammatic illustration of the A-range actuator drive only. It shows how the torque is limited to protect the actuator motor and the valve that is being driven. The physical linear movement of the motor shaft senses the torque developed. The pack of springs at the end of the driveshaft provides some mechanical resistance to the lateral/axial movement of the shaft when it is under load. A mechanical setting can be made in the combined torque/travel limit switches so that the motor switches off when the proportion of the torque, which had been previously set on the torque/travel limit switches, is reached. When the valve is being operated manually via the handwheel gearing, the electrical drive mechanism is disengaged. It is normal to provide this manual facility on all control valves so that under emergency conditions the process can be driven to a safe condition.

This type of actuator is normally supplied to fit on to valves that are manufactured by others, and therefore it is important to provide the following data to the actuator makers:

For gate- and globe-type valves

- The force that is required to drive the valve. This force must be defined in terms of torque and thrust necessary under the service operating conditions.

- The diameter of the valve stem.
- The operating speed.
- The valve stroke.

For quarter-turn valves—butterfly

- The torque required.
- The operating speed or travel time.

When the valves being driven are large gate types or penstocks, the thread efficiency must be taken into account. When the gate is a heavy one, long strokes
should be avoided, as the continuous energy dissipation during the stroke will cause
the drive motor to overheat, and will rapidly wear out the stem nuts. The total
thrust used for actuator sizing must not be less than three times the weight of the
gate.

    The valve speed required may not be matched exactly because gear ratios give
output shaft speeds that are in "steps." The operating speed will vary slightly while
the valve is under load, which is normal with all "cage" motor drives. Doubling
the speed of an actuator for a given torque requires the motor horsepower to be
doubled also. Hence, to avoid this situation and yet achieve nearly the same results, the "lead" of the screw is doubled instead. This has the effect of keeping the
motor rpm the same, but increases the torque and horsepower by only a small
amount. If the valve actuator is to be self-locking, that is, if the valve is to remain
in its last operating position with no signal applied, the stem thread or quarter-
turn gear box must be irreversible. Figure 17.8 is based on the experience gained
by Rotork over the years and shows the maximum thread lead that may be used
with a given stem diameter to prevent "overhauling" occurring in the gearing, i.e.,
keeping the gearing from being driven backward by the reverse thrust from the
valve.

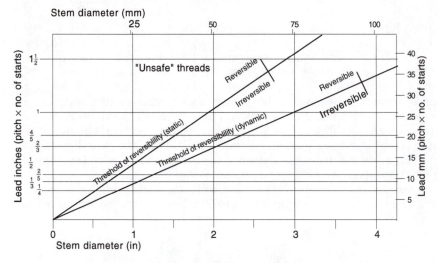

Figure 17.8: Guide to stem thread reversibility. The information shown applies to
both the A and IQ ranges of drives.

Cabling and Fusing

Only three parameters can be measured on the drive motor, and they are therefore important:

- The locked rotor current

- The locked rotor torque

- The stalling torque

However, the user is interested in only the first of the three. The other two are of interest to the actuator manufacturer. The locked rotor current on its own does not permit the selection of cables, fuses, or control switchgear, which require the full-load running conditions of the motor to be specified. Rotork publishes data on the locked rotor current, the seating torque, and the average load for all the actuators it manufactures. The following table is a sample of the data provided for single- and three-phase range of IQ and A actuators based on EE Type T, class Q1 HRC fuses to provide stalled motor protection on 380- and 415-V, 50-Hz supplies:

Actuator size	Actuator output (rpm)	Approximate locked current (A)	Fuse size (A)	Approximate disconnect time (s)
7	18, 24, 36	2.3	2	—
	48, 72, 96	4.6	2	4
14	18, 24	6.7	4	800
	36, 48, 72, 96, 144	11	6	200

Typical Actuator Specification

It is important that the design be able to withstand the elements, for most control devices are mounted in positions that are in the open and subject to the vagaries of the weather for all of their service life. A user's specification for a Rotork A range device could be drawn up from the following specification of the basic Rotork A range suitably enhanced if necessary, dependent on the application, from data on the options available that is listed after the basic Rotork A specification:

Main drive unit: (basic Rotork A range data) three-phase actuator with double-sealed enclosure conforming to the IEC IP68 standard (which means that the unit can be submersed in water to a depth of 3 meters for a period of 48 hours). The motor to be low-inertia/high-torque Class F insulated rated for 15 min and provided with thermostat protection, and complete with torque/limit switches for open and closed positions and auxiliary limit switches for use by others.

The unit to be fitted with an emergency handwheel: complete with a mechanism that permits it to be declutched and locked.

The drive to be fitted with a local three-position status indicator (this usually means that the indicator would be mounted on the drive itself).

The applications of the actuator are limited to physical locations that would subject it to vibrations not exceeding 0.5g over the frequency range 10 to 200 Hz.

The foregoing basic Rotork A drive can be enhanced by suitable selections from the following Rotork-supplied options to replace or add to those provided on the basic drive:

1. *Drive options:* The following are some of the variations that can be applied to the actuators:

- Flameproof enclosures to CENELEC or Factory Mutual (FM) specifications. **Note:** The safety aspects referred to here are detailed in Chapter 22 dealing with operational safety.

- Two-speed or low-speed operation.

- Integral reversing contactor starter mechanically and electrically interlocked.

- Instantaneous reversal protection.

- Pneumatic fail-safe system can be fitted to the three-phase actuator only.

- Local pushbuttons to open, close, and stop, and lockable selector switch.

- Indicator can be fitted with illumination (red, open; white, mid-travel; green, closed.)

2. *Remote control options*: The following enhance the pushbuttons available option stated in the foregoing in item 1.

   - Via powered pushbuttons to provide open/stop/close or open/close action of the device.

   - Via "volt-free" pushbuttons (e.g., Remote logic or local contactor or relay) located 600 m or further away.

   - Via an analog current signal in a general range of 0 to 50 mA although the signal need not be zero-based, i.e., it can have an elevated zero—e.g., 4 to 20 mA. Alternatively the signal could also be an analog voltage or resistance variation.

   - Via an On/Off switch contact.

3. *Valve status indication options*: The following are some available features to enhance the indicator option in the foregoing item 1:

   - Potentiometer for continuous remote valve position indication.

   - Retransmittable current signal representing valve position. The signal can be any range between 0 and 50 mA (including an elevated zero, usually—4 to 20 mA).

   - 24 V dc or 110 or 120 V ac supply to drive an external voltmeter or signal lamps. **Note:** "External" is meant not attached to the actuator itself.

   - Valve "running" (motor energized) and valve "available" indication (motor not energized).

4. *Interlock and monitoring facility*: Item 1 can be enhanced by arranging the device to be interlocked so that:

   - Opening or closing movement can be inhibited or permitted.

   - Manual operation can be inhibited or permitted.

   - An alarm can be initiated when either power supply is lost, the supply fuse is blown, the motor is overheated, or even when the control circuit is wired incorrectly or incompletely.

5. *Hazardous location option*: Item 1 drive motor safety compliance is elaborated upon with the following available safety certification.

   - CSA WT; and CSA EP, Class 1, Group D, Div. 1 (Canadian Standards Association).

   - FM Class 1, Groups B, C, and D, Div. 1 (Factory Mutual, USA).

- CENELEC EXd for EExd IIB T4; EXd Hydrogen for EExd IIB H2 T4; EXe for EExde IIB T4; EXe Hydrogen for EExde IIB H2 T4 (UK—BASEEFA). **Note** Under the CENELEC certification there are four categories of approval as shown previously.

Notes on vibration: If the drive is to be located in a seismic region and the device is to operate during and after a seismic event, 1.0 g in the frequency range 0.2 to 33 Hz is allowable, but if the drive is not to operate during or after the seismic event, and yet maintain its structural integrity, then the foregoing value can be increased to 5.0 g, and the device it will not be operable until fully tested. To avoid such limiting, due to "plant-induced" vibrations the actuator should be mounted remotely and the operating shaft extended and designed with vibration-absorbing couplings.

Or if an Rotork IQ actuator is more suitable, then: A user's specification for a Rotork IQ range device could be drawn up from the following specification of the Basic Rotork IQ suitably enhanced if necessary, dependent on the application, from data on the options available that is listed after the basic Rotork IQ specification:

> *Main drive unit:* (basic Rotork IQ data) three-phase Class F low-inertia/high-torque squirrel cage motor rated for 15 min for intermittent operation with double-sealed enclosure conforming to IEC IP68 standard. Burnout protection by embedded thermostat with facility of bypassing under emergency shutdown conditions. The power module has a mechanically and electrically interlocked reversing contactor, and a control supply transformer fed from two phases of the power supply.
>
> All torque, turns settings, and indicating contact configuration effected by nonintrusive handheld infrared setting tool. The actuator must not be used in applications where the physical location would subject it to vibrations exceeding 1.0 g rms (root mean square) over the frequency range of 10 to 1000 Hz.

The foregoing basic Rotork IQ drive can be enhanced by suitable selections from the following Rotork-supplied options to replace or add to those provided on the basic drive:

1. *Drive options:* The following are some of the variations that can be applied to the drive motor:

   - Flameproof enclosures to CENELEC or Factory Mutual specifications. **Note:** The safety aspects referred to here are detailed in Chapter 22, dealing with operational safety.

   - Four independently configurable latching contacts that can be selected individually to be normally open (NO) or normally closed (NC). Arranged to signal any one of the following valve conditions: opening, closing, moving, fully open, fully closed/intermediate positions, motor tripped in traveling, motor tripped on torque (open), motor tripped on torque (closed), preset torque exceeded, motor stalled, battery low, and actuator being driven by handwheel.

   - 'Opto isolators' provided and fed from an internal 24 V dc supply or 20–60 V or 60/120 V ac/dc external supplies for remote positive switching of open, close, stop, and ESD (emergency shut down) control signals and thermostat bypass, interlock functions. Negative switching is achieved by fitting an alternate module.

   - A lockable local/stop/remote selector operating internal reed relays.

   - A lockable open/close control selector operating internal reed relays

2. *Remote control options:* The following can be added to enhance the basic main drive unit to provide the means of driving the actuator to the required position by analog control signals proportional to current, voltage, or potentiometer position

are provided, the following being the operating control signal range:

- Current: 0 to 5, 0 to 10, 0 to 20, and 4 to 20 mA

- Voltage: 0 to 5, 0 to 10, and 0 to 20 V

3. *Valve status indication*: The following are some available features to enhance the independently configurable latching contacts of foregoing item 1:

   - Internally powered 4 to 20 mA analog signal proportional to valve position.

   - Internally powered 4 to 20 mA analog signal proportional to actuator output torque.

   - Three relays each with a volt-free contact to signal: (i) battery low, NO, rated 5 A, 250V ac/30 V dc; (ii) thermostat tripped, CO, rated 30 W, 62.5 VA, 110 V maximum; (iii) remote-selected CO, rated 30 W, 62.5 VA, 110 V maximum. Additional contacts are possible, but *not when* the Rotork *Folomatic* analog control or the Rotork *Pakscan* two-wire serial link is used.

   - Data logging is achieved by initially fitting or retrofitting an optional electronic card.

4. *Interlock and monitoring facility*: The device can be arranged to be interlocked in a variety of different ways via the contacts provided. The type of interlock is dependent on the application.

5. *Hazardous location*: Item 1 drive motor safety compliance is elaborated upon with the following available safety certification.

   - CSA WT; and CSA EP, Class 1, Div. 1, Groups C and D (Canadian Standards Association).

   - FM Class 1, Div. 1, Groups C and D to NEC Article 500 (Factory Mutual—USA).

   - CENELEC EExd IIB Norm EN50018 for EExd IIBT4; EExde IIB Norm EN50019 for EExde IIBT4; EExd IIC Norm EN50018 for EExd IIC T4; EExde Norm EN50019 for EExde IICT4 (UK—BASEEFA). **Note** Under the CENELEC certification, there are four categories of approval as shown here.

Notes on vibration: If the drive is to be located in a seismic region and the device is to operate during and after a seismic event, 2 *g* in the frequency range of 1 to 50 Hz is allowable, but if the drive is not to operate during or after the seismic event, and yet maintain its structural integrity, then the foregoing value can be increased to 5 *g*. To avoid such limiting, due to plant-induced vibrations, the actuator should be mounted remotely and the operating shaft extended and designed with vibration-absorbing couplings.

SUMMARY

1. On account of the sheer size and the weight of some of the final control devices used in industry, the normal diaphragm actuators are unsuitable, since these are unable to exert the necessary force to cause the device to move.

2. With the diaphragm actuator, the load-moving capability will depend on the strength of the diaphragm material when subjected to the pressure of the drive fluid.

3. The power cylinder is a device in which the movement is provided by the motion of a piston and piston rod that slide along the bore and longitudinal axis of the cylinder.

4. The greater driving force provided by a power cylinder is achieved by much higher pressures applied by the actuating fluid on to the moving parts of the device.

5. Since all power cylinders have circular bores, the force available is given by:

$$\text{Force} = \frac{\pi d^2}{4}(\text{operating pressure})$$

where d is the bore diameter.

6. The stroke of the piston depends upon the amount of movement required. The stroke length will determine the length of the cylinder itself.

7. The pneumatic version of the power cylinder uses air and the pneumatic positioner as a means of driving the piston and rod assembly. A feedback signal from the piston rod or "ram" is absolutely essential for stabilization and accurate positioning.

8. The feedback signal is derived from a mechanical link-and-lever mechanism. The link being firmly fixed to the piston rod and the lever made to act on a pneumatic relay, which makes small adjustments to the magnitude of the drive signal.

9. Pneumatic positioners receive control signals of 3 to 15 lb/in² (0.2 to 1.0 kg/cm²) and are usually only able to provide output pressures up to approximately 35 to 40 lb/in² (2.5 to 2.8 kg/cm²). A large driving force from the cylinder is obtained by increasing the cylinder bore. There are limits on the extent to which the pneumatic cylinder bore can be increased, and other means have to be found for those instances when the driving force required is so large that pneumatically actuated cylinders become inappropriate.

10. The hydraulic cylinder is very much more robust than the pneumatic version, because it has to be able to withstand much higher pressures, exert much higher forces on the equipment to which it is attached, and work with no cushioning effect of the motive fluid used.

11. One method of driving and positioning hydraulic cylinders, or directly coupled hydraulic actuators on quarter-turn valves such as butterfly valves or dampers or linear-stroke control devices such as plug-type globe valves, is by using an electrohydraulic system where the precision of electronic measurement is coupled to the precise means of transferring it to the control device itself. The mechanical drive of the actuator, powered by a compression spring, under emergency conditions moves the actuator to a predefined safe condition that is specifically designed into the actuator.

12. Control using an electrohydraulic system is effected by any one of the three types of actuator manufactured: the rack and pinion, the Scotch yoke, and the rolling diaphragm, which are differentiated by the type of motion (rotary or linear), the force needed to move, and the means used to transmit the drive developed in the actuator to the control device.

13. To minimize the number of conversions of the control signal in order to get the required process changes, it is possible to have final actuators that are electrically driven by control signals usually in the 4 to 20 mA range. There are differences

in the electrical actuators used. The electrohydraulic actuator is an electrically operated system with internal hydraulic positioning, specifically designed for continuous control duty under the dictates of the process controller output signal of 4 to 20 mA; the unit's rating is therefore 100%. Most other electric motor actuators are normally rated at 30% to 80%; the lower duty rating is brought about by the number of motor starts per hour allowed for these devices.

14. Electrical actuators are available for dc voltage and three-phase 50/60 Hz power supplies. The equipment gearboxes with worm drives transmit the drive to the control device, e.g., the valve.

15. A control drive is an assembly of pneumatic power cylinder with positioner and linkage mechanism enclosed in a protective housing and supplied as a complete piece of equipment.

16. Sizing of a control drive requires the following to be available or known:

    • The force to actuate the control element.

    • The angular distance through which the control element is to be moved.

    • The separation distance between the drive and the control element. This is the center-to-center distance between levers. The linkage between the two levers is usually made adjustable, by including a turnbuckle.

17. When using links and levers, it is recommended that:

    • The design should commence with the system at rest, and with all angles formed by the pin-jointed mating components at 90° to each other, to allow all succeeding movements to be directly related to each other.

    • The "line of action" for all members delivering mechanical power should avoid any twisting or bending of the components involved, and should be sufficiently robust to deliver the required power.

    • The mechanism should be kept as simple as possible and use as few components as needed to achieve the result required.

PART V

INSTALLATION OF INSTRUMENTATION

CHAPTER 18

Considerations in Proper Installation

While the thrust of all the foregoing chapters has been toward understanding the underlying principles of instrumentation, and control of the process, it is inevitable that, for the best results, very great dependence has to be placed on the signals related to the various process parameters that the measuring/transmitting and control instruments, located in the plant or control room, generate or receive. The following comments are made with the objective of providing both readers of this book and users of the instrumentation the most meaningful control system, and the information given is primarily concerned with the installation of instrumentation in the field, and, where necessary, the implications that control and computational functions have on the operation of the process plant. Without a good measurement—and by that is meant a truly representative and repeatable one—the control of the process will almost certainly always be poor. The topics discussed in this chapter are intended to help in developing an ability to recognize and avoid the hazards that could be encountered, and to be of assistance when designing a control system. This newly found expertise when coupled with correct installation would help produce meaningful signals.

Please be aware that the topics under discussion are *not exhaustive*, as they only cover the most common requirements and indicate some of problems that can be found and are associated with the instrumentation and its installation on-site.

GENERAL CAUTIONS

In this section we shall consider some aspects of instrumentation that do not appear to come under a specific category but nevertheless have to be considered when a system is being designed so that the final results are meaningful.

1. When "smart" or "intelligent" transmitters are used to provide the measurement in a control loop and are being reranged or recalibrated from the control room, be aware that the measurement signal to the associated controller(s) will go *inactive*. This results in a loss of control of the process for the duration. Appropriate steps to counter the situation must be taken.

2. Always make sure that the signal produced by the measuring device is truly representative of the actual process being measured and that if software is involved, then it, too reflects this. For an example, see the section on installing level devices later in this chapter.

3. When a start command or, say, a signal to open or close a valve is given to a plant-located device, it does not necessarily mean that the device has in fact (correctly)

carried out this instruction. The only proof that the command issued has been implemented is a feedback signal from the device itself. If the design of the scheme demands proof positive, then confirmation feedback is the only answer.

4. Some processes will demand a nonlinear characteristic in the controlled output. For an example, see the section on installing analytical devices later in this chapter.

5. The action of drawing cables (pulling them off the reel) in locations that are very dry, i.e., places that have a very low relative humidity, can cause a very high static charge to develop on the cable itself, which can sometimes reach a magnitude in kilovolts and can therefore cause considerable damage if proper precautions are not taken to protect sensitive electronic equipment and personnel. This phenomenon is called *electrostatic discharge*, sometimes abbreviated as *ESD*. However, be aware that most often *emergency shut down* is *also* abbreviated as *ESD*; therefore, read the document carefully to obtain the correct meaning when the acronym is used. To avoid electrostatic discharges, always make sure that cables drawn off a reel during installation are cleared of a developed static electric charge by grounding them before connecting up to any electronic equipment. Testing cables with a cable insulation tester can also result in high residual charges that can damage equipment, especially if one forgets to first isolate the instrument.

INSTALLING FLOW DEVICES

STRAIGHT PIPE RUNS

As a general rule, ensure that all flow primary devices have long, straight lengths of pipeline both up- and downstream of the instrument itself. As a rough guide, 15 to 20 pipe diameters upstream and 7 to 10 pipe diameters downstream should be sufficient, although (if possible) more are advisable and advantageous; see Figure 18.1. Slight reductions in these dimensions are allowable, but no less that 10 upstream and 5 downstream. In the event the requirements cannot be met, flow straighteners are recommended. The reason for using straight runs or flow straighteners is to eliminate the effects of turbulence on the measurement.

MINIMIZING PROCESS NOISE

If a flow primary device is to be installed in an existing plant, make every effort to eliminate process noise from all existing turbulence-creating equipment—i.e., pumps, control and isolation valves, pipe bends, and pipe fittings—whenever it becomes evident or is anticipated, by relocation and/or correction to obtain the recommended straight lengths of pipe run on the up- and downstream sides of the primary device. If the piping configuration will not permit suitable straight lengths up- and

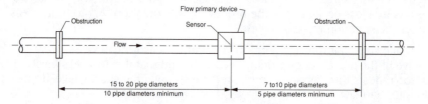

Figure 18.1: Definition of straight runs of pipe for flow measurement. For simplicity obstructions are shown as flanges; these could be valves, elbows, union connectors, or any other pipeline-mounted invasive devices.

downstream, install a flow straightener. In new installations, provide a stable measurement by keeping the straight runs of pipe work to the dimensions shown in Figure 18.1. Remember, flow is *always* a noisy parameter, even without turbulence-creating equipment upstream and/or downstream of the flow-measuring device.

Because the parameter flow is always noisy, derivative control action in the controller is not recommended. If derivative action was to be used, then the controller would respond to every fluctuation of the signal, and stable coincidence between setpoint and measurement would never be obtained.

ORIFICE PLATE INSTALLATIONS

The tapping points on orifice plate installations can be either D and $D/2$ (where D is the internal diameter of the pipe), flange, or carrier ring assembly, depending on the site requirements and different flow constants and tables apply to each type. Hence, use the correct data applicable to each type. See Figure 18.2 for details of each of the tapping points discussed.

Orifice plates may be mounted into the pipeline in one of several different ways. Two of the more common methods are shown in the upper part off Figure 18.2, which gives most of the salient detail to understand the methods used. Connections to the device are made via the ports shown in the diagram using "swan necks," which are really pipe nipples shaped with curves that are very gentle in the form of an S; the reason for the slow curves is to avoid generating the sediment or gas entrapment areas that would result if sharp bends, e.g., elbows, were used, and to provide separation for the isolation valves necessary to permit removal of the measuring device.

VALVE MANIFOLDS USED ON DIFFERENTIAL-PRESSURE INSTALLATIONS

The need to make up connections to the differential-pressure transmitter more complex than that shown in Figure 18.2 is eliminated by using proprietary valve

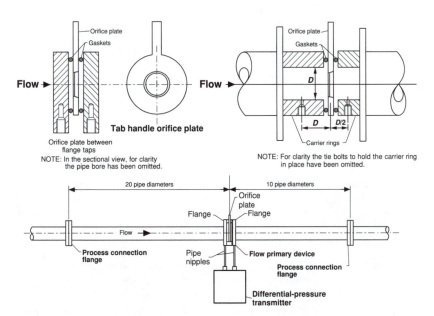

Figure 18.2: Typical orifice plate and flow metering section.

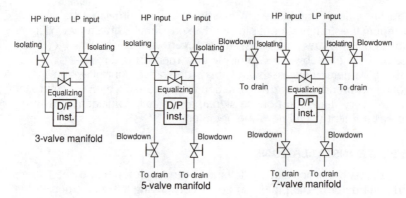

Figure 18.3: Schematic arrangements of valve manifolds.

manifolds, available from several manufacturers such as Mansfield and Green, Parker Hannifin, Unit Controls, and Anderson Greenwood. These manufacturers make the entire manifold as a single unit containing all the required valves, operating handwheels, and process connections. There are several configurations of the manifold valves, having typically two, three, five, and seven valves. These units are manufactured so that they can be either directly bolted to the body of the differential-pressure cell, or separated from the instrument in the impulse lines—i.e., connection pipes between primary device and transmitter (identifies as pipe nipples in Figure 18.2). The method of fixing the valve manifold directly to a differential-pressure cell has been standardized among all the major manufacturers, which means that the pitch of the fixing holes and process connections on the instrument are identical on D/P cells from any of the major instrument manufacturers.

Figure 18.3 shows schematic diagrams for some of the configurations of valve manifolds we have just discussed. In instances where proprietary items are unobtainable, Figure 18.3 could enable manifolds to be assembled using appropriate individual valves, pipe nipples, and/or lengths of pipe, but great care must be taken to ensure that every component used to do this is suitable and meets or preferably exceeds the entire specification of the process fluid and the material of the process line. It is most important that the assembled arrangement be thoroughly tested before installation and commissioning.

REQUIREMENTS FOR IMPULSE LINES

When using differential pressure transmitters on flow measurement, ensure that the impulse lines to the instrument conform to the following requirements, which are illustrated in Figure 18.4:

If the process fluid is a gas, make sure the tapping points are on the upper side of the pipeline or duct and the instrument located *above* the differential creating device with the impulse lines *sloping down* toward the primary device, as this will prevent solid particles or any entrapped liquid from entering the instrument and ensure that only the gas actuates the sensor diaphragms.

If the process fluid is a liquid and the pipeline is horizontal, the tapping points should be inserted at an angle either above or below the horizontal axis as shown in Figure 18.4. If the process line can run partially full, do not use the upper connections as this would allow any entrapped gas or vapor in the process fluid to enter the impulse lines, rendering the measurement unstable and inaccurate due to

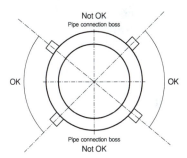

Figure 18.4: Cross-sectional view of a process pipe. Process connections for orifice plates can be made in the areas designated OK. Use the upper connections for gases, and the lower ones for liquids.

the compressible nature of the entrapped gaseous component of the fluid. The upper connections are most suitable for lines running full, as this avoids scale or other solid particles blocking the impulse lines. The very lowest locations of the connections are suitable for "solid free" liquids.

It is also imperative that both the impulse lines be on the same side of the horizontal axis and full of liquid only. To ensure that the impulse lines are full of liquid only, use the vent or bleed plug on each of the end caps on the transmitter body to vent the entrapped gas or vapor; see Figure 18.5. The procedure to be followed in this

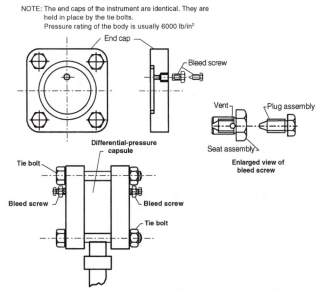

Figure 18.5: Bleed screw assembly on a differential-pressure transmitter. The cell shown is the electronic version; pneumatic types are also available.

exercise is identical to that used when bleeding the brake cylinders on an automobile. Do not forget to retighten the vent plug immediately after each line has been vented. Do not mount the instrument above the tapping point. For vertical process lines and liquid fluids, ensure that the flow is always upward.

On steam headers, takeoff positions of the tappings for the impulse lines should be truly horizontal, it is also vital that the impulse lines be full of liquid only from the outset, or measurements will be erratic for the time it takes to completely fill the impulse lines with condensate. When first commissioning steam lines, ensure that the impulse lines are full of water. This can be done prior to connecting the primary to the instrument. Condensate pots, though not strictly necessary with D/P cells, can also be used, since they make filling the impulse lines a lot easier, but the pots must be fitted via suitable nipples at the primary device end of the connection, and care must be taken to ensure that both the condensate pots are mounted in the same plane.

EQUALIZING STATIC PRESSURE HEADS

On all installations, it is essential that, when considered individually for each location, the two end connections (high and low pressure) of the orifice plate, venturi meter, or similar instrument and those of the differential-pressure instrument itself be coplanar; and this is especially important on liquids and steam services, for it means the static pressure heads from the outset are the same.

SEQUENCES NECESSARY WHEN OPERATING A THREE-VALVE MANIFOLD

For every installation where a differential-pressure instrument has been used, valve manifolds have to be included at both locations, primary device and transmitter; for without them it would be difficult to disconnect these from the process. There is a sequence to the actions that must be followed when operating the valves on the manifolds.

Let us consider the scenario where the measurement is flow, the loop operational, and a differential head type primary device together with the associated three valve manifolds and D/P transmitter have been installed. An operational loop such as this will have the isolating valves open and the equalizing valve on each manifold closed (see the manifold schematic of Figure 18.3 for the location of these valves). To remove the transmitter only from service, we shall have to:

First, open the equalizing valve to equalize the pressures on the capsule diaphragms. This will apply the higher of the two heads to both the high-pressure (HP) and low-pressure (LP) diaphragms. Second, close both the isolating valves. This will lock the line pressure (HP) in the transmitter. Third, vent each side of the transmitter in turn low side first, but slowly. This will allow the capsule to be subject to atmospheric pressure. Take suitable precautions when high-pressure or high-temperature fluid is involved. Fourth, remove only the instrument along with its manifold.

Remember, to remove the entire D/P cell assembly only, the manifold connections to the cell body must be broken and the transmitter mounting "U" bolts undone. If the manifold has also to be removed, remember to isolate the impulse lines at the primary device first, and follow the maker's torque recommendation for the manifold bolts on reinstallation.

Similar procedures will have to be followed when removing the primary device, but make sure the process line is shut down and drained before removal is carried out.

INSTALLING A CORIOLIS FLOWMETER

When Coriolis mass flowmeters are used, it is imperative that the instrument not be mounted on a pipeline subject to vibration, as this will introduce frequencies or harmonics that will interfere with the measurement being made, for, as explained in Chapter 1, all Coriolis mass flow meters use a technique that deliberately vibrates the measuring tube at its particular resonant frequency to effect the measurement, and therefore it is important that any secondary harmonics be excluded. Furthermore, with many instruments the meter should be mounted in a manner that will make the tube(s) self-draining, but with some tube geometries this is not possible; in these instances seek the makers recommendations.

INSTALLING A MAGNETIC FLOWMETER

When magnetic flowmeters are installed in nonconductive (e.g., plastic) pipelines, it is most important that *grounding rings* be installed and the device be fully grounded via the rings. (**Note**: Grounding rings are metallic rings provided with suitable grounding studs and fitted between the instrument and process flanges.)

Do not mount magnetic flowmeters in a manner that will allow the tube to be self-draining (i.e., with the fluid flow upward), as having fluid in the tube at all times will prevent the coils from overheating and burning out.

INSTALLING PRESSURE DEVICES

NEED FOR CONSISTENT ENGINEERING UNITS

Because there are a variety of engineering units for pressure to choose from, there are always problems with selecting the appropriate units to be used for this parameter. In the metric system, the *bar* is, strictly speaking, a gauge measurement of actual ambient pressure and equal to 10^5 N/m^2; hence, the atmospheric pressures in weather reports are always stated in the unit of bars, but since the actual pressures involved are so small, the millibar (mbar) is always quoted. The *atmosphere* (Atm) is another unit of pressure measurement, equal to 1.01325×10^5 N/m^2 (approximately 1013 mbar) and taken as the mean value of atmospheric pressure at sea level; it could be easier to use this unit to determine absolute pressures rather than the bar. Vacuum is measured in units of the torr, which is equal to 133.322 N/m^2 (1 torr = 1 mm Hg).

Consistency in the choice of units of pressure is essential, especially if using this parameter to carry out calculations for the compensation of one variable for changes due to other influencing ones—e.g., compensating the flow for pressure variations.

USE OF A SNUBBER

When the discharge pressure of a reciprocating pump is being measured, there is every possibility for the measurement to fluctuate, i.e., to be "noisy." One way of eliminating the fluctuations is to fit a pressure *snubber* between the pipeline and the pressure-measuring instrument. Figure 18.6 shows a piston-type snubber; there are several other designs available. This device can be subject to blocking if the fluid contains suspended particles, so in these situations the suggestion is to use an in-line

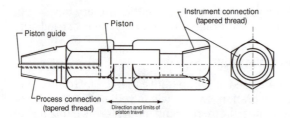

Figure 18.6: Typical pressure snubber assembly.

strainer mounted upstream of the snubber, which will protect the snubber and ease rectification when a blockage occurs.

COMPOUND-RANGE INSTRUMENTS

If the measured pressure can go negative, i.e., the measurement can make excursions from vacuum to positive pressure and vice versa, then a compound-range instrument will be required, and care must be exercised to ensure that the resolution is sufficient to meet the demands of the loop. Very large ranges always result in poor resolution—i.e., difficulty will be experienced in resolving (making visible small changes in) the measured variable. Modern "smart" devices can overcome this, since it is possible to rerange the instrument on-line, so that when a change occurs the best resolutions can be obtained.

MEASURING DIFFERENTIAL PRESSURE BETWEEN WIDELY SEPARATED POINTS

The pressure difference across a distillation column, say, is best measured with two pressure instruments, and the difference being obtained by subtraction of the two signals, rather than a single differential-pressure instrument. This procedure eliminates problems associated with reference legs, or thermal effects if capillary seals are used.

An alternative to using two pressure instruments is to use a differential-pressure instrument on the lower connection of the distillation column and a pneumatic *repeater* on the top connection, taking care to ensure that the range of the repeater is within the operating pressure range of the top of the column. (A repeater is a 1:1 pneumatic transmitter, which means that the measured pressure variable is directly converted into a pneumatic signal of identical pressure to the measurement.) The output from the repeater can then be piped to the low-pressure connection on the D/P instrument, and will save performing the subtraction externally.

There are limitations on the measurement range of the repeater, but when used on many distillation columns there are no high pressures involved and this method is viable. There must be an instrument air supply available for the repeater.

INSTALLING TEMPERATURE DEVICES

USE OF THERMOWELLS

For plant and instrumentation maintenance purposes, thermowells, sometimes also called *pockets*, are most often fitted to temperature sensors; see Figure 18.7. The thermowell is fitted to the process line or vessel either by screw threads, flange, or welding and therefore, to all intents and purposes, remains permanently in place.

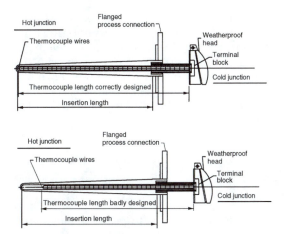

Figure 18.7: Typical thermocouple and thermowell assembly.

The sensor is fitted into the thermowell only semipermanently, as there may be a need to remove it for repair or replacement.

When installing a thermowell, be aware that a thermowell introduces a measurement lag into the system; this lag increases as the ratio of the well inner volume to the sensor volume increases. Also, the sensor outside diameter (O.D.) must always be less than the bore diameter of the well—as may appear obvious, but will be important when thermowell and sensor are obtained from different sources.

The design of the thermowell will have to take into consideration the following points:

- The maximum, design, and operating pressures of the process will determine whether a screwed, flanged, or welded design is to be provided.

- The material of construction for the thermowell must be compatible with the pipeline and the process fluid.

- The location and insertion length of the thermowell to be provided—e.g., in a stirred vessel—must be such that the agitator shaft and impellers are avoided; and consideration must be given to avoiding the bending of long unsupported lengths, and to the forces exerted by the liquid moving under the action of the agitator, or any other flow past the thermowell (e.g., when the element is inserted in a direct flow stream).

- The most representative temperature of a fluid is usually near the center of the containing pipe or vessel. Short insertion lengths in large cross-section vessels will not give accurate results. The sensed temperature in such cases will be that obtaining at or near the wall of the container.

- The insertion length must not be of a dimension that causes it to become an obstruction. Also remember that a thermowell when inserted in a flow stream behaves like a bluff body and will shed vortices with all the incumbent problems that these pressure fluctuations create.

INSTALLATIONS WITH THERMOCOUPLES

When using direct-wired thermocouples—i.e., without transmitters—always try to ensure that the interconnecting wiring between the thermocouple head and the associated equipment is a continuous run (no wiring breaks). This is necessary to avoid temperature gradients (the Thompson effect) between the point of measurement and the associated equipment. If breaks are unavoidable, then it is vital that the temperature at the junction of the break(s) be maintained constant with that at the thermocouple head, which is expensive to implement. A thermocouple with thermocouple-to-current converter mounted either in the head, or close by eliminates the problem, as the converters generate a proportional current signal in a standard range, e.g., 4 to 20 mA for transmission.

INSTALLING LEVEL DEVICES

NECESSARY TRANSMITTERS

When liquid levels in closed vessels are being measured, use differential-pressure or displacement transmitters. Direct-pressure transmitters are not suitable as level-measuring instruments in these applications, because:

1. The pressure head of the liquid is directly proportional to level only if fluid density and the pressure on the free surface of the liquid are constant, and in closed vessels this latter is usually not the case.

2. In closed vessels, the space above the free surface of the liquid will most often be filled with vapor or other compressible fluids, which exert a pressure on the liquid surface. The additional pressure at the liquid surface must be subtracted from the total pressure head in order to determine the pressure due to the liquid head alone, and this difference of pressure will then be proportional to the liquid level.

If the liquid temperature or composition changes significantly as part of a batch or recipe change, then the fluid density will change. It may be necessary to compensate for this change in density. Figure 18.8 shows a relatively simple way of accomplishing this compensation, but the important point is to ensure that the upper connection of transmitter A is always below the liquid level, as this will make the output from transmitter $A = \rho g h$, where h is the fixed liquid head, g is the acceleration due to

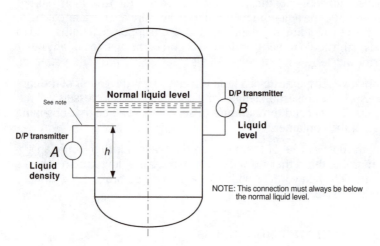

Figure 18.8: Liquid density correction.

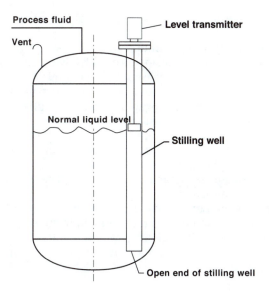

Figure 18.9: Stilling well installed to correct for surface fluctuations. The surface fluctuations shown have been exaggerated to show the effect of the stilling well.

gravity and ρ is the density and the only variable. The output from A can then be used to compensate the output from transmitter B, which is the measurement of the level in the vessel.

PRECAUTIONS IN MEASURING LEVEL

Measuring liquid level in vessels that have the free surface continuously disturbed, say, by the inflow of the process fluid can pose serious problems, especially when a float-and-cable or any surface-recognizing measuring instrument is used. Depending on the magnitude of the undulations, signal filtering may assist, but otherwise a *stilling well* will have to be installed and the measurement made within it. This arrangement is shown in Figure 18.9.

If the process offtake pipe protrudes into the vessel itself to form a weir or dam in the vessel and the level instrument connection is above the intruding pipe, then the zero level transmitted to the system by the measuring instrument is not zero level in the vessel; for because of the weir, there will still be an immeasurable quantity of liquid remaining in the vessel. If, on the other hand, the level instrument connection is on the vessel but below the top of the intruding offtake pipe, then again the weir will retain a quantity of the liquid, but this time the amount will be measurable, and there will never be a true zero liquid level signal produced.

INSTALLING ANALYTICAL DEVICES

Since pH is a highly nonlinear measurement, reagent addition in a pH control loop requires a controlled output that is nonlinear also. In the case of the pH loop, the most effective control is via a nonlinear controller and a valve with an equal-percentage characteristic.

In pH control, because of the characteristic of the titration curve, it may not be possible to achieve the desired reagent addition with a single control valve, due to the extremely wide rangeability and high resolution that would be required of the valve. In these instances it is suggested that two valves (a large and a small) be used in a split-range configuration (see comments further on regarding split ranging).

Almost every on-line analyzer will require a means of sampling the process fluid to allow an analysis to be conducted upon it, but since the majority of analyzers— e.g., chromatographs, pH with glass electrodes, oxygen—are designed to carry out the analysis at, or very slightly above, atmospheric pressure, these instruments are unable to cope with the high process pressures sometimes encountered. The sampling systems therefore usually involve some means of pressure reduction and are custom-made for each application. Process temperature can also cause problems in analyzers, and therefore, fluid coolers are also sometimes required and included in the design of the sampling system. The design of the sampling system requires specialist attention, and care must be exercised, since without a good sampling system there is always going to be a questionable measurement.

TIPS ON CONTROL APPLICATIONS

SPLIT RANGING

In those applications that demand a split in the range of the controller output (see Chapter 14 for a typical application) in order to drive two separate control valves, try to arrange the following:

- Range splitting to be carried out in the control room equipment rather than in the field

- The signal to the individual control valves, derived from each part of the split controller output, to cover the full standard range of 4 to 20 mA

When designed in this way, changing the ratio of the split is made relatively easy, if this is required, when commissioning the control system. Furthermore, this will also allow the current or pneumatic converters to be standard 4 to 20 mA or 0.2 to 1.0 bar instruments.

Figure 18.10 uses a distributed control system, described in detail in the following chapters, to illustrate how split ranging can be done within the system itself,

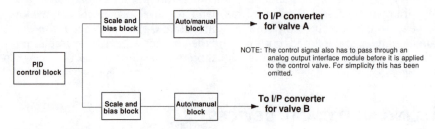

Figure 18.10: An arrangement for split-ranging a controller output in a DCS.

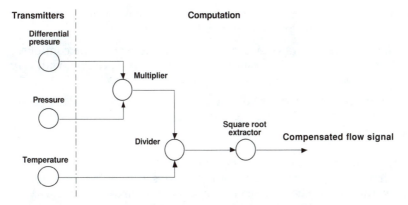

Figure 18.11: Typical scheme for flow compensation.

using as an example the Honeywell TDC 3000 Process Manager. This method leaves the output signal for each of the control devices in the field as the standard 4 to 20 mA. The Auto/Manual block arrangement in Figure 18.10 permits bumpless transfer when changing from Manual to Auto mode and vice versa; additionally, it also allows the valve positions to be read at all times, which is a great asset during plant commissioning.

COMPENSATING FLOW MEASUREMENTS FOR PRESSURE AND TEMPERATURE VARIATIONS

To obtain a pressure/temperature-compensated flow measurement using a differential-pressure transmitter, make the compensation calculation on the signal from the differential-pressure transmitter before extracting the square root to produce a measurement of flow. This is necessary because the square law Bernoulli relationship is applicable only to the measurement of differential head with respect to flow, and thus it is the signal from the differential-pressure transmitter that must be corrected for any variations in pressure and/or temperature. See Figure 18.11 for the arrangement.

AVOIDING DIFFICULTIES IN TUNING

Difficulties in tuning control loops may be due to any of the following :

* The design of the plant may be making the loop unstable, leaving both the control engineer and the control system few options to remedy the situation; the remedy is then in the hands of the plant owner. A self-tuning controller like the Foxboro Model 761/762 or the I/A Series with the "EXACT" (EXpert Adaptive Controller Tuning) algorithm may perhaps be of use here, but there is no guarantee of success in this particular instance, because the plant itself is questionable, and correction will be more difficult, especially if the loop is a fast one—e.g., flow or pressure.

* The sensor may be located too near the control device, so that the sensor is presented with the result of the control action as soon as it occurs and as a result the loop is unstable. Simply relocating either the sensor or control device, i.e., control valve or damper, can retrieve this situation.

- The control valve may not be sized correctly. In the following description it is assumed that the valve opens for increasing control signal:

  1. For installations where the valve size is too large for the range to be controlled, the valve is forced to operate at positions very near to the valve seat, with very serious consequences for the quality and sensitivity of control.

  2. For instances of an undersized control valve, the device will be forced to operate near its fully open position, which will also have similar serious consequences on the quality of control apart from the effects of maximum flow limitation.

The solution in both instances is to replace the valve with a correctly sized one. Sometimes valve oversizing may be rectified by fitting a reduced trim, but the pressure drop for the new trim will also have to be taken into account. In any event, however, the valve manufacturer must be consulted to ensure feasibility.

SUMMARY

This chapter makes no attempt to list all the possible difficulties that are to be found on a site. It does however tabulate some of the more common and therefore easily overlooked ones. It is recommended that readers think positively, and continue expanding the list for future benefit by recording and building on their own hard-earned experience, for their own benefit and their colleagues.

1. When "smart" or intelligent transmitters are used to provide the measurement in a control loop, reranging or recalibrating the transmitter from the control room will cause the measurement signal to the associated controller(s) to go inactive, and result in a loss of control of the process for the duration.

2. A start command or, say, a signal to open or close a valve is given to a plant-located device, does not necessarily mean that the device has in fact started. If the design of the scheme demands proof positive, then confirmation feedback is the only answer.

3. The phenomenon of electrostatic discharge is sometimes abbreviated as ESD; be aware that most often emergency shutdown is also abbreviated as ESD; therefore, read the document carefully to obtain the correct meaning when the acronym is used.

4. All flow primary devices have long, straight lengths of pipeline both up- and downstream of the instrument itself. As a rough guide, 15 to 20 pipe diameters upstream and 7 to 10 pipe diameters downstream should be sufficient, although (if possible) more are advisable and advantageous.

5. The tapping points on orifice plate installations can be either D and $D/2$ (where D is the internal diameter of the pipe), flange, or carrier ring assembly, depending on the site requirements and different flow constants and tables apply to each type; use the correct data applicable to each type.

6. When using differential-pressure transmitters on flow measurement, the impulse lines to the instrument should conform to the following requirements: If the process fluid is a gas, the tapping points are on the upper side of the pipeline or duct and the instrument located above the differential creating device with the impulse lines sloping down toward the primary device. If the process fluid

is a liquid and the pipeline is horizontal, the tapping points should be inserted at an angle either above or below the horizontal axis and completely full of fluid.

7. For every installation where a differential-pressure instrument with valve manifold has been used to remove the transmitter only from service: First, open the equalizing valve. Second, close both the isolating valves. Third, vent each side of the transmitter in turn low side first, but slowly. Fourth, remove only the instrument along with its manifold.

8. When Coriolis mass flowmeters are used, the instrument *must not* be mounted on a pipeline subject to vibration, as this will introduce frequencies or harmonics that will interfere with the measurement. With many instruments the meter should be mounted in a manner that will make the tube(s) self-draining, but with some tube geometries this is not possible and the makers recommendations must be followed.

9. When magnetic flowmeters are installed in nonconductive (e.g., plastic) pipelines, grounding rings must be installed and the device be fully grounded via the rings. Do not mount the flowmeters in a manner that will allow the tube to be self-draining.

10. Because there are a variety of engineering units for pressure, consistency in the choice of units of pressure is essential.

11. When a measured pressure fluctuates continuously, i.e., it is noisy, one way of eliminating the fluctuations is to fit a pressure snubber between the pipeline and the pressure-measuring instrument.

12. If the measured pressure can make excursions from vacuum to positive pressure and vice versa, then a compound-range instrument will be required. Very large ranges always result in poor resolution; modern "smart" devices can overcome the difficulty of resolution of the measured variable, since it is possible to rerange the instrument on-line.

13. The pressure difference across widely separated points is best measured with two pressure instruments, and the difference obtained by subtraction of the two signals, rather than a single differential-pressure instrument. An alternative to using two pressure instruments, say, on a distillation column is to use a differential-pressure instrument on the lower connection of the distillation column and a pneumatic *repeater* on the top connection, taking care to ensure that the range of the repeater is within the operating pressure range of the top of the column.

14. Thermowells, sometimes also called pockets, are most often fitted to temperature sensors. A thermowell introduces a measurement lag into the system; this lag increases as the ratio of the well inner volume to the sensor volume increases. The design of the thermowell will have to take into consideration the operating pressures of the process to select the appropriate process connection, the material of construction for the thermowell for fluid compatibility, and the location and insertion length to obtain a representative measurement and avoid interference.

15. When using direct-wired thermocouples, try to ensure that the interconnecting wiring between the thermocouple head and the associated equipment is a continuous run. If breaks are unavoidable, then the temperature at the junction of the break(s) must be maintained constant with that at the thermocouple head.

16. When measuring liquid levels in closed vessels, use differential-pressure or displacement transmitters. Direct-pressure transmitters are not suitable as level-measuring instruments in these applications.

17. Measuring liquid level in vessels that have the free surface continuously disturbed may require signal filtering, but otherwise a stilling well will have to be installed and the measurement made within it.

18. A nonlinear measurement in a control loop requires a controlled output that is nonlinear also. In the case of the highly nonlinear pH loop, the most effective control is via a nonlinear controller and a valve with an equal-percentage characteristic.

19. Almost every on-line analyzer will require a means of sampling the process fluid, which usually involves some means of pressure reduction and sometimes fluid coolers also. The sampling systems are therefore custom-made for each application.

20. Some applications demand a split in the range of the controller output in order to drive two separate control valves. It is preferable for the range splitting to be carried out in the control room equipment rather than in the field, and for the signal to the individual control valves, derived from each part of the split controller output, to cover the full standard controller output range.

21. To obtain a pressure/temperature-compensated flow measurement using a differential-pressure transmitter, make the compensation calculation on the signal from the differential-pressure transmitter before extracting the square root to produce a measurement of flow.

22. When control loops are unstable, a self-tuning controller could be of advantage. However, if the instability is due to the plant design, then no controller will be able to cope and a solution will have to be found elsewhere.

23. For incorrectly sized control valves, a sure solution to the problems encountered will be a replacement of the valves with ones that are correctly sized.

PART VI

MODERN CONTROL SYSTEMS

CHAPTER 19

Distributed Control Systems (DCS)

To appreciate the advantages of the modern control philosophy and technique of distributed control systems, we should take a look backward in time over a relatively short period of years to see how the ideas developed. Our review will commence from the period when there were only indicating and recording instruments available, and we shall take it as accepted fact that the need for such devices had already been established by the growing number of manufacturing processes. These industries were of limited scope, incapable of large-volume output and highly dependent on skilled operators, for there could not be any production at all without them. If the industrialists of the time were to be totally honest, process control was developed not only to increase earnings, but also to minimize the serious impact which the loss of a skilled operative would have on the company's profitability. By comparison with today, these very small-scale applications of mechanization not only increased output and stabilized the product, but to a large extent became the driving force, for money is a great motivator, even or especially today, when it goes under the pseudonym of "market forces." To achieve the profit objective, it is necessary to exploit more fully the capabilities that are available in modern control systems, and what is now being done is unleashing some of the power of the information that has always been available within the system but, up until the advent of new technology, could not be manipulated and used to advantage.

DEVELOPMENT OF MODERN CONTROL SYSTEMS

THE EARLY YEARS: THE FIRST CONTROL DEVICES

During the early period of instrumentation, the manufacturing processes using it were also in their infancy. Products were available and produced in small batches only, which were the result of the efforts of one individual, or in some cases a very few people. This limited scale of production was reflected in the cost of the product, which in many instances during that period could only be afforded by the more well-to-do members of society. To achieve consistency of the product, it was necessary to maintain constant and repeatable conditions in the manufacturing process itself, and to this end the available instrumentation, in spite of its limitations, was an invaluable asset. However, it did require the intervention of a human operator to physically manipulate the regulating devices fitted to the process lines in order to alter the process parameters that affected the product being made.

The invention of the mechanically actuated automatic control valve and its associated automatic controller made a profound change in the manufacturing process. These early devices were examples of ingenious mechanical engineering and amply reflected the quality of the engineers and craftsmen involved.

CONTINUED DEVELOPMENT: THE AGE OF PNEUMATICS

Human curiosity and inventiveness appear to have no barriers, and as a result technology moves relentlessly on. It was not long before the indicator was fitted with a transmitting mechanism that enabled the basic measured parameters of flow, pressure, temperature, and level to be read not only in situ, but also at a location some distance away. The medium used to achieve this breakthrough was air. The early mechanical controllers were redesigned so that they too could use the air signal being generated by the measuring and transmitting instrument.

It is not difficult to appreciate that each instrument maker used a pneumatic signal that best suited their design. Hence, there were a number of different pneumatic control signal ranges available—e.g., 5 to 25 lb/in^2 (psi), 3 to 27 psi, 3 to 15 psi—all non-zero-based. A number of years passed before there was a universally acceptable signal range, and this came about largely as a result of demands by the instrument users. The most common ranges at present are 3 to 15 psi, or 0.2 to 1 bar, or 20 to 100 kPa, all of which are very similar to each other and now accepted as standard.

The ability to locate the indicators, recorders, and controllers away from the point of measurement was the start to having specially designed mounting panels into which the instrumentation could be fitted, grouped to reflect the process they were controlling. The product manufacturers' objective for product consistency and repeatability in the production cycle was now slowly being seen as an achievable goal. The process designers exploited these newly found techniques, for they were now able to observe any effect that the alteration of one or two measured parameters had on the overall dynamic behavior of another, or a group of associated measurements, or even the whole plant. It was through these observations that a much better understanding of the product and manufacturing process was gained; the process designers were now able to show the operatives that it was possible to give desired qualities to a particular product by manipulating, at specific times during manufacture, the influencing process parameters. The results of this newfound ability were reflected in the product itself.

The ingenuity of the instrument designers produced a number of devices that performed mathematical calculations using the pneumatic signal, and these new devices, now available to the process designers, allowed them to further manipulate the manufacturing process according to established laws of chemistry and physics. These developments in instrumentation came together with the ability to handle larger amounts of material in the manufacturing process; and, provided the product itself fulfilled the demands of its purchasing clientele, the newly found economies of scale allowed it to be bought by a larger number of people, which assured its sustained production, improvements in manufacturing technique, and most important profitability.

Pneumatic instrumentation and control enjoyed great success for a number of years, and even today there are process plants worldwide that still use them for reasons of safety, tradition, or continued functionality versus replacement costs. It must be said, however, that the modern versions of the pneumatic measuring and control instruments are vastly different from the ones they replaced.

The Motive Power: The Air Supply

Although air-driven instruments are very good and in most instances very reliable, there are shortcomings, some not entirely concerned with the instrumentation itself. A few of the major issues when pneumatic instrumentation is used are:

- It is vital that there be a reliable source of compressed air, and this means that two compressors and air receivers have to be provided for an installation, the standard practice being to run the air compressors in a duty and standby manner. Sizing each of these has to provide for not only the running volumes, but also those periods when the other will not be available. Maintenance and running cost of the compression machinery, which is the responsibility of the user, represent an additional burden on financial resources.

- On installations of whatever size, the cost of the air supply equipment alone represents a sizable portion of the overall instrumentation cost.

- The air used has to be conditioned, for it must be clean, dry, and oil-free; this conditioning is most important, for the nozzles used in the instruments are very small indeed and therefore prone to blockage. To achieve the necessary quality, air dryers with either manual or automatic liquid draining and good air filters are required, all of which means additional purchasing, running costs, and maintenance.

- To have good response, the transmission distance for a single instrument may be up to 700 to 1000 ft (200 to 300 m); any greater distance requires insertion of boosters. This not only involves additional expense, but also some very small inherent loss of accuracy and an increase in transmission time, all of which, though not excessive, have to be considered.

- Since almost all signals are basically analog, computations performed with them demand that each signal used be scaled as a correct fraction of the whole, and subject to problems with the calculation(s), due to the inaccuracies of signal scaling involved. Correcting any errors made means either a complete "re-pipe," because the signals were applied to the incorrect instrument connections or possibly an instrument replacement, both of which involve additional time and expense.

FURTHER DEVELOPMENT: THE ELECTRONICS AGE

Electronic instrumentation received a massive boost after World War II—for it was during this conflict that the potential of electronics was increasingly exploited. The first instruments of this new technology were extremely large and heavy, mainly because of the size of their components. Many then state-of-the-art components such as transformers, metallic can capacitors, thermionic valves, ceramic-covered wire-wound resistors, finned rectifiers, and other components were so heavy that they had to be mounted on a metal chassis, which added to the weight of the equipment.

The electronic instruments performed the same functions as their pneumatic counterparts, but enabled much greater transmission distances, faster response, and a much reduced number of mechanical components requiring alignment and adjustment, as well as easier recalibration. In view of the advantages that electronics offered, much more attention was given to the development of newer and more sophisticated equipment using this technology, but the disconcerting, though understandable, part was that there again developed a situation where instrument makers took steps to protect their intellectual property. This resulted once more in equipment made in a multiplicity of signal ranges: e.g., 0 to 10 V, 1 to 5 V, 0 to 10 mA, 4 to 20 mA, or 10 to 50 mA, to state but a few. Subtleties were often designed into the circuits to ensure that the users of the equipment were forced to stay with the manufacturer of the instrument concerned.

This state of affairs could not be allowed to continue indefinitely for two very good reasons:

1. No single instrument maker could ever hope to meet all the requirements of every process industry, or for that matter even a single one.

2. The users of the instruments rightly insisted on exercising their right of choice to "mix and match" equipment for what they considered the best results at the most economical price.

To comply with these user demands, something had to give. Purely for the reason of staying in business, instrument makers were forced to give serious consideration to the compatibility of their products with those made by others, and one of the first things to be done was to arrive at a common transmission signal. A debate over voltage versus current transmission ensued; voltage eventually lost out in favor of current signals, the reason being that a current signal unlike a voltage signal can be transmitted over extremely long distances without significant loss in signal strength. (To be fair, voltage signals are still used, with the limitations of transmission distances recognized, accepted, and included if required, in the control system. In fact, there are instances where voltage is the most convenient signal to use. If transmission distances are long and the signal is a voltage, then a signal converter/booster can always be included in the loop, but there are penalties to be paid in loss of accuracy.)

Today the most widely transmitted signal in the process instrumentation industry is an electrical current of 4 to 20 mA dc, with the 4 mA usually representing either a low or "no" measurement and the 20 mA representing either a high or maximum measurement. Another reason for the choice and popularity of this signal range is the "live zero"; this means that when using any equipment with this signal range, one is confident that when a value of 0 mA is obtained at the input terminals of the receiving instrument, that not only is there a fault, but the fault is outside the receiver and could be either in the transmission line or in the transmitter itself—testing for these alternates being a relatively simple matter. Also, live zero simplifies the solution to low-signal linearity (the low-signal linearity a plus in pneumatics, but bottom-end stability due to the mechanics of the instrument can be a problem). However, signal loops operating entirely on current are a real difficulty and expensive to implement; voltage signal loops, by contrast, are much simpler to add to or alter. The other significant advantage of the 4 to 20 mA signal is that it can be generated at low voltage, typically from 18 to 24 V dc, so that it is suitable for use in those process industries subject to the risk of fire and explosion, such as petroleum oil refining or mining.

Electronics really made a breakthrough with the advent of the semiconductor; this miniature-sized element single-handedly caused a revolution in the whole field of technology. The transistors and circuits being developed and manufactured virtually made the thermionic valve obsolete, and true miniaturization a distinct possibility. Babbage's dream of an advanced *computing engine* was considered a real achievable goal, and instrumentation was another beneficiary of these advancements in solid-state physics. The electrical power to drive the semiconductor was also extremely small, and therefore most attractive, as it avoided the massive heat generation so predominant with electronic equipment in the past. The much reduced physical size and power requirements resulted in instrumentation becoming smaller, lighter, and therefore more easy to handle, with the first signs of advancement being a new range of analog instruments using solid-state electronics offered to the process industries. Equipment was being designed for high response speeds, much enhanced accuracy and repeatability, and much greater robustness though it was physically lightweight.

For the first time it was possible to have switches with no moving parts and capable of being operated at exceedingly high speeds, thus giving a tremendous boost to the application of binary arithmetic to computational problems and thereby opening the door to the world of digital electronics with all its advantages of versatility and ultimate computerization.

SPLIT-ARCHITECTURE CONTROL SYSTEMS: THE FORERUNNER OF DCS

While the capability of using digital electronics in a limited way for instrumentation was there, it was still considered a "novelty" by many. Decision making by the users was overshadowed by their skepticism, for digital systems were in their infancy, as yet untried and therefore viewed with apprehension. The trend at this time was to have sophisticated, yet flexible analog control systems, which gave rise to the concept of "split architecture" as shown in Figure 19.1. Under this design concept, the input, output, control, and calculation aspects of a system were separated from the display (visualization) of the process conditions. Since the measurement/control and display aspects of the system were separated, split architecture can be considered the forerunner of the distributed control systems (DCSs) of today.

A TYPICAL REAL SPLIT-ARCHITECTURE SYSTEM

One of the more popular systems using the split architecture concept was the then advanced Foxboro SPEC 200 series, the acronym SPEC standing for *Simplified*

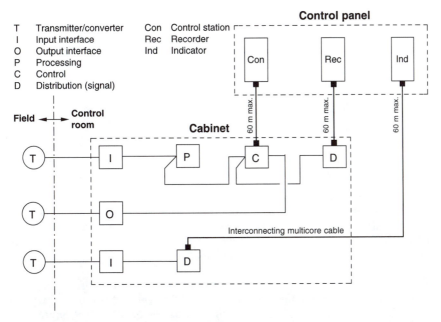

Figure 19.1: Block diagram of a split-architecture analog control system, based on the Foxboro SPEC 200 Series equipment.

Process of *Electronic Control*, and not specification, as incorrectly but popularly thought. We shall consider this system in some detail because of the general similarity it has to modern DCS designs. Under the split concept, the signals to and from the process plant were wired to input and output modules (IOMs), which acted as interfaces between the process plant and the controls. These interfaces were dedicated to specific types of measurement, e.g., current, voltage, resistance, pulse rate, or contact change. The incoming signals were converted to a standard internal operational signal in the range 0 to 10 V dc for display on indicators, control stations, recorders, and other internal functions. If computations and control were required, the scaling, mathematical manipulation, and control operations were acted upon by the system processing and control modules, respectively, and the results of these manipulations were then reconverted to produce outgoing signals in the range 4 to 20 mA and used to control the process. Direct current (dc) pulses to be displayed on electromechanical counters, and volt-free contacts or 24 V dc to drive signal lamps on a hand-painted "mimic diagram" (graphic), or to initiate alarm annunciators, sometimes directly from the 0 to 10 V signal were also provided.

The process input/output interfaces, the control (proportional, P + I, P + D, P + I + D), computation (arithmetic operations of addition, subtraction, multiplication, division, and square root), or processing (high, low, or median selection; averaging; characterizing; alarm functions; etc.) were individual and physically separate modules that could be mounted in cabinets located away from the display panel(s) for power sharing and mutual back-up; the separation distance between the control cabinets and the displays being up to 60 m (200 ft). Within the cabinets the location of the modules could, if so chosen, have been random, but on the contrary, the units were organized into areas, some containing the input/output interfaces only to make wiring to and from the field easy, and an assembly of control and processing modules that made up the control system; the control system being assigned to represent one or more related parts of the process. Each cabinet, for system integrity, had its own power supply, and if more than one cabinet was required, the individual power supplies of each cabinet would be integrated to form a composite whole for the full suite of cabinets, an arrangement allowing mutual backup within the cabinet group, and the integrated power supply to be backed up from either an uninterruptible power supply' (UPS) source on the power distribution grid or a standby dc battery system.

Within the confines of the cabinets, the signals used were voltage variations in the range of 0 to 10 V dc, as mentioned earlier, and manipulations were carried out on these signals. Voltage signals and voltage interconnection were chosen for the following reasons:

- Since there were only short distances between modules within the cabinet, the use of voltage as the manipulated variable was possible.

- The use of voltage signals within the cabinet completely eliminated the problems of current signal-to-load resistance that would have existed were this used instead, permitted the various electronic modules to be wired in parallel instead of in series as would be the case if a current signal were used, and thus made additions and alterations to the loops easier as mentioned earlier.

The input/output interfaces acted purely as converters to change the incoming analog or discrete signals (the latter could be digital variations that were volt-free, or carrying power supply voltage either ac or dc) to voltage variations of 0 to 10 V dc for use internally; and on the outgoing side analog current control signals of 4 to 20 mA, or discrete outputs that were either volt-free contacts, 0 and 24 V dc, 0 and 120 V ac, or 0 and 240 V ac for the "field" or other equipment in the control room. It can be seen that the system designers, with this clever arrangement, took best advantage of the strenghts of the different signal types: current for long-distance analog runs between control system and process plant, and voltage within the control system cabinet and panel itself.

DRAWBACKS OF ANALOG SYSTEMS

These analog distributed systems were a huge success, evidenced by the several thousands installed in process plants worldwide, and the many thousands that are still in full operation today. With the system described so far, one can see that, because it was analog, the following applied:

- Each system had to be custom-built for the task it was to perform.

- Since each function (control or processing) was achieved by a separate module, the connections between modules were made by discrete hard-wiring.

- When it was necessary to combine measurements for computation, each signal required scaling, and it was vital that each signal used in the calculation be an appropriate fraction of the overall signal range.

- If the process parameters were changed, then every computation and/or display scale associated with the change had to be individually modified, and this was an expensive task to implement.

- If a new or modified calculation or display were required, then either new wiring, or rerouting of existing wiring would be necessary. This involved additional implementation cost, and possible loss of revenue in lost production for the period it took to carry out the changes.

- The displays were instruments in their own right, and as such had to be mounted and readable by the operator, and so they could not be reduced below a certain size without sacrificing readability.

- Process graphics could be obtained only by actually, "sign painting" on the control panel, or by using a patent mosaic-type graphic system. Dynamic process data could be shown only via appropriately mounted signal lamps or miniature analog indicators inserted into the painted graphic.

- The cost of the control system display panels, module cabinets, and graphic represented a significant price tag and required considerable floor space in the control center.

In the world of commerce, profitability is always the "name of the game," and it therefore comes as no surprise that instrument users are forever looking at ways to reduce costs. For them, cost reductions could be obtained by being able to manufacture different products using the same process plant and operating the process at points within the product specification that gave the highest "return on investment" (ROI). The instrument makers, too, were being driven by a similar profit incentive, as manufacturing cost reduction applied to them also; but at the same time, meeting their customers' objectives demanded that they, the instrument makers, present to the instrumentation marketplace newer, cheaper, better, and more advanced products. The research and development budget of an instrument manufacturer takes a sizable slice of its corporate funds, and weighted consideration has therefore to be given to every technological advancement, for this is the only way of survival in the unforgiving, commercially driven technological world.

ADDITION OF DIGITAL COMPONENTS TO THE SPLIT-ARCHITECTURE SYSTEM

When the first integrated circuits were being introduced to the marketplace, the event was reflected in the control instrumentation industry by replacement of the hard-wired analog displays with graphic ones shown on a cathode-ray tube (CRT). The electronic display equipment was called the visual display units (VDU) and inevitably led to a reduction in the demand for control panels. The first systems were "hybrid," for the screen displays were digitally driven graphics but the controllers were still analog with digital interfaces that allowed the measurement and setpoint to be inputted to the computer digitally and the controlled output to be fed to an integrated circuit within which a setpoint calculation were performed. This calculated setpoint was downloaded to the analog controller, which manipulated the process accordingly. This method of control is called *setpoint* or *supervisory control*. Figure 19.2 shows the arrangement.

One should not get the impression that control of the process was left entirely to the values obtained by the calculated setpoint. Facilities were designed in to allow the process operator to intervene at any time to make adjustments to the calculated value or, if necessary, manipulate the controlled output directly, with the computing device tracking every operator-instigated change.

Since the VDU controller signals were digital, they were represented on the screen in a format that simulated the front face of the analog instrument they replaced, but with bar graphs instead of pointers and scales. In some instances these displays replicated the instrument face quite realistically (Figure 19.11 later in this chapter shows examples), but the instantaneous readout of values of each bar was presented

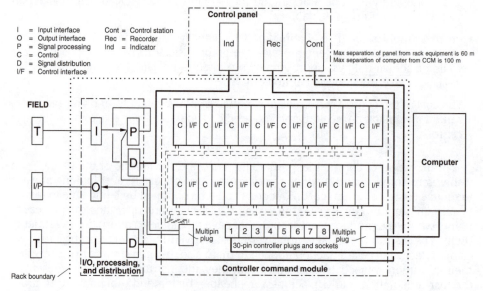

Figure 19.2: Block diagram of split-architecture analog system: Foxboro Spec 200 and Interspec. (Foxboro Spec and Interspec are registered trademarks of the Foxboro Company.)

numerically far more accurately than any operator could obtain with the analog version; while in other instances the displays were nothing more than three slender colored vertical lines with abbreviated descriptive script defining each line and the usual numeric readout of the instantaneous values of each bar; the simplicity of display outweighed their versatility and compact usefulness. The operators soon felt equally at home with the VDU displays as they did with the analog instruments, because of the possibility of having information at the touch of a button rather than having to read it off an instrument scale. All the process alarm functions and computations were now carried out in the integrated circuits, but in critical loops separate alarm trip units could be hard wired in parallel to provide enhanced safety and integrity, and selected process trends could be shown on hard-wired chart recorders that had at that time to be built into the system. The recording facility enabled any measurement from a predefined number of inputs to be selected and plotted; one system allowed any 16 from 100 input measurements to be graphed. The hard-wired chart recorders were necessary because the facilities for "live" VDU-based process graphics were not yet available.

The provision of both analog and digital controllers for a single loop continued for quite a while, the reason being that users of the system were still unsure or sceptical of the digital computer. Behind the suspicions the reason was mainly that business revenue depended on the long-term running ability of the computer, and users were therefore not prepared to take any exploratory chances.

BREAKDOWN OF THE PROCESS INTO MANAGEABLE SEGMENTS

The tried-and-tested method of dividing the process into specific parts developed very early on was enhanced, and resulted in displays that were allocated hierarchically to a particular "plant area," a "group of measurements within that area," and a "control loop within the group." These terms are defined a little further on. Since the system used microprocessors, the ability for the user and/or operator to "page" through the database of signals in an orderly way was built in. There was at least one system—e.g., Foxboro Videospec—on the market, which, through its design actually guided the operator into making the logical choice thereby removing the possibility of possible incorrect actions. The keys on the operations keyboard of this system carried no engraving at all, as their functionalities changed with each display on the VDU screen and were actually defined by the control processor itself, although the actual key functions' "engraving" was screen-simulated, relevant to the particular displayed page. Every system had a *display hierarchy* defined by the system manufacturer, but this did not mean that paging was a rigid sequential affair that had to be followed regardless, as there were facilities included that enabled the operator to access a specific display or control loop by merely keying in the *tag number* or the *loop designator* of the required instrument. (The tag number and loop designators are unique names assigned to each control, indicating, recording loop by the process and/or instrument engineer when developing the control strategy.) Assignment of each of the system inputs, controls, outputs, computations, and alarms for various parts of the process was the prerogative of the process and/or instrument engineers. The process operators were totally barred from making any changes to these, since it would have serious consequences on the process, product, or plant. This "exclusion" or security was a built-in feature of all systems offered and was implemented on the system either by software—i.e., a user-defined password—or hardware (a "key lock") and assigned to specified engineering or design staff.

To appreciate the technology just discussed, let us consider a fairly typical system. The system manufacturer divided the basic software representing the process plant into a maximum of eight manufacturing areas. Each of these areas was then subdivided into eight processing groups, and each group could have up to a maximum of eight control loops. The only option open to the control engineer was the number of loops contained within the group. This would give, by simple arithmetic, a total of 512 control loops and included assignable software indicators, recorders, alarm annunciators, motor control switches, etc., all of which, if required, could be accessed by a single operator, a formidable and daunting task if actually implemented in this way. Larger plants could have more than one such system, in which case the whole plant would be broken down into multiples of the eight areas, each covered by a single control system. Users, realizing the awesome responsibility that would be imposed on just one process operator if such a scheme were to be used, always assigned to their staff responsibility for a limited number of process loops. The result was that each process area was under the care of a team of operators whose job it was to ensure that the process within that area was uninterrupted and the product was held within its specification. In plants where more than one system was required, the whole large scheme was bound together by having the display(s) for the loop(s) in one area that, interacted with others in a different area(s), to have the displays of the loop(s) that caused the interaction repeated in each of the other affected areas, but leaving the control responsibility for the main loop(s) lie with only one operations team. Overall or selective plant-wide supervision and/or control was therefore possible from suitably configured "management consoles."

THE FIRST TRUE DIGITAL CONTROL SYSTEMS

With the enormous amount of money and development effort expended, computers became more reliable and therefore more secure, and from the instrument manufacturer's point of view, it was not such an enormous step to eliminate analog controllers from the system and replace them with easily "accessible on demand," easily reconfigurable control algorithms—i.e., software routines embedded in the system and forming an integral part of it. The first true DCS was designed and implemented by Honeywell and given the name TDC 2000. When Foxboro entered the same marketplace it coined the name *block* for the control algorithm, a very important part of the algorithm design being the provision of "hooks" to which signals—e.g., inputs, outputs, or other algorithms—could be connected by software commands. These algorithms were invoked every time a control action, computation, or alarm, was required, so now all that was necessary to make a control system was to configure the software to connect the input, output, control, calculation, alarm, and other functions together to form a composite whole, the only hard-wiring required being to connect the transmitters, sensors, and final control devices located in the plant to the appropriate input and output modules.

This elegant method of connecting the functions by software commands was achieved initially by a written code having an "English-type" grammar and syntax and referred to as a *high-level language* to distinguish it from the more fundamental codes of *machine language* or *assembly language* based on binary arithmetic of 0s and 1s at which the electronics operate. The reasons for making the system operate on binary arithmetic were that transistors when used in digital circuits are fundamentally two-state devices—i.e., they are either on or off—and more importantly, that they operate at extremely high speeds, they are less prone to signal *amplitude* fluctuations, and the circuits can be "stacked" or repeated and/or interconnected and signals reproduced ad infinitum.

The displays in the early DCSs (as yet, there were no process graphics, active or static) were still based on the CRT, as described earlier, resulting in a system to control a real process plant, which basically required input and output interfaces, the system intelligence being held in the integrated circuitry. The remainder of the hardware items—e.g., keyboard, printers, switches, alarm annunciators—were mounted conveniently nearby. The various basic functions, i.e., the fundamental system facilities and configuration capabilities held within the system are referred to as the *firmware* of the system, with the user designed system configuration, ranging, plant algorithms and constants, and data held on disk. It can now be appreciated that keeping abreast of developments in the world of algorithms is just a matter of making changes to the firmware and ensuring that the changes do not conflict with other algorithms or the hardware, but be more than aware that although, conceptually, the idea of implementing changes is simple, to actually achieve them can most often be quite involved.

Greater sophistication was continually being added to the system, and one of the first things to be included was a graphics capability, and with this enhancement came the ability to draw process mimic diagrams on-screen and to include in them "live," or dynamic, process data such as flow rates, pressures, temperatures, or levels at positions on the picture that more or less represented the actual points of measurement in the process together with switch and status information also.

By the late 1970s computers were becoming fashionable with the general public as well, and in the UK Sir Clive Sinclair played a very important part in making computers at an affordable price available to the masses in the early 1980s. These machines used a high-level language called *BASIC* (*B*eginners *A*ll-purpose *S*ymbolic *I*nstruction *C*ode) invented in the United States at Dartmouth College in New Hampshire. The language deliberately had an even greater bias toward English grammar and syntax, and was therefore available for almost immediate use by science undergraduates and others. The now famous Sinclair Model ZX 80 and the British Broadcasting Corporation (BBC) Model B did not use the original language, but a dialect of the original, hence they were usually referred to as Sinclair Basic and BBC *Basic*. Many of the old, but some of the latest as well, process control systems also use dialects of *Basic* suitably modified to reflect the requirements of the control and instrumentation industry. For example, Foxboro called its language FPB (Foxboro Process *Basic*) and Honeywell called its CL (Control Language). These modifications included some mnemonics—for example Honeywell Control Processor (CP) command "PO $PN" will output command prompts to a printer—which were very highly specialized and applicable to process control. Today's systems are not confined to using derivatives of *Basic*, but alternatively can, if required, be programmed in one of the following High Level languages: Fortran, Pascal, or 'C; for their speed and flexibility, C and its many derivatives such as C^{++} have considerable support among many manufacturers and users alike.

In any operational process, the physical parameters of flow, pressure, temperature, and level are continuously varying, and to maintain control a dedicated controller is assigned for each important parameter. In the distributed control system, controllers are configured when the system is being designed and the appropriate loop configurations, algorithms, scaling, and plant data are switched into by the processor in "sync" with the loop selection and field data in a loop/tag by loop sequence by the *processing unit*. However, there is a limit to the number of inputs, outputs, and controllers that one processing unit can handle, and when this is reached, another processing unit will be required.

In all "real" controlled processes a finite delay occurs between the making of a measurement and the controlled output's effect on the measurement, as shown earlier in Chapter 7. In a digital control system a dedicated controller is not monitoring the measurement continuously as in the analog version. What happens is that the system is arranged to look at each loop momentarily on a fixed rotation basis, perform the task(s) assigned within the period configured, and move on to the next. Because there is a continuous rotation, it is vital that the fast-acting loops are assigned more frequent "reviews" or *scans* so that any changes can be acted on. For example, suppose there are two loops, one a flow and the other a pressure; as we have seen before in Chapter 14 on steam generation, pressure is the faster of the two, hence, it is important that this loop is scanned more frequently. If, say, the flow loop is scanned every half second then, the pressure loop would be assigned a scan period of 0.1 s; in this way any change in pressure will be caught and acted on before the flow changes correspondingly. Usually the delay is greater than the scan period; therefore, any corrective action will be well within the time between successive measurements. It is clear then that a digital or analog control system will have similar effects upon the process, but in practice the digital system has a much greater effect, for the changes are acted upon with much greater precision.

Other advantages digital systems have over analog systems are as follows:

- All computations can be done in *actual* engineering units, and scaling becomes a thing of the past.

- The accuracy of control is limited only by the capacity of the display processor electronics, which is usually about 12 digits.

- Changes of control strategy involve no wiring changes, and when required, can be implemented by the engineer's programming as changes in block configuration.

- Sequence and/or logic control is a possibility with many systems.

- Any inputs can be trend-recorded if required. This facility is not usually implemented throughout for all measurement points as this would encroach on the processing time; although all critical points events and status are recorded.

MULTIPLE-PROCESSOR SYSTEMS: THE NEED FOR STANDARD COMMUNICATIONS

What has so far been discussed is a single distributed *control processor*. This is certainly not the end of the story, for the design has allowed for expansion, and we shall now look at how a single system can be enlarged. Suppose a process is so large that more than one processing unit is necessary; in this situation, some care is necessary in the configuration to ensure that all the control loops are self-contained within a single control processor, but this may not always be possible, and something has to be done to overcome the problem. The solution is to connect the processors together, but this is not as simple as it sounds, for remember, there also have to be a VDU to enable the engineers and operators to view the process, printers to provide logs and reports, and storage devices—i.e., hard and floppy disk drives—to hold long- and short-term data and system application information. It is therefore not difficult to visualize vast amounts of information passing along the interconnection cable, called the *data highway*, a name which, in fact, quite graphically describes the situation ensuing on the cable. As automobile drivers, we are all too familiar with the catastrophic consequences if discipline is not exercised when on a highway;

the same is absolutely true when dealing with data. To ensure orderliness, the data must conform to a specific format, or *protocol*, and must enter the highway at a time that will not cause a collision with other data that could already be there. In fact, the problem of communication can be solved relatively easily for equipment of one manufacturer, which can ensure that the same protocol is used for all its instrumentation, and would also take steps to prevent data collisions. However, when we have instruments of different manufacturers that have to be connected and "talking" to each other, we are at once faced with both differing standards and the question of protection of intellectual property. The situation became so bad for General Motors in the United States that, when it was found to be costing the corporation several hundred million dollars every year just to ensure communication among the different equipments it used, GM was forced to issue an ultimatum to suppliers, saying that unless compatibility of communication between equipment of all makes was guaranteed, the corporation would cease to do business with any defaulters. Thus, money was the lever used to force the issue. To facilitate implementation of its intention, GM agreed to put at the disposal of the equipment manufacturers all the information its engineers had designed, collected, and used over the years; this formed the basis of the *Manufacturing Automation Protocol*, or MAP, for short.

Today almost all DCS manufacturers use the MAP protocol and the ISO Seven-Layer Model for communications. This aspect of the system is a huge field in its own right, and we shall not go into it in any depth at this time, as we shall be discussing communications in more detail in Chapter 21. For the present, suffice it to say that layers 1 through 4 take care of the physical properties and transmission, and layers 5 through 7 the applications, with operators and engineers interfacing with the system at these upper layers. To give some idea of the data transmission speeds, the carrierband LAN (local area network) operates at 5 Mbit/s (megaBITS per second) and the broadband LAN at 10 Mbit/s—i.e., up to about 1 million characters per second or in computer users parlance about one megabyte (1 MB) per second.

To ensure that data from a control processor reaches the workstation processor (the operators VDU), a technique called *token passing* is implemented, which allows only one device on the network at any time and is applicable to single or multiple processors. The method operates in a manner analogous to the baton in an athletic relay race and is guaranteed by having one and only one "token" (baton) circulating on the network. The *communications processor* (Comms.) has a specific layer, which is the recipient and transferrer of the token, and the retention time of the token by the processor is predefined and invariable. During the period that the token rests with any specific processor, it can transmit or receive data, it is inhibited at all other times. A processor in possession of the token uses the allocated time and then hands the token to the next processor in the sequence. Communication between control processors on the same network takes place in an identical manner; in this case it is called peer-to-peer communication. Devices operating directly on the network are called *nodes*, and one should be aware of the differences in what the boundaries of a node encompass, as they do differ from one manufacturer to another. Of necessity, the length of an individual network is limited by data and signal integrity, but this really does not pose a problem, for with network extender interfaces, these lengths can be increased to cope with the largest plants. In order to keep the speed of communication as high as possible, the number of nodes on a network is limited, and it is recommended that the amount of peer-to-peer communications also be kept to the minimum.

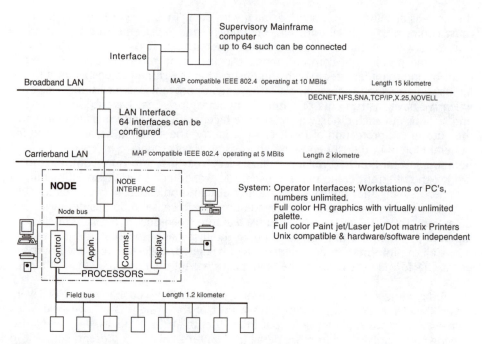

Figure 19.3: Simplified topology of the Foxboro Intelligent Automation system. Shown is a single node, which can be a whole control room for a particular process. The system can have 100 such nodes. The hardware can operate between the temperatures of −40 to +85°C. For simplicity, the system shown is nonredundant; full redundancy can be provided if required. For convenience the I/O has been shown remote; the I/O can be located along with the processors.

TOPOLOGIES OF REPRESENTATIVE DCSs

As a taste of the variety of DCSs available, the simplified block diagrams in Figures 19.3 to 19.8 have been provided to show the topology of some of the well-known major players in this field. These illustrate the concepts that have been discussed so far; it is hoped they will help consolidate the ideas.

The Foxboro I/A Series (Intelligent Automation) (Figure 19.3) was the system that broke the traditional practice of having the system located in a central control room and propriety operating system by using instead a Unix-based one. The arrangement is totally modular and capable of being field mounted, literally distributed throughout the plant; note the operating temperature limits of −40 to +85°C, also note the lengths of the carrierband and broadband LANs without interfaces being required and the fact that they are both MAP compatible, and the number of standard protocols that can be accommodated. Because of the operating system, the I/A Series is not application software dependent. Note the node arrangement.

Honeywell was the first company to present the DCS to the world. Their system (Figure 19.4) has undergone two major refinements since first being on the market. The system is modular also but requires a temperature and environmentally controlled control room to avoid problems with the equipment. Hardened (capable of working in higher temperature locations) workstations are available, and steps have been taken to obtain software independence. Note the different arrangement regarding nodes and the requirement of a propriety LAN protocol at some levels and the

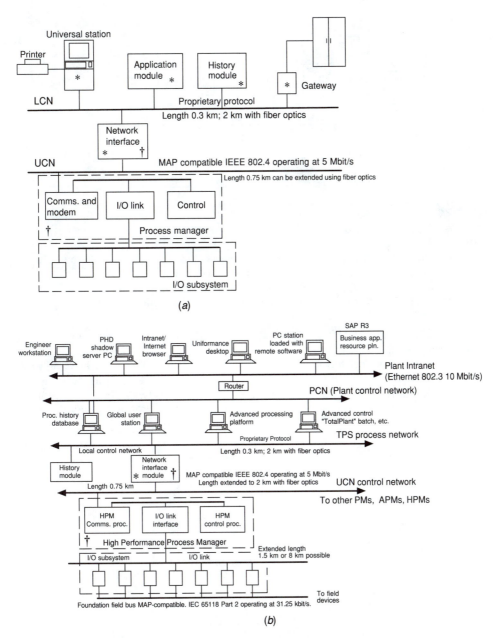

Figure 19.4: Honeywell's DCS system. (*a*) Simplified topology of the TDC 3000 system; (*b*) simplified topology of the TPS system. For both systems, items marked with an asterisk are nodes on the LCN (local control network). Items marked with a dagger are nodes on the UCN (universal control network). Maximum number of single devices on a UCN is 64. Maximum number of devices on an LCN is 40 without extenders, 64 with extenders. For the TPS, system operates on Windows 95, 98, or NT. All equipment on the PCN down to the foundation field bus is by Honeywell and runs on NT only; other equipment is by the vendor of choice. Most users of PHD access shadow server on main net.

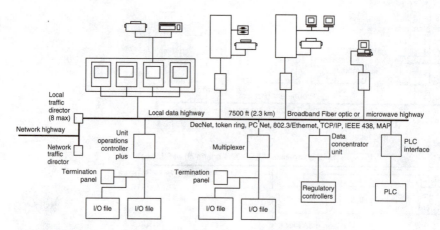

Figure 19.5: Simplified topology of the Fisher Provox system. There are three basic multiloop controllers: the UOC + for advanced sequencing and batch control, the IFC for mainly continuous control and limited batch control and sequencing, and multiplexers providing interface between Provox and process monitoring and final elements. Up to 30 devices can be connected to the local data highway.

use of the Foundation Fieldbus (one of the accepted standards) at the lowest level in the latest offering.

Honeywell has updated its TDC 3000 shown in Figure 19.4a system and now offers its Total Plant Solution (TPS) system shown in Figure 19.4b. Both systems are shown, for many TDC 3000 systems are still in use worldwide, and migration to TPS is available. There are changes in the upper levels of the LANs, which extend

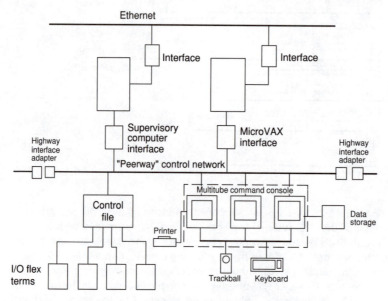

Figure 19.6: Simplified topology of the Rosemount System 3 control system. The control file contains eight multipurpose controllers/file. Maximum separation of I/O flex terms from control file is 5000 ft.

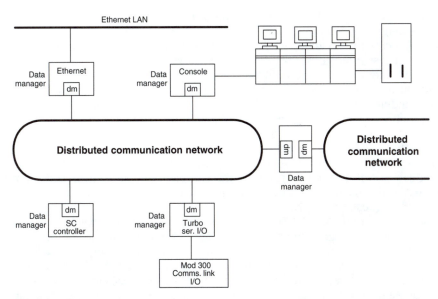

Figure 19.7: Simplified topology of ABB/Taylor MOD 300 system. Every node has a data manager structure (dm = data manager). Controllers: 6000 Series; 860 points using direct I/O. SC: Use remote I/O (TRIO) to control 2100 analog, 11,000 digital I/O. MOD 300 Turbo panel: 960 digital, 180 analog. Several distributed communication networks are connected to present a large network. Instrument communication network uses a 10-bit string (8 data, 1 start, and 1 stop).

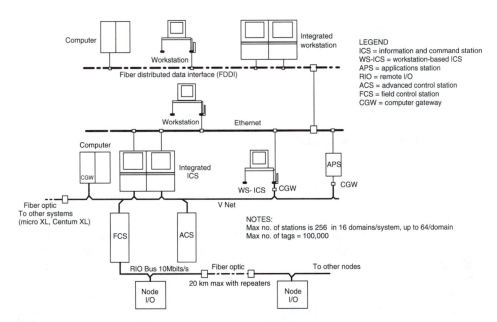

Figure 19.8: Simplified topology of Yokogawa Centum CS system.

the system in these areas. In the figure some of the acronyms not shown on the diagram are defined as follows:

PHD = process history database node; HPM = high performance process manager; PM = process manager; APM = advanced process manager.

With the Fisher Provox system (Figure 19.5), note the fact that there is only one LAN, which is MAP compatible and can also accept other standard protocols. Control is modular and achieved by using a selection of individual unit controllers (UOCs). Note the number of nodes permitted. In the diagram the undefined acronym PLC is interpreted as a programmable logic controller, and the acronym IFC is interpreted as an integrated function controller.

The Rosemount System 3 (Figure 19.6) is a relatively straightforward DCS using individual control files (name given by the company for the control algorithm) that connect to the plant input/output interfaces. Once again note the manner of connecting the nodes to the LAN.

The Taylor system (Figure 19.7) requires the configuration to be carried out in a selection of dedicated programming languages. These are: configurable control functions (CCF) for process control, Taylor ladder logic (TLL) for logic control processing, and Taylor control language (TCL) for complex arithmetic, serial interfaces to other manufacturers equipment, startups and shutdowns of the process, etc. Note the configuration of the LAN, which is quite different from all the others we have seen so far.

The Yokogawa system (Figure 19.8) is an advanced one. Because of the importance attached to the unit in the systems sense, the company has taken the initiative of including redundancy as a standard on the field control station (FCS) since regulatory, sequence, subsystem integration, user C language functions, and field instrument management functions are included in this unit. The networks follow the conventional layouts of those we have seen before.

OPERATION OF A TYPICAL DCS

Having now had the opportunity of seeing real system topologies, let us try to understand how these work. For the purpose of this exercise, let us concentrate on the Foxboro I/A Series (the author is most familiar with this system). Figure 19.9 should help in developing the ideas.

A TYPICAL CONTROL LOOP

We shall first consider a simple indication loop, and then a typical control loop. For this part of the exercise we should at first have a clear idea of what we want. To assist, let us prepare simple loop diagrams of the requirements; although as we become more familiar we may not need this step. However, experience has shown that diagrams are always helpful, especially when there are any complexities involved in the control strategy. These would look something like those shown in Figure 19.10.

This system uses actual engineering units throughout; hence, scaling is not a concern, except in those instances where conversion from one engineering unit to another is required, e.g., °C to K. For the indicator loop, the measurement signal—it could be an analog current or voltage—is transmitted to the system, where the first item of hardware encountered is the Field Bus Module (FBM), which is rack-mounted in the control room. At this point the signal is conditioned so that it is recognized throughout the system. The first software block encountered will be an Analog INput

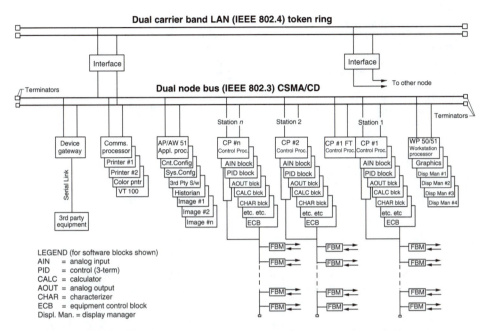

Figure 19.9: Detail of Foxboro Intelligent Automation system.

(AIN). The AIN is connected to an indicator (IND) faceplate (see Figure 19.11), which is a graphic that depicts an instrument showing the measurement as a variable vertical bar graph together with a display of its actual numerical value. The length of the bar corresponds to the value of the measurement at the time. Its unique assigned tag number and a brief description identify the measurement.

At this point it is also possible to configure alarms that are associated with the measurement; the alarms are in reality limits on the process that must be adhered to (not violated), as a violation could result in a product that will not meet specification, or the inception of safety or other problems that could be encountered in the production cycle. The operator will always be advised of an alarm condition by audible and visual means, and this state with details will be logged automatically. The visual indications could be either a change in color, or a flashing symbol, or text that changes color and flashes in the display or in a process graphic, if one is drawn and the measurement is included in it. Choice in the display of alarm alternatives is

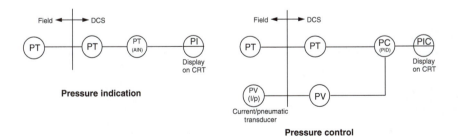

Figure 19.10: Typical indication and control loops.

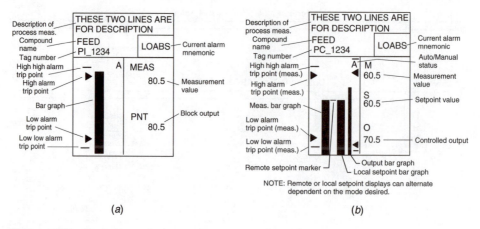

Figure 19.11: (*a*) Indicator faceplate. (*b*) Controller faceplate.

provided, and left to the discretion of the user. For convenience, Figure 19.11 shows typical single indicator and controller faceplate displays on the I/A Series system.

Alternatively, the IND faceplate can be configured as a multiple indicator of a maximum of three variables, which at a glance provides a means of comparison between the different measurements and have been found to be very useful. It is not recommended that multiple indicators have alarms, since it is not possible to segregate these to determine a clear alarm state for an individual measurement. For the control loop, the measurement enters the system via the FBM and then through an AIN block before it is applied to the PID control block—the acronym PID used defining the three control terms that are available if required and configured. The block diagram of the control block shown in Figure 19.12 is provided with all the functionality of a conventional controller, plus many additional embedded features not usually found on the equivalent hardware instruments.

The following partial listing of features provided in these powerful control algorithms should give a flavor of what can be expected:

• Alarms on the measurement

• Alarms on a preconfigured difference between the measurement and the setpoint (deviation alarm)

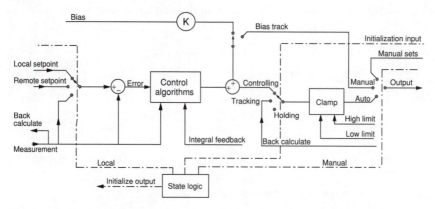

Figure 19.12: Simplified block diagram of control block.

- Alarms on the controlled output

- Choice of linear or nonlinear control

- Choice of self-tuning, i.e., the controller's ability to adjust the control terms without human intervention

- Possibility of inhibiting controller saturation (reset windup) when the output is disconnected

Very briefly, all that is required to implement a control scheme is to (1) make the appropriate software connections, (2) save the configuration, and (3) turn the scheme on.

Compare the diagram of the software controller in Figure 19.12 with the block diagrams of the classical controller that were given in earlier chapters, and see the similarities in the functionality provided. It is not possible, without complicating the diagram, to show the described enhancements available in the algorithms.

There are other blocks in the system that are capable of providing the control engineer with the means of implementing the most sophisticated control systems that may be needed. One of the latest additions to the control block structure is the *PIDA* multivariable control algorithm—the "A" in the acronym standing for "adaptive." This block allows implementing a controller in a loop that is influenced by other process variables to be continuously and adaptively tuned to bring stability to the loop, and hence to the process. The "disturbing" loops can be arranged as feed-forward and/or feedback systems to allow the best arrangement to be selected to form a complex yet stable controller.

PROCESS ALARMING

All process alarms that are tripped are automatically logged as they occur, and shown on the printers attached to the communications processor, with every entry date and time stamped so that a complete record is obtained. Every measurement, whether in alarm or not, can be stored in the "historian." However, because of the cost and storage size, one must use discretion on how much data is accumulated. The information stored in the historian can be configured to be presented in different ways; it can be printed out as individual records, retained and retrieved when required, or periodically issued locally or at specified locations as *reports* summarizing process conditions occurring within specific time intervals. The summarized data based on such timed periods is called a *reduction group*, which can also be archived in the historian. The math involved in producing the reduction group is freely configurable by the user and has to be designed specifically to give the most meaningful information to those needing it. The records of process parameters, materials used, etc., can also be presented as printed reports of product quantity over predefined time intervals; but the computations involved, and the location at which the report is to be printed out have to be designed and the system configured accordingly.

From the system overview of Figure 19.9, it will be seen that the control processor (CP) via the field bus modules detects any change in the process. The CP broadcasts any changes throughout the node via the node bus as soon as detected, so that every interested processor in the node is made aware of the change. It is in this way that any actuated process alarms are directed to the appropriate printer attached to the communications processor and the alarm history held in the application processor (AP).

The initial configuration, correction, or later modification of the control system is made using a display screen, keyboard, and pointing device—trackball or mouse—attached to the CP. Configuration is a process of specifying and setting the various parameters that form the operating boundaries of the algorithm within the processor. The method of accomplishing this task is via data entry into a predefined table appropriate to the algorithm used. As a very limited example of this task, each algorithm used is:

- Assigned a type—e.g., indicator (IND), calculation (CALC), control (P, PI, PID, etc.).

- Given a unique loop name in the form of a tag number by which it will be recognized throughout the entire system (tag duplication within the same loop will result in an error, of which the system will advise).

- Assigned to a *compound*, which is a collection of loops (a loop consisting of measurements, and/or control and/or computations) associated with a particular piece of plant or process. The compound is given a unique name by which it will be recognized throughout the entire system. Reference to a compound at any time will present, to the person making the reference, all the contents contained within it.

In addition,

- Each block is assigned an upper range value (urv) and a lower range value (lrv); these are the limiting values of the scale range for the measurement.

- If required, the measurement is assigned an alarm type (absolute, differential) and trip points—values that are within the scale range, but transgression of which will signal an alarm condition and clamp limits if appropriate.

- The block is given a *scan period*, which is the frequency at which the system will monitor the algorithm.

The implementation procedures will be explained in more detail in Chapter 20. Fair warning is given that system configuration is a long and sometimes boring chore, but one that must be done painstakingly to enable the system to perform its duty to the maximum of its potential.

A point worth remembering here, is that because the compound, loop, and group names are unique, access to them is immediate when any of the names are declared to the system. The final configuration is copied and held as an image in the appropriate AP using a procedure called *checkpointing*, which must be carried out on completion of the configuration. It cannot be emphasized enough that checkpointing must be carried out each time the configuration of the CP is modified in any way, for it is in this way that the most up-to-date image of the CP is always held in the AP, and that system integrity is guaranteed at all times. In the event of a CP failure—a rare occurrence, but it does happen—the system can be rebooted and restored fully directly from this image.

Another way of making the system less prone to catastrophic failure is to build in redundancy or fault tolerance, as it is commonly known. The control processor CP #1, shown in Figure 19.9 is fault-tolerant, as indicated by the acronym FT in the description, with both the CPs carrying identical configurations. Transfer of process information, which is totally transparent to the user, is made to both processors simultaneously. In the event of a processor failure, the operator is made aware as soon as it occurs, to enable steps to be taken to remedy the situation; but the control system continues uninterrupted with the task of maintaining the process and the overall plant in the meanwhile. The idea of fault tolerance can be implemented throughout the entire system, i.e., from the FBMs through the communications networks, right up to the supervisory computers, but there is a financial burden that the user must consider for this level of security.

Modern DCS control is so advanced that things, which a few years ago seemed impossible to achieve, are now an everyday occurrence. For example, it is possible for a CP located in one node to access data from any other relevant CP located in another node, but please be aware that this is peer-to-peer communication, which will always slow the data flow; hence, care is advocated to minimize any delays to the control of the process while this is occurring.

The process operator has the ability through the workstation processor (WP) to:

- View the entire process via a visual display unit (VDU), but intervention in an unassigned process area is restricted (i.e., "permit to access" has not been granted), for reasons of safety and product consistency.

- Alter setpoints within restricted limits via a keyboard, to provide smooth manufacturing operation.

- Freely pick a particular process aspect for detailed viewing via mouse or trackball.

- Acknowledge all process alarms. While alarm acknowledgment is freely allowed to the operator, in some important instances only supervisory or engineering staff must for safety or product consistency carry out resetting an alarm.

- Start and stop drive motors as necessary, but in some instances for safety and/or product consistency restart may have to be carried out by supervisory or engineering staff. The consequences of stopping a motor should be allowed for in the system design.

PROGRESSIVE TRENDS IN DCS TECHNOLOGY

EASE OF DATA MANIPULATION: MICROSOFT WINDOWS NT

One of the current topics is *system openness*, which means the ability of the system to use hardware and/or software that is not vendor (system supplier) specific. We shall be considering the implications of this when we discuss communications in Chapter 21. Another recent subject, and one that has gained importance is, ease of operation, i.e., the ease of data manipulation that PC (personal computer) users obtained when Microsoft's Windows and Windows NT entered the world of process control. Foxboro, with its I/A Series system, was the first to exploit this involvement, with a powerful partner who was up to that time unrecognized in the process control field. All the larger systems suppliers now have offerings with Windows NT—the multitasking version that is most suitable for process control applications. A totally unmanned plant will, in the author's opinion, remain just a pipe dream; but a process plant that is supervised and run by the very minimum of operators will one day be a very distinct and real possibility. Manufacturing processes are continuously growing more complex, and since process operators will always be needed, the type of person employed will be vastly different from those previously or even currently involved. The work will be much more technically demanding, and personnel will therefore have to be suitably qualified educationally, physically, and psychologically.

THE FIELDBUS CONCEPT

Before we leave this chapter let us look at how the process measurement is brought into the system. We know that all electrical equipment requires a minimum of two wires to provide the motive energy path to or from the device; the same is also true for all electrical and electronic instruments. In view of this, one can readily see that there will be many hundreds of pairs of conductors used in connecting the various measuring devices to the system I/O. All this cabling has a price tag attached; for not only are there cable costs, but also the cable trays, junction boxes, trenching, if required, and testing to be allowed for. In large plants the cabling costs alone run into several thousands of dollars; so any means of reducing this expenditure would be welcomed by users.

The incentive for attaining this objective has been great and pushed along quite hard by the large users, e.g., ICI, Shell, Exxon, and several others. The outcome of all this activity is the development of the *fieldbus* (this is the external bus that connects the system interface module, i.e., FBM to the field-mounted equipment for example, transmitters and final control devices and not the internal system fieldbus) idea, which operates on the same principle as the data highway we spoke about earlier basically requires only two conductors between the system and the multidrop junction. However, there are many other aspects to be taken into consideration before the fieldbus becomes a reality. Much work has been done in this area to date. In Europe a committee of interested parties under the name EUREKA has made significant progress; in the United States the Instrument Society of America ISA SP65 committee is also devoted to attaining the same goal. The task is not an easy one, for what is required and being developed is a universally acceptable specification. Some of the multitude of important considerations to be taken into account are:

- The communications protocol for the data

- The power-handling capability

- The intrinsic safety (IS) requirements

- The number of drops to individual instruments

- The type of cable to be used, i.e., coaxial or twisted pair

The main difficulty lies in obtaining a compromise that is feasible and acceptable to all in the short term, but at the same time capable of being developed further economically in terms of finance, time, and technology.

Many would say that programmable logic controllers (PLCs) do already have a fieldbus, but this is not totally correct, for measurements have to be brought into the machine by pairs of conductors. However, several PLCs can be joined together via a bus using an agreed-upon protocol, of which there are a few versions—e.g., Modbus, Profibus, FIP. Communications are possible between the PLCs, but only if the interconnecting protocols are the same. Once again, we can see that the same problems exist in the PLC scenario as they do in the distributed control system world. In fact, there is currently a dialog taking place between the PLC communications people and their European and US fieldbus counterparts. The concepts of a fieldbus have been demonstrated at a practical level and the principles generally accepted by all; but, in spite of this, there are still some points of disagreement. There is presently on the market at least one version (HART) of the technology, but the fieldbus project goes much further than that already available.

With such power available, the DCS has uses far beyond the mere business of process control. It is, in fact, a complete information-gathering, manipulating, and disseminating system that operates in real time. The DCSs of today have the power and capability of making use of all the data available. This data is nothing new—it has always been there, the truth being that never before have we had the vision or capability to exploit it. The limitations to our view of the possibilities lay squarely in the reality that we formerly simply did not have such enabling technological advantages at our disposal.

SUMMARY

1. To achieve repeatable consistency of product it is necessary to maintain constant and repeatable conditions in the manufacturing process itself.

2. Early instrumentation was limited to indications of flow, pressure, temperature and level. It did require the intervention of a human operator to manipulate the regulating devices to alter the parameters affecting the product. Later the indicators were fitted with a transmitting mechanism, which enabled the measured parameter to be read at a distance. The medium used to achieve this was air. Initially there were a number of pneumatic signal ranges available; it was some time before a universally acceptable signal range was adopted in response to user demands. The most common ranges today are: 3 to 15 psi or 0.2 to 1 bar or 20 to 100 kPa.

3. Some of the major issues to be considered when pneumatic instrumentation is used:

    • The need for a reliable source of compressed air, requiring two compressors and air receivers in an installation

    • The cost of the air supply equipment represents a sizable portion of the instrumentation cost

    • The necessity for conditioning the air so that it is clean, dry, and oil-free

    • A maximum transmission distance for any single instrument of the order of 700 to 1000 ft (200 to 300 m) to ensure good response.

    • Inaccurate scaling of the analog signals used in computations

4. Electronic instruments perform the same functions as their pneumatic counterparts, but allow much greater transmission distances, faster response, and far fewer mechanical components that require alignment and adjustment.

5. Initially, electronic instruments, like the pneumatic instruments, were offered in a multiplicity of signal ranges. Circuit subtleties were included to ensure that users were forced to stay with the instrument manufacturer for any required expansion of the control loop. The monopoly held by the instrument maker was broken for two very good reasons: instrument users insisted on their right of choice to "mix and match" equipment for what they considered the best results at the most economical price; and no single instrument maker could ever hope to meet all the requirements of every process industry.

6. To achieve universality of use for electronic instruments, one of the first things necessary was a common transmission signal. Voltage signals lost out in favor of current signals, because current signals can be transmitted extremely long distances without significant loss in signal strength. The most widely accepted electronic transmitter signal in the process control industry today is an electrical

current of 4 to 20 mA, where the 4 mA represents minimum value or no measurement and the 20 mA represents maximum value of the measurement. One of the reasons for its popularity is the live zero of this signal range. Also the 4 to 20 mA signal can be generated at low voltage, typically 18 to 24 V dc, making it suitable for use in industries prone to risk of fire and explosion, such as petroleum refining or mining.

7. The semiconductor caused a revolution in the whole field of technology. The transistors and circuits being developed and manufactured virtually made the thermionic valve obsolete and true miniaturization a distinct possibility. Instrumentation was a beneficiary of these advancements, and because the electrical power to drive the semiconductor was extremely small, it avoided the massive heat generation so predominant with electronic equipment in the past. The much reduced physical size and power requirements resulted in instrumentation becoming smaller, lighter, and therefore more easy to handle, with the first advancement being a new range of analog instruments using solid-state electronics offered to the process industries.

8. Analog control systems began to use the concept of split architecture; under which the input, output, control, and calculation aspects of a system were separated from the display (visualization) of the process conditions. Since the control functions could be separated from the displays, by a distance of up to 60 m (200 ft), it could be considered as the forerunner of the distributed control system. With split architecture, all incoming signals to the system were converted to a standard internal operational signal in the range 0 to 10 V dc for display on indicators, control stations and recorders, and other internal functions. If computations and control were required, the scaling, mathematical manipulation, and control operations were acted upon by the system processing and control modules, respectively, and the results of these manipulations were then reconverted to produce outgoing signals in the range 4 to 20 mA and used to control the process. Direct current (dc) pulses to be displayed on electromechanical counters, and volt-free contacts or 24 V dc to drive signal lamps on a hand-painted "mimic diagram" (graphic), or to initiate alarm annunciators, sometimes directly from the 0 to 10 V signal were also provided.

9. The early analog distributed systems were a success, but because they were analog, every system had to be custom built using separate modules for each function, and interconnections between modules had to be made by discrete hard-wiring; computations when required involved scaling for each measurement signal used, and every change could entail modified calculation and a lot of physical wiring or rewiring. Process graphics could only be obtained by sign painting on the control panel or by painting on a mosaic graphic system.

10. The early distributed control systems were usually configured as hybrid systems, because of user skepticism of the digital computer. The screen displays were digitally driven graphics; but the controllers were still analog with digital interfaces that allowed the measurement, and setpoint to be inputted to the computer digitally and the controlled output to be fed to an integrated circuit within which a setpoint calculation was performed. This calculated setpoint was downloaded to the analog controller, which manipulated the process accordingly. This method of control is called setpoint or supervisory control.

11. Every distributed control system has a display hierarchy defined by the system manufacturer, and means are included to enable access to a particular display

or control loop by keying in the required tag number, which is a unique name assigned by the process or instrument engineer when developing the control strategy. Process operators are excluded from making any changes that would have serious consequences on the process, product, or plant. Security is maintained through either a user-defined password or a physical key lock assigned to specified engineering or design staff.

12. In one typical control system, the process plant was divided into eight areas; each of these areas could then be divided into eight groups, and each group could be divided into a maximum of eight control loops. This would give a total of 512 control loops, all of which could be accessed by a single operator. This was not usually done; instead, responsibility for whole plant areas or parts of plant areas was assigned to several operators.

13. In the first true DCSs, the analog controllers were eliminated and replaced with control algorithms embedded in the system as an integral part. Each algorithm was provided with "hooks" to which signal inputs, outputs, or other algorithms could be connected. The algorithms were invoked every time a control action was required. By connecting the input, output, control, calculation, and alarm functions together, a complete control loop was assembled.

14. The required functions were connected by commands usually written in a high-level language, a language that had an English-type grammar and syntax. (The electronics operated at a much more fundamental level based on binary arithmetic.) Many older process control systems used dialects of *Basic*, modified to reflect the requirements of control and instrumentation. Modern systems are not confined to using *Basic*, but provide the choice of programming in a number of different high-level languages such as Fortran, Pascal, C, or C++.

15. There were no process graphics, active or static, in the early DCSs, but as greater sophistication was continually being added to the system, one of the first things to be included was a graphics capability. With this enhancement came the ability to draw process mimic diagrams on-screen and to include in them live (dynamic), process data such as flow rates, pressures, temperatures, and levels at positions on the picture that more or less represented the actual points of measurement in the process.

16. Scanning is the method by which a control processor is able to read in many measurements sequentially and provide associated controlled outputs. For a particular process parameter the time period assigned for this to be done is the scan frequency.

17. Digital systems have greater flexibility and power than analog systems. Some of the many advantages are:

   a. No scaling is required. All computations are done in actual engineering units.

   b. The accuracy is limited only by the capacity of the display processor electronics, which is usually about 12 digits.

   c. No wiring changes have to be made in the event of control strategy changes.

   d. Sequence and/or logic control is a possibility with many systems.

   e. All inputs can be trend-recorded if required. Although this feature is not usually implemented, critical points are recorded.

18. When a process is so large that more than one processing unit is necessary, the solution is to connect processors together, but vast amounts of information passing along the interconnecting "data highway" slows the transmission down. In such situations careful configuration is required to ensure that the control loops are not spread across different processors, but are self-contained within a control processor to minimize the amount of peer-to-peer communication. To ensure orderliness, the data must conform to a specific format or protocol and must enter the highway at a time that will not cause a collision with other data already there. To ensure that data, say, from an input interface reaches the control processor and from there proceeds to the workstation processor, one method is to allow only one device on the network at any time, using a technique called token passing, which operates in a manner analogous to the baton in an athletic relay race and is guaranteed by having one and only one "token" (baton) circulating on the network. The retention time of the token by the processor is predefined and invariable. During the period that the token rests with any specific processor, it can transmit or receive data, but it is inhibited at all other times. The communications processor (Comms.) has a specific layer, which is the recipient and transferrer of the token.

19. The *Manufacturing Automation Protocol* (MAP) standard communications protocol used today by almost all DCS manufacturers was forced on instrument makers by financial pressure exerted by GM. Within MAP is the ISO Seven-Layer Model for communications. Layers 1 through 4 taking care of the physical properties and transmission, and layers 5 through 7 the applications. Operators and engineers interface with the system at the upper layers. The data transmission speeds are: the carrierband LAN (local area network) 5 Mbit/s, and the broadband LAN 10 Mbit/s—i.e., up to about 1 million characters per second or about one megabyte per second.

20. Using a particular system for a control loop, the measurement enters the system via an interface, the field bus module (FBM), and then through an analog input (AIN) block before it is applied to the PID control block. The acronym PID used defining the three control terms is provided with all the functionality of a conventional controller, plus many additional embedded features not usually found on the equivalent hardware instruments. Some of features provided in these powerful control algorithms include: alarms on the measurement; alarms on a preconfigured difference between the measurement and the setpoint; alarms on the controlled output; choice of linear or nonlinear control; choice of self-tuning; and inhibiting controller saturation (reset windup). The latest addition to the system control block structure is a multivariable control algorithm (PIDA). This block allows implementing a controller in a loop influenced by other process variables to be continuously and adaptively tuned to bring stability to the loop, and the process. The disturbing loops can be arranged as feed-forward and/or feedback systems to allow the best arrangement to be selected to form a complex and stable controller.

21. In one system, all process alarms that are tripped are automatically date and time stamped, logged as they occur, and shown on the printers attached to the communications processor; thus, a complete alarm record is obtained. Every measurement, whether in alarm or not, can be stored in the "historian." However, the cost and storage size dictate how much data is accumulated. The stored information can be configured to be presented in different ways; it can be

printed out as individual records, retained and retrieved when required, or periodically issued locally or at specified locations as summarized *reports* of process conditions occurring within specific time intervals. The summarized data based on such timed periods is called a reduction group, which can also be archived in the historian. The math involved in producing the reduction group is freely configurable by the user and has to be designed specifically to give the most meaningful information to the recipient(s). Printed reports of product quantity over predefined time intervals using the records of process parameters, materials used, etc., can also be presented, but the computations involved, for example, parameter averages, departures from acceptable variations, etc., and the location at which the report is to be printed out have to be designed and the system configured accordingly.

22. The initial configuration, correction, or later modification of the control system is made using a display screen, keyboard, and pointing device—trackball or mouse. Configuration is a process of specifying and setting the various parameters that form the operating boundaries of the algorithm within the control processor. The method of accomplishing this task is via data entry into a predefined table appropriate to the algorithm used. In one system this consists of: assigning an algorithm (block) type, giving it a unique loop name in the form of a tag number by which it will be recognized throughout the entire system, assigning the block to a *compound*, which is a collection loop (the loop consisting of measurements and/or computations) associated with a particular piece of plant or process, and giving the compound a unique name by which it will also be recognized throughout the entire system. Additionally each block is assigned an upper range value (urv) and a lower range value (lrv), which are the limiting values of the range for the measurement; if required, the measurement is assigned an alarm type (absolute, differential) and trip points; and the block is given a scan period. The final configuration is copied and held as an image in the appropriate application processor (AP) using a procedure called checkpointing, which must be carried out on completion of the configuration and each time the configuration of the control processor (CP) is modified in any way, so that the most up-to-date image of the CP is always held in the AP, and that system integrity is guaranteed at all times. In the event of a CP failure, the system can be rebooted directly from this image.

23. Another way of making the system less prone to catastrophic failure is to build in redundancy or fault tolerance. The control processor is duplicated, and both the CPs carry identical configurations. Transfer of process information, totally transparent to the user, is made to both processors simultaneously. In the event of a processor failure, the operator is made aware as soon as it occurs, to enable steps to be taken to remedy the situation; but the control system continues uninterrupted with the task of maintaining the process and the overall plant in the meanwhile. Fault tolerance can be implemented throughout the entire system, i.e., from the FBMs through the communications networks, right up to the supervisory computers; but there is a financial burden, which the user must consider.

24. Electrical and electronic equipment requires a minimum of two wires to provide the energy path to and/or from the device. There will be many hundreds of pairs of conductors used in connecting the measuring devices to the system I/O. The fieldbus (this is the external bus that connects the system interface module, i.e., FBM to the field-mounted equipment, for example, transmitters and final control devices and not the internal system fieldbus), which operates on the same principle as the data highway, basically requires only two conductors

between the system and the multidrop junction. The fieldbus is not yet totally finalized, for what is being developed is a universally acceptable specification for:

- The protocol for the data
- The power-handling capability
- The intrinsic safety (IS) requirements
- The number of drops to individual instruments
- The type of cable to be used (coaxial or twisted pair) and a whole host of others.

CHAPTER 20

Implementing Control Schemes with a DCS

INTRODUCTION: USING THE FOXBORO I/A SERIES TO ILLUSTRATE DESIGN OF A CONTROL SCHEME

In this chapter we shall be considering the methods used to implement a control scheme using a distributed control system (DCS). We shall be using the Foxboro I/A (Intelligent Automation) Series as the vehicle, since it is the one most familiar to the author. The techniques involved however, do not change radically between the various system manufacturers. The Foxboro Company was the first to develop, and adopt *control blocks* as a means of implementing control schemes and also coined the term. As mentioned in Chapter 19, when we introduced distributed control systems, control blocks are really very powerful algorithms that have the means of being connected to process measurement devices located in the plant and to each other in a desired configuration to suit the application. The control strategies are limited only by the ingenuity of the control engineer; the system should be able to cope in almost all cases. It must also be remembered that the Foxboro I/A Series system is much more than a DCS; the built-in sophistication makes it a complete Management Information System (MIS).

The blocks and supporting literature are designed so that the person who inputs the data, or "implements the scheme," as it is usually put, is guided throughout the task. The blocks present themselves as a tabulated list of requirements against which the "implementer" has to assign appropriate answers. It must be stated from the outset that this job of data input is not without its measure of repetition, which some will undoubtedly find boring. To somewhat ameliorate the inevitable boredom, copying facilities have been included in the latest versions of software to minimize the onerous task of repeatedly entering similar data.

CONTROL SCHEMES

One or more control schemes dedicated to a control requirement that is designed to operate the process equipment within its capability and give a particular required result to the product brings about process control. The results obtained at the various stages of product manufacture are based on the overall process conditions prevailing at each stage, which in turn have their foundation in the laws of physics, chemistry, and mathematics. While we have in this text tried to limit the amount of theoretical work in the applicable scientific discipline drawn upon to produce the product, readers should investigate further the relevant subjects to appreciate the several engineering technologies responsible for the results and therefore behind the operation of the loops involved. In the I/A Series distributed control system, the control strategy is implemented by connecting together, via software links, the various blocks that are embedded as a fundamental part of the system. These blocks of software algorithms that perform the control functions for the scheme have been evolved and finely "honed" over several years of practical experience and are therefore highly stable. Irrespective of the actual DCS used, and before any attempt is made to code application specific software, the control block capability and its operation should be studied closely. Following this suggestion could result in saving considerable effort, time, and therefore money in control scheme implementation and furthermore produce a control loop that performs well and is very stable and, if necessary, easily modified.

DIAGRAMMATIC REPRESENTATION OF A FLOW CONTROL SCHEME

The flow control scheme in Figure 20.1 illustrates how a control strategy is developed and graphically depicted. In this control scheme we have deliberately derived the I/A Series loop in stages, starting from the standard Instrument Society of America (ISA) "bubble" diagram. In this bubble diagram, the early form of control loop drawing, it should be observed that, based on the application of the various technological precepts, the method shows in detail how the solution was arrived at. Analyzing the bubble diagram, we find that, having made the measurement of the pressure difference across the orifice plate, we have to relate this to the flow in the process line. We apply Bernoulli's law, which states, in simplified terms, that *the flow of the fluid squared is proportional to the pressure difference across a pressure-difference–creating element (orifice plate, venturi section, etc.)*, which we can write as:

$$Q_{vol}^2 \propto \delta p$$

where Q is the quantity of fluid flowing and δp is the pressure difference. As shown in the work we have done previously on process flow, this equation will have to be modified to include real-world and gravitational effects, so we can eventually rewrite it as:

$$Q_{vol} = K\sqrt{\delta p}$$

where K is the lumped constants. (Refer to the work done on flow in the first chapter of this book.) This is the reason for showing the square root extractor in the signal generated by FT (flow transmitter), the differential pressure cell connected across the actual pressure-difference creating device FE (flow element), the orifice plate. By doing this, the measurement signal applied to FIC (flow indicating controller), the flow controller, is in fact representative of the volumetric flow in the process line. It must be remembered that each bubble in the diagram represents an individual instrument performing the stated function. This methodology of an individual bubble for each instrument function does give one a "feel" for what is taking place in the control scheme. The limitations of the ISA bubble symbology are in the graphical representation of the functionality of the controller. For the author, the SAMA (Scientific Apparatus Manufacturers Association) symbol for the controller is by far the most descriptive of the graphical representations used, but requires a considerable amount of drafting time to reproduce as shown in Figure 20.2.

SAMA Method

The whole SAMA symbol applicable to a controller is divided into the three main functional sections—input, control, and output—as defined (in italics) in Figure 20.2a

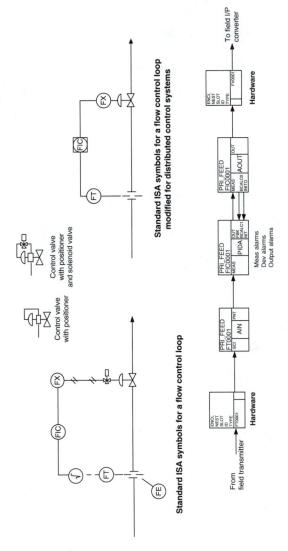

Figure 20.1: Implementation of the flow control loop in I/A.

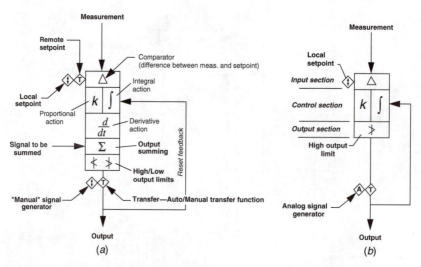

Figure 20.2: (*a*) SAMA three-term controller. (*b*) SAMA two-term controller with remote Auto/Manual station.

and *b*. In order to demonstrate the power of the graphical representation, the figure shows two quite sophisticated controllers, with three-term and two-term control action, respectively. In any real process application, the control actions necessary are selected to suit, clearly defined, and arranged in the allocated positions within the symbol by the user. In the interests of familiarity and continuity in using this graphical method, it is suggested that, the positions of the control terms within the symbol itself not be modified—for example, proportional and integral in the upper and derivative in the lower box (if, however, proportional plus derivative action is required, then the *k* will occupy the whole of the upper box, with the derivative symbol, as shown within the lower box). The option of *limiting* the controller output, if necessary, is user-selectable and shown at the output section of the symbol. These limit symbols are shown individually within the box. If, as shown for the three-term controller, both high and low limits are used, then the two limit symbols are located within the same box. On the other hand, if only one is required (as in this example a high limit shown for the two-term controller), then that symbol alone is detailed within the box. The three-term controller in Figure 20.2*a* has a local/remote setpoint transfer option—i.e., its setpoint is derived either locally (manually set by the operator at the controller itself) or from a source remote to the controller, and the transfer from one source to the other usually at the discretion of the operator in the case of a hardware controller, but the transfer in the case of DCSs can also be effected by commands from another block or a sequence program. However, some control schemes require either the setpoint to the controller, or the controlled output to the final actuator to be obtained from different sources dependent on attainment of specific process conditions within the process. In these instances the *transfer* symbol, with its two signal inputs and the *trigger*, which is usually a Boolean variable derived from some master initiator located elsewhere, could also be shown. The initiator, which could be a sequence program (an application-specific software routine), monitors the process and runs continuously while the process is in operation and produces the required trigger in the appropriate signal line that had been hard wired or software connected to the controller, when the predetermined conditions within the process are fulfilled.

When controlled outputs are switched in this way, it is absolutely vital that the reset feedback be taken only from the single signal drive signal to the final actuator. In control schemes where there are two controllers driving the same valve as in the case of the series compressors and a single recycle valve discussed in Chapter 13, the reset feedback is applied to the integral term of both the controllers involved, as only one controller can be in command at any time. This statement assumes that both the controllers have integral action. If there is no integral term, there is no need for a reset feedback signal, for the controller is unable to become saturated. Remember, integral control action will produce an output for as long as there is an error, i.e., difference between the controller measurement and setpoint. Note carefully, the reset feedback arrangement shown in the case of the two-term controller in Figure 20.2b. Why is this necessary? The answer is that since the controller output can be switched at any time, this arrangement will keep the controller from destabilizing the process when this is done. A full explanation of what actually occurs when outputs are transferred (switched) is to be found later in this chapter when Auto/Manual transfers are discussed. The explanations given for Auto/Manual transfers apply equally to any situation when the output only of a P + I controller is switched. In other control schemes the setpoint and output transfers could be linked so that there is always a one-to-one relationship between a setpoint and an output, with both changing in unison. In this instance, will it be necessary to provide reset feedback? The answer is no; the reason being that at no time is the controller output open circuited. However, under these circumstances valve failure modes become important considerations.

There are many more control loop symbols in the SAMA repertoire, and the reader is encouraged to investigate these symbols further.

The ISA Method

Using the ISA symbols for distributed control systems to depict the same control scheme will produce the result that was shown in Figure 20.1. Only one symbol with the appropriate tag number is made to convey the control intent of the loop. All the actual functionality to be implemented in a DCS is combined into this one graphical representation of the requirements. Putting the various control loop-illustrating methodologies for the same scheme together for comparison serves to illustrate how differently the various graphical techniques employed convey the same information. It is clear from the comparison that when a control scheme is drawn as a DCS symbol, it does require experience to interpret its full-intended meaning. It should be noted that the graphical symbols for automatic control supported by the BSI (British Standards Institute) follow very closely those of the ISA.

FoxCAE Method

When configured as an I/A Series loop the same control scheme (the lowest, i.e., the block diagram) of Figure 20.1 is shown as it would appear in the printout derived from the FoxCAE (Foxboro Computer Aided Engineering) package. In order to assist the understanding, additional notes (e.g., hardware, controller alarm capability) have been appended to the diagram; these will not appear in the printout from the FoxCAE design package. If we go carefully through this loop, it will be found that everything appearing in the original ISA bubble diagram also appears here. Note that the square root extraction takes place in the AIN (analog input) block in the form of the SCI (signal-conditioning index) parameter; this lines up with the Bernoulli requirement of having to take the square root of the differential-pressure signal before presenting it to the flow controller. A word of warning here: Always remember that when dealing with the process parameter flow, especially when differential-head devices are involved, there is usually a low flow cutoff built in. The reason for this is the parabolic (square law) nature of the relationship of differential-head vs. flow; the resolution at very low flow rates is not good, and therefore the signal is cut off to avoid cumulative errors developing when the measurement is down in this region.

TYPICAL DATA REQUIREMENT FOR A SINGLE INPUT TO A DCS

SPECIMEN OF DATA FOR AN ANALOG INPUT

Table 20.1 shows the standard capability of the Foxboro I/A Series AIN block. In this regard it should be noted that the block will support only a single input from an analog input type field bus module (FBM). Other blocks will have data requirements that differ from that shown in this sample.

Table 20.1 is given only as an example of the data input required, the type of data, the default value of the block data, and the connections that are possible between the block, hardware, and other blocks. This connection capability is shown in the last column of the table; these are the "hooks" spoken of in Chapter 19 and provided in the system to enable control schemes to be developed. The abbreviations *non-connect/non-setable* means that the parameter cannot be connected to, or changed by another block or routine; *con/setable*, on the other hand, means that the parameter could be used to connect to, or be set by, another block or routine. When *non-setable* appears on a row that specifies a choice of a value in a range, it means that the value is assigned when the system is configured initially and it cannot be changed without reconfiguring the block. The column headed "Parameter" in the table gives the name that appears on the configuration screen display of the system. Note that IOM is an abbreviation for input/output module. There are some items in the column headed "Units/Range" that need definition, and these are as follows:

• PERIOD refers to the block *scan periods*, which are set as any chosen index from 0, 1, 2, 3, 4, 5, 6, 7, 8

where $0 = 0.1$ s
        $1 = 0.5$ s
        $2 = 1.0$ s
        $3 = 2.0$ s
        $4 = 10.0$ s
        $5 = 30.0$ s
        $6 = 1.0$ min
        $7 = 10.0$ min
        $8 = 60.0$ min

• PHASE is an integer value that determines when the block will execute and is the period that can be assigned to prevent the block processor from overloading. It is best left at the default value unless processor-loading factors compel reconfiguration of the phase. There is a relationship between the period and allowable phase values, and it is suggested that this information in the Foxboro data dictionary be consulted when one is involved with actually testing or running the system.

• IOM is the hardware module that is mounted within an enclosure at a location determined by the system designer. The module itself is selected from the variety available to accept the signal provided by the measuring device, which could be a

TABLE 20.1

Data entry for an I/A Series analog input block in the "Control Configurator" (*reproduced with permission*). Items marked with * and ♦ are elaborated on.

Parameter	Description	Type	Default	Units/Range	System access
NAME	Block Name	String			non-connect/non-setable
TYPE	Block Type	Integer	"AIN"		non-connect/non-setable
DESCRP	Descriptor	String			non-connect/non-setable
PERIOD	Block Sample Time	Short	1	[0......8]*	non-connect/non-setable
PHASE ♦	Block Phase Value	Short	0	range varies	non-con/set
IOM_ID	Block Phase Value	Char	0		non-connect/non-setable
PNT_NO	IOM point No.	Integer	1	[1.......32]	non-connect/non-setable
SCI	Sig. Cond. Index	Integer	0	[0.......]	non-connect/non-setable
HSCO1	Hi Scale 1	Real	100.0	Specifiable	non-connect/non-setable
LSCO1	Lo Scale 1	Real	0.0	Specifiable	non-connect/non-setable
DELTO1	Change Delt.	Real	1.0	percent	non-connect/non-setable
EO1	Eng. Unit Out.	String	percent	Specifiable	non-connect/non-setable
OSV	Span Variance	Real	2.0	[0.......25] percent	non-con/non-set
MA	Manual/Auto	Boolean	false		con/set
INITMA	Initialise M/A	Boolean	true		non-connect/non-setable
LASTGV	Last Good Value	Boolean	true		non-connect/non-setable
INHIB	Alarm Inhibit	Boolean	false		con/set
MTRF	Meter Factor	Real	1.0		non-connect/non-setable
FLOP	Filter Option	Integer	0	[0......2]	non-connect/non-setable
FTIM	Filtr. Time constant	Real	0.0	[0.......] minutes	con/set
XREFOP	ext. Reference opt.	Boolean	false		non-connect/non-setable
XREFIN	ext. Reference input	Real	0.0	deg Celsius	con/set
KSCALE	Gain Scaler	Real	1.0	Scalar	non-connect/non-setable
BSCALE	Bias Scale Fct.	Real	0.0		non-connect/non-setable
BAO	Bad Alarm opt.	Boolean	false		non-connect/non-setable
BAT	Bad Alarm Text	String			non-connect/non-setable
BAP	Bad Alarm priority	Integer	5	[1......5]	con/set
BAG	Bad Alarm Group	Integer	1	[1.......3]	non-con/set
HLOP	Hi/Lo Alarm opt.	Boolean	false		non-connect/non-setable
ANM	Alarm name pnt. 1	String			non-connect/non-setable
HAL	Hi Alarm Limit	Real	100.0	LSCO1, HSCO1	con/set
HAT	Hi Alarm Text	String			non-connect/non-setable
LAL	Lo Alarm Limit	Real	0.0	LSCO1, HSCO1	con/set
LAT	Lo Alarm Text	String			non-connect/non-setable
HLDB	Hi/Lo Alm deadbd.	Real	0.0	LSCO1, HSCO1	non-con/set
HLPR	Hi/Lo priority	Integer	5	[1.....5]	con/set
HLGP	Hi/Lo Alarm Group	Integer	1	[1.....2]	non-con/set
BAD	Bad Status	Boolean			con/set
CRIT	Criticality	Integer		[0.....5]	con/non-setable
HAI	Hi Alarm Indic.	Boolean			con/set
LAI	Lo Alarm Indic.	Boolean			con/set
PNT	Point Output	Real		LSCO1, HSCO1	con/set
PRTYPE	Priority Type	Integer		[0......9]	con/non-setable

transmitter or a sensor (thermocouple, RTD, etc.). The enclosure and the module location are assigned unique identifications. The module itself is given a unique identity in the form of a six-character alpha and/or numeric name, which is made up by combining tiny hardware, electronic blocks that are associated with each alpha or numeric character. The fully assembled name, thereafter always known as a "letter bug," is plugged into the module circuitry to become part of it. The letter bug is an assignment made to posterity or for as long as the system exists.

• PNT/NO is in the range of 1 to 32 and corresponds to the number on the terminal board of the FBM associated with the block. A complete listing is not given here because the amount of relevant data available for each type of FBM. It is suggested

that this information contained in the text mentioned earlier be consulted when one
is involved with the actual configuration of the system.

• SCI (signal conditioning index) is a parameter that determines how the "raw
counts" from the FBM will be converted to engineering units and range. The mapping
of the index values of 0 to 49 is as follows

0 = no conditioning, pulse count

1 = 0 to 64,000 raw counts linear corresponding to signal range 0 to 20 mA

2 = 1,600 to 64,000 raw counts linear for signal range 0 to 10 V dc nominal

3 = 12,800 to 64,000 raw counts linear for signal range 4 to 20 mA

4 = 0 to 64,000 raw counts square root for signal range 0 to 20 mA

5 = 12,800 to 64,000 raw counts square root for signal range 4 to 20 mA

6 = 0 to 64,000 raw counts square root with low cutoff 3 to 4% for signal range 0 to
    20 mA

7 = 12,800 to 64,000 raw counts. Square root with low cutoff 3 to 4% for signal
    range 4 to 20 mA

8 = pulse rate

9 = unassigned

10 = IT2 support (Foxboro Intelligent Transmitter support)

    11 to 19 reserved for future use

20 = type 'B' thermocouple (°C)

21 = type 'E' thermocouple (°C)

22 = Unassigned

23 = type J thermocouple (°C)

24 = type K thermocouple (°C)

25 = type N thermocouple (°C)

26 = type R thermocouple (°C)

27 = type S thermocouple (°C)

28 = type T thermocouple (°C)

29 to 39 reserved for future use

40 = Copper RTD (°C)

41 = Nickel RTD (°C)

42 = Platinum RTD DIN standard (°C)

43 = Platinum RTD IEC standard (°C)

44 = Platinum RTD SAMA standard (°C)

45 to 49 reserved for future use

• HSCO1, LSCO1, and EO1 are the actual upper and lower limits of the scale
values and engineering units that are associated with the measurement being configured. For example these could be 500 to 50,000 GPH, 0 to 100%, etc.

- OSV is a real value that defines in percentage terms the amount by which the HOLIM and LOLIM clamps may exceed the measurement range specified by HSCO1 and LSCO1. It is expressed as a percent of the SPAN (defined by HSCO1—LSCO1) and permits a variation of up to 25%.
- FLOP provides input filtering that is specified as an integer in the range 0 to 2 where:

0 = no filtering

1 = first-order filtering

2 = second-order Butterworth filtering

- FTIM is a "real" value in minutes and represents the time for the output to reach 63% of its ultimate value after a step change in input, i.e., the time constant of the output.
- BAT is the alarm message that is printed out when the measurement is in alarm. It is a string that can be up to 32 characters long.
- BAP is an integer value in the range 1 to 5 that sets the priority of the output parameter BAD. 1 is the highest and 5 the lowest priority.
- BAG is an integer value in the range 1 to 3 that directs the BAD alarm message to any one of three groups of alarm devices chosen at will by the process operator. It can be modified to suit conditions and is changed when required via the workstation.
- BAD is a parameter of the block output that is set if either

  – The FBM device status has errors.

  – The channel status is bad.

  – The point output is being clamped.

- CRIT is a parameter of the block output and an integer in the range 0 to 5 that indicates the priority of a currently active alarm. 1 is the highest and 5 the lowest priority.
- PRTYPE is in general an indexed parameter of the block output and an integer in the range 0 to 9. It gives the alarm type of the highest priority active alarm. The following is a mapping of the index to the alarm types:

0 = no active alarm

1 = high absolute alarm

2 = low absolute alarm

3 = high/high absolute alarm

4 = low/low absolute alarm

5 = high deviation alarm

6 = low deviation alarm

7 = rate of change alarm

8 = bad alarm

9 = Boolean

For example, the AIN block has the PRTYPE 0, 1, 2, and 8 from the foregoing mapping for its output.

BLOCK DIAGRAM OF THE AIN BLOCK

A useful summary of the AIN block inputs and outputs can be derived from Figure 20.3, to which have been added a few notes shown in italics in the figure.

FURTHER DEVELOPMENT OF THE FLOW CONTROL SCHEME FOR DELIVERY OF A METERED QUANTITY

For the control loop of Figure 20.4 we have added some enhancement to the one in Figure 20.1 to show how schemes can be developed from simple beginnings. In this control scheme we wish to stop the flow after a predetermined amount has been delivered, and to achieve this, an accumulator block (ACCUM) has been inserted. The accumulated value is a single-precision floating-point value with seven-digit resolution and a user-specified range. A predetermined value is written to the PRESET input parameter of this block to provide a reference for the amount required. While the system is running, a comparison between the predetermined and actual values of flow is being made at every scan period. The limiting value HHAOPT, a Boolean variable, is set when the (predetermined) target value is attained. This Boolean variable (or bit) is connected to the MA (Manual/Auto) parameter of the control block to drive the block into the Manual mode and hold the output at its last value to prevent the integral term from saturating. In addition to manipulating the control block, the HHAOPT parameter also provides a cutoff signal to the flow control valve. This requirement is met by connecting a COUT (contact output) block to the HHAOPT. The discrete output from the contact out block is used to drive a solenoid valve placed in the pneumatic signal line to the control valve.

The following may initially appear trivial and simple, but it should probably be included at this point. The solenoid valve is a three-port device and operates as follows:

- A solenoid is placed around the movable valve stem and drives the stem through electromagnetic induction when it (the solenoid) is energized. The design of the valve is such that one of the ports is used as a "common" and the other two ports are switched with respect to it, thus ensuring that at any time the common and only one of the other two ports are in communication with each other. In other words, when the solenoid is *not energized*, the common and one of the two ports are connected together.

- When the solenoid is *energized*, the common and the other initially unused port are connected instead.

For this application we want a normally closed valve (no flow in the line); hence, the common port of the solenoid valve is piped to the connection on the control valve diaphragm and the port of the solenoid valve associated with the *deenergized* state is connected to the pneumatic drive signal to the control valve. The unused port is used as a "vent" to allow the air in the diaphragm motor to be exhausted to the atmosphere when the solenoid is *energized*. This simulates a drive signal failure to force the valve shut (normal state). Other schemes may require a normally open valve, in which case the valve failure mode will be open, and the solenoid valve must be connected appropriately. In every case, a failure in the air supply will force the valve to its failure mode; so choose the mode carefully.

When the system is in operation, the electronic controller output signal is converted to a standard pneumatic signal, which is applied to the valve actuator to

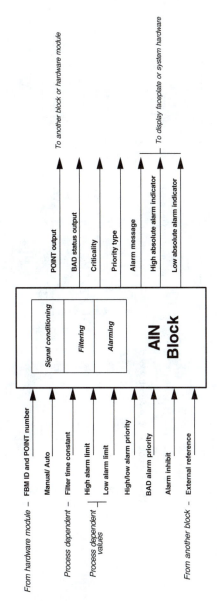

Figure 20.3: AIN block inputs and outputs.

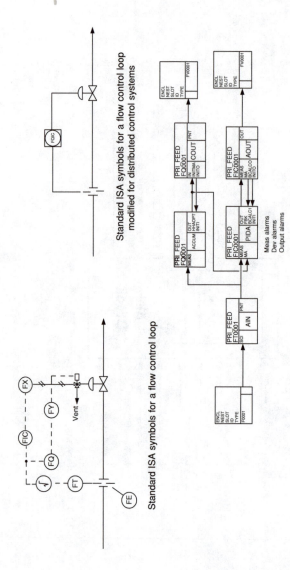

Figure 20.4: Implementation of the flow integration and control loop in I/A. The output from the contact output block is used to initiate the solenoid valve.

regulate the flow of the process fluid. The accumulator integrates the quantity passed, and when the amount actually delivered is equal to the preset amount, a cutoff signal from the COUT block is applied to the solenoid. The solenoid switches the valve ports to:

- Inhibit the controller signal to the valve
- Cause the pressure in the valve actuator to escape through the now selected vent port
- Cause the control valve to be driven to its failure mode, which in this instance must be closed; and as a result stops the fluid flowing

Before restarting the system it is vital that the integrating counter of the accumulator block be reset to zero and the controller be returned to the Auto mode. The reset is accomplished by initiating the CLEAR input on the ACCUM block with a 0 to 1 transition.

THE "STANDARD" CONTROL BLOCK

The PID block algorithm allows any one of five, selectable, control actions that form part of the block design and provides the controller PRI_FEED FIC-0001 a PIDA block in Figure 20.4. (**Note:** This block is chosen for no other reason than it offers all the advantages of the PID, plus many more that do not have to be used immediately but could prove useful later if required, and then avoid reconfiguring the loop.) The control action to be used is configured under the MODOPT parameter of the block and requires a numeric entry in the range 1 to 5. The meaning of each numeric entry is defined in the following list:

1. PO (Proportional only)—default configuration
2. IO (Integral only)
3. PD (Proportional + Derivative)
4. PI (Proportional + Integral)
5. PID (Proportional + Integral + Derivative)

Once the selection is made the parameter is set. Change to another type of control action can only be obtained by reconfiguring the block. Since we are controlling process flow, the type of control action required will have to be the numeric 4 (Proportional + Integral), as stated earlier. There are several other features associated with the PID block that have to be specified, or first selected and then specified, when configuring this block, in a manner similar to that used for the AIN block.

The controller setpoint is configured under the block parameter SPT and always represents the active setpoint of the controller. It can be either derived locally or set remotely when the block parameter LR (a Boolean variable) is configured. By setting the Boolean parameter LR true (= 1), a remote setpoint is connected. The default for LR is false (= 0), i.e., local setpoint. The configured setpoint can also be supplied as a value to other blocks. In Figure 20.4 because the setpoint is local, this parameter has not been shown in the diagram.

Three connections are vital between each primary and its secondary control block and must be configured for correct operation:

1. INITI—INITO

2. BCALCI—BCALCO

3. OUT—RSP

The parameter INITI is the acronym for *INITi*alization In and defines the source block and parameter that drives the block into initialization that could be the parameter INITO, the acronym for *INITi*alization Output, of a downstream (AOUT) block. The parameter BCALCI (Back *CALC*ulation *In*) provides the starting value of the output before the PID block enters the controlling state. The source for this parameter is the parameter BCALCO (Back *CALC*ulation *Out*), of a downstream (AOUT) block. It provides the means of having bumpless return to the controlling state of the block.

The parameter RSP (Remote *Set*Point) connects from an upstream block in a cascade control scheme. It is the selected source of the controller setpoint when the parameter LR is set to true (1). The parameter OUT (output) is the manipulated variable of the control block. In the automatic mode of operation it is the result of the calculation performed by the block control algorithm based on the difference between the setpoint and measurement within the limits of any configured output clamping.

A useful summary of the PID block inputs and outputs can be derived from Figure 20.5, to which have been added a few notes shown in italics. As shown for the AIN block, there are many other parameters than those shown in the diagram that have to be completed when configuring the block.

THE "ENHANCED" CONTROL BLOCK: INCLUSION OF ADAPTIVE CONTROL

Despite the power contained in the PID algorithm, the desire to improve the control capability of the system resulted in the design and inclusion of the PIDA (PID Adaptive algorithm) block in the system. This system innovation brought benefits to users also, for it improved the quality of their products and return on investment (ROI). As far as the author is aware, this algorithm is the most advanced and sophisticated available on any distributed control system. It brings to the user adaptive control capability as a standard feature of the system. It should be appreciated that adaptive control is not necessary for simple control loops, of which there are several in any manufacturing process. However, there are instances where a particular loop is difficult to control because of the interaction of other process loads with the control scheme. In such cases the PIDA block comes into its own. The PIDA block enables the control engineer to take account of up to four other load variables that his or her operating experience has shown disturb the controlled process variable, and to automatically and continuously tune the controller to meet the variations that these disturbances impose. The result of this enhanced control algorithm is to radically improve the quality of control of the affected variable and bring stability to the process as well. This algorithm is designed to take account of the difficulty in controlling a process with deadtime (distance/velocity lag we discussed in Chapter 7) by implementing either the PITAU or PIDTAU algorithms as required, and also provides the users with the

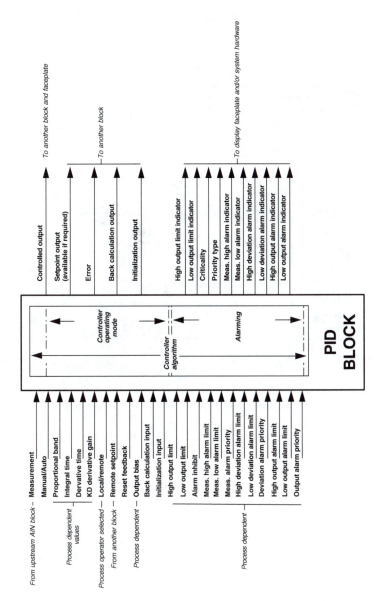

Figure 20.5: PID block inputs and outputs.

399

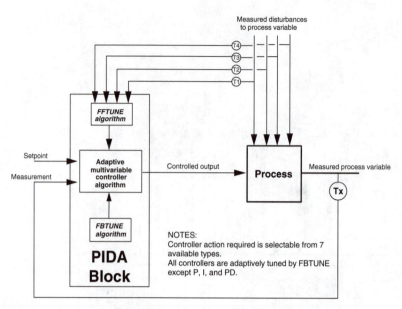

Figure 20.6: Simplified schematic diagram of the PIDA block.

option of selecting either an interacting or noninteracting controllers, both of which we discussed in Chapter 10.

Figure 20.6 is a simplified schematic of the PIDA block, which has the following capabilities:

- Interacting P, I, PI, PD, PID, and PITAU (PI with deadtime) control actions. The most suitable one from the six being selected for the task.

- Noninteracting PID and PIDTAU (PID with deadtime) control actions. These two further control algorithms increase the selection ability.

- The local setpoint can be ramped at a specified rate, or within a specified ramp time, and have limits imposed on it. The path (with respect to time) traced by the setpoint is variable, for example, (1) the setpoint can be made to move at, say, a specific rate and stop when the target limit is reached, or (2) the setpoint can ramp a specific rate to a target value, and when the target is reached, ramp to a new target value within a specified ramp time instead of a ramp rate. When the controller is transferred to the Manual-operating mode, not only will the controller freeze at the last value, but also the ramp will stop.

When choosing controllers, interacting controllers are preferred, for they permit an integral feedback signal to prevent the integral term from saturating (reset windup) when cascade loops are involved. Noninteracting controllers are valuable when the process contains a secondary lag that is twice as large as the effective process deadtime. The effective process deadtime is considered to comprise the true deadtime plus the sum of tertiary lags minus negative leads. The noninteracting control actions of the PIDA block benefit from the inclusion of an integral feedback signal to bring the advantage of the interacting controller within its design remit.

FURTHER DEVELOPMENT OF THE SCHEME TO INCLUDE FLOW RATIO CONTROL

Continuing the development and complexity of the control scheme to show how the system copes, we have used the same process parameter flow to produce a control loop that is in reality a combination of two control schemes. We have modified it to provide *ratio control*, where ratio, when two separate numerical terms are considered, is defined exactly as it is in arithmetic. In this instance we want one term to be a fixed percentage of the other at all times. The loop in Figure 20.7 is also an example of a "cascade" control scheme. This means that one loop dictates how the other should respond. The loop that dictates the behavior of the other is called the *primary*, while the one dictated to is called the *secondary*. (**Note:** Some people refer to these as "master" and "slave," the choice of nomenclature is a matter of personal preference.) In this instance note that control loop F0001 is the primary, because it provides the setpoint for control loop F0002, thereby dictating how loop F0002 should respond. Furthermore, *the required ratio is derived from the measurement of the controller in loop F0001*, and this is most important. Hence, we can conclude that in any ratio control scheme the objective is to ratio the amount in one process stream to the amount in another process stream, in other words we are only concerned with the measurements of the two process streams, and to ensure that the ratio between the two is obtained we must regulate only one of the two streams to the ratio of the other. We have used the word "amount" in the foregoing deliberately to signify its independency of a measurement parameter, e.g., we can have a ratio of one temperature to another, or one pressure to another. Following on from this we can modify the foregoing statement and say, that, if there are controllers in both process loops, then we must only heed the measurement of one controller (ignoring its output, but only for the purposes of ratioing) and regulate the other to give the required ratio. To explain the reasoning behind this recommendation, we shall use the ratio scheme presented here as an example. What is required is the ratio between the actual flow in loop F0001 and the actual flow in loop F0002. The measurement of each of the controllers in the two loops represents the actual flow occurring in the associated process line and, provided the controllers are tuned correctly, will coincide with the desired value (setpoint) of the controller. If the output of the controller in loop F0001 is incorrectly used as the setpoint of controller F0002, then certainly the results will be nonsense, as will be now be explained. During operation, the controller in loop F0001 will change its output just sufficiently to maintain a constant flow in the process line, at a value determined by its setpoint. In common with all process controllers the controlled output in F0001 includes (inseparably) some proportion of the gains of the control actions of the controller and of the control valve involved. Thus, it is not a measure of the actual flow in the process line; it is more correctly visualized as a measure of the amount of flow change necessary to bring about coincidence of the controller's measurement and setpoint values. In this application, because we want F0002, the secondary loop, to respond to any change in F0001, the primary, the output of F0001, the primary, is set to change in the same direction as its measurement—i.e., increasing measurement giving increasing output (the valve requires air to close, i.e., fail open) to ensure measurement/setpoint coincidence. The measurement of F0001 is taken through a ratio block FR0001 and on to the setpoint of secondary controller F0002, which will manipulate its valve to bring about coincidence between its measurement and the dictated ratioed setpoint.

A point to remember here: Many process control engineers are confused when talking of the direction the controlled output is to take. There are two differing points

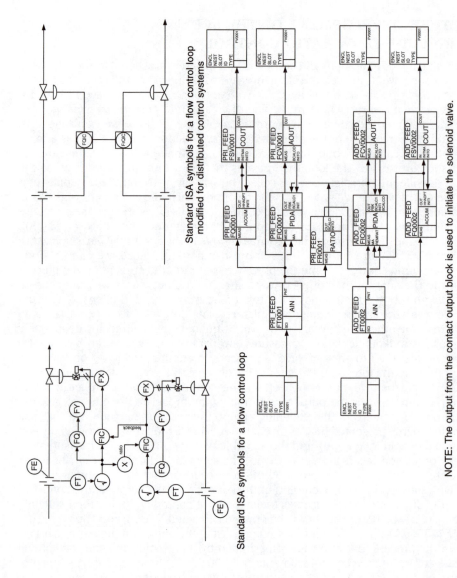

Standard ISA symbols for a flow control loop
modified for distributed control systems

Standard ISA symbols for a flow control loop

NOTE: The output from the contact output block is used to initiate the solenoid valve.

Figure 20.7: Implementation of the flow ratio control and integration loop in I/A Series.

of view:

- Some refer to *increasing measurement giving increasing output* as *direct action* and *increasing measurement giving decreasing output as reverse action.*

- Others refer to *increasing measurement giving increasing output* as *reverse action* and *increasing measurement giving decreasing output as direct action.*

As can now be seen, one statement is the exact opposite of the other. The recommendation is: Avoid all confusion, and state in full the output action required.

AUTOMATIC/MANUAL TRANSFER—THE IMPLICATIONS

We now consider the practical aspect of the control loops while they are in operation. The *main assumption* here is that *both the controllers involved are correctly tuned*, which means that any movement of the setpoint is instantly reflected in the same amount of movement in the measured value. The behavior just described can very easily be demonstrated and observed occurring in an extremely fast loop that can be set up on the workshop bench, using a fully operating hardware controller. To do this we connect the controller output back through a very small delay to its measurement input, to simulate a real process, and then proceed to finely tune the instrument. When the tuning is correct, the slightest shift in the setpoint will result in an equivalent measurement change; the two instrument pointers will begin to move as one, with the barest of difference. Since FIC0002, the secondary, is a controller, it is possible for the process operator, at any time, to change this controller's operating mode from Automatic to Manual and drive the output from the controller to a value other than that produced by the control algorithm. This Automatic to Manual transfer facility is provided to the operator for safety reasons, so that under dangerous conditions the process can be forcibly driven to a point that eliminates the hazard. By deliberately changing the secondary controller's operating mode from Automatic to Manual, the operator effectively open-circuits the output of FIC0001 the primary controller. This operator intervention causes FIC0001, the primary controller, to drive its output beyond its limits and, dependent on the duration, very possibly into saturation. If the operator, having overcome the hazard then returned FIC0002, the secondary controller, back to the Automatic mode without first attending to FIC0001, the possibly saturated primary, then there would for certain be a large bump in the process instigated by FIC0002, the secondary controller's output change. The reason for the unacceptable bump is that the output of FIC0001, the saturated primary controller, would be at one extremity (far removed from its previous operational setting) of its output scale and similarly reflected in the setpoint of FIC0002, the secondary controller (remember, in this simple case, the primary output is connected to the secondary setpoint). The now large disparity between the measurement and setpoint of FIC0002, the secondary controller, would cause its output to change a considerable amount from where it had been deliberately put previously by the operator, upsetting once more the considered balance of the system.

Such an occurrence would be a classic example of the effects of integral saturation or reset windup. This is a situation to be avoided at all cost and must be eliminated. Two possible methods to surmount the problem are:

- (The assumption here is that FIC0001, the primary controller, is saturated or very nearly so and still in the Automatic mode of operation.) The operator first transferrs FIC0001, the primary controller, to Manual mode to stop integral action from having any further effect, and then drives its (FIC0001) output, visible as the setpoint of FIC0002, the secondary controller, to a value coincidental with the measurement of FIC0002, the secondary controller. When, and only when, this is achieved, transfer FIC0001, the primary controller, back to the Automatic mode of operation to allow the controller to keep the process on control at the operator-set safe condition. This procedure ensures that the control valve associated with FIC0002 the secondary controller does not alter its position and maintains the safety of the process. This procedure is called bumpless auto/manual transfer. The method, being operator driven, is therefore entirely manual, requiring great manual dexterity and near-perfect visual/manual coordination to manipulate the functions of the two controllers involved, and considerable plant experience, time, and effort. All these present difficulties to a busy operator.

• The reader might like to consider the interconnections to specific block parameters in a real-life loop situation as described, and thereby attain an appreciation of the phenomenal capability of the present state-of-the-control-industry-art systems equipment that is presently available to systems designers and engineers. The whole procedure can be automated by making a connection from the secondary controller output to the FBK (feedback) input on FIC0001, the primary controller.

Implementing the latter method will ensure that there will never be a saturated primary controller, for under all conditions the smallest change in one controller is instantly reflected in the other.

Having described the hazards of allowing the integral term of a controller to saturate raises the question, What happens if one does not heed the advice given? To answer this, let us assume that we have a situation as described earlier, in that the entire control loop was in the Automatic mode and the operator transferred FIC0002, the secondary controller, to Manual and started to adjust its output to eliminate the cause of a problem. This has to be done slowly, for the effect of every change made has to be carefully observed before any further changes are made. Once satisfied that the action taken eliminates the problem, the operator returns FIC0002 back to the Automatic mode. While the adjustments to FIC0002 were being carried out by the operator, neither the manually set desired value (setpoint) nor the operating mode (Auto/Manual switch) of FIC0001, the primary controller, was touched, and therefore remained unchanged, but its output had been deliberately open circuited and as a result its integral term was allowed to begin gradually saturating, which finally resulted in a "wound up" controller. As soon as the operator put FIC0002, the secondary controller, back to the Automatic mode, FIC0001, the primary controller, was now placed in command of the whole loop, but it was saturated and with its output at one extreme of its output scale. The whole system responds immediately to the dictates of FIC0001, the primary controller; the valve associated with FIC0002, the secondary controller, will be driven hard to a position demanded by its setpoint, which is really the output of FIC0001—results being the complete unbalancing of the system and therefore an unsafe plant situation. The process oscillates wildly as the controllers struggle to gain order; this upheaval will continue for some time. Since the process operating point that initially had been manually set by the operator as the setpoint on FIC0001, the primary controller, was the parameter not touched during the whole traumatic episode, it will be the focal point to which the system will endeavor to correct itself. Correction will eventually be achieved, but at a rate determined by the integral action time set initially on the controllers, and if these are large values it will be a considerable amount of time (in some cases this could be the major part of an hour) before the situation will ease and some semblance of control begin to be observed. This return to some sort of normality at a rate determined by the integral action time is usually referred to by control engineers as "returning to control at the reset rate." To avoid such situations occurring and, almost certainly, affecting other parts of the process—and as well, due to the integrated nature of most modern manufacture—it is advisable, and highly recommended, that the preventative means available in the system are not disregarded, but be included from the outset of the control scheme design.

In implementing the scheme of Figure 20.7, use has been made of the I/A Series RATIO (ratio) block to provide the setpoint of the secondary controller. However, it is also possible for a CALC (calculation) block to be used instead. The penalty in using the latter is a greater computational load because of its greater and more complex facilities (in terms of block equivalents) imposed on the CP (control processor). The function of FR0001, the ratio block, is to multiply the input (which is also the measurement to FIC0001, the primary controller block) by a user-specified constant to provide the setpoint of FIC0002, the secondary controller block. In view of this arrangement (setpoint input is the primary measurement multiplied by a constant), it will be clear that the secondary loop will always be regulated in a fixed relationship to the primary. The actual amount of process fluid flowing being dependent on the value of the multiplying factor set in FR0001, the ratio block.

We have included an "end of metered run" shutdown facility also, which is shown as the circuit included with the solenoid-operated valves placed in the drive signal lines to the control valves. The initiators for the solenoids are the COUT (contact output) blocks shown. The accumulator block triggers these contact output blocks when the predetermined amounts have been passed. One could, if desired, enhance this further by including some logic that shut both loops down at the same time.

It has been found from experience that the ISA bubble diagram is a very helpful tool in control loop design. In view of its simplicity and the ability to construct individual functions for each aspect of the overall loop, it offers a good (in the author's experience, when combined with the SAMA recommendation for controller depiction, unrivaled) basis upon which to build ideas for most control schemes.

BLOCK TYPES AVAILABLE ON THE I/A SERIES SYSTEM

As has now been shown that when using the I/A Series system, the assembly and correctly connecting together of a number of appropriate blocks produce schemes to form the required control strategy. The control strategy normally referred to in the system as a *compound* is associated with a part of the process or particular piece of plant equipment, and is assigned an appropriate and unique name. The compound name, usually task-descriptive, can be made up from 12 alphanumeric characters and is the means by which the compound is known throughout the whole system, no matter how large it may be. The compound name must be entirely in capital letters; no lower case letters are permitted. Several (entire) compounds can be implemented and made to execute within a station and can be interconnected across station boundaries. It is not possible to split a compound across two stations. Every compound assignment made to the system is incorporated into a true relational database. Hence, invoking the compound name presents to the user all the control blocks associated with it. Details of the compounds are held in the *control configurator* the first screen of which is the CSA (Compound Summary Access), a very powerful utility that allows:

- Creating new compounds, and then copying these to an appropriate configurator

- Finding and selecting a compound anywhere in the system

TABLE 20.2

Block types available in the I/A Series System.

Name	Function	Capability	Remarks
AIN	Analog input	1 input	Type of input depends on SCI parameter chosen
AINR	Analog input redundant	1 measurement up to 2 transmitters	
MAIN	Multiple analog input	Up to 8 inputs	
CIN	Contact input	1 input	
MCIN	Multiple contact input	Up to 32 inputs	
AOUT	Analog output	1 output	
AOUTR	Analog output redundant	1 output redundant FBMs	
COUT	Contact output	1 output	
MCOUT	Multiple contact output	Up to 16 outputs	
PID	P, I, PI, PD, PID control	1 output	
PIDA	P, I, PI, PD,PID adaptive control		Adaptive control
PIDE	As PID + EXACT tuning	1 output	Self-tuning
PIDX	PID add on	1 output	Extended function sampled data, batch, tracking
PIDXE	PID add on + EXACT	1 output	Extended function sampled data, batch, tracking
PTC	Proportional time controller		Pulse duration On/Off control
DGAP	Differential gap + PID control	2 Boolean outputs	
FFTUNE	PIDA add-on	Feed-forward	For multivariable control
FBTUNE	PIDA add-on	Feedback	For multivariable control
BIAS	Bias sum of 2 inputs		Each input scaled independently
DPIDA	PIDA controller		Adaptive control algorithm located in FBM
OUTSEL	Selects 1 from 2 outputs		Controller output selection
LIM	Position and velocity limiter		
GDEV	Controller		2- or 3-wire motor circuits
MTR	Controller		2- or 3-wire motor controls
VLV	On/Off controller		Motor-driven valve actuators
MOVLV	On/Off driver		Incremental drive for motorized valves
MDACT	Driver		Motor-driven actuators, single-directional/bidirectional motors
ACCUM	Accumulator		Integrating/predetermining counter
RAMP	Ramp function	Up to 5 segments	Multisegment ramp sequence
SIGSEL	Signal selector	Up to 8 inputs	Any one from 8 on a high or low specifiable basis

(Continued)

TABLE 20.2
(*Continued*)

Name	Function	Capability	Remarks
DTIME	True delay	Specified in minutes	"Follow" mode bypasses delay, appears as output lead, lag, or lead/lag the input
LLAG	Dynamic compensation		
SWCH	Input selection	Any one of 2 inputs	
CALC	Calculation in 50 steps	8 analog, 16 Boolean inputs	Full arithmetic, Boolean and logic functions, 4 analog, 8 Boolean outputs
CALCA	Calculation in 50 steps	8 analog, 16 Boolean, 2 integers, 2 long integers.	Full arithmetic, Boolean and logic functions + control functions
MATH	Programmable calculation in 20 steps	8 real inputs, 4 real outputs	
CHAR	Characterizer	20-segment linear piecewise	Unidirectional only (i.e., rising or falling not both)
RATIO	Scaled multiplication		
ALMPRI	Alarm priority change		Reassigns dynamically priority of the point
MEALM	Measurement alarm	1 input	For "intelligent" field device
PATALM	Pattern alarm	8 Boolean inputs	Compares 8 Boolean inputs with 8 unique specified patterns
REALM	Real alarm	1 input	3 types of alarms: absolute high or low, d/dt measurement, high or low deviation
STALM	State alarm	1 input	Event change alarm for "intelligent" field device
MSG	Message generator	8 inputs	State change for "intelligent" field devices
DEP	Dependent sequence	50 long variables, 127 subroutines	User-defined sequence algorithm works with monitor, timer, exception blocks
IND	Independent sequence	50 long variables, 127 subroutines	User-defined sequence algorithm works with monitor, timer, dependent blocks
MON	Monitor	16 Boolean Expressions	User-defined expression works with exception, monitor, independent, dependent blocks
EXC	Exception sequence	50 long variables, 127 subroutines	User-defined sequence algorithm works with independent, dependent blocks
TIM	Timer	4 independent timers	Works with any block requiring timed periods
EVENT	Event	32 Boolean inputs and outputs	Event reports with time stamp messages

(*Continued*)

TABLE 20.2
(*Continued*)

Name	Function	Capability	Remarks
STATE	State block		16-bit packed Boolean state to drive the output
DataVar	Data variables		Separate blocks for Boolean, real, long, pack, string
PLB	Programmable logic block	32 contact inputs, 16 contact outputs	User defined ladder logic in FBM

Notes: The following are explanations of some of the abbreviations and descriptions used in the table.

SCI	signal-conditioning index
EXACT	EXpert Adaptive Controller Tuning—(Foxboro registered name)
Batch	batch switch feature added to a PID controller on batch operations to hold control when output is left open for extended periods as required in batch applications
Tracking	tracking the output of any control or analog block
Sampled data	data collected over time
True delay	real-time delay
Integers, long integers	integer length depends on the number of digits
d/dt meas	rate of change of measured variable

- Finding and listing all compounds in the system

- Finding and listing all compounds in the system of a particular name

- Finding and listing all compounds in the system that contain a particular type of block

and several other functions.

There are other blocks that act as gateways to programmable logic controllers (PLC) by Allen Bradley, using the ABSCAN block; Modicon, using the MDSCAN block; and equipment manufactured by others (Foreign Devices) using the FDSCAN block. Since the Foxboro Company has been making and installing control instruments and systems for many years, there are blocks (*gateways*) that enable the still-operating older systems to be included in the latest I/A Series system offering. Those in the control industry refer to this idea as *backward compatibility*. Backward compatibility is of interest to users of older control equipment who now have the opportunity to maintain production of their product and gradually update the control systems with modern technology at a pace of their choosing to meet their needs, and without the immediate financial burden that wholesale replacement of the entire system entails.

As we all know, advancement in technology moves on continually; therefore the list of blocks in Table 20.2 could already have changed.

SUMMARY

1. Each bubble in the ISA diagram represents an individual instrument performing the stated function. However, the SAMA symbol for the controller is by far the most

descriptive of the graphical representations used, though it requires a considerable amount of drafting time to reproduce.

2. In the ISA symbols for distributed control systems, only one symbol with the appropriate tag number is made to convey the control intent of the loop. The full control functional control of the entire loop previously shown in a bubble diagram is encapsulated in this one graphical symbol. It requires experience to implemented in a DCS all the actual functionality required.

3. In any ratio control scheme the objective is to ratio the amount in one process stream to the amount in another process stream; in other words, we are only concerned with the measurements of the two process streams, and to ensure that the ratio between the two is obtained we must regulate only one of the two streams to the ratio of the other. Following on from this we can modify the foregoing and say, that, if there are controllers in both process loops, we must only heed the measurement of one controller (ignoring its output, but only for the purposes of ratioing) and regulate the other to give the required ratio.

4. To avoid misinterpretation of the required output action of a controller, always define the action in full, because, some refer to *increasing measurement giving increasing output as direct action* and *increasing measurement giving decreasing output as reverse action* and others refer to *increasing measurement giving increasing output as reverse action* and *increasing measurement giving decreasing output as direct action*. And the two have directly opposite meanings.

5. Integral saturation or reset windup must be avoided and eliminated at all cost, for the damage to product loss in revenue and time it entails.

6. The reader might like to consider the interconnections to specific block parameters in a real-life loop situation as described, and thereby attain an appreciation of the phenomenal capability of the present state-of-the-control-industry-art systems equipment that is presently available to systems designers and engineers.

CHAPTER 21

Communications, Signal Processing, and Data Handling

It is probably true that the whole of mobile creation depends on some means of communication to let other members of the species know their intentions or desires. *Homo sapiens*, being further up the intellectual scale, have carried this a whole lot further, for humans have devised methods of establishing dialogue not only between themselves and their companions, but also the various inanimate machines of their design. This chapter is devoted to providing a very general overview of how communications are established so that, as engineers involved with systems, we have a reasonable appreciation of what is involved. We shall be considering how man has achieved this objective of man/machine dialogue and try to see what the future holds.

THE NEED FOR A STANDARD PROTOCOL

If it is necessary to open a dialogue between human and machine, or between machine and machine, then designers will tend to protect their intellectual property by using a specific language unique to their equipment. If this thinking is applied across the whole industry, it is not difficult to envisage the chaos that will ensue. We shall be replicating in this age, incredulous as it may sound, the pandemonium that must have surrounded the Tower of Babel in ages gone by, but it is a fact that this was the case up to only a few years ago, and as mentioned earlier, General Motors in the United States took a stand against the way things were and decreed that if suppliers wished to continue doing business with the corporation they would have to ensure that all their devices in addition to communicating with each other, would also be capable of dialogue with equipment provided by other manufacturers, and that such a rule would not be confined to all the supplied equipment, but would also apply between such equipment and the corporation's data management systems.

DEVELOPMENT OF MAP AND TOP

In view of the amount of money involved, GM's ultimatum to suppliers was, in a manner of speaking, an "offer they could not refuse," and it became the starting point for the manufacturing automation protocol (MAP). MAP is concerned with the manufacturing side of industry; the administrative environment has an equivalent in the technical office protocol (TOP). The Boeing Corporation—the airplane makers—also in the United States pioneered this latter protocol. To accelerate the development GM agreed to make available the specifications it had developed during its years of direct involvement. To link one item to the another is not just a matter of correctly wiring the different equipment together, for, in the joining together, problems at different levels are manifest in the interconnecting, such as:

- Spotting errors

- Correcting the errors

- Correctly routing messages

- Synchronizing the conversation among the equipment

and these are only some of the problems with the communication system alone; there are several more. Remember that the difficulties in communication are not confined to the manufacturing industry only; in fact, they span the whole realm of business.

What do these two protocols of MAP and TOP do? To begin to understand them, we must first appreciate the difference between *comprehension* and *connectivity* and the best way is probably by analogy.

Connectivity: Suppose we wish to make a telephone call to Italy, say, and in today's situation we dial direct. The telecommunications people in both our country and Italy have ensured that the equipment used is designed so that both parties can be linked together. In other words, they have used an agreed manufacturing standard to ensure that this is possible, even though the networking systems behind the two connection interfaces are quite different. Hence, the reliable end-to-end connection between dissimilar systems is a kind of definition of connectivity.

Comprehension: Having now made the connection to the Italian party, we are not at the end of the story. We are now faced with the problem of actually talking to each other. If we knew and spoke only English and the other party knew and spoke only Italian, we would certainly not be able to say anything comprehensible to each other. This, then, is a kind of reverse definition of comprehension.

The MAP and TOP protocols provide only *connectivity* to make a full working network. They do not provide comprehension, but contain *protocols to support comprehension*. The difference between MAP and TOP is the method by which data is transmitted between the various linked devices, even though the communication path is the same—i.e., the broadband or baseband.

THE PROCESS INDUSTRY SCENARIO

Before we proceed any further, let us stop and consider the manufacturing process industry, and see how data from the sensor is sent to the equipment that will manipulate it. At present, the most common method for those sensors not capable of direct connection to the receiver is for the very small changes discerned by the sensor to be converted into an analog electrical current signal (i.e., a signal that varies in a continuous manner) usually in the range 4 to 20 mA, which represents the minimum and maximum values of the sensor's measuring range, for connection to the receiver. The receiving equipment senses this signal and processes it. If the receiving equipment was digital, then the transmitted analog sensor signal would first have to be converted into a digital one, within a predefined digital range, so that the equipment would be able to recognize it, before any action could be taken on it. This conversion from analog to digital takes place in an interface module called an analog-to-digital converter (ADC), and the result of the conversion could be sent to either a digital indicator or to a visual display unit (VDU) for viewing by the operator; or if the signal is to be manipulated by a control algorithm, then manipulation takes place in digital form, and the output of the algorithm could be either cascaded onto another control algorithm or directed to a control device that regulates the process itself. Most of the common final control devices are analog, and therefore the digital signal will have to be converted back to an analog in an interface module called a digital-to-analog converter (DAC) before it can be used to initiate the control device.

At present, the most common way to connect the sensors and control devices to the control system is by running individual pairs of conductors between the two. There is considerable financial cost involved in doing this; so, it should not be a surprise to learn that a data link to achieve the connections is actively under consideration, and some data links are actually working at the moment.

THE COMMERCIAL SCENARIO

In commerce today it is more than likely that, provided the distances are short, the inputs will be digital, i.e., via a keyboard; and the outputs will also be digital and directly to a VDU, printer, or mass-storage device—floppy disks/hard disks, etc. However, if the data are entered from locations separated by long distances, then modems—the name obtained from MOdulatorDEModulator—will be employed, and analog current signals will be used in the field, with telephone lines carrying the data from one point to the other. Modern technology has now made possible digital transmission on telephone lines, but to avoid complete failure, care is needed to ensure that the connected equipment—e.g., modems—is compatible with the transmission signal. Most commercial and business data are transferred to a computer on a batch basis, and, because there is usually a very large volume of information to input to the system, the data transfer is carried on at night. The reasons for this are:

1. Data transfers could take a very long time, sometimes measured in hours.

2. Rates for nighttime use of the telephone lines are significantly cheaper than the normal daytime rates.

It will be appreciated that since there is a very great difference between the requirements of commerce and manufacturing industry, there is also going to be a significant difference between the architecture of the two systems. In commercial dealings "on-line status" is generally limited mainly to, for example, stocks trading, currency values, and the like; but, in manufacturing, most information must be direct on-line all of the time. In commerce, pure science and technology have little if any relevance in the day-to-day running of the enterprise, except in providing the means to an end. However, unlike commerce, the manufacturing industry does have scientific and technological requirements, and also needs to incorporate a great many of the features of the world of commerce into its systems, for proper management of its finances, material resources, and marketing; all necessary for being a viable enterprise.

THE PROTOCOLS

Modern systems generally do not have a multiplicity of individual circuit wiring connectors to each individual equipment, but they instead have a single set of conductors (pairs, if electrical) that services all the equipment as a group. Information is carried along conductors that are either coaxial (i.e., like TV aerial cable), a twisted pair of insulated copper wires, or, more recently, a fiber-optic cable, where the carrier medium is light. Since a single cable is used, it is clear that because all devices are connected to it, they could all try to transmit data simultaneously and along the same path. If this were allowed, then bedlam would surely prevail, as the messages would definitely collide, with intercorruption of the data. To bring order and avoid collisions, two different systems are adopted:

1. Token passing

2. Collision detection

What do these terms mean?

Token Passing

For an explanation of token passing, we shall again resort to an analogy. Consider an athletic track and one team of relay athletes (each analogous to a device) spread around the track; but the only runner who may start is the one who has the baton, and this runner stops (becomes inactive) on handover. Let us assume the running track is the network formed by the single interconnecting cable that has been designed so that the baton, called a *token*, or a permit to transmit or receive data, passes around the network on a set route continuously, and stops at each athlete (device) for a fixed period during which the device is allowed to receive or transmit information. Just as the athletes have to wait their turn, data receipt or data transmission is inhibited at all other times. If a device has no data available to transmit or it receives no instructions, then the athlete (device) passes the baton (token) on to the next athlete (device). The system is designed so that there is only one baton (token) in circulation on the network at any given time. From this it is clear that there will not be any data collisions, and furthermore, entry to the communications path (running track) is guaranteed. The protocol MAP 802.4 uses the token-passing procedure for the following reasons:

1. Every device is guaranteed to be able to receive data, or transmit instructions for action to be taken.

2. Token passing is *deterministic*, i.e., it is possible to calculate the time for the token to complete the circuit, and this is important when we consider the system response time to a change in the process conditions.

Both these aspects are vital to the manufacturing process industry, where speed of transmission and guarantee of receipt within a set period are of prime importance.

Collision Detection

The commercial world is not so fussy, since no business can, or does, react to changes in milli- or microseconds, as is the case in the manufacturing industry; hence, a different data transmission system called *collision detection* is employed. This system is much less costly than token passing and works as follows. Every device on the network "listens" for any other device that is transmitting, and if the network is clear, any device that has data to transmit starts its transmission and continues until it is completed. If, during the time a device is transmitting data, another device wishes to send information, it is blocked from doing so, and will have to wait until the network is clear again before it can start to send its data. The waiting (listening) period is short and of random time duration, and in this way the possibilities of gaining access to the network are increased. Delays with this system are inevitable, and of the order of one-half second, and therefore nondeterministic, i.e., it is not possible to calculate the time the device is on the network—but the waiting period is of little concern to the user, because the administrative side of the industry is unable to react to a change in fractions of a second, it can afford to wait one-half second, which is more than fast enough before making a decision, and for this reason TOP uses the collision detection procedure.

THE OPEN-SYSTEM INTERCONNECTION MODEL (ISO SEVEN-LAYER MODEL)

Both MAP and TOP use a layered strategy to achieve their objectives, as shown in Figure 21.1. The International Standards Organization (ISO) developed this concept, which works as follows: The model is referred to as the *open system interconnection*, or OSI, under which layers 1 to 4 are designed to achieve *reliable communications*, and layers 5 through 7 are to service *applications programs*. We shall now generally consider each of the seven OSI layers and their relation with MAP and TOP.

LAYER 1: THE PHYSICAL LAYER

The *physical layer* (1) in both MAP and TOP is based on the IEEE 802 Standard; MAP chooses 802.4, and TOP, 802.3. This offers three modulation techniques:

1. Single-channel continuous FSK—frequency-shift keying

2. Single-channel continuous FSK (carrierband)

3. AM/PSK (broadband)—amplitude modulated phase-shift keying

MAP uses AM/PSK at 10 Mbit/s, two-channel operation on a directional bus with re-modulator, 75-Ω trunk and drop cables. **Note:** A *bit* (binary dig*it*) is a single character that could be either 0 or 1, and M = mega or 1 million; hence, Mbit is 1 million bits, also abbreviated to Mb. A *byte* is a collection of bits, most commonly 6 or 8, but the number depends on the transmission packaging, etc. Bytes are usually measured in millions (mega) and abbreviated to MB, or Mbyte. Transmission is usually timed over a one-second period; therefore, abbreviations are Mbit/s, Mbps, or Mbyte/s.

The physical layer provides the physical means (i.e., the cables, modems, etc.) of transmitting data between two nodes, organizes the connection or disconnection of the network and its hardware, and is responsible for the electrical and mechanical

7	Application layer
6	Presentation layer
5	Session layer
4	Transport layer
3	Network layer
2	Data link layer
1	Physical layer

Figure 21.1: The International Standards Organization model.

operations that accompany this activity. In addition, it modulates all outgoing and demodulates all incoming data and puts the outgoing data on to the network.

MAP V3.0 allows for two transmission mediums, both conforming to IEEE 802.4:

1. Broadband technology operating at 10 Mbit/s

2. Carrierband technology operating at 5 Mbit/s

What is the difference between the two? *Broadband* transmits one way only (i.e., it is unidirectional) and has low attenuation and high immunity to electrical and magnetic interference. The most important thing is that it can be multiplexed, making it eminently suitable for carrying a great many different signals simultaneously. *Carrierband*, on the other hand, is bidirectional and is more reliable than broadband; and, because of its two-way communication capability at coincident times, data transmission is not subject to great delay. Carrierband is also a lot less costly; hence, it is most often used as the link at the manufacturing floor level.

LAYER 2: THE DATA LINK LAYER

The data link layer provides for media access. It is here that data is "queued" if collision detection is used, or where the token is held if token passing is implemented. Managing the token is the task of a sublayer called *media access control* (*MAC*), and linking to the communication path is the function of another sublayer called *logical link control* (*LLC*), with three options in the ISO 8802.2 Standard:

1. Type 1: connectionless

2. Type 2: connection-orientated

3. Type 3: acknowledged connectionless

There are four choices of services to support these three types of connection:

1. Class 1: supports type 1

2. Class 2: supports types 1 and 2

3. Class 3: supports types 1 and 3

4. Class 4: supports types 1, 2, and 3

Let us try to visualize what is happening at this level:

For type 1, connectionless: The data is labeled and is sent on its way when the link is available, it is really going off "blind," and therefore *no data acknowledgement, sequencing, flow control, or error recovery* is possible. A good analogy here is what happens when we mail an ordinary letter: the address can be absolutely correct, but once the letter is in the mailbox the sender has no more control over what happens to it; there is no certainty that the communication will even reach its destination.

For type 2, connection-orientated: Once again the data is labeled, but the *connection* between the sender and the recipient is established *before any data is transmitted*. With the connection made, the sender and the recipient have full control, and all segments of the data can be validated *before communication is terminated*.

For type 3, acknowledged connectionless: The data is labeled and sent on its way "blind," but in this case *the recipient has the ability to acknowledge receipt of the data*. A telephone call is a good analogy here, because when we dial a number we do not know whether the path is OK or not, but when the person dialed acknowledges, we know the link has been established and information can be exchanged.

MAP chooses type 1, class 4, for its speed and arranges for layer 4, transport, to fill in the missing functions.

LAYER 3: THE NETWORK LAYER

The network layer has the task of routing messages between two nodes, irrespective of whether the nodes are on the same or different interconnected subnetworks, and regardless of the physical distance between the two subnetworks. To get a feel for the complexities involved, let us see what does happen. At a given location we could have a local area network (LAN) that is providing a communication path for data exchange. Suppose additional data is required from another location that is being served by another LAN, and/or X.25—which will be discussed later in this chapter—telephone or radio links; then some means will have to be provided to link the different networks together and ensure data exchange. It is the task of the network layer to establish the best communications path, even if the path crosses fundamentally different subnetworks. How it does this is by constructing a consistent set of services from the inconsistencies of the different subnetworks. It also makes sure that the outgoing data packet is of the correct size for transmission, and that the physical address of the next recipient is correct; this is necessary because the data makes a sort of leapfrogging path to its final destination, all this activity being transparent to the layers above it. A reasonable analogy, perhaps, is a diner at a restaurant: all he sees is the waiter giving him the meal he ordered; he is totally unaware of the frantic hustle and bustle in the kitchen and the serving bay.

LAYER 4: THE TRANSPORT LAYER

The transport layer provides the session layer, the next layer up, with a network-wide data transfer service. The data path is error-free because of the work done by the network layer, thus allowing the layers above to carry out their tasks without having to concern themselves with achieving end-to-end data transfer. This layer marks the end of the communications ladder, and ensures that everything that needs to be done to connect one node to another and permit the free flow of data has been accomplished. Because MAP chooses type 1 (the simplest) for the data link, it chooses class 4 (the most comprehensive) to make up for the shortcomings. By doing this MAP allows data segmentation and reassembly, flow control, out-of-sequence data checking, error detection, and resynchronization to be available after error; these are functions that could be obtained if type 2 were chosen for the data link layer (layer 2).

The last three layers are concerned with the way the data is to be organized—in other words, the data has to be applicable to the process, and also has to be managed.

LAYER 5: THE SESSIONS LAYER

The sessions layer is the first of the layers that are applications-oriented. Layer 4 ensures the structure and synchronization of the data, but layer 5 is the one that coordinates the procedure. For example, when two devices—they could, for example, be processors—are interdependent, then each device must know the status of the other before dialogue can commence. A very loose analogy would be the function of the chairperson of a committee. It is the task of this person to ensure that the attention of all committee members is focused on the matter to be discussed, authorize the opening and closing of the discussion (session), be responsible for how the session is conducted, and manage the way the discussions proceed.

Very early on, we outlined the difference between comprehension and connectivity, and now can begin to see how the connectivity part has been implemented in MAP, and begin get a feel for what is to follow—i.e., comprehension, to which the last two layers of the model are devoted.

LAYER 6: THE PRESENTATION LAYER

Every computer manufacturer represents the data it handles in a way best suited to its own design, and because of this an *open communication system*—i.e., a communication system that will suit all—is very difficult to achieve. One very obvious way out of the difficulty is to have all the data handled in the same way, but this is something that will be almost impossible to implement, for not everyone will be prepared to do this, or possibly the data handled may not lend itself to universal standardization. Hence, the only way out of the dilemma would be some kind of "translator" that would make one system comprehensible to another.

MAP and TOP solve the problem by having a translator based on two types of notation:

1. Abstract syntax

2. Transfer syntax

The translator works as follows: Using *abstract syntax* notation on the data, strips it (data) of all its "personality traits" (analogous to converting all local dialects to "standard American speech") that make it specific to a particular system, and turns it into a format that is capable of being encoded at the lowest (i.e., bit) level, with the set of rules for encoding the abstract provided by the *transfer syntax*, but it must be appreciated that the transfer syntax depends on the particular type of data formats being translated. The sequence of the complete transmission is:

1. Apply abstract syntax

2. Encode into transfer syntax

3. Transmit data

4. Decode into abstract syntax

In view of the fact that data types differ from each other, selection of the abstract syntax is something that must be considered very carefully. All the conversions are carried out in this layer.

LAYER 7: THE APPLICATIONS LAYER

The applications layer is the uppermost of the layers, and is the one through which the network user gains access to the network. The applications layer deals with information in any form: data, text, graphics, video, or voice, whereas the layers below deal only with data.

In view of the fact that applications programmers use this layer exclusively, it is important that it should be versatile, and versatility is made possible by the association control services element (ACSE), the old acronym being CASE. This element is a "provider," for through it, and in collaboration with an applications service element (ASE), common communication parameters can be established. At the present time only four ASEs have been developed:

1. File transfer access and management (FTAM)

2. Directory services

3. Manufacturing message standard (MMS)

4. Network management

FTAM provides for the transfer of information from application programs and file stores, and guarantees the ability to work with text or binary files. To do this it has the usual attributes of a filing system, i.e., the ability to create, delete, read, write (create), change, and locate specific records. It supports random access, sequential, and single-key indexed sequential files; but because of the differing structure and organization of the various network user systems, the problems involved with providing the services mentioned are enormous. A solution is obtained by inserting a software device, which intervenes when file access is required and that operates on the basis of what all the files have in common; but to arrive at this commonality requires an abstract model, and in this case it is the *virtual file store*. This model defines the file in terms of its structure, its attributes (name, access mechanism, size, etc.) and the other facilities offered by the file. FTAM is transparent to the user; i.e., it works, but the user is unaware of it.

Directory services operate in the same way as telephone directory assistance does, in that every accessible object on the network is given a unique address. This address is made known to the directory services and recorded in the directory information base (DIB), and the directory service agent (DSA) is an interface through which the directory user agent (DUA) accesses the DIB. It is not difficult to visualize the DUA, which could be a person or a device, accessing the DSA with a query, and the DSA in turn interrogating the DIB, which conducts a search and returns with the answer; this is much the same sequence that a telephone operator, in the same circumstances, would take.

Network management performs the tasks of

• Obtaining information on the network usage by the various network devices

• Ensuring that the network functions correctly

• Producing reports on the network performance

This very complex functionality can be broken down very broadly into three roles, which can simply be described as:

1. The *network manager application*, which resides in its own node and operates through the *network administrator* (a person), who reads, alters, and controls the network. The administrator can also access the reports.

2. *System management application process* and *application entity*. These are present in all nodes. The management application process manages the system and collects

information on *all* the layers. The management application entity acts as a communications element.

3. *Manager/agent protocol* enables the network management function on a manager/agent basis using manufacturing message standard (MMS). This is a communal language that has its roots in the one invented by General Motors, and is not based on any programming language, but is a list of mnemonic instructions and messages to get shop floor devices to act or interact. The standard for numerically controlled machines is the EIA specification; for programmable controllers, it is the NEMA specification; for robot controllers, it is the RIA specification; and for process control systems, it is the ISA specification.

PACKET SWITCHING

We hear a lot of talk about X.25. What is it, and what does it do? To give an answer we shall look at the history of how it all came about. About 25 years ago the telephone lines began to carry data in addition to speech. Many subscribers of data communication found that the circuits then being used were inadequate for the purpose, due to (1) poor error control, (2) long delays in setting up a call, and (3) very high cost, especially for long-distance calls. In response to these concerns, the communications companies brought out *packet switching*. With this method, messages or "data" are broken into blocks with a label or *header* attached to each block; this label gives details to each switching node to enable (to make possible) the routing of the data, and the combination of header and data is called a *packet*. Since each packet is defined uniquely, it is possible to interleave packets of different data on the same paths, in that they *do not need end-to-end dedication*, i.e., predefined paths or routes in order to provide communication across the network. The switching node takes care of the routing, via the routing algorithms, which are very complex and outside the scope of this book. This new technology enabled the charge structure to be radically changed, from one based on distance to one based on the volume transmitted.

All data links are subject to interference from external sources, which can cause corruption of the data, and the data packet is no different. Many methods have been tried to reduce the effects of the interference, and so far the most effective, *high-level data link control (HDLC)* keeps the error rate to approximately 1 in 10.

Figure 21.2 shows a typical frame format of a data packet: Opening and closing flags of a single 8-bit word having the unique sequence 01111110 define *frame boundaries*. Therefore, the close flag of one frame can be the open flag of the next in a continuous sequence. If the transmission is discontinuous, then the interframe time is filled with a stream of continuous flags. To prevent the data within a frame from being corrupted by the flags, the transmitter checks the bit stream between two consecutive flags, and after 5 consecutive flags inserts a 0 (zero). The inserted zeros are deleted when the transmission is received.

Start					Finish
Flag	**Address**	**Control**	**Information**	**FCS**	**Flag**
01111110	8 bits	8 bits	*X* bits	16 bits	01111110

NOTE: FCS = frame checking sequence (check sum).

Figure 21.2: Composition of one data packet.

In addition to the opening and closing flags, the frame contains the following fields (sections):

The *address field* of 8 bits identifies the source of the data and its destination.

The *control field* of 8 bits is used for recording, sending and receiving sequence numbers, mainly; it is here also that the receipt of error-free frames and their correct sequence is recorded for intelligible reconstitution, and acknowledgment of the receipt of data is made to the sender.

The *information field* has as many bits as necessary to suit the message.

The *information field* is followed by a *frame check sequence (FCS)*, which is a standardized checking algorithm and results in a 16-bit word that is appended in this field. On receipt at the node, the procedure is repeated and another 16-bit word produced for the FCS. If there is no error between the two words, then the check will result in a known 16-bit word. If, on the other hand, there is a difference, then the frame is discarded and a request made to the sender to retransmit the data.

The X.25 protocol is defined in terms of a three-level structure:

1. *The physical level:* the mechanical, electrical, functional, and procedural characteristics to initiate, maintain, and deactivate the data terminal equipment (DTE) and the data circuit terminating equipment (DCE)

2. *The link level:* the access procedure between the DTE and DCE

3. *The packet level:* the packet format, control procedures for the exchange of packets, control information, and user data.

Figure 21.3 summarizes this structure.

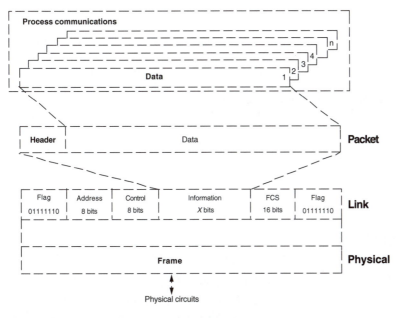

Figure 21.3: X.25 communications protocol.

SIGNAL PROCESSING

We have seen previously that analog signals can be generated by numerous different sensors, but these measurement signals, though very useful, can only be used directly and processed in very costly analog computers dedicated to specific tasks, and because of the dedication any changes required are effected only through a redesign of the relevant circuits, which is an extremely costly affair. Digital computers, on the other hand, manipulate digits, specifically 0s (zero) and 1s (one) to perform their tasks; and at this level we speak of encoding in *machine language*, reconfiguring the machine to fulfill another function is simply a matter of defining how to manipulate the digits. It is therefore little wonder that digital computers are so popular. As we have previously discussed, the majority of *process measurements* use an analog signal for transmission; therefore, to allow these to be used by the digital machine, a conversion has to take place.

FREQUENCY-SHIFT KEYING

For data transmission, one of the simplest techniques is to represent each digit by either of a unique pair of tones (frequency), one tone dedicated to represent a zero and the other to represent a one. This process is called *frequency shift keying (FSK)*, and it is not difficult to visualize that for very low bit rates the two tones approximate two spectral lines. As the bit rate increases, the spectrum of a data sequence spreads on each side of the mean frequency. For a given bit rate, FSK signals have and require a wider bandwidth than other methods, but the main advantages are:

- Constant power level

- Good signal-to-noise ratio (SNR)

- Relative simplicity of generating and detecting signals

FIELD/PROCESS SIGNALS

At present, the majority of signals going back and forth between the process plant and the control system are analog in the range 4 to 20 mA. This, by the way, is not the ideal signal; however, that is another subject and not covered in this text. Suffice it to say that there are other, more suitable means of communication. The first piece of systems hardware that the measurement signal encounters is the interface module, or an input device; other interfaces take the signal output from the control system to the field- (process-) mounted control actuator. Collectively, the input and output interfaces are called I/O. The input/output interfaces are designed to accept the usual signals available in most plant, which are typically:

Input

- 4 to 20 ma; 0 to 20 mA dc

- 0 to 10 V dc; 1 to 5 V dc

- 10 to 70 mV dc (typical) from thermocouples or analyzers

- 0 to 320 Ω (typical) from RTDs or other resistance elements (load cells)

- 1.25- to 12.25-kHz pulses at 24 V ac and 30 mA (maximum)

- 24 to 125 V dc
- 120 V ac (voltage monitoring)
- 240 V ac (voltage monitoring)
- Contacts (volt-free)

Output

- 4 to 20 mA; 0 to 20 mA dc
- Digital (transistor or contact/switch closure)

In the very latest designs the interfaces have microprocessors that converse with the system control and applications processors and so never need to be calibrated by an instrument engineer or technician.

The control processor scans each input at predetermined intervals, the intervals being set by the systems or process engineers. This enables "slow-changing" process parameters to have long scan intervals while "fast-changing" process parameters can be assigned shorter intervals. The scan periods are changeable on demand (via security) thus allowing the system to be adjusted to suit operating conditions; and short enough for "signal following" in *each* case, but not unnecessarily so to collectively overload the processor capability.

ANALOG-TO-DIGITAL CONVERTERS

We shall now consider analog-to-digital converters, which all work on the basis of sampling a continuous input and obtaining instantaneous values at discrete time intervals. Each of the instantaneous values are acted upon and converted into a series of binary digits (bits) representative of the instantaneous value, and an output is produced that is representative of it. Figure 21.4 explains the idea pictorially.

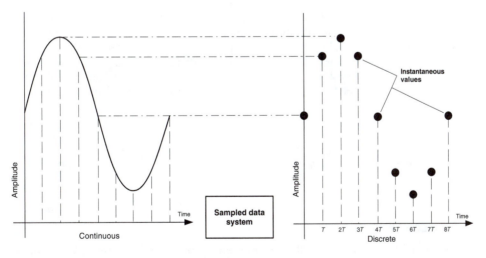

Figure 21.4: Analog-to-digital conversion.

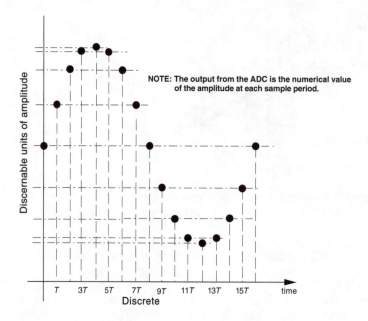

Figure 21.5: Schematic diagram of technique of analog-to-digital conversion.

The converter operates as follows. The analog signal is viewed at fixed intervals of T, which is the sampling period, and so it follows that $1/T$ is the sampling frequency, and at each interval the instantaneous amplitude of the signal is obtained and stored as a digital number; implicit in this is the fact that there must be a fixed range of units for the amplitude. It is the discrete value of the quantified amplitude for each sampling period that is stored, and because processors recognize only discrete numbers, the stored values are made available to it. Figure 21.5 is the sine wave of Figure 21.4, but shows that we can replicate the curve from a set of numbers alone and explains the idea. From the figure it is clear that the sampling frequency and accuracy of the value of the quantified amplitude will affect how well the conversion from analog to digital has been done, and this will affect the accuracy of the controller output. Data sampling rate affects the behavior of the system; as sooner or later an analog signal will be produced from this variable frequency pulse train, which may or may not appear smoothed, and the degree will depend on the sampling frequency.

What we have just explained is how individual signals are converted, and by using this, how a large variety of individual or separate signals can be "seen" by the process computer. But if this method was used, then each signal would require its own interface. There is a way, which uses only one ADC, and this method is called *multiplexing*.

MULTIPLEXING

In multiplexing systems all measurement signals are converted to a standard voltage variation, say -10 to $+10$ V dc, which is generally acceptable as a range, and then passed on to an ADC, which changes each analog voltage to digital pulses in the range, say, 0 to +4000, which corresponds to the associated input range of ±10 V. We can visualize the multiplexer as a large rotary switch (by large is meant a great number of inputs and not the physical size). The switch rotates continuously, stopping for a brief period at each input, during which the input signal instantaneously on that point is read by the processor in the analog-to-digital converter, but from this it should be clear that

1. The processor in the analog-to-digital converter will not see *all* the undulations of the measurement, because of the time interval of the switch rotation and the A/D conversion itself.

2. The smaller the time interval at each successive scan of each specific signal input point of the multiplexer, the better the input value to the processor in the analog-to-digital converter.

The output from the analog-to-digital converter processor can then be used where required throughout the system, as a stream of digital information.

AMPLITUDE MODULATION

Amplitude modulation is another method of transmitting data, and to see how this works, let us consider a signal of fixed relatively high frequency and amplitude on which we superimpose another varying signal. The signal of fixed amplitude and frequency is called the *carrier frequency*. The effect of the imposition of the varying signal on the carrier frequency will be to alter the amplitude of each wave in unison with the varying signal, giving the modulated carrier as in Figure 21.6. On arrival at the receiver, a unit called a *demodulator* extracts the modulation signal (envelope) from the modified carrier and so obtains the transmitted data. Theoretically, the carrier frequency can be eliminated and just the sideband(s) transmitted, for it does not itself contain any data; however, in practice, a pilot carrier is usually transmitted to facilitate signal recovery at the receiver by synchronizing, i.e., frequency-locking the demodulator. By multiplexing the modulation signal content and/or modulating onto subcarriers multiple signals can be sent. Figure 21.6 may help visualize the idea.

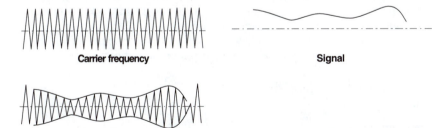

Carrier frequency   Signal

Amplitude-modulated carrier

Figure 21.6: Amplitude modulation.

PHASE-SHIFT KEYING

A more common type of modulation is *phase-shift keying* (*PSK*), in which the data signal changes the phase of the carrier by 0 or π radians (180°) corresponding to 0 or 1, and comparison against a reference carrier at the receiver decodes and reconstitutes the absolute values directly. In practice, the system is very susceptible to interference, leading to phase ambiguity (spurious phase generation); hence, a modified PSK is used. In this system each data element has a phase change corresponding to that of the preceding element; decoding is made possible by determining the relative phase changes. This method increases the noise immunity significantly and is called *differential phase-shift keying* (DPSK).

SIGNAL CONDITIONING USING OPERATIONAL AMPLIFIERS

We are aware that sensors can be classified either by the principle they use to measure the parameter or by the parameter measured. Signals are categorized generally according to the following terms:

- *Type of signal:* alternating, unidirectional, bidirectional
- *Kind of output:* current, voltage
- *Measurement/output relationship:* linear, nonlinear
- *Information content:* amplitude, frequency, phase
- *Output impedance:* load into which the signal will drive

Signal conditioning is used where required to make the signal suitable for use within other equipment, for example:

- To change current signals to voltage variations
- To change the output impedance capability
- To change voltage signals to current variations for data transmission
- To provide amplification, isolation, linearization

In modern electronic equipment, an *operational amplifier* usually accomplishes signal conditioning. This is an integrated circuit that behaves as a high-gain dc coupled amplifier with so much negative feedback that the closed-loop performance is defined primarily by the passive feedback components and not by the active device or amplifier itself. An *ideal op-amp* (operational amplifier) would provide:

- Infinite gain and bandwidth or frequency response with no noise
- Infinite input impedance
- Infinite common-mode rejection
- Zero output impedance
- Zero bias currents
- Zero input offset voltage

However, no practical op-amp has all of these characteristics, but they do approximate them. From a signal-processing point of view, the most important specifications for the device are:

- Common-mode rejection ratio
- Input bias current
- Input offset voltage and its drift with time and temperature

Let us see the effect that real op-amps have on closed-loop performance, in contrast to the effects of the ideal op-amps:

1. Infinite gain and bandwidth are replaced with finite gain and bandwidth, making the closed-loop system no longer dependent on the feedback components only, and there could also be some instability.

2. Infinite input and zero output impedance are replaced with finite input and output impedances, which can and do modify the feedback factors and alters the closed-loop input and output impedances from the ideal values.

3. Infinite common-mode rejection ratio is replaced with finite values, which makes the amplifier respond to the common-mode signals applied to its differential inputs. The result is an output component due to these signals.

4. Zero bias currents are replaced with definite currents, because the input stages of the differential amplifier have to be provided with dc bias paths, which cause bias currents that interact with the feedback resistors and produce additional offset.

5. The closed-loop amplifier adds noise to that already present in the signal.

6. An open-loop amplifier produces a finite dc output for zero difference of the input, resulting in an output offset voltage when the loop is closed.

Figure 21.7 shows the two basic methods of applying feedback to op-amps. The first is the *inverting* mode using a shunt feedback; and the second is the *noninverting* mode, using series feedback. These amplifiers are generally not used as

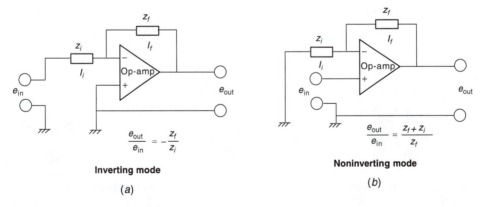

Inverting mode

(a)

$$\frac{e_{out}}{e_{in}} = -\frac{z_f}{z_i}$$

Noninverting mode

(b)

$$\frac{e_{out}}{e_{in}} = \frac{z_f + z_i}{z_f}$$

Figure 21.7: Basic feedback applied to operational amplifiers. (a) Inverting mode. (b) Noninverting mode.

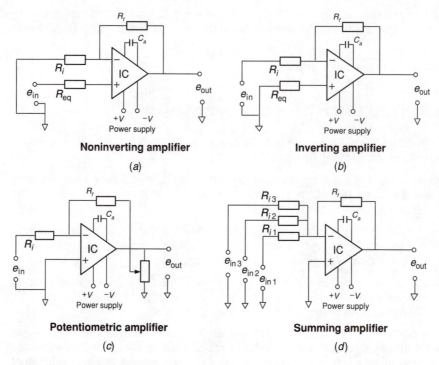

Figure 21.8: Op-amp circuits: (*a*) noninverting amplifier, (*b*) inverting amplifier, (*c*) potentiometric amplifier, (*d*) summing amplifier.

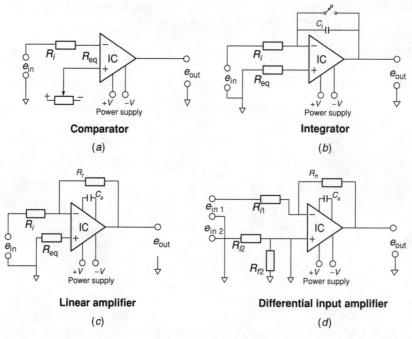

Figure 21.9: Op-amp circuits: (*a*) comparator, (*b*) integrator, (*c*) linear amplifier, (*d*) differential input amplifier.

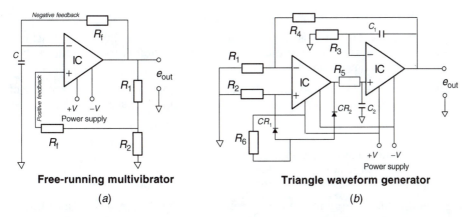

Free-running multivibrator

(a)

Triangle waveform generator

(b)

Figure 21.10: Op-amp circuits: (a) free-running multivibrator, and (b) triangle waveform generator.

instrumentation amplifiers, since measurement of the signal causes amplification of the common mode as well as the transducer signal.

The circuits shown in Figures 21.8 through 21.11 are a collection of op-amps circuits applicable to a real series of instrumentation that have been proven over a number of years of active service in the process control industry.

In Figure 21.9 the capacitor C_a in the two amplifiers shown is used for stability and not directly for frequency response; the reset switch shown in the integrator is used to reset the output voltage, for closing it during integration or at any time will instantly make the output voltage return to zero.

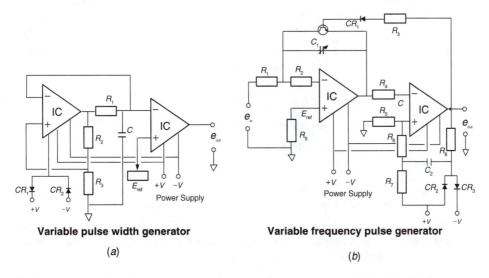

Variable pulse width generator

(a)

Variable frequency pulse generator

(b)

Figure 21.11: Op-amp circuits: (a) variable–pulse-width generator, and (b) variable-frequency pulse generator.

SUMMARY

1. Dialogue between humans and machine or between machine and machine, requires a code or protocol. If the machines are of different manufacturers then the code must have a standard format.

2. The two accepted protocols are MAP, concerned with the manufacturing side of industry; and TOP, concerned with the office or commercial environment.

3. Connectivity may be thought of as reliable end-to-end connection between dissimilar systems. Comprehension may be thought of as the means of enabling two different systems to understand each other. The MAP and TOP Protocols provide only connectivity to make a full working network; they do not provide comprehension, but contain protocols to support comprehension.

4. Since all devices are connected to the same network cable, they could all transmit data simultaneously and along the same path. If this were allowed, then the messages would collide. To bring order, two different systems are adopted: token passing and collision detection. The token, or permit to transmit or receive data, passes around the network on a set route continuously, and stops at each device for a fixed period during which the device can transmit or receive data. In collision detection, every device on the network "listens" for any other device that is transmitting; if the network is clear and a device has data to transmit, then it starts its transmission and continues until finished.

5. MAP uses token passing for the following reasons:

    a. Every device is guaranteed to receive or transmit data.

    b. It is possible to calculate the time for the token to complete the circuit, and this is very important when we consider the system response time to a change in the process.

   TOP uses collision detection, because its users are mainly administrators who are able to wait a short time (delays of the order of one-half second are inevitable with this system) before making a decision,.

6. Both MAP and TOP use a layered strategy, developed by the International Standards Organization (ISO) and referred to as the open-system interconnection, or OSI. Under this model:

    a. Layers 1 to 4 are designed to achieve reliable communications.

    b. Layers 5 through 7 are to service applications programs.

7. The physical layer (layer 1) provides the physical means (cables, modems) of transmitting data between two nodes and organizes the connection or disconnection of the network and its hardware and the electrical and mechanical operations that accompany this activity. It also modulates all outgoing and demodulates all incoming data and puts the outgoing data onto the network.

8. The data link layer (layer 2) provides for media access. It is here that data is queued, if collision detection is used, or where the token is held if token passing is implemented.

9. The network layer (layer 3) has the task of routing messages between two nodes irrespective of whether the nodes are on the same or different interconnected subnetworks, and regardless of the physical distance between the two subnetworks.

10. The transport layer (layer 4) provides the session layer with a networkwide data transfer service. The data path is error-free because of the work done by the network layer, thus allowing the layers above to carry out their tasks without having to concern themselves with achieving end-to-end data transfer. This layer marks the end of the communications ladder and ensures that everything that needs to be done to connect one node to another has been accomplished to permit the free flow of data.

11. The sessions layer (layer 5) is the first of the layers that is applications oriented. Layer 4 ensures the structure and synchronization, but layer 5 is the one that coordinates the procedure.

12. The presentation layer (layer 6) provides a "translation" service to make one system comprehensible to another, since every computer manufacturer represents the data it handles in a way best suited to its design.

13. The applications layer (layer 7) is the uppermost of the layers, and is the one through which the network user gains access to the network. The applications layer deals with information in any form: data, text, graphics, video, or voice; whereas the layers below deal only with data. Applications programmers use this layer exclusively; therefore, it is important that it be versatile.

14. MAP V3.0 allows for two transmission mediums, both conforming to IEEE 802.4:

    a. Broadband technology, operating at 10 Mbit/s. Broadband is unidirectional and has low attenuation and high immunity to electrical and magnetic interference. Most important, it can be multiplexed, making it eminently suitable for carrying a great many different signals simultaneously.

    b. Carrierband technology, operating at 5 Mbit/s, is bidirectional and more reliable than broadband. Because of its two-way communication capability, data transmission is not subject to great delay. Carrierband is also less costly; hence, it is most often used as the link at the manufacturing floor level.

15. Media access control (MAC) is a sublayer having the task of managing the token. Another sublayer called logical link control (LLC) has the function of linking to the communication path. There are three options given for this in the ISO 8802.2 Standard:

    a. Type 1, connectionless. The data is labeled, and when the link is available it is sent on its way, "blind"—no data acknowledgement, sequencing, flow control, or error recovery is possible.

    b. Type 2, connection-orientated. The data is labeled, but the connection between the sender and the recipient is established before any data is transmitted. With the connection made the sender and the recipient have full control, and all segments of the data can be validated before communication is terminated.

    c. Type 3, acknowledged connectionless. The data is labeled and sent on its way "blind," but in this case the recipient has the ability to acknowledge receipt of the data.

There are four choices of services to support the three types of connection:

    a. Class 1: supports type 1

b. Class 2: supports types 1 and 2

c. Class 3: supports types 1 and 3

d. Class 4: supports types 1, 2 and 3

MAP chooses type 1 (the simplest) for the data link and class 4 (the most comprehensive) to make up for the shortcomings—thus allowing data segmentation and reassembly, flow control, out-of-sequence data checking, error detection, and resynchronization after error.

16. The MAP and TOP translators are based on two types of notation:

a. Abstract syntax, used on the data, to strip the data of all its personality traits and turn it into a format that is capable of being encoded at the lowest level (bit) with the set of rules for encoding the abstract provided by the transfer syntax. Because data types differ from each other, selection of the abstract syntax must be considered very carefully.

b. Transfer syntax, which provides the set of rules for encoding the depersonalized data. The transfer syntax depends on the particular type of data formats being translated.

The sequence of the complete transmission is:

a. Apply abstract syntax

b. Encode into transfer syntax

c. Transmit data

d. Decode into abstract syntax

17. The versatility of layer 7 is made possible by the association control services element (ACSE; previously CASE). This element is a provider, for through and in collaboration with an applications service element (ASE), common communication parameters can be established. At the present time only four ASEs have been developed:

a. File transfer access and management (FTAM) for transfer of information from application programs and file stores. It guarantees the ability to work with text or binary files and has the usual attributes of filing systems. It supports random access, sequential and single-key indexed sequential files. It inserts a software device, which will intervene when file access is required and operate on the basis of what all the files have in common. To arrive at this commonality requires an abstract model, the virtual file store. This model defines the file in terms of its structure, its attributes (name, access mechanism, size), and the other facilities offered by the file. FTAM is transparent to the user.

b. Directory services, similar to telephone directory assistance. Every accessible object on the network has a unique address, known to the directory services and recorded in the directory information base (DIB). The directory service agent (DSA) is an interface through which the directory user agent (DUA) accesses the DIB. The DUA, which could be a person or a device, accesses the DSA with a query; the DSA interrogates the DIB, which conducts a search and returns with the answer.

c. Manufacturing message standard (MMS) is a communal language with its roots in that invented by General Motors. It is not based on any programming language, but is a list of mnemonic instructions and messages to get shop floor devices to act or interact. For numerically controlled machines, the standard is the EIA specification; for programmable controllers, it is the NEMA specification; for robot controllers, it is the RIA specification; and for process control systems, it is the ISA specification.

d. Network management performs the tasks of obtaining information on the network usage by the various network devices, ensuring that the network functions correctly, and producing reports on the network performance. This very complex functionality can be broken down very broadly into three roles: the network manager application, which resides in its own node and operates through the network administrator (a person), who reads, alters, and controls the network and can also access the reports; the system management application process, which manages the system and collects information on all the layers; and the application entity, which acts as a communications element, both of them present in all nodes; and the manager/agent protocol, which enables the network management function on a manager/agent basis.

18. X.25 is a protocol for data communications that is also known as packet switching. The packet is a combination of header and data. Messages or data are broken into blocks with a label or header attached to each block. This label gives details to each switching node to enable routing of the data. Since each packet is defined, it is possible to interleave packets of different data on the same path, with the switching node taking care of the routing. Packet switching paths are very different from circuit switching paths, in that they do not need end-to-end dedication to provide communication across the network.

19. A data packet consists of a six-section frame, the boundaries of which are defined by opening and closing flags of a single 8-bit word having the sequence 01111110; therefore, the close flag of one frame can be the open flag of the next in a continuous sequence. If the transmission is discontinuous, then the interframe time is filled with a stream of continuous flags. To prevent the data within a frame from being corrupted by the flags, the transmitter checks the bit stream between two consecutive flags, and after five consecutive flags inserts a 0 (zero). The inserted zeros are deleted when the transmission is received. The four intervening sections between the opening and closing flags of the frame in order of sequence from the opening flag are: the 8-bit address field that identifies the data source and its destination; the 8-bit control field mainly used for recording, sending, and receiving 'sequence numbers,' and also recording receipt of error-free frames and its correct sequence, and acknowledgement to the sender of data receipt; the information field has as many bits as necessary to suit the message; and the frame check sequence (FCS) that follows the information field, and is a standardized checking algorithm that results in a 16-bit word that is appended in this field, on receipt at the node the procedure is repeated and another 16-bit word produced, if there is no error between these two words then a known 16-bit word will result; if not, the frame is discarded, and the sender requested to retransmit the data.

20. The X.25 protocol is defined in terms of a three-level structure: the physical level, the mechanical, electrical, functional, and procedural characteristics to initiate, maintain, and deactivate the data terminal equipment (DTE) and the data circuit terminating equipment (DCE); the link level, the access procedure between the DTE and DCE; and the packet level, the packet format, control procedures for the exchange of packets, control information, and user data.

21. For data transmission, one of the simplest techniques is to represent each digit by a unique pair of tones (frequencies), one tone dedicated to represent a 0 and the other to represent a 1. This process is called frequency shift keying (FSK). For a given bit rate, FSK signals have a wider bandwidth than other methods; but the main advantages are constant power level, good signal to noise ratio (SNR), and relative simplicity of generating and detecting signals

22. In multiplexed systems all measurement signals are first converted to a standard voltage range, say, +10 to −10 V dc, and then passed on to an ADC, which changes the voltage to digital pulses in the range, say, 0 to +4000, which corresponds to the associated input voltage range of ±10 V.

23. Amplitude modulation is another method of transmitting data. The varying signal is superimposed on another signal of fixed frequency and amplitude called the carrier frequency. The effect on the carrier signal is an alteration of the amplitude of each wave corresponding to that of the varying signal. On arrival at the receiver, a demodulator strips away the carrier frequency to recover just the original data.

24. Phase-shift keying (PSK) is a more common type of modulation. The data signal changes the phase of the carrier by 0 or π radians (180°) corresponding to 0 (zero) or 1. A reference carrier at the receiver decodes the absolute values directly. The system is very susceptible to interference, leading to phase ambiguity; hence, a modified PSK is used. The alternative method of differential phase-shift keying (DPSK) increases the noise immunity significantly. In this system each data element has a phase change corresponding to that of the preceding element, and decoding is carried out by determining the relative phase changes.

25. Signals are generally categorized according to *type of signal* (alternating, unidirectional, bidirectional), *kind of output* (current, voltage), *measurement/output relationship* (linear, nonlinear), *information content* (amplitude, frequency, phase), or *output impedance* (load into which the signal will drive). Signal conditioning is used to make the signal suitable for use within other equipment, for example, to change current signals to voltage variations, to change the output impedance capability, to change voltage signals to current variations for data transmission, or to provide amplification, isolation, linearization.

26. Signal conditioning is usually accomplished by an operational amplifier, which is an integrated circuit that behaves as a high-gain dc coupled amplifier with negative feedback, so that the closed-loop performance is defined by the passive feedback components and not by the active device or amplifier itself. An ideal op-amp would provide infinite gain and bandwidth or frequency response with no noise, infinite input impedance, infinite common-mode rejection, zero output impedance, zero bias currents, and zero input offset voltage.

27. Real op-amps can only approximate the ideal characteristics. The most important specifications for an op-amp from a signal processing point of view are

common-mode rejection ratio, input bias current, and input offset voltage and its drift with time and temperature. Because practical op-amps have real characteristics and have the following effects on closed-loop performance:

- Infinite gain and bandwidth are replaced with finite gain and bandwidth, making the closed-loop system no longer dependent on the feedback components only. There could also be some instability.

- Infinite input and zero output impedance are replaced with finite impedances. These modify the feedback factors and alter the closed-loop input and output impedances from the ideal values.

- Infinite common-mode rejection ratio is replaced with finite values, making the amplifier respond to the common-mode signals applied to its differential inputs. The result is an output due to these signals.

- Zero bias currents are replaced with definite currents because the input stages of the differential amplifier have to be provided with dc bias paths, and these cause bias currents that interact with the feedback resistors and produce additional offset.

- The closed-loop amplifier adds noise to that already present in the signal.

- An open-loop amplifier produces a finite dc output for zero difference of the input, resulting in an output offset voltage when the loop is closed.

CHAPTER 22

Electrical Safety: Power and Grounding

THE BASIS OF ELECTRICAL SAFETY DESIGN

Safety in modern process plants has an important bearing not only on the people employed in manufacture, but also on users of the products. We shall confine our interest to the manufacturing side of the business alone and in particular to the avoidance—through proper electrical safety measures—of explosive risk to both plant and personnel. The risk due to fire generated by means other than electrical are equally important to deal with, but this aspect is beyond the scope of this text. As we all know most plant processing materials based on crude petroleum pose an explosive risk because of the highly inflammable nature of the material. These plants provide us with all grades of gasoline, industrial spirits, and the many chemicals we use everyday. However, even though we may not realize it, we should not forget that the distilleries (whisky/whiskey, gin etc.) are also susceptible to the risk of explosion. This short-listing shows the extent of coverage of process plants subject to strict safety measures. The main cause of explosions in such plants has been narrowed down to risks brought about by electrical wiring and services to equipment, for as we said before much of the motive, processing power, and all of the lighting is provided by this most useful, but at the same time unforgiving if mishandled, source of energy. The risk of explosion is categorized according to the susceptibility of a flammable atmosphere to igniting either from the energy released by a spark, in the main, when an electrical circuit is opened either deliberately or inadvertently (there are other causes, e.g., sparks generated by tools or footwear, but these are not of concern here and will therefore not be pursued), or because of a rise in surface temperature of the containing enclosure following a release of heat energy within it due to abnormal conditions i.e., a fault.

This chapter is only a guide; all applicable national and local regulations and practices concerning these aspects must be strictly followed and adhered to.

APPROACHES TO SAFETY DESIGN

A safe environment, in the context of what has just been said, is achieved in general by either of two methods:

1. Limiting any energy release into an explosive atmosphere to such a value that it is incapable of causing an explosion

2. For any hazard that may develop in an item of plant equipment, designing the equipment itself to contain the hazard and thus prohibit its spread

The first carries the name *Intrinsic Safety* (IS), and the second *explosion proofing* or *flameproofing*.

EXPLOSION- VERSUS FLAMEPROOF

"Explosion-proof" and "Flameproof" are two very different concepts as far as design specification and manufacturing construction is concerned. The commonality is in the fact that both *accept* a hazardous condition developing, but strictly within the confines of the equipment itself only. To briefly summarize the requirements:

In explosion proofing, the results of any explosion will be mechanically contained entirely within the very substantial equipment housing and any conduits connected to it.

In flameproofing, the fire resulting from any combustion will be entirely contained within the equipment housing. This requires all flame paths to be of a length that prohibits its escape from the affected equipment. The miner's Davy lamp is a good example of this.

In both methods the resulting surface temperature rise of the housing must be constrained to predefined limits, resulting in the design and use of very substantial metal enclosures.

In view of the complexity of the specifications, it is suggested that *any further detailed information be obtained directly from the published national standards codes of practice of the relevant countries.*

The items most likely to pose a hazard are mainly electrically powered. This statement assumes that naked flames, internal combustion engines, electrical storage batteries, and equipment using any of these, together with items that can be ignited by friction alone (e.g., matches), or otherwise without a source of ignition are by their very nature prohibited from the hazardous areas anyway. In "hazardous" plants all electrically driven motors must be flameproof or explosion-proof.

BODIES RESPONSIBLE FOR SAFETY IN EUROPE AND THE UNITED STATES: COMPARISON OF STANDARDS

The safety authorities in most European countries are government-sponsored bodies, whereas in the United States the responsibilities for safety are mainly with two private nonprofit organizations, Underwriters Laboratories (UL) and Factory Mutual (FM), these two companies are responsible for drawing up specifications and codes of practice to ensure safety.

In the European Community (EC) most member countries have their own national certifying authority, and an overall central body, CENELEC (Comité Europeén de Normalisation Electrotechnique) made up of delegates who are concerned with safety from each member country, that coordinates the safety specifications and codes of safety practices throughout the EC. CENELEC does not have any legal powers, but exerts considerable influence because of its membership. Any equipment used in hazardous locations within the Community, must carry certificates of approval by the relevant safety authority.

In the United States and Canada the author has found some difficulty in obtaining information on a "central body" equivalent to that in Europe, but the codes of practice to which both UL and FM comply in their testing and certification are based on specifications in the NFPA (National Fire Protection Association) publication 493 dealing with standards for intrinsically safe and associated apparatus; in Canada the Canadian Electrical Code publishes data for safety in industry. In the United States it appears that local regulations apply in the use of certified safe equipment in the manufacturing processes and when involved in work in any plant; it is therefore recommended that these regulations be consulted.

The IEC (International Electrotechnical Commission) is the world authority for standards in electrical and electronic engineering and composed of national committees in 43 countries, including the United States. Most countries with the exception of the United States are inclined to follow IEC recommendations. The IEC produces some very good work, but because of its international composition suffers from the inertia that dominates such organizations and results are therefore very slow in coming to fruition. CENELEC follows the IEC standards as far as possible, but has the advantage (as it is not as "international") of producing results faster.

We shall now draw a comparison between the IEC and US specifications for I.S. equipment to see where the differences lie. Most West European countries adhere to CENELEC standards, which are in turn are based on the IEC codes of practice.

LOCATIONS OR AREAS

To enable the correct equipment and protocols to be supplied and implemented, the authorities concerned with safety divide the plant into categories of locations or areas with appropriate and specific standards for each. The subdivision of classes in the United States is similar to that used in Italy.

IEC Zone 0: Explosive atmosphere present continuously or for long periods

IEC Zone 1: Explosive atmosphere likely during normal operation

IEC Zone 2: Explosive atmosphere not likely during normal operation

US Division 1: Flammable atmosphere present continuously, intermittently, or occasionally during normal operation

US Division 2: Flammable atmospheres present, but confined in closed containers from which it is possible to escape under abnormal operating conditions

OPERATIONAL CATEGORIES FOR EQUIPMENT

To provide protection to the plant and personnel, it would normally be acceptable for electrical wiring to equipment to be of such a design that it would protect the whole system within a hazardous area; but since intrinsically safe equipment is of such a design that energy release to the surrounding atmosphere is so small in the event of a circuit component breakage, short-circuit to earth, or between the power supply lines, mechanical protection to the explosion-proof/ flameproof specification is unnecessary. However, the insulation between the cable conductors must conform to certain standards. The following defines the allowable fault conditions permitted for safe operation of the certified equipment. The certifying authority specifies the allowable faults on the equipment, for example, capacitors in a circuit that can fail with serious consequences on energy release.

IEC Ex ia: Protection maintained with a single fault or up to two independent faults and a safety factor of 1.5 to relevant ignition data for normal operation or single fault condition, and a safety factor of 1.0 for a two fault condition. The equipment may be located in Zone 0, 1, or 2.

IEC Ex ib: Protection maintained with a single fault and a safety factor of 1.5 to relevant ignition data for normal operation, unless the equipment has no exposed switch contacts involved and the fault is one where it is apparent to the user and has to be corrected before plant operations can continue, in which case the safety factor is 1.0. The equipment may be located in Zone 1 or Zone 2.

United States (only one category): Protection maintained with up to two component faults or other faults. Equipment may be located in Division 1 or 2. The American category is similar to the IEC Ex ia, but IEC Ex ib is not recognized in the United States.

GAS CLASSIFICATION

The following listing gives the gas classification groups assigned by the certifying authorities in descending order of ease of ignition in air. To use the information provided in this list of IEC (European) and United States equivalents, look at the gas involved and the accompanying gas classification, e.g., for acetylene (the most susceptible)—it would be Class I, Group A in the United States, whereas in Europe it would be Group IIC. The United States has groupings for which there are none similar in Europe

IEC Group IIC Acetylene	US Class I, Group A	Acetylene
IEC Group IIC Hydrogen	US Class I, Group B	Hydrogen
IEC Group IIB Ethylene	US Class I, Group C	Ethylene
IEC Group IIA Propane	US Class I, Group D	Propane
IEC Group I Methane*	US Class II, Group E	Metal dust
(*known in the mining	US Class II, Group F	Carbon dust
industry as *firedamp*)	US Class II, Group G	Flour, starch, and grain
	US Class III	Fibers and flyings
	US Unclassified	Methane (mining industry)

TEMPERATURE CLASSIFICATION

The IEC has one additional classification, temperature, broken down into the following. There is no similar categorized specification in the United States

T1 450°C
T2 300°C
T3 200°C
T4 135°C
T5 100°C
T6   85°C

These values represent, under fault conditions, the maximum allowable surface temperature the equipment can attain from an ambient of 40°C.

A TYPICAL INTRINSICALLY SAFE LOOP

Having now outlined the standards to which intrinsically safe equipment has to conform mainly for Europe and the United States, we shall see how the results are achieved in a typical instrument loop, and we shall look at a typical interface device later in this chapter. The basic circuit of an intrinsically safe system can be broken down into the parts shown in Figure 22.1.

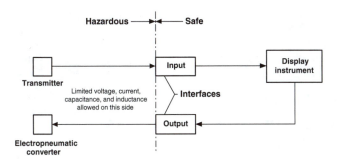

Figure 22.1: Basic requirement of an intrinsically safe circuit.

Let us assume that the system shown is to conform to IEC requirements, and therefore the instruments in the hazardous location must be certified for use in the area, which means that they must be suitable for either Zone 0 or Zone 1. (**Note:** Standard, uncertified transmitters and converters and those certified to Zone 2 requirements that are most often field mounted are generally not considered suitable for the plant and process shown in Figure 22.1. Instruments to Zone 2 certification are usually located in areas considered nonhazardous during normal operation).

When working in Europe, it must not be assumed that because a suitable interface is provided, the circuit is intrinsically safe irrespective of the instrumentation connected within it. At one time the whole circuit, including the instruments in the safe location, had to be certified by the appropriate authority, but this procedure was found to be unnecessarily constrictive and later abandoned. *Present requirements* allow the designer to assemble the circuit from instruments obtained from different manufacturers, *provided they are to the same standard of safety certification.* The instruments must be used only within the surface temperature limits of the gas classification group for which they are designed and certified.

SIMPLE APPARATUS

In both the United States and Europe, the only field-mounted apparatus that does not require certification is defined as "simple apparatus." The devices included are: thermocouples, resistive sensors, LEDs, and switches. However, it is imperative that, when these items are used, they must not be capable of generating, carrying, passing, or storing more than 1.2 V, 0.1 A, 20 mJ, and/or 25 mW.

PROCESS PLANT CABLING REQUIREMENTS

Multicore cables, when run between the hazardous and safe areas, must carry only noninterconnected intrinsically safe circuits, unless specifically permitted; additionally, in Europe, no intrinsically safe circuit will be permitted to operate at more than 60 V rms. When single wiring runs are used, they are considered as 'separated circuits'; the insulation thickness on each core must be 0.2 mm (in Europe) or 0.25 mm (in the United States and Canada), and the cables must also be protected against damage. Any overall screen must be grounded at one point only.

SHUNT DIODE BARRIERS

Shunt diode barriers are usually mounted in the safe location, although they can be mounted in the hazardous location, provided only that they are contained in an enclosure that is flameproofed or pressurized to Zone 1 requirements. The barriers must be grounded to the equipotential bonding system for the plant via an insulated, identified conductor, having a resistance of no more than 1 Ω, for reliability. A duplicate conductor is recommended. In the UK, a key or index to the system documentation for the protected circuits must be provided at the barrier location.

CATEGORIES OF INTERFACES: ZENER BARRIERS AND GALVANIC ISOLATORS

Let us now consider the interfaces shown in Figure 22.1 in a little more detail. These can be broken into two distinct technologies, Zener barriers and galvanic isolators. How do these operate? The following is a description.

THE ZENER BARRIER

The *Zener barrier* shown in Figure 22.2a uses a combination of resistances, fuses, and Zener diodes to provide a physical path between the safe and hazardous locations. The electrical power to drive the field device is generated in the safe area, and is fed via series resistances and fuses to limit the current going to the field device. The Zener diode regulates the voltage and limits the output terminal voltage to a predetermined value (the Zener voltage) and diverts the excess energy to ground. By these means the energy to the hazardous location is limited under all circumstances, while a through path from the safe to the hazardous locations is still provided. The field device uses this power to excite its circuits and controls the measurement signal back to the safe area on the same pair of conductors that brought the power to it; this is known as *two-wire transmission*, and the barrier device is called a *passive barrier.* A secure equipotential grounding system is required.

THE GALVANIC ISOLATOR

Galvanic isolators, shown in Figure 22.2b, provide a signal path between devices in the safe and hazardous locations without a direct physical connection between the two. They are very useful in situations where equipment with floating circuits in the hazardous area is required; such a device is called an *active barrier*. The isolation between the equipment in the hazardous and safe areas as shown in the figure is achieved using a transformer between the safe and hazardous locations; however, other isolating devices such as relays, or optical couplers, can also be used instead. In view of this it will be appreciated that this type of isolation is the only one that can meet the requirements of Zone 0. A secure, maintained safety ground is not a

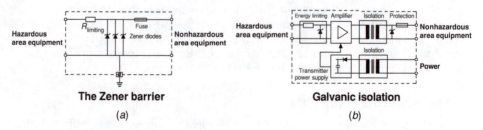

Figure 22.2: (a) Typical passive Zener barrier. (b) Typical active galvanic isolation.

fundamental requirement, but should be provided in the event that Zener barriers might also be connected at sometime in the same "area."

COMPARISON BETWEEN THE ZENER BARRIER AND GALVANIC ISOLATOR

Zener barriers and galvanic isolators each have their advantages and drawbacks:

Zener Barrier

1. Low cost.

2. Requires a secure equipotential grounding system, which must be tested on a regular basis to ensure safety.

3. No protection on short circuit; the safety fuses blow.

4. Some errors of measurement are possible, due to dependence on temperature of the Zener leakage current.

5. Can only be used on hazardous location devices that have 500 V isolation from ground.

6. If the voltage on the series resistance changes, then the length of the connection between the field and safe area equipment will be affected.

Galvanic Isolators

1. High initial cost

2. No need for secure grounding system

3. Short circuit protected

4. Isolated 4 to 20 mA signal output

5. Can be connected to grounded field devices

6. Supply voltage to field devices not limited by series resistor volt drop

INSTALLATION OF INTRINSICALLY SAFE CABLING AND EQUIPMENT

Wiring from and to the hazardous locations have to conform to specific regulations, the most striking being that the color of the cable insulation and terminals and wiring enclosures must be easily identifiable, light blue is the usually chosen color. There are other requirements also and in general these are as follows.

CONDUCTORS AND CABLES

The following listing should not be considered exhaustive; it gives those items considered to be of general importance. The national and local regulations will have to be complied with in all circumstances.

- The minimum conductor size is 0.017 mm^2.

- The minimum insulation thickness on the conductor is 0.2 mm (0.25 mm in the United States and Canada) polyvinylchloride (PVC) or equivalent.

- The cable must be capable of withstanding an insulation test of 500 V rms 50/60 Hz to ground for one minute.

- If screened conductors are used, then the screens must be grounded at one point only. In the field, screens must be isolated from ground and from each other.

- Multicore cables must carry intrinsically safe circuits exclusively; i.e., no mixtures of circuits with "non-IS" are permitted (i.e., wiring connections to instrumentation that do not carry safety certification).

- Multicore intrinsically safe cables should be laid in paths that are subject to the least mechanical damage. The cable should be fixed throughout its run length. When more than one IS cable is required to be laid, each IS cable should be run adjacent to the next, and none should carry a voltage exceeding 60 V peak.

- The overall cable armor of multicore cables should be solidly grounded via the junction box to the plant structure at one point only.

Figure 22.3 summarizes these requirements.

All cables can store energy, and the amount stored is affected by the capacitance and the inductance of the cable. If we consider the energy stored in a capacitor, then:

$$\text{Energy} = \tfrac{1}{2}cV^2$$

where c is the capacitance and V is the voltage. Hence, we can say that: *for a defined amount of energy the capacitance limit is determined by the voltage.*

And now, if we consider the energy stored in an inductor, then:

$$\text{Energy} = \tfrac{1}{2}LI^2$$

where L is the inductance and I is the current. Hence, we can say that: *for a defined amount of energy the inductance limit is determined by the current.* These principles are important for ensuring that the intrinsic safety requirements are not violated.

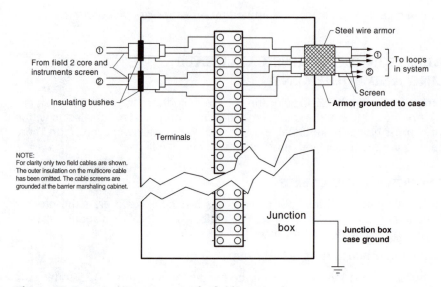

Figure 22.3: Typical arrangement of a field junction box.

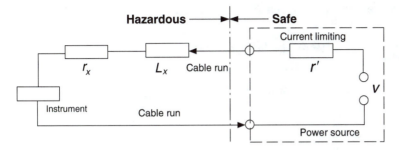

Figure 22.4: Typical cable run for an instrument.

WIRING CONNECTION TERMINALS

Terminals used on intrinsically safe wiring have to be of a distinctive color; light blue is generally the accepted color. There are other requirements also, but these concern the separation distance between the conductor holding screws, although the necessary distances have been taken care of by the terminal manufacturer in the design of the product.

An important aspect for consideration is the effect that the cable, which connects equipment in the safe location to the instruments in the hazardous location, has on safety. There are curves published that show the relationship between the inductance and capacitance of a circuit and the energy available for release for various voltages and currents. Very often the restriction in terms of the L/r ratio on cable inductance, i.e., on energy release rate, is quoted (as developed in the next section). The circuit shown in Figure 22.4 gives the detail of a typical configuration and the explanation below which generally follows that by H. G. Bass, in *Intrinsic Safety Instrumentation for Flammable Atmospheres* (Quartermaine House, 1984), gives good reasons for using it.

CABLE L/r RATIO

In the circuit of Figure 22.4, we assume that the inductance and the resistance of the cable can be shown as single-lumped electrical items; this is not strictly correct, as they are known as "distributed components," but it enables a visualization. Let the values of these be L and r per unit length, respectively, let the total length of the cable be x, and power source provide V volts through an internal resistance of r' and a current of I. Then, considering the power in the circuit, we can say:

$$W = \tfrac{1}{2}(Lx)I^2$$

$$= \frac{1}{2}Lx\left(\frac{V}{r'+rx}\right)$$

since

$$I = \frac{V}{R}$$

where Lx is the total inductance.

Collecting terms containing x, we can rewrite this equation as

$$W = \frac{1}{2}LV^2\frac{x}{(r'+rx)^2}$$

Differentiating the energy W with respect to x using the mathematical method of dy/dx when $y = u/v$, we have:

$$\frac{dW}{dx} = \frac{(r' + rx)^2 \frac{1}{2} LV^2 - \frac{1}{2} LV^2 x [2r(r' + rx)]}{(r' + rx)^4}$$

For a turning point value, dW/dx will $= 0$ (and the turning point will be at W_{max}) then:

$$0 = \frac{(r' + rx)^2 \frac{1}{2} LV^2 - \frac{1}{2} LV^2 x [2r(r' + rx)]}{(r' + rx)^4}$$

$$= (r + rx)^2 \frac{1}{2} LV^2 - \frac{1}{2} LV^2 x [2r(r' + rx)]$$

Dividing the right-hand side by $\frac{1}{2} LV^2$, we have:

$$0 = (r' + rx)^2 - 2rx(r' + rx)$$

$$= r'^2 + 2r'rx + r^2 x^2 - 2r'rx - 2r^2 x^2$$

$$= r'^2 - r^2 x^2$$

Taking the square root and transposing, we have

$$rx = r'$$

(i.e., the familiarly—from (electronics, etc.)—maximum power to load if load $=$ source). Use the equation:

$$W = \frac{1}{2} Lx \left(\frac{V}{r' + rx} \right)^2$$

(already shown) and rewriting it using the relationship $r' = rx$, we have

$$W = \frac{1}{2} Lx \left(\frac{V}{r' + r'} \right)^2$$

$$= \frac{1}{2} Lx \left(\frac{V}{2r'} \right)^2$$

$$= \frac{1}{2} Lx \frac{V^2}{4r'^2}$$

$$= \frac{1}{8} \left(\frac{V^2}{r'} \right) \frac{Lx}{r'}$$

$$= \left(\frac{V^2}{8r'} \right) \frac{L}{r}$$

since

$$\frac{x}{r'} = \frac{1}{r}$$

where also $V^2/r' =$ short-circuit current.

This shows that the energy is dependent on the ratio L/r only, provided that the supply voltage is constant, and independent of the cable length. To obtain the maximum value of the ratio L/r we must equate the above to the maximum power that is available in the circuit in other words:

$$\frac{1}{2}L_{max}I_{max}^2 = \left(\frac{V^2}{8r'}\right)\frac{L}{r}$$

where L_{max} is the maximum cable inductance for the maximum current I_{max} that can occur in the circuit. Therefore

$$\frac{L}{r} = \frac{1}{2}L_{max}I_{max}^2\frac{8r'}{V^2}$$

$$= \frac{4r'L_{max}I_{max}^2}{V^2}$$

Maximum current I_{max} will flow under short-circuit conditions, we can then write the above in terms of the current-limiting resistance r' of the source:

$$\frac{L}{r} = \frac{4L_{max}}{r'}$$

since

$$\frac{I^2}{V^2} = \frac{1}{r'2^2}$$

That is, L_{max} depends on source resistance r', and L/r of the external cable.

In this calculation it is assumed that all the inductance is in the cable and the instrument has zero inductance. If there is inductance in the instrument, then this should be subtracted from the cable inductance.

PLANT GROUNDING

Grounding and earthing are terms used to describe that part of an electrical circuit having a common voltage and serving as a datum or reference point for all voltage measurements that are to be made in the circuit. This datum point acts as the zero and does not have to be connected to earth ground, but an earth connection is normally used for the following reasons:

1. To fix the common point to earth potential, which is theoretically at zero volts

2. To provide a low resistance path to earth for dangerous currents, thus providing personnel protection and safe equipment operation.

The grounding system must include all power available at a site—i.e., distribution, lighting, equipment and accessories power, instrumentation, and signaling equipment. It is in this context that a grounding system cannot be considered for only one aspect, but has to take into account all the others as well, but in such a way that interaction and mutual interference cannot occur, and the integrity of one set can have no adverse effects on the others.

The flow of electrical currents in the earth is dependent on so many variables that it is impossible to predict or calculate the magnitude and distribution at any time.

It must, however, be said that there are mathematical expressions that have been developed for both ac and dc currents that use the idealized conditions of constant resistance and a perfectly homogeneous earth structure, assumptions that are rarely, if ever, encountered in practice. Notwithstanding this statement, observations of real situations allow some general statements to be made:

- The distribution of dc current flow in the earth depends on the resistivity of the earth. If it is assumed that the earth is homogeneous in depth between two electrodes spaced apart over a large area, then a dc current entering by one electrode and leaving by the other will cause the current to spread over a great distance and depth in seeking a path of least resistance between the two points.

- The distribution of ac current flow in the earth is influenced strongly by the inductive effect of the changing magnetic field, the inductive effects, sometimes ac and dc effects, etc., overriding the resistivity of the earth. The earth currents will distribute themselves to minimize the energy in the associated magnetic fields. Considering the earth return current beneath an ac power or transmission line, the current flows in a broad band on both sides of the line closely following the route, and restricted both in depth and width. For an overhead ac transmission line, the relation between current density and lateral distance, assuming homogeneous earth, is a very complex Bessel function. The current is a maximum directly below the line, gradually falling to zero at the extremity of the zone. The width of the zone varies directly with the soil resistivity, and inversely with frequency.

- The main variable that precludes accurate prediction is the resistivity of the earth, as no soil is perfectly homogeneous. In most instances, even though the surface may be relatively uniform, the subsoil is stratified, having layers of varying moisture content and absorbing capability, which makes the resistivity vary seasonally or even daily.

- Long-buried conductive material, such as metallic pipelines or uninsulated steel wire-armored cable, alters the distribution very considerably. Items such as these collect the earth currents over a very large area and carry them to vast distances.

- No satisfactory method has been developed to directly measure the current density or stray earth currents at a particular point. The ground currents arising from a specific source can be measured at the point where the current enters or leaves the earth—e.g., the grounded neutral on a transformer. The potential difference between two points on the earth's surface can be measured, and this will give an idea of the magnitude of the earth currents in an area.

COMMON-MODE VOLTAGE

Common-mode voltages are those that appear equally on each side of a signal line pair to a common reference point, which is usually the system ground. These common voltages are caused by magnetic induction, and by capacitive and resistive coupling, and of these the resistive couple between two separate ground points is the most troublesome. When this is the case, the stray ground potentials establish a voltage difference between the two ground points and act as a voltage generator, which affects the accuracy of transmitted signals. Figure 22.5 will clarify the situation.

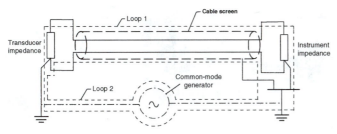

Figure 22.5: Common-mode voltage. Loop 2 is a low-impedance ground loop. A certain amount of total loop current will develop an *IR* drop across the transducer in loop 1.

Common-mode voltages can be divided in three categories:

- Earth common voltages
- Transducer common voltages
- System common voltages

The circuits in Figure 22.6 show typical examples of each.

EARTH CURRENTS

Earth currents are always present in areas where there is heavy electrical equipment in use. When instrumentation is to be placed in such areas, it is most important that the effects of these earth currents be minimized. One easy, but not always practicable, way is to place the instrumentation far away from the electrical gear so that the resistance of the earth itself reduces the magnitude of the earth current. When high-power voltage lines are buried, the earth currents are still associated with the cable

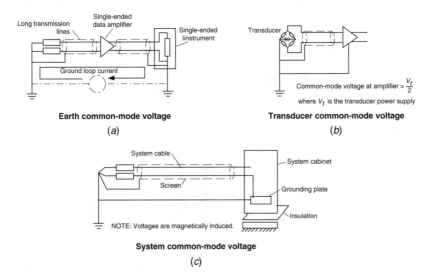

Figure 22.6: Common-mode voltages: (*a*) earth, (*b*) transducer, (*c*) system.

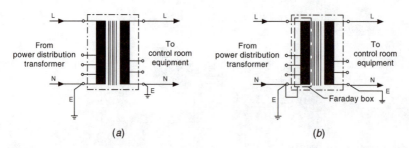

Figure 22.7: (*a*) System power isolation transformer. Primary neutral is grounded at source. (*b*) System power isolation transformer with Faraday box on primary. Primary neutral is grounded at source.

and follow the direction of the transmission line. It is recommended that a perpendicular relationship, i.e., to cross or run at right angles to each other, between instrumentation lines and power lines be maintained, and care must be exercised to avoid any instrumentation conduit passing under or parallel to such power transmission lines.

Within the control room itself, the grounding of the low-level, low-frequency instrumentation cabinets should be separate from the normal ac power equipment circuits and grounding. The grounding of high-frequency circuits such as radar stations is vastly different from that of low-frequency instrumentation. As a general rule, in high-frequency installations, grounding should consist of several unconnected paths, each originating in a particular module and terminating in a common ground plate. The paths themselves must have a very low impedance and be as short as possible, because as the frequency increases, the inductance and impedance in the center of the wire increase very greatly, to the point where they affect the current in this area, causing it to flow at the surface of the conductor; thus, the useful current-carrying area of the conductor is reduced and the effective impedance of the conductor increased—the well-known (in electronics) "skin effect." Having several ground paths leads to a lowering of the impedance offered to the current and divides the total current among the several conductors.

THE USE OF ISOLATION TRANSFORMERS

To minimize the effects of circulating earth loops, the arrangements in Figures 22.7 and 22.8 give details of the connections to isolation transformers. It should be noted that protection of the transformer windings and core as shown in Figure 22.7*b*

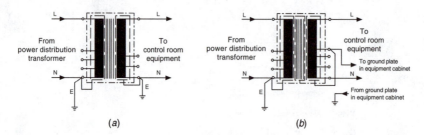

Figure 22.8: (*a*) System power isolation transformer with box shields on primary and secondary windings. (*b*) System power isolation transformer with box shields on primary, secondary, and core.

and Figure 22.8*a* and *b* from the influence of circulating ground currents is part of the transformer design and achieved during manufacture by the inclusion of (often multiple) screens located round section of the windings and core, each being provided with appropriate connections to internal and/or external (common) grounding points, and therefore cannot be added at a later stage.

UK REQUIREMENTS

In the United Kingdom, the IEE regulations require all safe area equipment powered from the electrical distribution grid, irrespective of whether it is are protected by barriers or not, to be supplied with power from a double-wound isolation transformer. For intrinsic safe systems, the output from the secondary of the isolating transformer must be restricted to 250 V rms maximum. It is normal for the instrumentation power to hazardous locations to be dc and normally 30 V maximum, preferably lower.

TYPICAL POWER AND GROUNDING ARRANGEMENT FOR AN INSTRUMENTATION SYSTEM

Figure 22.9 shows a typical power and grounding arrangement for an instrumentation system. *All national and local regulations must be strictly adhered to, and this may alter the arrangement shown.* Also note the conductors for the power distribution have to be

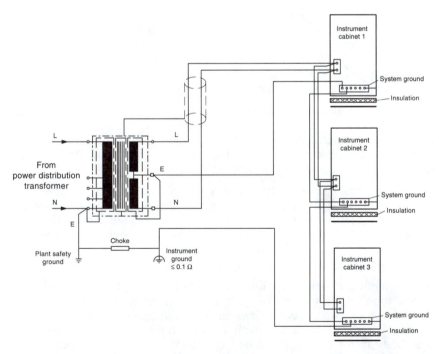

Figure 22.9: Typical power and grounding system. In this example only three cabinets are shown additional cabinets can be included if required. An alternative is to have a common instrument ground plate with each cabinet wired to it. This common point is called a star point and is grounded as shown.

of adequate rating and size to carry the current that is drawn by the circuits without excessive volt drop (regulation) or power heating. The ground conductors between cabinets and/or to *star point* should be a minimum of 4 mm^2, preferably larger. The ground plates in each cabinet must be isolated from the body of the cabinet, which must be grounded to the safety ground in order to protect the personnel using it. The choke since it connects the instrument and plant safety grounds together provides a back-up safety path for personnel under fault conditions, while preventing a further earth loop with its high impedance as far as high-frequency interference is concerned and for other valid reasons must be one that will take at least 100 A.

Within the cabinet, every care should be taken to segregate ac power and dc signal cables. This precaution will minimize the possibility of induced power line interference in the signals. When safety barriers and/or isolators are used, these should also be mounted on isolated mounting rails and easily accessible. Signal wiring to and from the safety barriers and isolators must, as mentioned earlier, be carried out in cable having a distinctive insulation color and be contained in separate cable trunking, also in the same distinctive color. It must be remembered that the path taken by the current under fault conditions will depend on the design of the circuits in the equipment and the nature of the fault. There will be a return path through the 0 V line to the secondary winding of the isolation transformer, and this will be grounded at the barrier.

SUMMARY

1. A safe environment is, in general, achieved by either limiting the energy release into an explosive atmosphere to such a value that it is incapable of causing an explosion or designing the equipment itself to contain the hazard and thus prohibit its spread. The first of these methods ensure intrinsic safety, whereas the containment approach is a matter of either explosion proofing or flame proofing, depending on the housing design criteria used.

2. In the United States, definitions of hazardous plant conditions, as well as intrinsic safety standards for equipment, are set by Underwriters Laboratories, Inc., and Factory Mutual Research Corporation. In the European Community, the safety authorities are government-sponsored bodies; most countries follow the CENELEC standards, which are based on IEC codes of practice.

   - Areas are classified according to presence of flammable atmospheres: in the United States, as Division 1 (flammable atmosphere present to any degree during normal operation) or Division 2 (present only under abnormal conditions); in Europe Zone 0 (explosive atmosphere continuously present), Zone 1 (explosive atmosphere present during normal operation) and Zone 2 (not likely under normal conditions).

   - Equipment is classified for intrinsic safety according to number of faults permitted. In the United States, protection must be maintained up to a maximum of two faults for Division 1 or 2. In Europe, equipment falling into IEC category Ex ia a single fault or up to two independent faults and a safety factor of 1.5 to relevant ignition data for normal operation or single fault condition, and a safety factor of 1.0 for a two fault condition is acceptable for use in Zones 0, 1, or 2; and that in category Ex ib a single fault and a safety factor of 1.5 to relevant ignition data for normal operation, unless the equipment has no exposed switch contacts involved and the fault is one where it is apparent to the user and has to be corrected before plant operations can continue, in which case the safety factor is 1.0 only in Zones 1 and 2.

- Gases are grouped according to the ease of ignition in air: in the United States, in ascending order of class and group, with acetylene listed in Class I Group A, and air containing fibers in Class III; in Europe the gases are arranged in descending order of group and there are only two groups, acetylene and hydrogen listed in Group IIC ethylene and propane in Group IIB, and methane in Group I.

- The IEC also has a safety classification based on temperature.

3. In Europe, the only field-mounted apparatus that does not require certification is defined as simple apparatus, comprising thermocouples, resistive sensors, LEDs, and switches. Items in use must not be capable of generating or storing more than 1.2 V, 0.1 A, 20 mJ, or 25 mW; this is also applicable in the United States.

4. Multicore cables run between hazardous and safe areas must carry only noninterconnected intrinsically safe circuits unless specifically permitted. In addition, in Europe no intrinsically safe circuit is permitted to carry more than 60 V rms. Single wiring runs are considered separated circuits, and the insulation thickness on each core must be 0.2 mm (Europe) or 0.25 mm (United States, Canada) and protected against mechanical damage. Any overall screen must be grounded at one point only.

5. Shunt diode barriers are usually mounted in the safe location, but they can be mounted in the hazardous location provided that they are contained in an enclosure that is flameproofed or pressurized to Zone 1 requirements. The barriers must be grounded to the equipotential bonding system for the plant via an insulated, identified conductor having a resistance of no more than 1 ohm; a duplicate conductor is recommended. In the UK, a key to the system documentation for the protected circuits must be provided at the barrier location.

6. The Zener barrier also known as a passive barrier uses a combination of resistances, fuses, and Zener diodes to provide a physical path between the safe and hazardous locations. The Zener diode clamps the output voltage to a predetermined value and diverts the excess energy to ground. Hence, the energy to the hazardous location is limited under all circumstances.

7. The galvanic isolators, also called an active barrier, provides a signal path between devices in the safe and hazardous locations, but without a direct physical connection between the two, achieved by using transformers, relays, or optical couplers between the safe and hazardous locations. This type of isolation is the only one that can meet the requirements of Zone 0.

8. Wiring from and to the hazardous locations has to conform to specific regulations. Most visible is color of the cable insulation, terminals, and wiring enclosures, which must be easily identifiable. Light-blue is the usual choice. Other requirements of general importance: The minimum conductor size is 0.017 mm^2; minimum insulation thickness is 0.2 mm (0.25 mm in the United States and Canada) polyvinylchloride (PVC) or equivalent; capable of withstanding an insulation test of 500 V rms. to earth for one minute; multicore cables must carry intrinsic safe circuits exclusively and not carry a voltage exceeding 60 V peak; screens must be grounded at one point only; the overall cable armor of multicore cables should

be "solidly earthed" to the plant structure. However, national and local regulations on installation will have to be complied with in all circumstances.

9. All cables can store energy. The amount stored is effected by the capacitance and the inductance of the cable. The energy stored in a capacitor is:

$$\text{Energy} = \frac{1}{2}cV^2$$

where c is the capacitance and V is the voltage. The capacitance limit is determined by the voltage. The energy stored in an inductance is:

$$\text{Energy} = \frac{1}{2}LI^2$$

where L is the inductance and I is the current. The inductance limit is determined by the current.

10. The maximum value of the ratio L/r is given by:

$$\frac{L}{r} = \frac{1}{2}L_{max}I_{max}^2\frac{8r'}{V^2}$$

$$= \frac{4r'L_{max}I_{max}^2}{V^2}$$

Maximum current I_m will flow under short-circuit conditions; therefore, in terms of the current limiting resistance r' of the source:

$$\frac{L}{r} = \frac{4L_{max}}{r'}$$

since

$$\frac{I^2}{V^2} = \frac{1}{r'^2}$$

11. Grounding and earthing are terms used to describe that part of an electrical circuit having a common voltage and serving as a datum point for all voltage measurements that are to be made in the circuit, and acts as the zero. It does not have to be connected to earth ground, but an earth connection is normally used to fix the common point to earth potential, which is theoretically at zero volts, and to provide a low resistance path to earth for dangerous currents, for personnel protection and safe equipment operation. The grounding system must include all power available at a site—i.e., distribution, lighting, equipment and accessories power, instrumentation, and signaling equipment.

12. The flow of electrical currents in the earth is dependent on so many variables that it is impossible to predict or calculate the magnitude and distribution at any time. The distribution of dc current flow in the earth depends on the resistivity of the earth, and that of ac current flow in the earth is influenced strongly by the inductive effect of the changing magnetic field, the inductive effects overriding the resistivity of the earth. The main variable that precludes accurate prediction is the resistivity of the earth, as no soil is perfectly homogeneous. Long-buried conductive material alters the distribution very considerably. No satisfactory method has been developed to directly measure the current density or stray earth currents at a particular point.

13. Common-mode voltages appear equally on each side of a signal line pair to a common reference point, usually the system ground. These common voltages are caused by magnetic induction, and by capacitive and resistive coupling; the

resistive couple between two separate ground points is the most troublesome, as the stray ground potentials establish a voltage difference between the two ground points and act as a voltage generator, which affects the accuracy of transmitted signals. Common-mode voltages can be divided in three categories: earth common voltages, transducer common voltages, and system common voltages

14. Earth currents are always present in areas where there is heavy electrical equipment in use, the effects of these earth currents must be minimized when instrumentation is to be placed in such areas. An easy, but not always possible, way is to place the instrumentation far away from the electrical gear or maintain a perpendicular relationship, i.e., cross or run at right angles to each other, between instrumentation lines and power lines, and avoid any instrumentation conduit passing under or parallel to power transmission lines.

15. In the United Kingdom, the IEE regulations require all safe area equipment powered from the electrical distribution grid, whether protected by barriers or not, to be supplied with power from a double-wound isolating transformer. For intrinsic safe systems, the output from the secondary must be restricted to 250 V rms maximum. It is normal for the power to hazardous locations to be dc and normally 30 V maximum, or lower.

CHAPTER 23

Process Alarm Systems

THE PHILOSOPHY BEHIND PROCESS ALARMING

In event of a malfunction, all manufacturing processes require some means of advising the operating personnel that the plant, or the process or both are moving toward a condition where either the product specification, product quality, or safety of plant or personnel could be affected. Notice that in the foregoing statement we have said "moving toward"; this is very important. The reason for raising an alarm *prior* to reaching an unwanted state is to enable safe corrective or evasive action to be taken in order to avoid the aftermath of not being prepared, or of learning about a malfunction when it is already too late to correct it.

The process engineer and the process chemist are the most qualified people to stipulate the conditions to avoid to ensure good product and a least hazardous environment. In matters of plant equipment and personnel safety, the plant maintenance engineer, safety engineer, process engineer, and process chemist are those who must be consulted to determine the precautions and procedures in the event of any malfunction in the process or the processing equipment in use.

Let us visualize the considerations that go into reaching such decisions. Later in this chapter, we shall see how a process and instrument diagram may be of assistance; for now, however, we must look at the scenario first so that we can understand the philosophy. Alarms can be initiated by:

- A measured variable outside the band of acceptable variation.

- Failure or malfunction of an item of plant equipment involved in the manufacturing process, including control valves.

- Failure or malfunction of the measuring device itself, which includes the sensor and the transmitter as a single entity. Device failure or malfunction therefore means that the cause could be in either the sensor or the transmitter or both.

- Failure or malfunction of the interconnections (transmitter, signal cable, or receiving instrument) between the field-mounted and control room-mounted devices' input and output.

- Low level or loss of motive power, which could be electrical, steam, hydraulic, or pneumatic, for example.

Most distributed control systems have alarms built in to signal the failure of any item included in the system; and if redundancy is designed in as well, then changeover to standby units should be seamless, but the operating personnel will always be made aware of the situation so that corrections or replacement can be effected and the level of protection maintained at all times. Priority of actions to be taken in the event that an alarm is triggered depends on the severity of the consequences.

Critical measured/monitored parameter(s)	Will warrant immediate action
Process in critical events	Will require an immediate response
Not so critical measured/monitored parameter(s)	Can tolerate some delayed action, but with an estimate of the time allowable
Process in not so critical events	Can tolerate some delayed action, but with an estimate of the time allowable

In the foregoing, if the alarms triggered in the process are continually repeated then not only will they have to be dealt with at the priority shown, but also personnel procedure(s) to be followed must be reviewed at the relevant priority level as process changes, present alarm settings, product specification margins, or plant design versus operating conditions may be involved. The following are some of the more common considerations to take into account when planning for hazardous conditions arising during plant operation, but remember, there could be, and usually are, several more specific ones:

1. Can initiating some process action, e.g., automatic addition of a substantial quantity of reagent to counteract the initiating condition and restore equilibrium?

2. Can the countermeasures be called upon immediately, i.e., *without* operator intervention, or, if allowable, with minimum delay when required?

3. Can the hazardous condition affect the safe working of the personnel, site property, and/or the site equipment in the immediate surrounding or adjacent areas of the plant?

4. Can the hazardous condition affect the environment, property, and persons adjacent to the plant?

5. Can product in mid-process be saved after the hazard has been contained?

6. If not, can it be disposed of with minimum risk to personnel, property, and environment?

To indemnify the plant owners plans must be in existence for the foregoing items 5 and 6, and invoked for remedial actions and procedures, so that when these situations arise they can be dealt with expeditiously.

PROCESS ALARM SETTING

Figure 23.1 shows how process alarms might be set out, but be aware that the configuration of the alarm levels and the assignments are only typical; there are many other arrangements that can be tailored to the individual process and operational requirements. The inclusion of suitable alarms applies equally to the storage of raw materials and to fully or partially processed products.

For the example shown in Figure 23.1 (which concerns the process alone, not how these are interpreted as process alarm settings on the instrument), the operating point to give quality product has been chosen to lie midway within the band of "marketable product," which is the product's specification limits and cannot change if the product is to be viable. The process operating point can, on the other hand

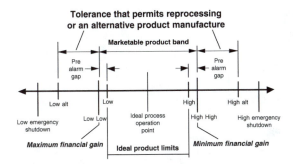

Figure 23.1: Typical process alarm settings. Alarms do not have to be symmetrical.

be allowed to change—it needn't fall midway—and we shall assume that a product within specification could equally well be obtained with a process operating point having a lower measured value than shown in the figure, provided it did not violate the product's lower specification limit. This, we should agree, would be acceptable as far as product quality is concerned. However, let us consider another point. If we first look at the production cost of the product with the operating point as shown in the figure, this will give a particular monetary value. Let us now lower the operating point to a position more toward the low alarm and close up the gap between the instrument's low and high limits, to make sure that the modifications carried out are still within the original specification limits, and operate the process at this point to assess the operating cost. We shall hope to find that the production cost has dropped because we are able to produce more within the same time scale and at less support expense— e.g., fuel cost—but the important point is that the product is still within the specification limits and the profit margin on it has increased. This very simple exercise shows that lowering the control point and closing up the alarm settings achieves "tighter" control, facilitates staying within specification, and increases profitability. To do this, we will need the enabling instrumentation, and what has been said is only by way of example, but it is important to appreciate that each case has to be treated on its merits.

PROCESS ALARM INITIATION

TWO-WIRE TRANSMISSION CIRCUITS

Let us now consider how the various alarm points are obtained and initiated, and for the moment let us look at an analog system to get an appreciation of how it has all been brought about. Consider the instrument loops as shown in the Figure 23.2, depicting the two ways in which power supplies can be connected to produce a current measurement signal in the range, say, 4 to 20 mA. Of the two arrangements, the externally powered loop can be used only in nonhazardous environments, whereas the internally powered can, with suitable barrier devices (not shown), be installed in hazardous areas not classified as Zone 0 (as defined in Chapter 22). The important things to appreciate are the limiting value of the load— i.e., the readout instrument that can be driven by the measurement signal—and the fact that signal and power, which are separately derived, but superimposed, are carried on the same pair of interconnecting conductors, the transmitter *controlling* the power or current from the source with the 4 mA still enough to ensure that the transmitter is able to operate properly within its specification, and the range

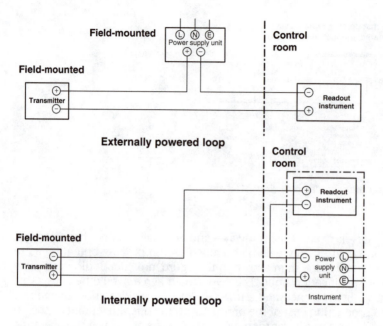

Figure 23.2: Typical two-wire transmission circuits (4 to 20 mA).

(swing) of 16 mA from 4 to 20 mA being the measurement signal, i.e., measured variable constituent, within the combination. Since the conductors have resistance, usually stated by the cable manufacturers in ohms per foot or meter, the total resistance of the cable run must be included in determining the load into which the signal is to drive, plus any series contribution by the barriers for reasons defined later.

Figure 23.3 correlates the power supply with the current signal produced by the sensor/transmitter, and the total load into which the signal can drive. It is usual for all instrument manufacturers to provide a graph similar to the one shown in the figure to allow the user to calculate the resistance of the loop and so choose the appropriate instruments to connect. From this graph it can be seen that, with a supply voltage of approximately 18 V, it is possible to drive into an extra load of approximately 250 Ω, and as stated earlier, this includes the resistance of the interconnecting wire. If this were to be the case, then very few practical loops could be implemented. Increasing the supply voltage on the basis of the graph to, say, approximately 26 V allows much more to be accomplished—e.g., an indicator, a recorder, or a controller can be connected. Furthermore, because of the higher load capability (approximately 625 Ω), it may be possible to connect both a recorder and a controller in the circuit, assuming that each instrument presents a load of no more than about 250 Ω and to allow sufficient capability for the interconnecting wire as well. For the instrument shown in the figure, the limit of the output load is approximately 1500 Ω, with a supply of approximately 42 V. **Note**: The power source may have some effective resistance at its output terminals, including any current-limiting items, *and* the loop will have to include effective series barrier elements where appropriate. Be aware that most if not all modern transmitters are two-wire devices (power and signal on the same conductor pair) having a 4 to 20 mA current output. *In all current loops, the loads, which form the loop, must be connected in series.*

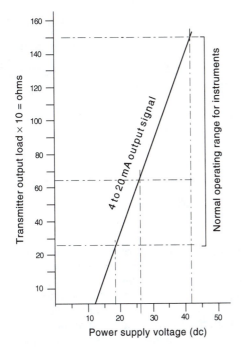

Figure 23.3: Transmitter output load capability.

SETTING THE ALARM TRIPS

We shall now discuss how alarms are set and initiated, and to do this we shall go back in time to see how they developed, for in many instances the principles are still valid. Figure 23.4 is a diagram of a design for an electrical current-actuated indicating alarm instrument; there are also designs of temperature alarms with a completely mechanical (bimetallic) sensor that initiates electrical contacts—e.g., the popular room thermostat used with central heating systems. In both of these designs the alarm trip point setting is achieved mechanically.

In the indicator shown in Figure 23.4, the measurement pointer and the two mechanically operated setters with electrical contacts one for each trip point move

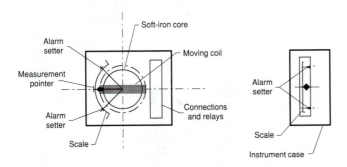

Figure 23.4: Typical indicator with alarm contacts.

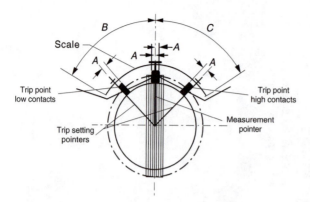

Figure 23.5: Enlarged view of indicating alarm.

across the whole length of the scale (instruments with a single alarm are also obtainable). Because of the mechanics of the arrangement, the setters for the trip point will have a restricted movement; and when using a large measurement span, together with a small alarm span, problems will be encountered, i.e., by mutual interference between the two setters due to the mechanical design of the instrument. Figure 23.5 shows what could occur when dual alarms with a very small alarm span are set up (**Note**: For explanation purposes, the size of the alarm contacts has been exaggerated and assumed to travel on the same arc; the problem disappears if the arcs are different.); since it is the contact on the measurement pointer that actually initiates the alarm. In Figure 23.5 the dimension A is the thickness of the electrical contact material, which will vary depending on the electrical current to be handled; one assumes that the current is low as relays within the case are usually involved. The dimensions B and C are the length of the arcs to be swept by the measurement pointer taken from the position shown on the diagram, which for clarity is shown mid-scale.

To overcome the limitation of such an arrangement, other methods have been developed that permit the use of very small ranges for each type of alarm, and furthermore allow the high and low ranges to be set very close together. In these designs separate electronic alarm trip amplifiers have to be assigned to each alarm function, and each amplifier allowed to function over the whole of the transmitter signal range. The trip point of each alarm is capable of being set anywhere in the measured signal range; Figure 23.6 is a simplified block diagram of a single-channel instrument such as we have been discussing. It is possible for such instruments to have an indicating meter to permit the setpoint to be seen, and/or an instrument panel-mounted lockable 10-turn potentiometers with a rotation counter, to permit the

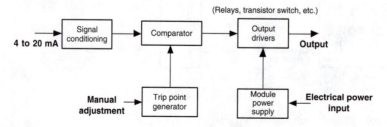

Figure 23.6: Block diagram of a typical trip amplifier.

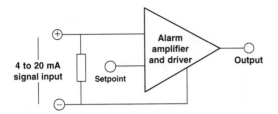

Figure 23.7: Typical trip amplifier circuit.

trip point to be set. In this application the potentiometers are usually linear devices, we can generally say that the resolution for the alarm setting is better than 0.1% of the full-scale resistance of the potentiometer.

The block diagram of Figure 23.6 can be converted very easily into the electronic schematic of Figure 23.7. See Chapter 21, Figure 21.9a (comparator), for a single-trip point amplifier input that is the front end of that that shown in Figure 23.7. These alarms can be made as multiple units so that they are able to accept two separate inputs having a common negative, or they can have the common linked to provide dual alarms on the same signal, or they can be made as difference alarms. The difference alarms operate on the basis that the difference between two signals should not exceed a predetermined value.

Having now understood how hardware alarm units operate, it is easy to understand how alarms in a distributed control system would function. In this case the alarm functionality is contained in an algorithm provided with "hooks" to which the measurements and setpoints can be attached. Figure 19.11 in Chapter 19 showed typical faceplate displays with their integral alarm/trip point presentation.

VISUALIZING THE ALARM CONDITION VIA A PROCESS ALARM ANNUNCIATOR

Before we get into the protection of process plants in the event of abnormal process conditions, let us discuss what can be done with the alarm outputs that have been generated by one or another of the methods already discussed. To get a feel for what this topic is about, we shall consider a slightly older type of instrument, still in use with instrumentation and systems other than DCS, or even with a DCS as a independent supervisory safety overview—the *alarm annunciator*, for use with separate contact/trip units—and then look at how the philosophy has been implemented in modern systems. As an example, let us consider only one small part of the process, say, the water and steam circuit of a steam generator. Figure 23.8 (which should be familiar from Chapter 14) shows this, and for the purpose of the exercise we shall discuss only a simple framework of an alarm system. In practice this framework is usually elaborated upon; specifically, the present illustation *excludes* any alarms associated with the combustion air/fuel systems, which there will certainly be on the actual plant.

To start, let us list the parameters of interest to us:

Feedwater flow

Drum water level

Steam pressure

and for the present describe the process as briefly as possible to recall the dependence each has on the other. (A much more detailed study of steam generation

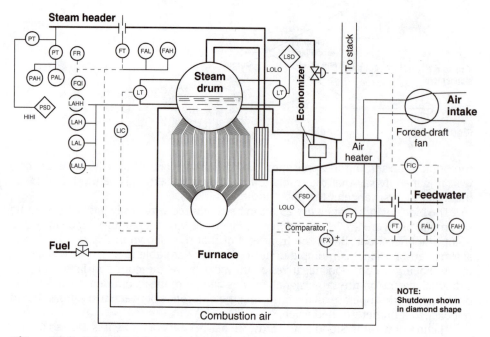

Figure 23.8: Schematic of typical steam generator with alarms.

control is given in Chapter 14.) In steam generation, feedwater is the raw material and steam the finished product, with the conversion from one to the other taking place in the steam drum under the influence of the heat supplied by the fuel in the combustion chamber. The object of the control system is to detect the changing demand for steam, balance the demand with an equal quantity of feedwater, and at the same time maintain the water level in the steam drum throughout the changes. For the present we are considering only the protection of personnel and plant by giving timely warning. If the warning is left unheeded, or counter actions are too slow, then a process shut down can be arranged, but for the present a shut down has been shown on the diagram; but the implications are not discussed.

We start with the feedwater, and since there is a measure of the quantity being pumped, we use the signal developed from the flow measurement as the basis for setting the trip points, which, for simplicity are just one high, and one low alarm (FAH and FAL on the figure), and use these to inform the process operator of the onset of unwanted conditions. Since steam is required continuously in a real process, there will always be demand for feedwater to replenish that which has been used, which means that there is a point on the flow scale that can be referred to as the *normal* flow rate; and is the usual operating point of the process. Please bear in mind that we are not considering how efficiently the generated steam is being used, but rather only that there is a normal consumption rate. The low and high flow alarm trip points are set at the maximum and minimum values the measured variable is permitted to take, but these will lie well within the safe operating limits of the pump and drum, thus allowing the generator to continue operating so long as the flow does not exceed these limits, although the process operator is always in attendance monitoring the performance. The alarms are wired to two separate inputs of an alarm annunciator that provides both audio and visual indication of the condition attained.

THE ALARM ANNUNCIATOR

The audio indication is common to both alarms, external to the annunciator, and is usually a bell, horn, or siren depending on the location on the plant; the visual indication, on the other hand, is individual to each alarm, and comprises a back-lit light box called a "window" that has a translucent screen carrying an engraved message such as: "FIC-xxxx FEEDWATER LOW FLOW" and "FIC-xxxx FEEDWATER HIGH FLOW," or something similar with appropriate tag number, etc. When either of the alarms is initiated, both the audio and the visual alarms are triggered simultaneously; the appropriate window is illuminated and made to provide a flashing display. When, and only when, the process operator acknowledges the alarm, the flashing light turns to a steady light, and the audible alarm is silenced; the steady light continues for as long as the alarm condition persists. When the alarm condition is rectified, either the lamp is automatically extinguished and the system reset (extremely rare), or, more usually, the operator has to manually reset the alarm circuit, in readiness for the next time. The translucent screens can be provided in a variety of colors and a code used to signify the severity of the alarm. Since each alarm window is unique to an input, the windows can be grouped together to form a meaningful set.

Figure 23.9 summarizes the alarm annunciator operating sequence. In the figure the alarm contacts are made (closed) at all times when the process is "healthy" and broken (open) on alarm conditions, this sequence of alarm contact action being called *fail-safe*, for the reason that should the contact be broken (say, mechanically, due to wear, bad connection, or unit or relay coil failure), the alarm will be initiated and corrective action can be taken by the maintenance engineers.

A further enhancement of the alarm annunciator allows a group of interdependent and process-associated alarms to be collected together in such a way as to permit any single alarm, whichever of these to be actuated first, to initiate the audio and visual display with a fast flash rate, which when acknowledged, resets the audible function but makes the display continue to flash until the fault has been corrected. Any other subsequent alarm in the group retriggers the audible alarm and makes the triggering alarm also visually discernible and to be acknowledged in the usual way or sequence as already described. Such an arrangement is called a *first-up* system.

Since, as we have said earlier, the system we have just described is a stand-alone version, each of the alarm inputs is hard-wired to a dedicated window, with the initiating contact acting only on the associated alarm window and common audible alarm.

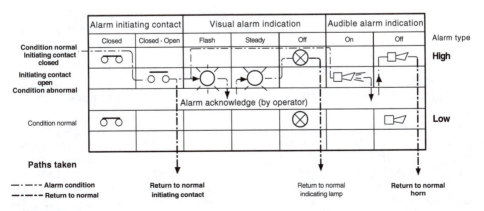

Figure 23.9: Sequence of alarm functions.

DCS PROCESS ALARM SYSTEMS

Modern distributed control systems, also use for their alarm systems the concept of an audible alarm, which is eventually silenced and a flashing visual that turns steady when acknowledged. However, the first-up sequence is not available as standard; custom software logic and sequences have to be provided if this is required, and are a very costly addition. The advantage of the DCS alarm system is that separate trip amplifiers are not needed on any measurement if it is part of the system and thus internal signals such as results of calculations or output signals can also be alarmed. However, they are frequently included as independent, hard-wired, overcritical backup alarms. The trip points of such signals can be set or adjusted when the system is being configured or at any time thereafter, subject to the system security and authorization arrangements enforced by the user. DCS alarm systems have extremely versatile additional functionality available, such as:

1. Alarm history reporting, which includes date and time stamping, both as VDU screen display and hard-copy printout. Some systems offer two-color alarm printouts, usually red for critical and black for noncritical alarms.

2. Most, if not all, DCS alarms are shown on-screen in a display called an *alarm summary*. This display will contain many of the following data for each alarm point:

   • Type of alarm—high, low, high-high, low-low

   • Tag number of the alarm

   • Reason for the alarm—e.g., tag number of the initiating contact

   • Short description of the alarm—e.g., "Temp. R-xxxx High"

   • Reference to the associated process unit—e.g., "Reactor R-xxxx"

   • Date and time

   When a point goes into alarm, the line of data associated with that point will start to flash to advise the process operator of the situation.

3. Alarm history archiving is achieved by writing alarm data to magnetic media (usually hard disk). Some systems also provide an automatic tape streaming facility; this keeps disk space always available so that writing process data to disk is uninterrupted.

4. Historical alarm data can be directed to printer at predetermined intervals and/or on demand and presented as shift, weekly, and monthly logs or recalled on-screen.

5. Alarm reports usually follow a DCS system manufacturer's standard format, but as an alternative it is also possible to have custom report formats. If a custom presentation is chosen, then it is to be expected that there will be additional costs, and the possibility of compromise in presentation of data and layout is also very real.

None of the foregoing features are available or can be obtained on the stand-alone alarm annunciator systems described earlier. Let us look at how this capability is used to provide other features that form part of the versatility of the DCS alarm system we are discussing. As we have seen, it is possible to set the trip point anywhere on the scale of the measured variable input and, by means of the trip, set *flags* that can be used to initiate other procedures which could involve logic, sequential routines, sending messages to the process operators, changing the mode of regulatory controllers or initiating a ramp of the setpoint on a regulatory controller, initiating or stopping plant functions such as motors (e.g., pumps or fans) opening or closing safety-related valves, and/or other functions too numerous to state.

Figure 23.10, which is based on the Foxboro I/A Series DCS control system, is used as an example to describe the operation of a modern process control alarm system. The diagram shows only as many items as required to meet the basic demands of the very small and necessarily simple system description. The system node can actually contain many more items of equipment and can be very complex in operation. For example, there can be as many as 64 Control Processors (CPs) on the node shown.

The process is monitored by sensors coupled to transmitters and provides measurements of the required process parameters for the system. The measurements enter the system via the field bus modules (FBMs) assigned to particular control processors (CP). Since the node bus links the control processors, every CP in the system has the capability of obtaining data from any transmitter in the system, but unfortunately this peer-to-peer capability results in a slowing down of the system, for it takes time to get the information across a working node bus. Careful configuration of the system CPs, to avoid or limit the number transactions across the node bus and yet have all the data necessary to meet the control strategy, can minimize the delay.

Specification of all the required alarms, alarm messages, and the trip points at which these will initiate, is carried out during the configuration of the control blocks in the CP. The trip point on the measurement scale is where the system will either have to take some control action automatically, or warn the operator of any impending violations. An alarm message is text that describes the alarm state and its origin, and a typical message could read "6-LAH-1234 Condensate Vessel T-567 Level High." This message tells the process operator that the condensate level of vessel T-567 in process area 6, monitored by level instrument with alarm tag number LAH-1234, is at the high level and something ought to be done. If it were serious, the trip could automatically start a discharge pump; but, before that is allowed, the pump outlet path must be checked to see that it is clear, and a receiving vessel is available. From this, one can now begin to appreciate that there will be logic, and more measurements necessary, adding to the complexities involved in interlocks and status checks. As mentioned before, all process measurements, controlled outputs, and points of predetermined difference (deviation) between the measurement and setpoint, can have alarms applied and control actions initiated from these alarms. There are many types of alarm. The standard types of high, low, high-high, and low-low are *absolute* alarms; this means that actions are initiated at very specific values on the measurement scale. *Deviation* alarms, another standard type, are those that actuate when the alarm parameter moves higher or lower by more than a specifiable amount. *Difference* alarms, yet another standard type, actuate when the difference (plus or minus) between two inputs exceeds defined limits. Finally, *rate-of-change* alarms are initiated when a parameter changes (usually) faster than a specified rate.

The alarm system shown in Figure 23.10 operates as follows. A comparison between the conditioned incoming signals and the configuration of the associated blocks determines the alarm state. If the signal is in alarm, the system broadcasts the condition over the node bus. The application processor (AP) receives the message and displays it on the CRT. At the same time, through the communication processor (Comms), the data prints out as an alarm log on the preassigned printer, and since the alarm being discussed originated in the process, the condition prints out as a process alarm. The historian date- and time-stamps and records the alarm data. The workstation processor (WP) simultaneously receives the data and distributes it to the process graphics and all display managers on the system, and, depending on the requirements of the display, the alarm can even alter a process graphic on the

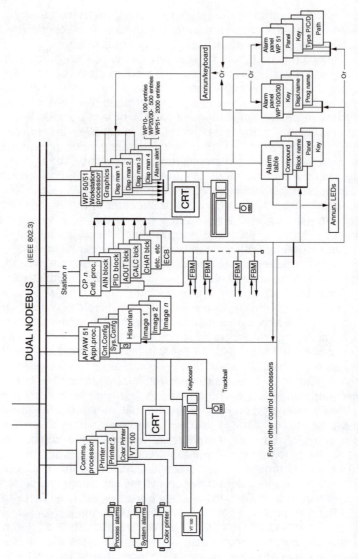

Figure 23.10: Schematic of I/A Series system under alarm conditions. A thin, continuous line depicts an alarm signal path. The alarm data requirements are shown under each alarm configurator header.

operator workstation. The alarm annunciator panel is initiated and the appropriate LED (and window) illuminated to signify the alarm condition via the alarm alert subsystem, which has a handling capacity dependent on the WP bought by the user. For example, a WP10 has a maximum of 100 alarm points, a WP20 or WP30 a maximum of 500 alarm points, and a WP51 a maximum of 2000 alarm points.

This completes a very simplified overview of the concept of process alarms, the functionality of which can be, and is often, exploited in continuous processes enabling the protection of personnel, plant, and product, and in batch process operations where the trips not only initiate alarms to perform the safety and protection functions, but can also trigger some of the process changes required during product manufacture. Since the monitoring is carried out continuously by machines, it adds to peace of mind, for, assuming that the design is securely based, equipment fully maintained, and operating personnel vigilant and responsive, we can be reasonably sure that hazardous conditions can be avoided or at worst contained. Environmental considerations in today's world will demand that greater attention be paid to the setting of the alarm trip points to ensure that damaging emissions, are minimized, where they exist, or avoided completely. Such requirements will force processes to operate within smaller (tighter) processing limits.

EMERGENCY SHUTDOWN SYSTEMS (ESDSs)

In principle, emergency shutdown systems (ESDSs) operate much like the process alarms we have just described. However, the point at which they come into effect is that which the experts (process engineers and/or process chemists) decide will be a hazard to both life and manufacturing plant itself. In this respect ESDSs are really safety-related, and therefore subject to legislation and routine government inspection in some countries. It is imperative that ESDSs be stand-alone systems and never incorporated within the process control system itself; this separation of the two should extend even to the power line supplies to the two systems. It is most important that the shutdown system be provided with an uninterruptible source of power (UPS) even if the control system is not. There is ground for debate as to whether the control system should also be fed via an uninterruptible power source, but the usual reason for not doing so in every case is financial. This is the one line of reasoning that, in the author's opinion, can in most cases be very shortsighted.

Since the ESDS is the last line of defense (last-ditch precaution), it is vital that all information taken regarding the condition of the process and equipment be absolutely secure. This does not mean that the system is never going to fail; rather it does mean that the system designers have taken measures to reduce to a minimum the consequences of a major disaster occurring. It is usual for there to be separate multiple measuring devices, and a selection routine to be configured to ensure that the ESDS always has the most reliable and indicative data upon which to operate. Any failure in the measuring system is advised to the operating personnel, via its own display and alarm system, for action to be taken to reinstate or restore the integrity of the measuring system as quickly as possible.

Within the shutdown alarm system, the logic is also designed on a multiple redundancy basis, and we shall consider Triconex "Tricon" Fault Tolerant Control system as an example of multiple redundancy. As the name implies the system is 'triple redundant' and microprocessor-based. Figure 23.11 is a block schematic of the system only, the voting logic for the multiple set of transmitters and final element actuators do not form part of Triconex offering and has to be designed separately. For the purposes of

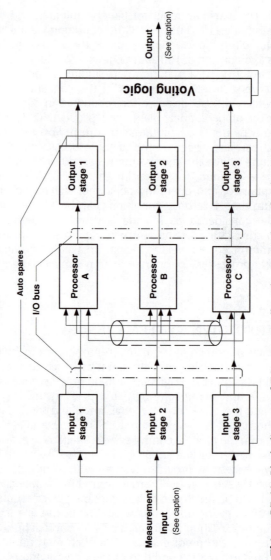

Figure 23.11: Block diagram of Tricon controller. The voting logic for the multiple transmitters and final element actuators is omitted for clarity. This will have to be designed and supplied separately.

the diagram, we refer to the input and output modules as stages, and as can be seen the inputs, processors, and output stages are triplicated, with each of the input and output stages fully independent of the other, in addition to which both the input and output stages are backed up with spare stages see Figure 23.11. The measurement from the selected field device is fed in parallel to each of the three input stages, and from there to one of the three main processors. The "Tribus" interprocessor performs a two-out-of-three majority vote on the data; this ensures that each of the main processors uses the same voted data to execute its program. The outputs from the main processors are sent down three paths to the output stages, where once again they are vote-selected; this maintains the integrity of the signal output. There are extensive diagnostics on each of the independent stages to detect and report any faults immediately, with the results of all the diagnostics stored as system variables and/or displayed by LEDs and alarm contacts. The information can also be used within the system to modify the control actions or to direct maintenance or to manipulate the various items of plant equipment to ensure personnel and plant safety.

With this information, it is hoped that a feel has been given for what process and shutdown systems are and what they can do. However, there is something to be said concerning a phrase, "graceful degradation," so often used in the context of shutdown systems. What this means is that the system is to be designed to position the process and the associated equipment in such a condition that there is no chaotic halt, to either the system or the process and its equipment, so that recovery can be swift and as trouble-free as possible. To achieve this predetermined ordered sequence of events and to implement them calls for very carefully designed loops, sequences, and logic, but mostly for cooperation and joint agreements between the process engineers, process chemists, instrumentation and control engineers, and the control system suppliers and builders to ensure that all foreseeable eventualities have been provided for.

SUMMARY

1. All manufacturing processes require some means of advising operating personnel of malfunction of the plant, process, or both, and warning them that some parameter is *moving toward* a condition that could compromise the product specification or quality, plant or personnel safety, or the environment. The reason for raising an alarm prior to attainment of the unwanted state is to enable safe evasive action to avoid the aftermath of being unprepared, or of finding out about the malfunction when it is too late to correct it.

2. The most qualified people to state what conditions to avoid to ensure good product and a least hazardous environment are the process engineer and the process chemist. Those who must be consulted to determine the appropriate precautions and procedures for safety of personnel or processing equipment in the event of malfunction are the plant maintenance engineer, safety engineer, process engineer, and process chemist.

3. Alarms can be triggered by measured variable outside its acceptable band; failure or malfunction of a piece of plant equipment involved in the manufacturing process, including control valves; failure or malfunction of the measurement device itself, which includes the sensor and the transmitter as a single entity; failure or malfunction of the interconnections (transmitter, signal cable, or receiving instrument) between the field-mounted and control room-mounted devices' input and output; or low level or loss of motive power, whatever the energy source.

4. Most distributed control systems have alarms built in to signal failure of any item in the system. If redundancy is designed in as well, then changeover to standby units should be seamless, but operating personnel will always need to be made aware of the situation. Critical process conditions, measured or monitored parameters warrant immediate action, including an immediate response. Not so critical process conditions, measured or monitored parameters can tolerate some delayed action in implementing but with an estimate of the time allowable. If the alarms triggered in the process are continually repeated, personnel procedure(s) to be followed must be reviewed at the relevant priority level as process changes may be involved.

5. Some of the more common considerations to take into account when planning for hazardous conditions arising during plant operation (there could be many more) are:

   • Can initiating some process action, e.g., automatic addition of reagent, to counteract the condition, restore equilibrium?

   • Can countermeasures be called upon immediately, without operator intervention or minimum delay?

   • Can the hazardous condition affect the safety of personnel or equipment in the immediate surroundings or adjacent areas of the plant?

   • Can the hazardous condition affect plant personnel, property, or equipment, or the environment, property, and persons adjacent to the plant?

   • Can product in mid-process be saved after the hazard has been contained? If not, can it be disposed of with minimum risk to personnel, property, and environment?

6. Configuration of alarm levels and assignments must be tailored to individual process and operational requirements. Including and configuring suitable alarms are as important to the storage of raw materials as to processed products.

7. Choosing a process operating point that gives an acceptable product (within the specification) for the lowest manufacturing cost will increase profitability. Alarm settings that permit this advantage are invaluable.

8. Fail-safe alarm contacts are electrical contacts that are energized to close in order to signify the "healthy" state of the measured parameter and deenergized to open on an alarm condition.

9. When alarm contacts are grouped together so that when any one member's into alarm signifies that the associated process could be entering an alarm condition, the designed alarm system is termed a "first up" alarm. Once initiated, the alarm condition is held until the cause of the alarm is rectified, after which the alarm is reset and the system returned to normal. Any other subsequent alarm in the group retriggers the audible alarm and makes the triggering alarm also visually discernible to be acknowledged.

10. Functions in DCS alarm systems not found in alarm annunciators are available: Alarm history reporting includes date and time stamping both as VDU screen display and hard-copy printout. Some systems offer two-color alarm printouts, usually red for critical and black for noncritical alarms. Most, if not all, DCS alarms are shown on-screen in a display called an alarm summary, which will contain many of the following data for each alarm point:

- Type of alarm—high, low, high-high, low-low

- Tag number of the alarm

- Reason for the alarm—e.g., tag number of the initiating contact

- Short description of the alarm—e.g., "Temp. R-xxxx High"

- Date and time

When a point goes into alarm, the line of data associated with that point will start to flash. This advises the process operator of the situation. Alarm history archiving is achieved in DCS alarm systems by writing alarm data to magnetic storage media (usually hard disk). Some systems also a provide an automatic tape streaming facility; this method keeps disk space always available so that writing data to disk is uninterrupted. Historical alarm data can be directed to printer at predetermined intervals and/or on demand and presented as shift, weekly, and monthly logs or recalled on-screen. Finally, alarm reports usually follow a DCS system manufacturer's standard format, but it is also possible to have custom report formats. If a custom presentation is chosen, then it should be expected to cost more, and the possibility of compromise in presentation of data and layout is also very real.

11. ESDSs (emergency shutdown systems) are safety-related and therefore subject to legislation and routine government inspection in some countries. The point at which they come into effect is that which process engineers and/or process chemists decide will be a hazard to both life and manufacturing plant itself.

12. It is imperative that ESDSs, which are the last line of protection, be stand-alone and hard-wired, i.e., wired separately and individually, and never incorporated within the process control system itself. This separation of the two should extend even to the power line supplies to the process control and shutdown systems. It is most important that the shutdown system be provided with an uninterruptible source of power even if the control system is not.

13. To ensure that the ESDS always has the data upon which it is to operate, it is usual for there to be separate multiple measuring devices and a device selection routine configured.

14. Within the shutdown alarm system, the logic is designed on a multiple redundant basis.

15. In the context of shutdown systems, the phrase "graceful degradation" means that the system positions the process and the associated equipment in such a condition that there is no chaotic halt to either the system or the process and its equipment.

CHAPTER 24

Boolean Logic

Boolean logic is the method by which computations and decisions are carried out in digital computers. The manipulation is based on a number system that has a base of 2, as opposed to the base of 10 of the more familiar one that we all know. The base 2 system is called the *binary* and the system with base 10 *decimal*. We are going to introduce an algebra that uses the binary system and thereby show how the digital system is made a practicality, and one that we nowadays almost take for granted.

The reason for choosing the binary system for digital computers is that numbers can be represented very easily using the two states of an electrical (electronic) switch. We can all appreciate that a switch can be only ON or OFF, and if we call the ON state 1 and OFF state 0 then we can have a basis for doing arithmetic or making logical decisions. The latter assumes that we consider the initiators and the results of a decision to have possible values of either true or false, yes or no, with 1 synonymous with true or yes and 0 synonymous with false or no; *no* intermediate or mitigating circumstances are permitted.

NUMERICAL REPRESENTATION IN ELECTRICAL FORM

We shall consider how decimal numbers can be represented in an electrical form by using a set of wires, one assigned to each separate digit where each digit, if available, carries a logical 1 (which could represent a value of either voltage or current) on its designated wire and, if not available a 0. It follows that in order to accomplish this we must have 10 wires to represent the decimal number 10, remembering that zero is a valid digit. Table 24.1 will result.

For the representation in the table to work, each wire has to be dedicated to represent a unique decimal value. Hence, with this arrangement, it is necessary to have as many as groups of 10 wires as there are digits in the number we wish to represent; e.g., for numbers of "weight" 1 (where weight refers to a multiplier of a number to indicate the importance of the number's contribution to the total), we need 10 wires for decimal numbers of a single digit, and 20 wires for numbers of two digits, and so on.

To reduce the number of wires required to represent a decimal number, we can allow more than one wire to carry a logical 1 at the same time, and "weight" the value of each logical 1 dependent on its *position* as shown in Table 24.2. The weighting values as shown in Table 24.2 are signified by a factor that is a power of 2.

Let us see how such a number system works by using some examples. Suppose we want to represent the number 1 in binary, we would write this as a *binary number* 1, if the number were 2 then we would have to write this as a binary number 10, and if the number were 3 then we would write this as a binary number 11, and if the number were 8 we would have to write this as a binary number 1000. The reason for this is we have only a base of 2 to work with and every time we come to a number that is 2 or more we have to "move" to the left just as we do in the decimal system

TABLE 24.1
Binary representation of individual decimal values

Decimal value	Wire no.									
	0	1	2	3	4	5	6	7	8	9
0	1	0	0	0	0	0	0	0	0	0
1	0	1	0	0	0	0	0	0	0	0
2	0	0	1	0	0	0	0	0	0	0
3	0	0	0	1	0	0	0	0	0	0
4	0	0	0	0	1	0	0	0	0	0
5	0	0	0	0	0	1	0	0	0	0
6	0	0	0	0	0	0	1	0	0	0
7	0	0	0	0	0	0	0	1	0	0
8	0	0	0	0	0	0	0	0	1	0
9	0	0	0	0	0	0	0	0	0	1

where every time the units position of the number reaches 9 (remember 0 is a valid number). When we weight the digits in the binary system, as we have done in Figure 24.2, then using the same examples, as before number 1 would be represented as the "coded" binary number 0001, number 2 would be represented as the "coded" binary number 0010, number 3 would be represented as 0011, and number 8 would

TABLE 24.2
Binary-coded decimal (BCD)

Decimal value	Wire no.			
	1	2	3	4
	(Weight)			
	8	4	2	1
0	0	0	0	0
1	0	0	0	1
2	0	0	1	0
3	0	0	1	1
4	0	1	0	0
5	0	1	0	1
6	0	1	1	0
7	0	1	1	1
8	1	0	0	0
9	1	0	0	1

Wire no. 1: decimal value 0–15
Wire no. 2: decimal value 0–7
Wire no. 3: decimal value 0–3
Wire no. 4: decimal value 0–1

TABLE 24.3
Interpretation of the binary number
11100111

1	1	1	0	0	1	1	1
R^7	R^6	R^5	R^4	R^3	R^2	R^1	R^0

be represented as the "coded" binary number 1000. The maximum number we could represent by this method would be the "coded" binary number 1111 which in decimal would be obtained by adding $8 + 4 + 2 + 1$, which would give 15 as a result. The system is called binary-coded decimal (BCD).

The decimal and binary number systems have similarities. Both systems have a *radix*, or *base*; for the decimal, it is 10, and for the binary, it is 2. The first (i.e., the rightmost) digit in a number in either system is multiplied by the radix raised to the power 0 (R^0), the second digit is the value multiplied by the radix raised to the power 1 (R^1), etc. In general, we can say the nth digit is obtained by raising the radix to the power $n - 1$ (R^{n-1}). A decimal equivalent of a binary number is obtained by addition of the decimal equivalent of all the digits carrying a 1; any leading zeros will not change the value of the number in either systems as shown in the earlier examples given.

CONVERSION OF ONE SYSTEM TO ANOTHER

Binary to Decimal Conversion

Suppose we wish to convert the binary number 11100111 to its decimal equivalent. The meaning of which is given in Table 24.3, where R is the radix, in this case 2. Table 24.4 shows the decimal equivalent of the binary number. Adding the values together gives the decimal number 231.

Decimal to Binary Conversion

Let us convert the decimal number 97 to binary representation. In Table 24.5, we begin by placing the decimal number at the top of the left-hand column. We then successively divide the last number (97) in the left-hand column by 2 writing the whole number quotient (48) in the next row of the left-hand column and the remainder (1) in the right-hand column. The process is continued until there is a zero quotient in the left-hand column. For the decimal number 97, we obtain the binary number 1100001 *reading the right hand-column from the bottom of the table upwards*.

LOGIC BLOCKS

All logical manipulation within a digital machine is carried out by means of *logic blocks*; these blocks should be considered for their functionality, and for the purpose

TABLE 24.4
Decimal equivalent of the binary number in Table 24.3

1×128	1×64	1×32	0×16	0×8	1×4	1×2	1×1
128	64	32	0	0	4	2	1

TABLE 24.5
Decimal to binary
conversion

97	
48	1
24	0
12	0
6	0
3	0
1	1
0	1

of design this is all that is necessary. However, it is sometimes necessary to be aware of the circuits involved, and we shall consider the circuits later, but shall for the present concentrate on the design of the logic itself.

There are a number of criteria that apply to the assembly of a logical function, all of which have to be provided through the use of logic blocks arranged in a particular order to accomplish the task required. Initially we shall consider only the three basic and main logical functions of AND, OR, and NOT conventionally and frequently referred to as *gates*. Logic blocks for the moment should each be visualized as a "black box" into which we can put different conditions to obtain from it one result governed by the rules for which the black box has been designed. Summarizing this idea, we can say that the black box is able to accept either one input or more than one input, but can produce only one output, all signals being in the binary form of 0s and 1s.

DEFINITIONS OF THE LOGIC FUNCTIONS

AND   This function requires *all* inputs to the box to be true (logical 1) before the output will be true. Putting this another way, if the inputs were numbered 1 through n, then input 1 AND input 2 AND input 3 AND... input n *must* be true for the output to be true.

OR   This function only requires *any* one (one or more than one) of the inputs to the box to be true (logical 1) before the output will be true. Putting this another way, if the inputs were numbered 1 through n, then either input 1 OR input 2 OR input 3 OR... input n must be true for the output to be true.

NOT   This function, also referred to as *inversion* (INV), acts on a single input only. In operation, the box receives an input, which could be either true or false, and operates on it to produce the inverse of the input as the output. Putting this another way, if the input is true, i.e., 1, the output will be false, i.e., 0, or if the input is false, i.e., 0 the output will be true, i.e., 1.

These definitions are shown in circuit diagram form in Figure 24.1
Note: The symbols shown in the center and right-hand columns of Figure 24.1 are very frequently found on diagrams of logic circuits. However, the author has found the symbols shown in the left-hand column most convenient for writing Boolean expressions, as will now be shown.

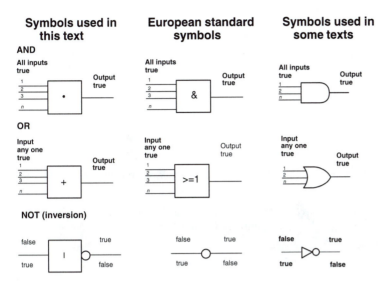

Figure 24.1: Symbols for logic functions.

ALGEBRAIC SYMBOLS FOR THE LOGIC FUNCTIONS

There are variations on the symbols used for the algebraic manipulation of Boolean logic, but for this part we shall standardize on a "·" (center point) to denote an AND and a "+" (plus sign) to denote an OR. Adding a bar over the appropriate input denotes inversion, in the same way as a negative characteristic is denoted in logarithms.

TRUTH TABLES

A *truth table* is a tabulated list of values of a Boolean expression for each possible combination of the variables in the expression. The writing of a truth table can be reduced to a systematic method of construction that can be applied to any number of Boolean variables. To completely specify the condition of a single variable A, we can write the truth table shown in Table 24.6. To completely specify all the conditions of two variables A and B, we can write the truth table shown in Table 24.7.

CONSTRUCTING A TRUTH TABLE

The method for constructing a truth table for three or more variables (Table 24.7 defines the table for two variables completely) is as follows:

1. Give each variable a name. Start with a single variable A and define it completely as shown in Table 24.6.

TABLE 24.6
Single variable (Stage 1)

A
0
1

TABLE 24.7
Two variables (Stage 2)

B	A
0	0
0	1
1	0
1	1

2. Add the second variable B, to the left of the first variable name. Then add two zeros to align with the first two conditions of variable A as shown in Table 24.7.

3. Repeat the two conditions of variable A in the same order directly under the initial two conditions of variable A as shown in Table 24.7.

4. Add two 1s under the 0s of variable B to align with the repeated initial conditions of variable A as shown in Table 24.7. This completes the truth table for two variables A and B.

5. Add the third variable name C to the left of variable B, and add four zeros to align with the four conditions of variable B; this is not shown in Table 24.7, but steps 6 through 8 following are in principle the same as steps 2 through 4.

6. Repeat the four conditions of variable A in exactly the same order, but directly under the initial four conditions of variable A.

7. Repeat the four conditions of variable B in exactly the same order, but directly under the initial four conditions of variable B.

8. Add four 1s under the 0s of the third variable C to align with the repeated conditions of the variable B.

The procedure is repeated one at a time for as many variables as required, always adding the new variable to the left of the existing one.

Figure 24.2 is a diagram of the method for the first two variables, with the step numbers of the method shown in parentheses.

BASIC OPERATIONS

Using values of the logic levels for two variables we obtain the following relations by ANDing the two:

$$1 \cdot 1 = 1 \qquad 0 \cdot 1 = 0 \qquad 1 \cdot 0 = 0 \qquad 0 \cdot 0 = 0$$

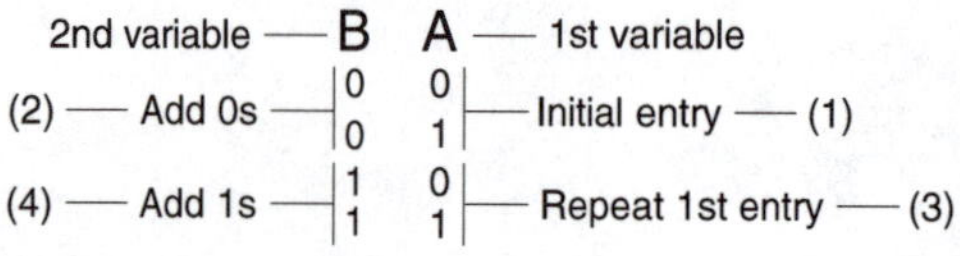

Figure 24.2: Diagram of the method of constructing a truth table (first two variables only).

Now, next, by INVerting the *first* of the logical variables, we have:

$$0 \cdot 1 = 0 \qquad 1 \cdot 1 = 1 \qquad 0 \cdot 0 = 0 \qquad 1 \cdot 0 = 0$$

If the *second* of the logical variables is INVerted, we have:

$$1 \cdot 0 = 0 \qquad 0 \cdot 0 = 0 \qquad 1 \cdot 1 = 1 \qquad 0 \cdot 1 = 0$$

Similarly, we derive the following relations by ORing two variables:

$$1 + 1 = 1 \qquad 0 + 1 = 1 \qquad 1 + 0 = 1 \qquad 0 + 0 = 0$$

Now, next, by INVerting the *first* of the logical variables, we have:

$$0 + 1 = 1 \qquad 1 + 1 = 1 \qquad 0 + 0 = 0 \qquad 1 + 0 = 1$$

If the *second* of the logical variables is INVerted, we have:

$$1 + 0 = 1 \qquad 0 + 0 = 0 \qquad 1 + 1 = 1 \qquad 0 + 1 = 1$$

From what we have seen so far, it will not be difficult to establish the validity of the following theorems to determine the equivalence of two expressions:

AND:	$A \cdot 0 = 0$	$A \cdot 1 = A$	$A \cdot A = A$	$A \cdot \bar{A} = 0$
OR:	$A + 0 = A$	$A + 1 = 1$	$A + A = A$	$A + \bar{A} = 1$

CONVENTIONS IN BOOLEAN ALGEBRA

The algebra with which we are all familiar has an operational priority that carries the acronym BODMAS (*B*rackets *O*f *D*ivision, *M*ultiplication, *A*ddition, *S*ubtraction). Boolean algebra, too, has an order of priority in carrying out operations, which is Inversion, AND, OR. In Boolean algebra, the "·" signifying AND is usually omitted, such that A·B·C (*A* AND *B* AND *C*) is written as ABC; it is also normal to use parentheses around terms that are ORed; e.g., $A \cdot B + C$ can be written as $A(B + C)$; $A + B \cdot C + D$ can be written as $(A + B)(C + D)$.

Before we list another set of theorems, let us see how we manipulate an expression. Suppose we want to prove the identity $B + AB = B$. Let us look at Table 24.8; we can see that:

- The column headed AB combines the results of the truth table of the two variables A, B following the rule of ANDing for each of the states of the variables.

- The column headed $B + AB$ combines the results of the states of variable B and the states of the ANDed variables A with B according to the rules for ORing variables.

- A comparison of the column headed $B + AB$ and the column headed B shows full agreement within each row, which proves the identity.

When variables are ANDed, the effect is the same as if an algebraic multiplication is taking place, with the exception that when a variable is ANDed with itself the result

TABLE 24.8
Meaning of Boolean
expressions

B	A	AB	B + AB
0	0	0	0
0	1	0	0
1	0	0	1
1	1	1	1

is not the square, but the variable itself—e.g., $AA = A$. And if the variable is ANDed with the inverse of itself, the result is 0; e.g., $A \cdot \bar{A} = 0$. The following theorems shall also be very useful in further work.

$$A + AB = A \qquad A + \bar{A}B = A + B$$
$$A(A + B) = A \qquad A(B + C) = AB + AC$$

There could be instances when the number of variables involved makes the solution of a problem appear difficult; however, substitution for a group of combined variables with a single new variable may allow the problem to be reduced to one of the standard forms just shown, thus providing the possibility of easy manipulation to obtain the result. One should not overlook this method of working, as shown in the following example. Suppose we have an expression

$$E(AB + CD) + \bar{E}$$

Substituting X for $(AB + CD)$, we have the expression

$$EX + \bar{E}$$

This is equivalent to

$$X + \bar{E}$$

Prove this for yourself, using the method shown in Table 24.9. Resubstituting, we have

$$AB + CD + \bar{E}$$

as the result.

TABLE 24.9
Inversion of two variables
ORed together

A	B	A + B	$\overline{(A + B)}$
0	0	0	1
0	1	1	0
1	0	1	0
1	1	1	0

TABLE 24.10
Identities of two variables inverted
ORed; inverted ANDed

A	B	$A + B$	$\overline{A+B}$	$\bar{A}$	$\bar{B}$	$\bar{A} \cdot \bar{B}$
0	0	0	**1**	1	1	**1**
0	1	1	**0**	1	0	**0**
1	0	1	**0**	0	1	**0**
1	1	1	**0**	0	1	**0**

DE MORGAN'S THEOREM

Earlier we showed that inverting a variable resulted in changing a true (1) to a false (0); if we now invert the false (0), we shall revert to a true (1) again. This gives a double inversion theorem $\bar{\bar{A}} = A$. Some interesting consequences result as we invert expressions; they lead us to important theorems in Boolean algebra. Let us consider two variables A and B and write down the truth table, OR them first, and then invert the ORed result. This will give the results shown in Table 24.9. If each variable is inverted first, and then ANDed we get the results shown in Table 24.10. Examining the results of the inverted OR and the results of the inverted AND, we see that they are identical. This result will be obtained irrespective of the number of variables involved and gives the generalized theorem:

$$\overline{A + B + C} = \bar{A} \cdot \bar{B} \cdot \bar{C}$$

or alternatively,

$$\bar{A} \cdot \bar{B} \cdot \bar{C} = \overline{A + B + C}$$

This theorem is called *De Morgan's theorem*. An equivalent electronic circuit demonstrating it is shown in Figure 24.3.

MINIMIZATION METHODS

We can apply Boolean logic not only to mathematics, but also to determining the truth or falsehood of a combination of language (spoken or written) statements that have only two possible interpretations each.

ALGEBRAIC MINIMIZATION: THE QUINE-McCLUSKEY METHOD

In order to obtain the most economical and efficient logic, it is necessary to combine and eliminate those variables that are repeated, and thus form a set of terms

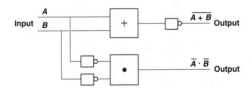

Figure 24.3: Equivalent circuit proving of De Morgan's theorem.

called *prime implicants*. A minimization method using binary notation called the *Quine-McCluskey method* is used. This method is a simplified one that depends on the repeated application of the theorem $AB + A \cdot \bar{B} = A$ (**P**roof: **A**lgebraically, $AB + A \cdot \bar{B} \equiv A(B + \bar{B})$; **B**ut $(B + \bar{B}) = 1$; hence, $AB + A \cdot \bar{B} = A$: In this theorem the important point to notice is that one term appears once in both the true and inverted form.), together with rules that determine the number of times the theorem is applied.

In algebraic manipulation, it is usual to assign to the variable a value 1 for its true form and the value 0 for its inverted form. This gives the following as examples:

1. For the combination ABC, we obtain 111.

2. For the combination $A\bar{B}C$, we obtain 101.

3. For the combination $\bar{A}B\bar{C}$, we obtain 010.

4. For the combination AB, we obtain 11-.

Note: In item 4, since we started the sequence with three digits, we have written a hyphen in place of the variable C; this is necessary to keep the number of characters of all expressions the same.

Two terms can be combined if they differ in *only* one digit position (e.g., as in items 1 and 2 of the preceding example); however, this is not the only means of determining that the terms could be combined. The number of 1s in a term is important and is called the *index* and it is written in decimal form. The index is the sum of all the 1s in the binary form of the term. Using the same examples, we have:

1. For the combination ABC, we obtain 111, index = 3.

2. For the combination $A\bar{B}C$, we obtain 101, index = 2.

3. For the combination $\bar{A}B\bar{C}$, we obtain 010, index = 1.

4. The combination AB cannot be included, as it has only two variables.

We can now restate the foregoing index definitions as a *rule: The indices of two terms must differ by exactly one for the same digit position.* This is necessary to allow us to combine terms in the second list. The following should clarify what is meant. Suppose we have two terms $ABCDE$ and $ABCD\bar{E}$. These can be written as 11111 and 11110, respectively. Comparing the two terms on a component-by-component basis, we find that component E is the only one that differs, which means that it is possible to combine the two leaving 1111— as the result. If, on the other hand, we had $ABC\bar{D}E$ and $ABCD\bar{E}$, these two could not be combined, for if we wrote them in binary form we would get 11101 and 11110, respectively, where the difference is in *two* digit positions, even though the index for each term is 4.

For the Quine-McCluskey minimization technique, we follow a particular method, as follows:

1. Convert the problem into the binary notation of the standard sum form; remember, any problem can be converted into this form.

2. Sort the terms into groups of the same index.

3. Tabulate the groups and separate each by a horizontal line.

4. To ensure identification, write the decimal equivalent of each term alongside the associated term.

5. Consider pairs of terms to see if combination is possible. Two terms can be combined if they differ in only one digit position—i.e., the theorem stated earlier $(AB + A\bar{B} = A)$.

6. As a result of step 5, write a hyphen (-) at the position of the difference in digit.

This procedure will result in a list of binary-coded digits separated by horizontal lines that group codes of equal decimal weight together. An example of the procedure is as follows:

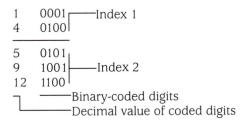

```
1    0001 ┌──Index 1
4    0100 ┘

5    0101 ┐
9    1001 ├──Index 2
12   1100 ┘
     ┐ ──────Binary-coded digits
     └──────Decimal value of coded digits
```

Having done this, we shall have to compare the coded digits and tabulate the results of the comparison using hyphens where appropriate; this is shown in the example below.

```
1   0001      1, 5    0 - 0 1
4   0100      1, 9    - 0 0 1
              4, 5    0 1 0 -
5   0101      4, 12   - 1 0 0
9   1001         ┐ ──────Results of the comparison (2nd list)
12  1100         └──────Decimal values of the code that gave the results
     ┐ ──────── Binary-coded digits (1st list)
     └──────── Decimal value of coded digits
```

We repeat the procedure of tabulation and comparison until no more terms can be combined, always being sure to append the decimal values that give the reduced binary code shown in the results column.

Note that one term can be used as often as necessary, but once it is used an asterisk must be appended to it to indicate that this is the case, as follows:

```
1   0001*     1, 5    0 - 0 1
4   0100*     1, 9    - 0 0 1
              4, 5    0 1 0 -
5   0101*     4, 12   - 1 0 0
9   1001*        ┐ ──────Results of the comparison (2nd list)
12  1100*        └──────Decimal values of the code that gave the results
     ┐ ────────Binary-coded digits
     └────────Decimal value of coded digits
```

Using the example, we can group the binary terms of the second list and separate

TABLE 24.11
Chart of prime implicants for the
worked example

	1	4	5	9	12
$A\bar{C}D$	*		*		
$\bar{B}\bar{C}D$	*			*	
$AB\bar{C}$		*	*		
$B\bar{C}\bar{D}$		*			*

them by a horizontal line and make a comparison as before:

1	0001*	1, 5	0-01
4	0100*	1, 9	-001
5	0101*	4, 5	010-
9	1001*	4, 12	-100
12	1100*		

The comparison does not give any further terms that can be combined; therefore we shall have to convert back to the original variables to give:

$$X = \bar{A}\bar{C}D + \bar{B}\bar{C}D + \bar{A}B\bar{C} + B\bar{C}\bar{D} \tag{24.1}$$

This expression shows the prime implicants, comprising all terms that cannot be combined any further by the Quine-McCluskey minimization method.

We can now plot a chart of the prime implicants of expression 24.1 as follows:

1. Write down all the decimal values in ascending order of magnitude.

2. Write down all the terms in the expression in a column on the left-hand side.

3. Write an asterisk alongside each decimal value of the term.

4. Select the smallest set of terms of prime implicants that contain all terms of the expression

From Table 24.11 we can see that there are no redundant terms; hence, the expression is the minimum that can be obtained. This would not have been the case if a different reduction expression had yielded the prime implicant chart of Table 24.12. In the *assumed* chart of prime implicants, it is clear that the term $A\bar{B}$ is redundant, for the other two terms contain all the prime implicants as indicated by the positions of the asterisks of the second term, which occupy decimal values 4 and 5, which also appear as components in the other two terms.

TABLE 24.12
Assumed chart of prime
implicants

	0	4	5	7
$\bar{B}\bar{C}$	*	*		
$A\bar{B}$		*	*	
AC			*	*

GRAPHIC MINIMIZATION: THE KARNAUGH MAP

The next minimization technique, a relatively straightforward graphical representation of the truth table, is called the *Karnaugh map*. This is a graphical method and is suitable for a small number of variables. Four variables can easily be handled, but numbers above six become very involved and difficult, and because of this we shall not be illustrating these. It is suggested that for five or more variables the Quine-McCluskey method shown previously be used instead.

We have seen that a single variable can take two possible values, 0 and 1, and in a truth table we listed the variables as columns with the variable name at the head. In a Karnaugh map, we allocate an area for each value of the variable instead, as in the following:

A 0 1 ——The independent variable is shown here.

☐☐ ——The dependent variable is shown here.

Suppose we had to show the relationship:

$Y = 0$ when $A = 1$

and

$Y = 1$ when $A = 0$

This would be written as:

A 0 1

Y $\overline{1\ 0}$

For two variables we make the chart as follows:

A 0 1——This is one variable with its two possible states.

B 0 ☐ ——This is the area in which can be shown the combinations that
    1 ☐   the two variables can take.

– ———This is the second variable with its two possible states.

What we will notice is that we actually have the space to show the eight possible combinations that the two variables A and B can take. That two variables have eight possible combinations we determined when we constructed the truth table for them. The open area will enable the results of the combinations—i.e., ANDing, ORing—to be written down. As an example, let us AND the two variables and see what the table will look like in this case:

A	0 1	
B 0	0	| Result of ANDing $A = 0$ with B
1	0	

A	0 1	
B 0	0	| Result of ANDing $A = 1$ with B
1	1	

A	0 1	
B 0	0 0	| Full result of A ANDed with B
1	0 1	

TABLE 24.13
Four-variable   Karnaugh
map

A B	00	01	11	10
C D				
0 0				
0 1				
1 1				
1 0				

The map for three variables will appear as follows:

A	0 1
B C	
0 0	Note the order of the rows for the variables B and C.
0 1	This order has been chosen so that when changing
1 1	from one row to the next only one variable changes.
1 0	

As before, the tabulation allows us to fully define the conditions that the three variables can take up.

For four variables, we shall obtain a map that will look like the one in Table 24.13. A five-variable map will look like the one in Table 24.14.

So far we have shown the plot for variables in the sum form only. If the maps are to be of any use, we shall also have to account for the nonstandard forms as well. To show how this is done, we shall use an example again. Suppose we had to show:

$$Y = A\bar{B}CD + \bar{A}BC$$

This will require a four-variable map and will appear as follows:

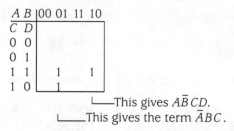

          └──This gives $A\bar{B}CD$.
          └──This gives the term $\bar{A}BC$.

What one should notice is that only 1s are used to indicate the selected conditions of the variables and the 0s are omitted; this is necessary to avoid confusion that would result with too many characters within the area that defines the condition.

TABLE 24.14
Five-variable Karnaugh map

E	0				1			
A B	00	01	11	10	00	01	11	10
C D								
0 0								
0 1								
1 1								
1 0								

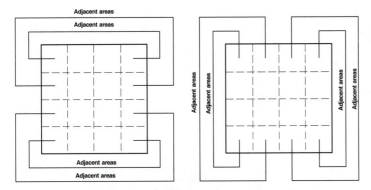

Figure 24.4: Adjacent areas on a Karnaugh map.

The properties of a Karnaugh map are:

1. Each area on the map corresponds to a specific value of a variable.

2. Each adjacent area differs by one variable.

3. Adjacent areas have a common side.

Adjacent areas are defined as shown in Figure 24.4.

It must be pointed out that a Karnaugh map is in fact a visual representation of the Quine-McCluskey minimization method and therefore should give the same results. In the following map we shall construct, we can see the relationship between the Quine-McCluskey and the Karnaugh methods of minimization. As shown, the terms differ in one variable only (A) and therefore can be combined according to the Quine-McCluskey rule. There is a similar rule in the Karnaugh method, which says that *two adjacent areas can be read as one term by including only those variables whose values are the same in both areas.*

Taking the two terms shown in the map of Figure 24.5, we can write the following. Using Quine-McCluskey symbols,

$$\bar{A}B\bar{C}D \equiv 0101$$

$$AB\bar{C}D \equiv 1101$$

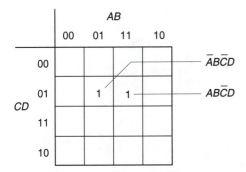

Figure 24.5: Equivalence of Karnaugh and Quine-McCluskey expressions.

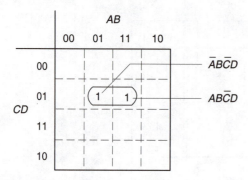

Figure 24.6: Method of depicting combined terms.

This gives the combination of the two as:

-101

which is the same as obtained from the Karnaugh map.

To simplify the process of identifying the terms that are combined, it is suggested that they be looped together as shown in Figure 24.6. A 1 on a map can be used several times in loops that are to be combined, as shown in Figure 24.7 and when the terms are being read and noted down it does not matter in which order this is carried out. When we read the map of Figure 24.7 as a combination, we get the following:

For the two horizontal loops: $X = B\bar{C}D + BCD = BD$

For the two vertical loops: $X = \bar{A}BD + ABD = BD$

This will yield a result for all the four loops of BD as shown in Figure 24.7 after applying the theorem $AB + A\bar{B} = A$ to both the horizontal and vertical loops in turn, since for the horizontal loops the only differing variable is C and for the vertical loops the only differing variable is A.

There are some points worth noting with regard to loops:

The number of 1s in a 1-loop $= 2^0$

The number of 1s in a 2-loop $= 2^1$

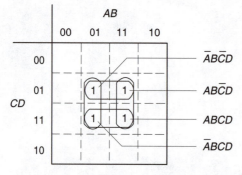

Figure 24.7: Method of identifying terms that are combined.

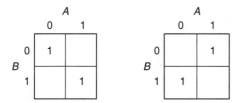

Figure 24.8: Inversion in a Karnaugh map.

The number of 1s in a 4-loop $= 2^2$

The number of 1s in an 8-loop $= 2^3$

Inversion in a Karnaugh map is carried out by changing all the 0s in the areas to 1s; even though we have stated that only 1s are shown on the map, it is understood that 0s are there even though they are not shown. We shall give an example of this using two variables as in Figure 24.8. The map represents:

$$y = \overline{A\bar{B} + \bar{A}B} \qquad y = A\bar{B} + \bar{A}B$$

This relation can be verified by writing a truth table of Table 24.15.

The Don't Care Situation

We are sometimes faced with a situation where the inputs are interdependent, which means that they are such that certain combinations of the inputs cannot occur. Under these circumstances we really do not care what the output will be when the combinations are realized. The combinations will have to appear on the map, but it will not be possible to leave the area that contains a "don't care" blank, for that would indicate that a zero was present in that location, which is really not the case. To overcome the difficulty, a symbol is used, Φ, the Greek capital letter phi and has the physical appearance of a 1 superimposed on a 0, or in other words a don't care.

There is a general rule which requires that for each 1 in the Karnaugh map we should try to configure the largest possible loop, with the loop containing either 1s or don't cares. Then the smallest number of loops that contain all the 1s is chosen, using the don't cares if necessary, the don't cares within the loops being converted to 1s, and all the others outside the loop being converted to 0s. If we wish to invert the function that contains don't cares, then it should be clear that it does not matter what the state of the don't care component is: the 1s will change to 0s, the 0s will change to 1s and the Φs will remain as Φs.

TABLE 24.15
Verification of Figure 24.8

A	B	$A\bar{B}$	$\bar{A}B$	$\overline{A\bar{B} + \bar{A}B}$	$A\bar{B} + \bar{A}B$
0	0	0	0	1	0
0	1	0	1	0	1
1	0	1	0	0	1
1	1	0	0	1	0

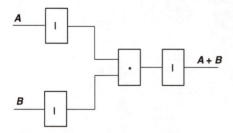

Figure 24.9: Implementation of OR function with AND and Inverter blocks only.

Other Logic Functions

Thus far, we have considered only the AND, OR, and INVERTER functions. There are other blocks (functions) that can be used for implementing logical operations, and these are EXCLUSIVE OR, NOR, NAND, MAJORITY, MINORITY, and IMPLICATION. Before proceeding further, we shall describe in words the functionality of each of these blocks in turn so that an idea can develop.

In the following listing, the output of the block will be 1 when:

AND	all inputs are 1
OR	at least one input is 1
INVERTER	if input is 0 (and vice versa) can have only one input (see note)
EXCLUSIVE OR	(has two inputs) only one of the inputs is 1 but not both
NOR	all inputs are 0
NAND	at least one input is 0
MAJORITY	more than half the inputs are 1 (has odd number of inputs)
MINORITY	less than half the inputs are 1 (has odd number of inputs)
IMPLICATION	has two inputs called A and B, except when A is 1 and B is 0, $A \cdot B = 0 =$ don't care

Note: An inverter will produce a 0 if the input is a 1.

A set of blocks is said to be functionally complete when the combination is able to implement any function. As an example, we can implement an OR function for two inputs using only AND and INVERSION functions, as shown in Figure 24.9.

RELAY LOGIC SYSTEMS

Relay logic, the last topic we shall cover in this chapter, is an important one in that the configuration of many of the programmable controllers in use is based on this technique. When such a method is used, it is referred to as *ladder logic*, the reason being that the configuration is graphical, in essence with one line on the diagram

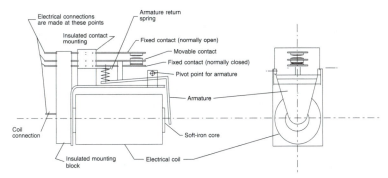

Figure 24.10: Structure of a relay with changeover contacts. The relay shown is an open type; but, more commonly, a polycarbonate cover is used for dust protection. Sometimes the entire relay assembly is hermetically sealed.

representing (in most cases) a logical operation; hence a series of operations give the appearance of the rungs of a ladder. In recent times, however, ladder logic has been enhanced through combination with *sequence logic* using a technique called *Grafcet* that has its origins in France as a method to configure complete programmable logic controllers (see Chapter 25).

STRUCTURE OF A RELAY

Before we proceed further, let us define what a *relay* is so that we can comprehensively visualize practical switching. A relay is an electromechanical device used to mechanically operate or initiate one or several electrical contacts, the drive for the contacts being provided by an electromagnet. Figure 24.10 shows a relay from which the principle of operation can be deduced and which for simplicity is kept to a single change over contact. (Changeover contacts change position when the electromagnet is energized and, normally, change back when it is deenergized.)

To energize the relay coil, it is necessary to provide an electrical source of power to it. This electrical source of power can be provided at definite times dependent on the logical requirements of the circuit, the implication being that one relay can be the initiator, via its contact set, of another or several other relays, with the overall effect of producing a logical set of operations.

THE RELAY AS A BOOLEAN DEVICE

When applied to relay circuits, the Boolean constants, variables, and operations that we have been discussing take on a different meaning from when we are considering logic blocks or logic networks. When applied to a relay circuit, the logic 1 and the logic 0 could actually apply to the single circuit or contact; hence, they represent a closed or open circuit, respectively. Or they could apply to a complete, complex circuit.

The typical motor control circuit in Figure 24.11 will give a good idea of what is involved. The start and stop pushbuttons are under the control of the operator and are initiated only momentarily. The figure also shows the following:

1. Each circuit has a "ladder" configuration.

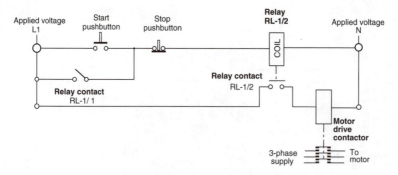

Figure 24.11: Motor control circuit.

2. One relay can drive another—e.g., to start the motor, as shown.

3. Discrete logical conditions are present on components—e.g., the start and stop pushbuttons—in the total logical system.

4. A relay operating on a low-voltage source can manipulate components on another, much higher energy level via a slave contactor (particularly advantageous if the controls are distant from the motor).

We shall now describe the circuit operation.

A TYPICAL MOTOR START/STOP CIRCUIT

Let us assume that the operating (control) voltage is 24 V dc (we can use low-voltage ac instead, say, 50 V ac at 50 Hz). In fact the Institute of Electrical Engineers (IEE) recommends that all control voltages be low, something of this order. When the normally open-contact start pushbutton is initiated, there will be a current flow from terminal L1 to N. This current flows through the normally closed-contact stop pushbutton to one end of the relay coil, through the coil, and on to terminal N. In passing through the two-contact relay (RL-1) coil, it "energizes" (i.e., magnetizes) it, an action that we have seen causes the associated contacts to change state. The contacts are labeled RL-1/1 and RL-1/2, and these names show that they belong to relay coil RL-1 and are contacts 1 and 2 of that relay. The relay and contact naming protocol shown can be extended to cover as many contacts as are involved, with each contact numbered sequentially.

Contact RL-1/1 is normally open and closes as soon as the coil is energized; this action provides an alternative source of power to the coil to cover for the time when the start pushbutton is released. Since the time to energize the coil is of the order of milliseconds, the transition from pushbutton to contact is seamless. The coil will remain energized as long as the alternative path, also called the *hold-on* path, is uninterrupted. The hold-on path is also made through the stop pushbutton; this is necessary in order to break the supply of electrical energy to the relay coil, causing it to deenergize or "drop out" and thus stop the motor. The relay that initiates the three-phase motor is called a *contactor*—to give it its full description a direct in-line (DIL) contactor. The contactor is designed to carry the very heavy currents involved with such motors, which are far beyond the electrical current-handling capabilities of the control relay itself.

One can imagine several contacts—normally open (NO) or normally closed (NC)—placed between supply terminal and the start pushbutton: e.g., NO contacts in parallel with the start pushbutton for remote or auto start to stipulate the conditions under which a motor would be inhibited or locked out, or would be allowed to start; and several NC contacts either in series with contact RL-1/1 or, more usually, in the circuit position directly after the stop pushbutton to stop the motor. These latter additional NC contacts could, for instance be initiated by a value of the motor current or motor winding temperature, associated with the equipment, or by some process parameter. For example, in a batch control system with which the motor is linked, the NC contacts would stop the motor under abnormal conditions.

This is just a brief and simple illustration of the many capabilities of relay systems. It also briefly outlines the vast field of logical actions that are associated with process control that use not only relays, but also solid-state devices and programmable controllers in pursuit of control objectives.

SUMMARY

1. The system called binary-coded decimal (BCD) uses weighted binary digital values to represent decimal numbers.

2. The first (the rightmost) digit in a number in any number system is multiplied by the radix raised to the power 0 (R^0), the second digit is the value multiplied by the radix raised to the power 1 (R^1), etc. In general, we can say the nth digit is obtained by raising the radix to the power $n - 1$ (R^{n-1}). Any leading or place-holding zeros will not change the value of the number.

3. As an example of binary to decimal conversion, the binary number 11100111 can be represented as follows:

1	1	1	0	0	1	1	1
R^7	R^6	R^5	R^4	R^3	R^2	R^1	R^0

In this case the radix is 2. Substituting decimal equivalents for all the digital values, we get:

1×128	1×64	1×32	0×16	0×8	1×4	1×2	1×1
128	64	32	0	0	4	2	1

Adding these together gives the decimal number: 231.

4. As an example of decimal to binary conversion using the decimal number 97, place the number 97 at the upper left of an empty two-column table. Successively divide by 2, and writing the remainder (if any) on the right-hand side of the vertical line and the whole number quotient in the next row on the left. The process is continued until there is a zero quotient in the left-hand column. The table will appear as follows

```
97 |
48 | 1
24 | 0
12 | 0
 6 | 0
 3 | 0
 1 | 1
 0 | 1
```

The binary number is 1100001 and is obtained by reading the right-hand column upward from the bottom.

5. AND is a function that requires all inputs to be true (logical 1) before the output will be true. If the inputs were numbered 1 through n, then input 1 AND input 2 AND input 3 AND ... input n must be true for the output to be true.

6. OR is a function that requires any one of the inputs to be true (logical 1) before the output will be true. Put another way, if the inputs were numbered 1 through n, then either input 1 OR input 2 OR input 3 OR ... input n must be true for the output to be true.

7. NOT is a function that is also referred to as inversion (INV) and made to act on a single input. In operation an input that could be either true or false is conditioned to produce the inverse of the input as the output. Put another way, if the input is true, the output will be false, or if the input is false, the output will be true.

8. To construct a truth table for three or more variables:

    (a) Give each variable a name. Start with a single variable, A, and define it completely as shown in Table 24.6.

    (b) Add the second variable name, B, to the left of the first variable name. Then add two zeros to align with the first two conditions of variable A as shown in Table 24.7.

    (c) Repeat the two conditions of variable A in the same order directly under the initial two conditions of variable A as shown in Table 24.7.

    (d) Add two 1s under the 0s of the variable B to align with the repeated initial conditions of variable A. This completes the truth table for two variables A and B.

    (e) Add the third variable name C to the left of variable B, and add four zeros to align with the four conditions of the variable B. Steps 6 through 8 following will then in principle be the same as steps 2 through 4.

    (f) Repeat the four conditions of variable A in exactly the same order, but directly under the initial four conditions of variable A.

    (g) Repeat the four conditions of variable B in exactly the same order, but directly under the initial four conditions of variable B.

    (h) Add four 1s under the 0s of the third variable C to align with the repeated conditions of variable B.

The procedure is repeated for as many variables as there may be.

9. Boolean algebra, like the familiar algebra, has an order of priority in carrying out operations, in this case Inversion, And, OR, AND is written as ABC. It is also normal to use parentheses around terms that are ORed; for example, $A \cdot B + C$ can be written as $A(B + C)$; $A + B \cdot C + D$ can be written as $(A + B)(C + D)$.

10. The following theorems will be useful. For AND

$$A \cdot 0 = 0 \qquad A \cdot 1 = A \qquad A \cdot A = A \qquad A \cdot \bar{A} = 0$$

For OR,

$$A + 0 = A \qquad A + 1 = 1 \qquad A + A = A \qquad A + \bar{A} = 1$$

Also

$$A + AB = A \qquad A + \bar{A}B = A + B$$
$$A(A + B) = A \qquad A(B + C) = AB + AC$$

11. The results of inverting OR and inverting AND, we have seen, are identical. These results will be obtained irrespective of the number of variables involved, and gives the generalized De Morgan's theorem:

$$\overline{A + B + C} = \bar{A} \cdot \bar{B} \cdot \bar{C} \qquad \text{or alternatively} \qquad \overline{\bar{A} \cdot \bar{B} \cdot \bar{C}} = \overline{A + B + C}$$

12. For the Quine-McCluskey minimization technique, we follow a particular sequence of operation:

    (a) Convert the problem into the binary notation of the standard sum form, remembering that any problem can be converted into this form.

    (b) Sort the terms into groups of the same index.

    (c) Tabulate the groups and separate each by a horizontal line.

    (d) To ensure identification, write the decimal equivalent of each term alongside the associated term.

    (e) Consider pairs of terms to see if combination is possible. Two terms can be combined if they differ in only one digit position, i.e., by the theorem $(AB + A\bar{B} = A)$.

    (f) As a result of the previous step, write a dash or hyphen (-) at the position of the difference in digit.

This procedure will result in a list of binary-coded digits separated by horizontal lines that group codes of equal decimal weight together. We repeat the procedure of tabulation and comparison until no more terms can be combined, always making sure to append the decimal values that give the reduced binary code shown in the results column. Note that one term can be used as often as necessary, but once used, it must be marked with an asterisk for identification. When the comparison does not give any further combinable terms, we convert back to the original variables to give the prime implicants comprising all terms that cannot be further minimized by the Quine-McCluskey method.

13. The Karnaugh map is a graphical method of showing a truth table suitable for a small number of variables. Four variables can easily be handled, but numbers above six become very involved and difficult. It is suggested that, for five or more variables, the Quine-McCluskey method be used instead.

14. The properties of a Karnaugh map are:

    (a) Each area on the map corresponds to a specific value of a variable.

    (b) Each adjacent area differs by one variable.

    (c) Adjacent areas have a common side.

15. A Karnaugh map is a visual representation of the Quine-McCluskey minimization method and therefore should give the same results. When two terms differ in one variable only, they can be combined as one term according to the Quine McCluskey rule. There is a similar rule in the Karnaugh method that says that

two adjacent areas can be read as one term by including only those variables that are of the same value in both areas.

16. Where the inputs are interdependent, or are such that certain combinations of the inputs cannot occur, we do not care what the output will be when these combinations are realized. The combinations will have to appear on the Karnaugh map. However, it will not be possible to leave the area that contains a don't care blank, for that would indicate that a zero was present in that location, which is really not the case. The difficulty is overcome by using a Φ (Greek capital letter phi) instead of a blank space.

17. For each 1 in the Karnaugh map, we should try to configure the largest possible loop. The loop can contain either 1s or don't cares. The smallest number of loops that contain all the 1s is chosen, using the don't cares if necessary. The don't cares within the loops are converted to 1s; all the others outside the loop are converted to 0s.

18. When applied to a relay circuit, the logic 1 and the logic 0 could actually apply to the single circuit or contact (hence, they represent a closed or an open circuit, respectively), or they could apply to a complete, complex circuit.

CHAPTER 25

Sequential Logic

CONTINUOUS VERSUS BATCH (SEQUENTIAL) PROCESSES

Process industries can be divided into two very broad segments: continuous process operation and batch process operations. In the continuous processes, raw materials are continuously taken in at one end, operated on during their passage through the various stages of manufacture, and turned out as a continuous stream of finished product(s) at the other end of the operation. Once set in motion, there is, in principle, no break in the manufacturing cycle. Batch operations, on the other hand, take in very specific quantities (batches) of raw materials in an ordered sequence, operate on these in an ordered sequence, and, at the end of processing, produce a batch of finished product. In batch operations there are by definition a start and a finish point for the batch, but under no circumstances should a batch operation be considered a shortened version of a continuous one. As examples of batch operations, we can consider almost all pharmaceutical products and many of the "coating" products (e.g., paints, varnishes, photographic emulsions).

A TYPICAL BATCH (SEQUENTIAL) OPERATION

Since batch production always uses operational steps that repeat in the course of manufacturing each batch of the product, each step will be associated with a definite state of the product at any given point in the manufacturing cycle. The transition from one stage to another is determined through specific indicators inherent in the process material; the transition could be, for example, a rapid change in pressure, temperature, viscosity, or some other associated parameter. Figure 25.1 is a block diagram of a typical overall batch process scheme. Note that like most batch control scheme diagrams, the control loops shown in the figure only cover some of the overall controls, there are several others that for clarity are not shown, and those that are shown, are not detailed, but give only the barest information; this is deliberate because in reality the process is most often very complex and this method avoids overcrowding the diagram. The full control loops are usually defined in separate documents that accompany the overall scheme. In this instance, as an example only, a full flow control scheme is shown in Figure 25.18, the other loops will of course have separate diagrams. Looking at the full flow loop diagram and the information given should go a long way toward explaining the reason for the practice.

THE STARTING POINT OF A BATCH CONTROL SCHEME

The Process Analysis

The starting point for the design stage of a control system to automate any batch operation is a complete, written, detailed description of the manufacturing

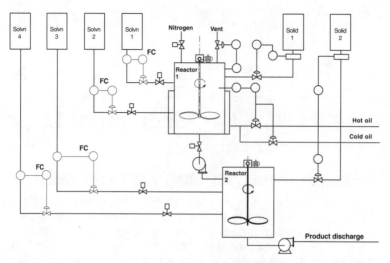

Figure 25.1: Typical batch process diagram. (Note: Loops marked FC are expanded in Figure 25.18.)

cycle, usually referred to as the *process analysis*. The process analysis is usually prepared jointly by the process engineers, process chemists, and production personnel employed by the user or licensee of the process, since it is only these individuals who would be fully conversant with the total manufacturing requirements, including the conditions that have to be alarmed, avoided, or handled by set operating procedures.

It is worth noting that batch processes, because of their complexity and secrecy of processing, are most often subject to patents or copyrights, and thus personnel associated with them are usually required to sign a secrecy agreement binding them not to divulge any part of the process.

Design Considerations—"Inerting"

To give some idea of the complexity of the requirements, let us consider a simple but typical batch operation. Let us assume that the product has to be produced in two vessels, called reactors, which are usually equipped with a means of agitation and, depending on the process, with heating and/or cooling facilities, and that the product is made in the presence of an inert atmosphere to eliminate unwanted reactions between the ingredients and oxygen, of which air contains approximately 23%. Since oxygen is to be avoided, and to ensure that this requirement is met, it is important that the reactors initially be purged of air, and filled with an inert gas instead, a process commonly referred to as *inerting*. Most often the gas used to achieve this is nitrogen (N_2), which, it must be emphasized, is a lethal gas. Although the air we breathe contains approximately 77%, this amount is finely balanced, anything much above that will cause discomfort, and when the concentration approaches 100%, survival under these conditions is impossible, no human or animal exposed to it will live; and particular care therefore has to be exercised to ensure a safe environment. In our example, once the reactor is filled with the inert gas, the first of the raw materials are discharged into it. The method employed in determining when the vessel is full of the inert gas is to monitor the vented gas for oxygen using an O_2 analyzer of the type described in Chapter 5.

PRODUCT INGREDIENT ADDITION

Let this first material be a liquid, and since a specific quantity is required, this will have to be metered (measured) in, a typical flow measuring, integrating, and control loop as shown in Figure 20.4. By introducing a mass of liquid into the reactor, it is inevitable that the "blanket" gas is going to be compressed unless some regulation is applied. We shall assume that the pressure within the reactor has to be approximately atmospheric, or preferably just slightly higher to maintain O_2 exclusion, and any material additions will have to be forced into the reactor and any reaction carried out at this pressure. Let us now assume that the next raw material to be added is a solid as shown in Figure 25.1. This is also metered into the reactor, but this time the measurement is most often based on ingredient weight; however, the principles of control and integration are those given earlier in Figure 20.4, but the final control element will have to be considered carefully and will depend to a large extent on the process and site equipment. Usually the added solid is in granular or powder form and fed into the reactor from a hopper suitably N_2-locked to exclude air. Before the solid ingredient is added, it is necessary to start the agitator to ensure that the added solid is dispersed quickly throughout the liquid. The addition of the second ingredient could result in either an endothermic or an exothermic reaction. Note for endothermic reactions to occur heat will have to be applied to the vessel, whereas for exothermic reactions cooling will be necessary; heating or cooling is also initiated by the batch control system and shown in generalized form in Figure 25.1; the principles of reactor temperature control and mixing were discussed in Chapter 12. In both these circumstances it is usually important that specific temperatures of the contents be maintained or limited or controlled. Throughout all the material addition and the other operational procedures, the slightly positive pressure in the reactor is maintained with a blanket of N_2 gas to avoid product contamination. This pressure control loop is indicated in Figure 25.1, and while only the vent valve is shown as being manipulated, it should be obvious that the N_2 supply line will also have to be regulated as well, but this is quite a complex loop and too much to be shown in the figure presented.

Product Processing

It may be that the combined ingredients have to be held in the reactor for a specified time or until the material being processed exhibits particular characteristics. During this period of retention the product within the reactor could be giving off vapors that when condensed produce valuable by-products that can be used elsewhere in the process or sold directly. To obtain the condensed product will require a reactor that has a distillation column built at the top of the vessel, as shown in Figure 25.2. This distillation process can give rise to a means of controlling the reaction, for by reintroducing small quantities of the *reflux* (the term for the liquid formed by the condensed vapor) back into the distillation column, the desired conditions can be maintained. To do this requires a level control loop (not shown in Figure 25.2 for simplicity and clarity) on the reflux vessel, and more than likely a differential-pressure control loop across the column itself, the two loops working in unison. The distillation column could be part of the reactor itself, as shown in the figure or it could be a separate piece of equipment; quite often the column associated with a batch reactor is a *packed* distillation column—i.e., the column has a material filling instead of the more customary "trays." When distillation is involved, means are often provided via suitable valved bypasses to isolate the column so that a faster startup can be achieved. The vapors on start up are initially made to recirculate through the reactor

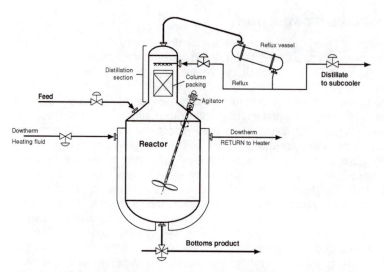

Figure 25.2: Typical batch reactor with distillation column. Not shown, for simplicity and clarity, are the nitrogen feed and vacuum system; all control loops; and pigging, which could be employed on the bottoms product.

without passing through the reflux vessel, and at the appropriate time the process lines are reset so that the reflux vessel is included in the circuit to allow the reflux to be collected. The level of the material in the reactor as it varies is an important parameter, and some control of the processing sequence is usually based on it.

Compliance Testing of Product During Manufacture

The product in the reactor is periodically sampled for analysis and its composition determined, either by tests in the laboratory by a process chemist, or automatically by some analyzer instrument. In any case, while the testing is being carried out, the product has to be maintained at the conditions prevailing when the sample was removed. If the product is found to be acceptable after testing, then the next processing steps of the *recipe* (holds the same interpretation as in food making) can be implemented. Let us assume that a liquid has to be added first, followed by a much smaller amount of solid material than was initially introduced. Once acceptance of the material by the test has been obtained, the second liquid can be metered in. Metering the liquid must usually be accomplished as quickly as possible, or at a specified rate (recipe dependent) if necessary. The process is monitored continuously to make sure that the product is meeting the specification at each stage.

All through this operation the agitator is kept moving to ensure complete mixing, and it may be useful to consider the requirements of the agitator itself at this point. Some agitators should not be operated without a load applied to the blades, to avoid mechanical damage to the agitator, vessel, and motor drive assembly; but it is recommended that every agitator be considered on an individual basis and suitable steps be taken to ensure its safe operation. It is often the case that the current drawn by the agitator motor provides a good indicator of the condition of the material being processed and can serve as a basis for initiating the required control actions, (i.e., current reflects torque, which in turn provides a measure of fluidity, viscosity, consistency of paper pulp, or density).

Product Processing—Continued

At the end of liquid addition, there may be a waiting period to ensure complete reaction or mixing, and product sampling can then follow, subject to similar criteria as described earlier. On completion of this stage of processing and product acceptance, addition of the second solids can be made. Once again, any addition of an ingredient could give rise to an exothermic reaction, so that suitable precautions have to be taken to prevent thermal runaway (an uncontrollable temperature rise); or, if the process demands it, further heating may have to be applied, to sustain an endothermic reaction. Let us take this second addition of solids to be the end of the first part of batch processing of the product defined in our assumption of two reactors, which results in, say, a viscous fluid that needs further processing, but not in the same reactor. Note that for some other, less complex product, reaching the end of this first stage of material processing could mean that the required product had been obtained, and the batch processing would be terminated.

In the event of further processing, the product in the first reactor has to be discharged to the second reactor where the next phase of the processing is started. Pumping out the partially complete product from the vessel is not as simple as it sounds, for while the initial processing is being carried out in the first reactor, the second (receiving) reactor must be set ready to receive the product. It could be that this second reactor had been in use for a previous batch, which could mean that cleaning and purging of the vessel will have to be completed prior to receipt of the soon-to-be-discharged product from the first reactor. To achieve this implies that the processing time in the second reactor has to be shorter than that in the first, and if several batches are being made consecutively, then synchronizing batches becomes very important; this synchronization is referred to as *batch scheduling*. The *batch scheduler* (an algorithm in the system) has to be kept informed of each stage of the processing by signals generated through the process, or those developed during the run of the *sequence* program.

Wash Cycle and Solvent Recovery

Emptying the first reactor is not the end of the cycle, for the vessel has to be made ready for the next batch, which means that it has to be washed clean and dried and its atmosphere made inert once again. This is sometimes referred to as the *wash cycle* and could involve one or more chemicals or water as the solvent with all the attendant pump and valve-sequencing. The agitator is usually kept moving throughout the wash cycle and will have to be stopped only when the wash solvent is discharged. In many instances the solvent is also discharged into the next vessel, assuming compatibility, and hence the quantities have to be carefully metered so that the batch processed subsequently is not spoiled. Sometimes after the wash cycle, the solvent, if it is not usable in the subsequent processing, is pumped to another part of the process, where the solvent is recovered and the remnants of the product, if appropriate, even recovered for subsequent reuse, conversion, sale, etc., or discarded; and this part of the operation is called *solvent recovery*. The waste has to be treated before it can be disposed of, or it can be incinerated, but in any event the effect of waste disposal on society and the environment has to be considered most carefully.

SECOND REACTOR PRODUCT PROCESSING

In the second reactor, let us assume that additional ingredients are to be added and the product made to undergo a controlled temperature change. In some cases it may be that the combination of both additional ingredient and temperature change is not required, but a close watch on the product is necessary in any event. The addition of ingredients has been dealt with earlier, and similar procedures could be used, although these may have to be modified dependent on the actual requirements and ingredients or materials involved. Temperature control will follow the good practices used in continuous control loops, but care must be exercised, since the thermal characteristics and the inertia involved can be very different compared with those of continuous processes.

In batch operations it is quite normal for the operator to intervene at any stage, as it is not always possible, or in some instances advisable, for the process to proceed without human intervention, because some products are so sensitive to small variations that if these were not attended to immediately, the results could have a severe impact on personnel, plant, or the environment, as well as the product. To address these circumstances the instrumentation may have to be configured with duplicates for all automated loops implemented on the sensitive parts of the process, with one continually tracking the other during the production process and ready to take over at any time if required; the operator also being allowed to take over the automated loops at any time. Such arrangements are referred to as *redundant* systems with each suite of equipment being called a *level of redundancy*; it is not uncommon to find two or sometimes even three levels of redundancy in critical processes. These ideas of redundancy are also relevant and applicable to continuous processes.

Process Line-Pigging Operation

When the processing in the second reactor is complete, the product has to be pumped away. However, by this time the monetary value has increased quite enormously, and therefore as much of the product as possible has to be collected from the containing equipment and pipework. When required, a method of scouring the vessel with the inert gas or solvent and the pipelines with a mechanical piston called a *pig* is resorted to, with the procedure of cleaning the pipeline referred to as *pigging the lines*. To give some idea of what happens during a pigging operation, let us first try to visualize the pig, which is a roughly dumbbell-shaped piston made from a PTFE block, steel-reinforced for strength and stability, that fits the internal bore of the pipe. Since the pig is a solid of fixed dimensions, it is necessary to predetermine the minimum allowable pipe bend radii through which it can be freely moved and so traverse the length of pipe. The pig is driven through the pipe under the exerted force of a compatible gas or solvent at high pressure, often of the order of 2 bars, applied to the face of the dumbbell in the direction of travel. When not in operation, the pig is retained in a specially designed section called a *keeper*, which is a cylinder having a bore the same size as the pipeline fitted with a solid transverse metal gate on one end, pneumatic (gas) or hydraulic (for solvent when used) connections along its length, which is sufficient to accommodate two pigs, and at the other end a sliding flange operated by a lever that allows the pig to be inserted or removed. All the moving parts have seals that maintain the fluid at the operating pressure; the arrangement enables the pig to be inserted in a position behind the gate. Solenoid-actuated valves that are initiated at the appropriate time in the processing cycle control both inlet and exhaust of the motive gas or solvent. During product processing operation the closed gate holds the pig in the keeper and blocks its entry into the pipe line, and when the line is to be pigged, both the retaining gate, and the high-pressure gas supply solenoid valve are initiated, causing the pressure behind the pig to rise to the full gas or solvent supply pressure and the gate to open fully. This forces the pig into the pipeline, and because it fits the pipe bore all material within the pipe is scraped together and forced forward in front of the pig, leaving a clean pipe in its wake. The scrapings are diverted to the product receivers just before the completion of the pipe traversal and the capture of the pig in another keeper located at the far end of its

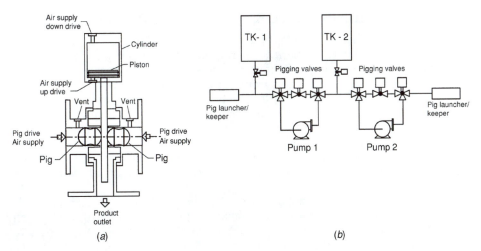

Figure 25.3: (*a*) Sectional view of a typical pigging valve. A three-way valve has been shown; there are other alternatives. (*b*) Typical line pigging arrangement. The lines to be pigged should offer unobstructed passage to the pig. For clarity, the gas lines and valves have not been shown, and the electrical signal lines and initiating logic circuits have also been omitted.

travel. The motive gas is then vented or the solvent diverted to the solvent recovery unit, and the starting position gate opened. The gas flow is reversed to drive the pig back to the keeper at the starting position. When the pig is returned to the keeper once again, the gate is closed, making the line ready for the next time. Figure 25.3 shows a typical pigging system.

It must be emphasized that this description of the batch and pigging system is extremely simplified. In practice, each of the measuring devices, controllers, solenoid valves, pumps, actuators, and other equipment has to be specified in detail with initiating and shutoff times precisely defined. It is also vital that the quantities required, and the complete and detailed set of operating procedures at each and every stage, be precisely specified.

Since it is the duty of the process engineers, safety engineers, and chemists to ensure that the plant, personnel, and the environment will be protected under all operating conditions, manual or automated, hazardous or nonhazardous, they must give detailed descriptions in clear, precise language of the procedures and actions to be followed under any hazardous conditions during manufacture—whether routine stoppages for, say, product analysis, operator intervention to correct a process deviation, or putting production on hold while some function is carried out. The procedures can and will vary dependent on the part of the manufacturing cycle in which the conditions occur. It is the duty and responsibility of the control system designers to insist on being given these vital specifications and to ensure that the crucially important instructions given in the manufacturing and safety procedures are followed and the control system built to meet all the requirements. Furthermore, control system designers must ensure that the process follows logical steps and is easy to understand, and that the process operators are given complete operating instructions for plant, personnel, and environmental safety. All this must appear as a coherently written and presented manual called the process analysis referred to earlier; it is a very much-elaborated form of operating manual, for it sets out in the most minute detail the precise operation for all eventualities. All process operating personnel must be aware and conversant with every aspect of the procedures. Contravention of this could have serious consequences, including legal.

THE NAMUR STANDARD FOR BATCH OPERATIONS

In 1949 a number of large German, Dutch, and Swiss chemical companies decided to pool their information and experience in process measurement and control, for in the opinion of these founding organizations it was necessary to make the practices in process industries uniform. It was from this beginning that the NAMUR (Norme-narbeitsgemeinschaft für Mess-und Regeltechnik in der Chemischen Industrie zur Interessengeimenschaft Prozessleittechnik der chemischen und pharmazeutischen Industrie) standard was born. Success saw the group expand to include other European companies and professional bodies, and in 1979 a subcommittee was formed to investigate the possibilities of applying the standards to batch processes. This resulted in a report and recommendations published in 1985, which set out guidance for software development and control system evaluation.

Since the computer today controls virtually all batch operations, standard software routines and documentation permit changes to be implemented quickly—an important advantage in a competitive marketplace. The following is a summary of the NAMUR batch committee recommendations:

• A hierarchical control system structure that reflects the process

• A terminology that is standard and applicable to batch control

• Use of *recipes* to produce and manage a batch and modular software that relates the recipe to a *unit operation*

• Use of configuration techniques—e.g., as used in DCS and other block-configurable systems—rather than custom-written software to achieve the application requirements

Meaning of some of the terms used in the recommendations: A *unit operation* is a set of instructions for process operations that covers:

• The process equipment

• Materials (raw and finished)

• System communications

• Operator interactions

Such elements in the manufacture of a specific product are generally repeatable from one process plant to another, though the plants and processes may differ in some minor details. A *recipe* in batch terms, as in common usage, specifies:

• The quantities of each material to be used

• The operating conditions for each part of the manufacturing process

• The sequence of operation for each manufacturing stage

• Each item of plant equipment required to produce the batch

According to the NAMUR recommendations, a *recipe manager* should form part of a batch control package. The object of the recipe manager is, in general, to divide recipes into *master* and *control* and to provide additional functionality to handle the separate and specific requirements of each. The difference between the two forms of recipe is that the *master* recipe is process equipment independent whereas the *control* recipe is not. Control recipes are modified masters in which:

- The process equipment to be used is specified.

- Unit operations of the master recipe are replaced with control phases, which are used to produce the batch of product.

- Facilities are provided for the operator to modify those parts of the recipe that have not yet been executed, without interrupting the process.

- Facilities exist to modify the recipe to allow for variations in material quality or composition.

- Facility is available to manually reassign process equipment if required.

The ISA SP88 committee of the Instrument Society of America was formed in 1988, and it also has a recommendation for batch process operations, which in many ways reflects those of NAMUR. There are differences, especially in the scope of the control recipe, but the trend in batch process control is toward an internationally acceptable standard that is applicable throughout the world.

SPECIFYING A BATCH OPERATION: GRAFCET

Having described the operations involved in batch manufacture of a product, the question now arises as to how these descriptions can be converted into detailed information that can be readily understood by a suitably capable reader and converted into "plant comprehensible" instructions. This proviso of a "qualified" readership is necessary, for we are dealing with systems that can be highly complex and therefore not really understandable by the casual first-time reader. However, it must also be stated that the acquisition of the necessary qualification should not be made so complex or specialized that it acts as an impediment to persons willing to get involved.

We shall now present a sequence language covered by IEC 1131.3 that is accepted as an international standard. The basis for this standard is *Grafcet*, a symbolic graphical method developed in France, initially for organizing and regulating numerically controlled machines, e.g., manufacturing robots.

As the language is graphical, the system *model* is constructed from a set of graphic "primitives" or elements. These primitives are assembled in accordance with a *syntax*, and are elaborated on by succinct descriptions that define the *function*. The syntax will be given later in the chapter, since it involves definitions that must be stated beforehand. The Grafcet is always read from the top downward, and the evolution from one state to another can proceed only in accordance with the *five rules of evolution*, which will also be given later in this chapter. The reason for this delay is that some basic definitions must be given prior to a statement of the rules so that readers can clearly understand the implications of these rules.

THE PRIMITIVES

There are three graphic primitives:

The step

The transition

The directed link

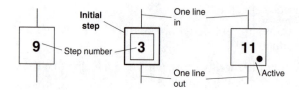

Figure 25.4: Typical Grafcet steps.

The step can be of three forms:

Basic

Initial

Active

In a Grafcet, there can be more than one initial step, and there can be multiple links to a step. However, each step can only have one entry line and one exit line as shown in Figure 25.4. To meet the requirement of single entry to and exit from a step, the convention demands that the graphic representation be as shown in Figure 25.5.

A *link* connects one *step* to another, and in doing so goes through a *transition*. It is normal for a link to be shown as vertical line segments. However, if circumstances demand, a link may be "bent" through 90° to one side in order to accommodate another part of the Grafcet. In the top-down—i.e., top-to-bottom—representation of the sequence, directional arrows are not shown, but if a loop, or a link from a step lower down in the evolutionary path of the sequence to a step higher up in the evolutionary path, is required, then directional arrows are permitted. It is recommended that all connecting links be perpendicular to each other even though diagonal links are allowed. The representation is shown in Figure 25.6.

It is normal to define actions associated with a step in either symbolic form or text, and in both cases these actions are noted directly alongside the relevant step. If the actions are in text, then the instructions are enclosed within quotes and written alongside—for example "Activate relay 5" as shown in Figure 25.7b; if on the other hand, the action is in symbolic form, then a suitable table must be given defining the actions associated with each symbol used. For example, the symbolic form RL5 shown in Figure 25.7a is defined in the accompanying table (Figure 25.7c) as "Activate relay 5".

An action can carry a qualification such as Delayed, Stored, Limited, Pulsed, Conditional, and Combined as defined in Figure 25.8. To minimize the text involved, these actions are contracted to their first alphabetic character; e.g., D is equivalent to Delayed as shown in Figure 25.9. For a Conditional action, a C will be written in

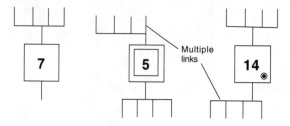

Figure 25.5: Typical links in Grafcet.

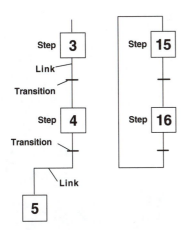

Figure 25.6: Typical Grafcet action sequence.

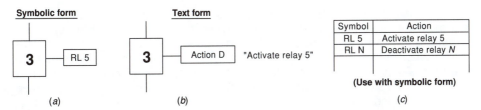

Figure 25.7: Typical definition of Grafcet actions. (*a*) Symbolic form. (*b*) Text form. (*c*) Table of symbols for symbolic form.

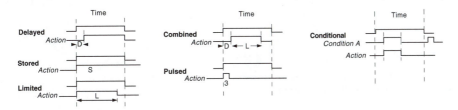

Figure 25.8: Definition of actions associated with a Grafcet step.

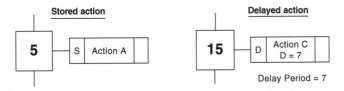

Figure 25.9: Typical actions associated with Grafcet steps.

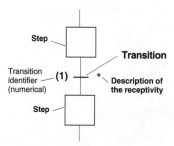

Figure 25.10: Typical method for transition identification.

place of the S shown in the stored action space of Figure 25.9, and the condition itself written in two lines within the action space, e.g., "Action E // if cond c." It is possible to combine the actions—e.g., DL is equivalent to Delayed Limited—and each action is defined in terms of the time period involved; the DL is written in place of the D shown in the delay space of Figure 25.9, and action itself defined in two lines within the action space, e.g., "Action F // D = 4, L = 6."

The *transition* is a method for bringing control to a Grafcet, and it is cleared—i.e., the action is completed or processed, when the preceding step is active and the transition condition is satisfied, i.e., made "true." The clearance of the transition condition is termed the *receptivity*. The transition is identified by a numeral in parentheses—e.g., (1)—with the description of the receptivity (shown by an asterisk) on the opposite side of the numerical identifier, as shown in Figure 25.10. Receptivity is expressed either literally (text) or as a Boolean expression, which can be composed of Boolean operators and/or *edge* symbols; the expression can also be *time-dependent*. The active edge symbol can be either rising or falling. Figure 25.11 illustrates these definitions.

Either a single or multiple steps can clear a transition, as shown in Figure 25.12. In both cases the transition will be cleared when the condition defined by the user—e.g., "Pump is stopped and valve closed"—and denoted by the asterisk turns true.

SEQUENCES

A sequence is a logical series of operations that are to be implemented to achieve the result(s) required. A *linear sequence* comprises a series of steps that are activated

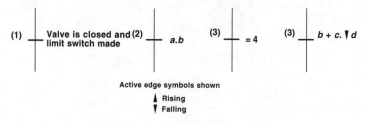

Figure 25.11: Typical examples of alternatives to express receptivity. Transitions can be Boolean expressions—e.g., $a \cdot (b + c)$. Any of the Boolean operators are valid.

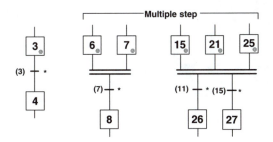

Figure 25.12: Typical examples of defining transition clearance.

one after the other; in practice, each step is followed by a transition, and each transition by a step. An active sequence is one in which any one step is active. The sequence is inactive only if all steps are inactive as illustrated in Figure 25.13.

It is possible in a particular instance to *jump* steps in a sequence, but only when the steps jumped over are not required in that *particular* instance, as shown in Figure 25.14. *Selection* among several sequences from one or more steps can be made as shown in Figure 25.15, the *single horizontal line* indicating an OR function, the *double horizontal lines* indicating an AND function.

To ensure that only one sequence is selected for a particular application, the *receptivities of the associated transitions must be made mutually exclusive*, as illustrated in Figure 25.16. Sometimes it is necessary to repeat a sequence operation—in other words create a loop of the sequence, as shown in Figure 25.17. Here the loop will be repeated for as long as the condition *p* is not true.

SYNTAX AND THE RULES OF EVOLUTION

The *syntax rule* states that *the order step/transition and transition/step must be respected for every sequence*. This means that two steps or two transitions must never be connected by a link; therefore, it is mandatory that a directed link connect:

- A transition to a step.

- A step to a transition.

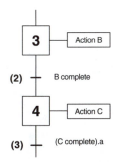

Figure 25.13: Typical linear sequence.

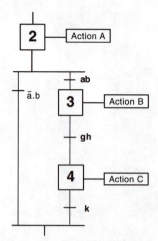

Figure 25.14: Typical jump function.

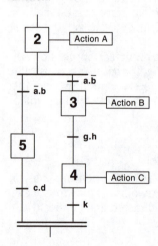

Figure 25.15: Sequence selection (AND).

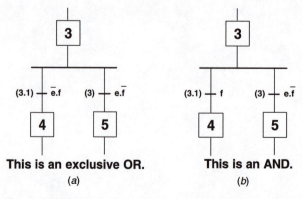

Figure 25.16: (a) Exclusive OR function. (b) AND function.

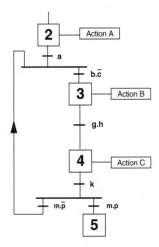

Figure 25.17: Typical loop functon.

The rules of evolution referred to earlier:

1. One or more initial steps determine the initial state of a Grafcet, and the rest state of the system is described by the actions associated with the initial step(s).

2. A transition will be cleared when:
   - All immediately preceding steps are active. The receptivity associated with the transition is true.

3. A cleared transition results in simultaneous:
   - Deactivation of all immediately preceding steps
   - Activation of all immediately following steps

4. All transitions that can be cleared simultaneously will in fact be cleared simultaneously.

5. If a step during the course of operation is logically activated and deactivated at the same instant, then the result will be that the step will become active.

THE CONTINUOUS CONTROL ASPECTS OF A BATCH CONTROL OPERATION

As previously stated, a batch process operation will have some continuous control features that will need to be brought into play during the run of the batch. In the process described in the beginning of this chapter, there are a few continuous control loops that have to be allowed for; however, we shall not investigate all of them, but will choose one and describe it in fair detail to obtain a feel for the involvement.

Let us for the present consider how a required quantity of a solvent or "raw process liquid" is metered into the reactor; Figure 25.18 illustrates what is taking place. This diagram is based on the use of a DCS, since it is the most currently used means of process control; and, for clarity, the controls shown in the diagram are

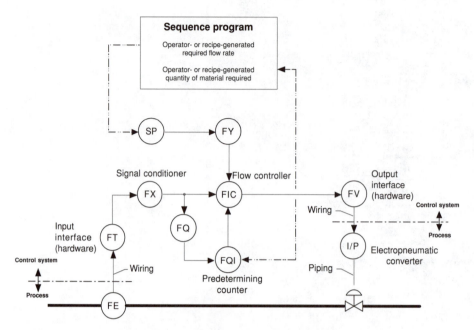

Figure 25.18: Functionality of a flow-metering control loop. All equipment associated with the process is hardware; the remainder is software. This system has been drawn to show the functionality of a DCS control block structure.

broken down to explain the individual functionality required. The majority of the sequence of actions will in fact be contained either in a sequence control block or in some of the other blocks used.

Referring to Figure 25.18, we shall systematically describe the functions starting with the liquid-flow input interface; this provides facilities for adjusting the gain, bias, filtering, and alarms of the incoming signal and is shown collectively as "signal conditioning." In practice, the signal conditioning is part of the analog input block (AIN), as we have seen in Chapter 20, attached to interface hardware module.

The conditioned signal forms the measurement input to the flow controller, which compares it with the setpoint to produce the error signal; and dependent on the magnitude (i.e., value) and sense or polarity of the error (for the error is a vector quantity) an appropriate output, both in direction and controlled magnitude, is produced. The actual workings of a process controller have been given in previous chapters and should be referred to if required.

The setpoint is the flow rate at which the material must be delivered to the vessel and will either be known to the process operator or form part of the product recipe. The sequence block can be programmed to allow either of these options. The setpoint is most often ramped up to the required value in order to avoid a sudden step change and resultant oscillatory swing of the controller output. The ramp function and the "rate of ramp" of the setpoint can, in the newer PID block, be specified during configuration or programmed as a subroutine, usually held in the sequence control block; and because the setpoint is being ramped, it is possible to achieve a very smooth change, which effectively has the measurement and setpoint move in unison.

The actual amount of material required is entered, either by the operator or automatically from the product recipe, into a set of two registers arranged to form a *predetermining counter*, which could be an accumulator block or be configured in the sequence block. The principle upon which the counter operates is as follows. The required quantity of liquid material will be downloaded to one of the two counters (or registers as they are called) arranged to accept a value from an operator or from a command in the software routine. The second register is arranged to receive the measurement signal and made to increment for each measured unit of material that has passed through the flowmeter. Both counters (registers) are initially set at zero by the sequencing logic, in order to form a reference datum, and a comparison between the two registers is made on each scan. When the quantity of material passed is equal to the required quantity initially set on the counter, a flag (trigger) is set, and this flag is used for:

- Closing the liquid supply valve directly

- Closing the line select solenoid valve (see Figure 25.1) after a delay period, to ensure that the full quantity required by the recipe is in fact delivered

- Putting the flow controller into manual mode to hold the control action ready for the next time

- Driving the setpoint to zero

- Initiating other functions required by the sequence

SUMMARY

1. Process industries can be divided very broadly into continuous process and batch process operations. Continuous processes are in principle, no-break manufacturing cycles. Raw material is continuously taken in at one end, operated on during its passage through the stages of manufacture, and turned out in a continuous stream of finished products. Batch operations take in very specific quantities (batches) of raw material in an ordered sequence, operate on these in an ordered sequence, and, at the end, produce a batch of finished product. Thus batch operations have definite start and finish points.

2. The process analysis is the starting point for the control system design function to automate any batch operation. Since the operation is one that uses raw ingredients of the product in strict sequence in definite quantities and at specific times during the manufacture, this data together with all the operating conditions, procedures, and precautions, must be made available to the system designers to enable them to produce the process analysis document. In practice the completion of one batch of product follows another. To obtain an ordered sequence of production the synchronizing of batches of batch scheduling is a very important function that is handled by an algorithm called a batch scheduler. The batch scheduler is continually updated by live data from the system, so that at all times the scheduler is fully aware of the manufacturing status, so that in the event of a unscheduled occurrence recovery is possible.

3. It is quite normal in batch operations for the operator to intervene at any stage, for it is not always possible or in most instances advisable for manufacturing to proceed without human control input. Some products are so sensitive to small variations that the results could very quickly have severe impact on personnel, plant, or environment. To address these circumstances instrumentation may have to be configured with duplicates for all automated loops implemented on

the sensitive parts of the process, with one continually tracking the other during the production process and ready to take over at any time if required; the operator also being allowed to take over the automated loops at any time. Such requirements are referred to as redundant systems. Each collection of equipment being a level of redundancy; it is not uncommon for two, or sometimes even three, levels of redundancy to be present on critical processes. These ideas of redundancy are also relevant and applicable to continuous processes.

4. When the processing of a batch is complete the product has to be pumped away, and because the monetary value has increased quite enormously, as much of the product as possible has to be collected from the containing equipment and pipework. The vessel is scoured with the inert gas or solvent, and the pipelines, with a mechanical piston called a pig. This procedure is referred to as pigging the lines.

5. Control system designers must give consideration to the procedures that must be followed when the process enters an alarm condition through some sequence malfunction; when a fault in the process equipment causes the processing to stop; when the operator intentionally puts the system on hold; when recovering from such interruptions; and, in general to ensure that the plant, personnel, and the environment will be protected under hazardous operating conditions.

6. Today almost all batch operations are controlled by computer; hence, standardization of software methods and documentation permits changes to be implemented quickly. The NAMUR batch committee report of 1985 recommends:

   • A hierarchical control system structure that reflects the process

   • A terminology that is standard and applicable to batch control

   • Use of recipes to produce and manage a batch and modular software that relates the recipe to a unit operation

   • Use of configuration techniques, e.g. as used in DCS and other block-configurable systems rather than custom software to achieve the application requirements.

As used in these recommendations, a unit operation is a set of instructions for process operations that cover the process equipment, materials (raw and finished), system communications, and operator interactions; these are generally repeatable, allowing for some variation, from one process plant to another. A recipe specifies the quantities of each material to be used, the operating conditions for each part of the manufacturing process, the sequence of operation for each manufacturing stage, and each item of plant equipment required to produce the batch.

7. A recipe manager, according to the NAMUR recommendations, should form part of a batch control package. The object of the recipe manager is in general to divide recipes into master and control and to provide additional functionality to handle the separate and specific requirements of each. The master recipe is process equipment—independent whereas the control recipe is not.

8. Control recipes are modified masters in which:

   • The process equipment to be used is specified

   • Unit operations of the master recipe are replaced with control phases to produce the batch of product.

- Facilities exist for the operator to modify the recipe to allow for variations in material quality or composition, and those parts of the recipe not executed, without interrupting the process

- Facility is available to manually reassign process equipment if required

9. GRAFCET uses a graphical method for a sequence language that is accepted as an international standard covered by IEC 1131.3. As the language is graphical, the system model is constructed from a set of graphic primitives, assembled in accordance with a syntax and elaborated on by succinct descriptions that define the function. The Grafcet is always read from the top downward.

10. The five rules of evolution are:

  1. One or more initial steps determine the initial state of a Grafcet. The rest state of the system is described by the actions associated with the initial step(s).

  2. A transition will be cleared when all immediately preceding steps are active, or the receptivity associated with the transition is true.

  3. A cleared transition results in simultaneous deactivation of all immediately preceding steps and activation of all immediately following steps.

  4. All transitions that can be cleared simultaneously will in fact be cleared simultaneously.

  5. If a step during the course of operation is logically activated and deactivated at the same instant, then the result will be that the step will become active.

11. The syntax rule states that the order *step/transition* and *transition/step must be respected for every sequence*. This means that two steps or two transitions must never be connected by a link. Therefore, it is mandatory that a directed link connect a transition to a step and a step to a transition.

12. There are three graphic primitives: the step, the transition, and the directed link.

13. The step can be of three forms: basic, initial, or active. There can be more than one initial step. The step can have only one entry line and one exit line.

14. A link connects one step to another and in doing so must pass through a transition. Links are normally shown as vertical line segments. However, if required, then the link may be "bent" through 90° to one side in order to accommodate another part of the Grafcet. Directional arrows are not shown, except if a loop or a link from a step lower down in the sequence to a step higher up is required, when directional arrows are permitted. It is recommended that all connecting links be perpendicular to each other even though diagonal links are allowed.

15. Actions associated with a step are defined in either symbolic or text form; in both cases the actions are noted directly alongside the relevant step. If the actions are in text, then the instructions are enclosed within quotes and written alongside; if the action is in symbolic form, then a suitable table must be given defining the actions associated with each symbol used.

16. An action can carry a qualification such as Delayed, Stored, Limited, Pulsed, Conditional, or Combined. These above actions are contracted to their first alphabetic character; of the above; e.g., D is equivalent to Delayed. It is possible to

combine the actions—e.g., DL is equivalent to Delayed Limited—and each action is defined in terms of the time period involved.

17. The transition is a method for bringing control to a Grafcet. It is cleared when the preceding step is active and the transition condition is satisfied, i.e., made "true." Either single or multiple steps can clear a transition, and the clearance of the transition condition is termed the receptivity. The transition is identified by a numeral in parentheses, with the description of the receptivity shown by an asterisk on the opposite side of the numerical identifier.

18. Receptivity is expressed either literally (text) or as a Boolean expression, which can be composed of Boolean operators and/or edge symbols, with the active edge symbol shown either as rising or falling. The expression can also be time-dependent.

19. A linear sequence comprises a series of steps that are activated one after the other, and in practice each step is followed by a transition, and each transition by a step. An active sequence is one in which any one step is active. The sequence is inactive only if all steps are inactive.

20. In a particular instance, it is possible to jump steps in a sequence, but this is allowed only when the steps jumped over are not required in that particular instance.

21. Selection between several sequences from one or more steps can be made. A single horizontal line joining the sequences indicating an OR function; double horizontal lines joining the sequences indicating an AND function.

22. To ensure that only one sequence is selected for a particular application, the receptivities of the associated transitions must be made mutually exclusive.

23. The continuous aspects of control follow those that had been discussed previously for DCS systems. However, when metering liquid fluids especially in batch operations, it is recommended that the setpoint and measurement move in unison. This is obtained by carefully setting the ramp rate of the setpoint either in the PID control block or in the subroutine of the sequence program. Be aware that it is sometimes necessary to meter the fluid within a specified time; in these instances, the ramp rate will have to be chosen accordingly.

Bibliography

Bass, H.G. *Intrinsic Safety Instrumentation for Flammable Atmospheres.* 1984, Quartermaine House Ltd., Sunbury Middlesex. UK.

Baumann, H.D. *Control Valve Primer—A Users Guide.* 1994, ISA Publications, NC USA.

British Standards Institution. *Methods for the Measurement of Fluid Flow in Pipes BS 1042 Part 1.* 1943, London UK.

Crane Company. *Flow of Fluids through Valves, Fittings and Pipes.* Technical paper 410, Engineering Division Crane 1969 Illinois USA.

deSa, D. & Maalouf, S. "The Operating and Control Philosophy of Turbo Compressors" 1993 *Measurement + Control*, pp 69–72. Institute of Measurement & Control, Vol 29, No 3, London UK.

deSa D. "Signal Processing, Data Handling and Communications" Institute of Measurement and Control London, Residential School "Process Analysis: A Road to Safety and Efficiency," University of Warwick; April 29 to May 4 1990, Lecture Notes pages 33 through 56.

Dukelow, S.G. *The Control of Boilers.* 1991, ISA Publications, NC USA.

Elcon Instruments, Inc. *Introduction to Intrinsic Safety.* 1992, Maryland USA.

Fisher Rosemount, Ltd. *Control Valve Handbook* Fisher Controls 1965 Iowa USA.

———. "PROVOX® Process Management System" Form 8570/20M/10–97 1997, Texas USA.

———. "PROVOX® Operator Console" Form D350562X012/3MAQ, 1MA4/10–98 1998, Texas USA.

———. "2390–249 Series Level Transmitters & Level Sensors" Bulletin 11.2:2390–249, 1996, Iowa USA.

———. "2500–249 Series Pneumatic Controllers & Transmitters" Bulletin 34.2:2500, 1997, Iowa USA.

Foxboro Great Britain, Ltd. *I/A Series® Systems*, Book 021 compilation of Product Specification Sheets for Intelligent Automation Systems, 1997. Ma. USA.

———. *I/A Series® Systems, Integrated Software Concepts*, 1989, Ma. USA.

———. *I/A Series® Systems, Integrated Control Block Descriptions*, 1989, Ma. USA.

———. *Control Valve Engineering Handbook* Publication PUB 237D 10M 3/85,1985, Ma USA.

———. "Pressure Drop through Control Valves" Publication TI 31-4M, 1970, Ma. USA.

———. *M & I Devices*, Book 010 Vols 1 & 2 compilation of Product Specification Sheets for Process Measurement Instruments and Stand Alone Controllers 1997, Ma. USA.

———. *SPEC 200®*, Book 200 compilation of Product Specification Sheets and Instruction Manuals for SPEC 200 Analog Control Systems.

———. *INTERSPEC® systems*, Book 210 compilation of Product Specification Sheets and Instruction Manuals for Spec 200 Computer Interface Devices.

Francis, J.R.D. *A Textbook of Fluid Mechanics.* 1958, Edward Arnold Ltd., London UK.

Hoernes, G.E. & Heilweil, M. F. *Introduction to Boolean Algebra and Logic Design.* 1964, McGraw Hill, New York.

Honeywell Control Systems, Ltd. "TotalPlant® Solution System" Bulletin EN/IAC/99–024 1999, AZ USA.

Jones, E.B. *Instrument Technology*. Vols. 1–3, 1960 Butterworth & Co Ltd., London UK.

Kocache, R. "The Measurement of Oxygen in Gas Mixtures" Journal of the Institute of Physics E: Science Instrum. (1986) Vol 19 pp 401–412 London UK.

Maihak Aktiengesellschaft "Series 600 Oxor Oxygen Analyser," Product Specification Sheets 21.043 E, 07 1994, Hamburg Germany.

———. "Series 600 Multor NDIR Gas Analyser for Multi-component Gas Streams," Product Specification Sheet 21.001 E, 02 1995, Hamburg Germany.

———. "Unor NDIR Gas Analyser for Single Component Gas Streams," 22.001 E, 24.006 E, 01 1995, Hamburg Germany.

Masoneilan-Dresser *Handbook for Control Valve Sizing* Bulletin OZ1000E 09/97. 1997, Texas USA.

———. "Camflex II the Second Generation of All Purpose Automatic Control Valves" Bulletin BF 5000 E, 1997, Texas USA.

———. "SVI® The Smart Valve Interface" Publication BW 1000 E, 1998, Texas USA.

Miller, J.T. *A Revised Course in Industrial Instrument Technology*. 1952, United Trade Press Ltd., London UK.

Noltingk, B.E. *Instrumentation Reference Book*. 1988, Butterworth & Co. Ltd., London UK.

Perry's Chemical Engineers Handbook. 1984, McGraw-Hill New York.

Rogers, G.F.C., & Mayhew, Y.R. *Engineering Thermodynamics* 1992 Longmans Scientific & Technical, Harlow UK.

Rotork Controls Ltd., "Rotork 'A' Range Double Sealed Electric Actuators," Publication E210E 1997 Bath UK.

———. "Sizing Guide for Electric Actuators," Publication AE2/0.2 1993, Bath UK.

———. "Rotork IQ Range Intelligent Non intrusive 3 Phase Electric Valve Actuators," Publication E110E, 1998, Bath UK.

SDS Data Systems. *Grounding and Noise Reduction Practices*, SDS Data Systems, CA USA.

Seborg, D.E. Edgar, T.F. Mellichamp, D.A. *Process Dynamics and Control*. 1989, Wiley & Sons New York.

Shinskey, F.G. *Process Control Systems*. 1988, McGraw-Hill, New York USA.

Shinskey, F.G. *Energy Conservation through Control*. 1978 Academic Press, New York USA.

Skil Controls, Ltd. "Electro Hydraulic Actuators," Product Specifications for Models 301/2, Models 405/6/7, Model 413, Models 416/7, Model 411, Models 413, 481, 1999, Skelmersdale UK.

Ward, Brice *Computer Technician's Handbook*. 1971, Foulsham-Tab Ltd., Slough UK.

White, M.H. *Surge Control for Centrifugal Compressors*, Chemical Engineering, July 18, 1988, pp. 39–42.

Wika Instruments, Ltd. *Chemical Seals Technical Handbook*, Klingenberg, Germany.

Worcester Controls, "Series 44/45 Full & Reduced Bore Ball Valves" Publication PB105 Iss KI, Sussex UK.

———. "Series 40R (pneumatic) actuator" Publication PB29 Iss HF, Sussex UK.

———. "Series 75 (electric) actuator" Publication PB25 Iss MH, Sussex UK.

———. "New Ball Control Valves for High Performance" Publication PB30 Iss MD, Sussex UK.

Yokogawa Electric Corporation, "Centum CS Integrated Production" Bulletin 33G1A12-01E, 1993, Tokyo Japan.

———. "Centum CS" Bulletin 33A0B0-E-H, 1993, Amersfoot The Netherlands.

———. "Concentral Solutions" Bulletin OA2A1-E 1993, Tokyo Japan.

Index

IOM—DCS systems, **19**-360; **20**-390
Ionization, **5**-96
IPST, **3**-61
ISA—DCS symbol, **20**-387
ISO—7 layer Model, **21**-415
Isolation Trasformer—power distribution,
  22-450

K

Karnaugh map Minimisation, **24**-487
Katharometer—*see* Shakespear
  katharometer detector
Keeper—Pig, **25**-504
Kelvin—*see* Absolute Temperature
Key lock, **19**-363

L

Ladder diagram, **24**-492
Lag—definition, **7**-139
Lag—Distance/velocity, **7**-140
Lag—Measurement, **7**-139
Lag—Output **Intro**-xxx
Lag—Transfer, **7**-142
Law of Intermediate
  metals—thermocouples, **3**-57
Law of Intermediate
  temperatures—thermocouples, **3**-57
Lead-lag—*see* cross limiting—steam
  generators
Letterbug, **20**-391
Level Instruments—criteria, **4**-89
Level switch, **4**-84
Linear—*see* characteristic curves—control
  valves
Link—Grafcet, **25**-508
Live zero, **19**-358
LLC—*see* logical link control
Logic Blocks, **24**-477
Logical Link Control, **21**-416
Loop designator/name, **19**-363
Low flow cut-off, **20**-390
lrv—lower range value—*see*
  range—measurement

M

MAC—*see* Media access control
Machine Language, **19**-364; **21**-422
Magnetic Flow meter, **1**-17
Management consoles, **19**-364
Manifolds—*see* valve manifolds

Manufacturing Message
  Standard—communications, **21**-419, 420
MAP protocol, **19**-367; **21**-415 to 420
Mass Flowmeter—*see* Coriolis Mass
  flowmeter
Master Pressure control—steam generators,
  14-253
Master Recipe—batch, **25**-506
Measurement lag—definition, **7**-139
Measurement lag—*see* lag—measurement
Measurement span—*see*
  Span—measurement
Measuring cell—katharometer, **5**-106
Measuring cell—NDIR, **5**-110
Measuring unit—*see* input unit—controller
Media Access Control—communications,
  21-416
Metering Section—flow measurement,
  18-339
Minimisation techniques—Boolean logic,
  24-483, 487
MMS—*see* manufacturing message standard
Modes-control **Intro**-xxviii
Mol equivalent, **5**-97
Multi step Control, **8**-151
Multiple Compressor control, **13**-231, 233
Multiplexing, **21**-424

N

NAMUR Batch ops, **25**-506
NDIR gas analyser, **5**-110
Negative feedback—*see* feedback—negative
Network layer—ISO model, **21**-417
Network Management—communications,
  21-419
Node—*see* DCS node
Nondispersive Infrared analyser—*see* NDIR
  gas analyser
Normal Water Level—boiler drum—steam
  generator, **14**-247
Normality—pH, **5**-97
Nucleonic Level—*see* radiometric level
  detector
NWL—*see* normal water level—boiler
  drum—steam generator

O

On-Off Control, **8**-150, 151
Op-Amp—typical circuits, **21**-427 to 429
Open Channel Flow measurement—*see*
  Parshall and Venturi flume